Algebra I

NEXT GENERATION

Course Workbook
with Regents Questions

2024-25
Donny Brusca

ISBN   978-1-952401-37-4

www.CourseWorkBooks.com

The CourseWorkBooks Catalog

CourseWorkBooks has been publishing high school workbooks since before the introduction of the Common Core curriculum and remains an essential resource for New York schools transitioning to the Next Generation Regents standards.

New to our catalog for 2024 is **CWB's Course Workbook for AP® Statistics**, the first CourseWorkBooks publication aligned to an AP® curriculum. Also new this year is the supplementary **Statistics Calculator Guide, TI-83/84+ Edition**.

Answer Keys are available as free PDF downloads or as paperbacks.

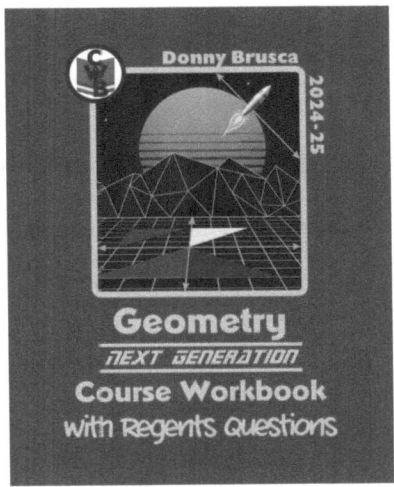

AP® is a trademark registered by the College Board, which is not affiliated with, and does not endorse, this product.

www.CourseWorkBooks.com

Digital Site Licenses

Purchase a Digital Site License to grant teachers and students access to the online editions of every book in the CourseWorkBooks catalog.

The License includes our proprietary e[Work]Books software for writing and drawing directly on the workbook pages and saving them as image files.

The License also includes access to our NEW instructional videos, which are being created for each section of the Next Generation workbooks.

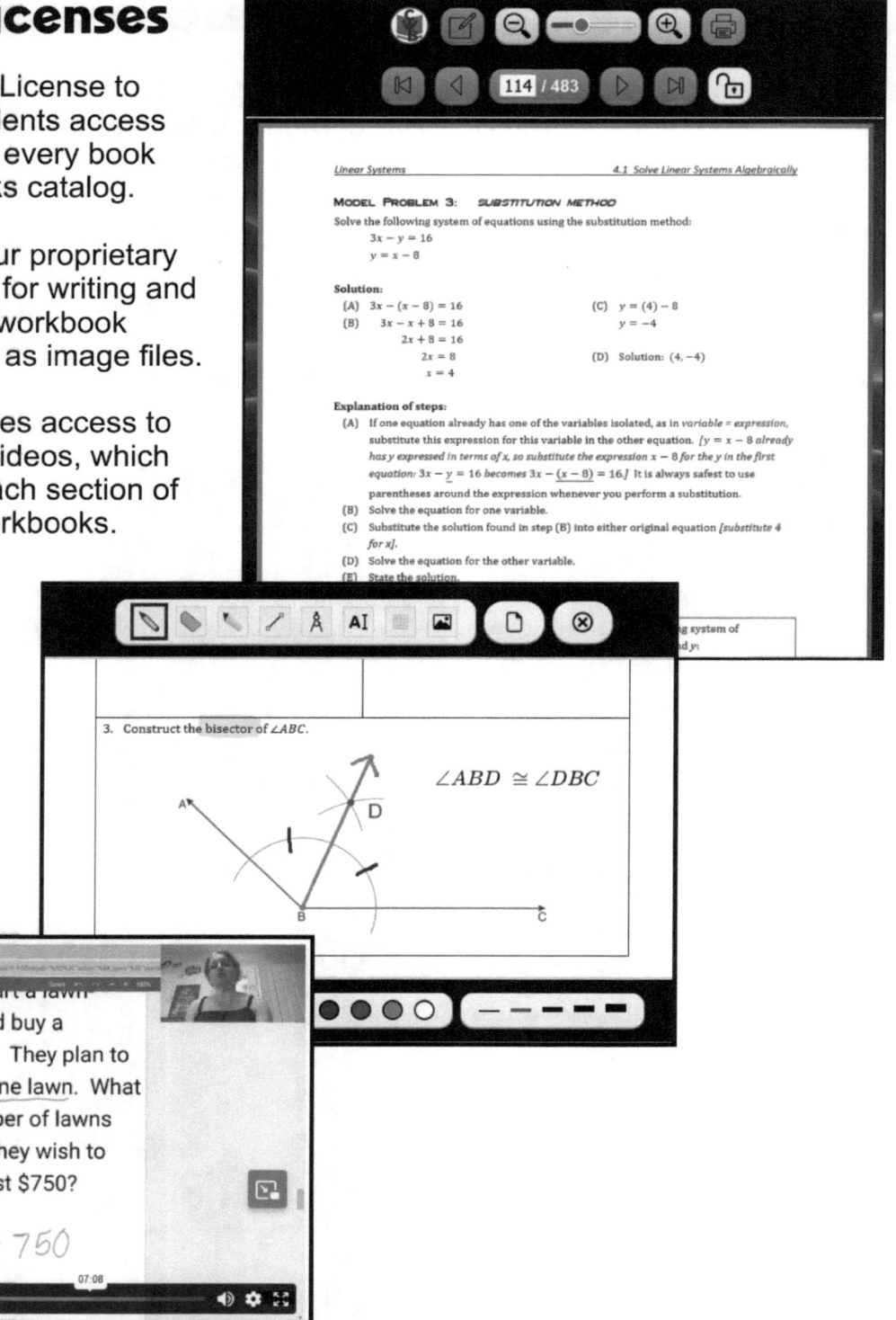

Table of Contents

ABOUT THE AUTHOR

Donny Brusca founded CourseWorkBooks in 2010.

He has retired from teaching and school administration after about 30 years of employment, mostly on the high school and college levels. He has a B.S. in mathematics and M.A. in computer and information science from Brooklyn College (CUNY) and a post-graduate P.D. in educational administration and supervision from St. John's University.

His college-level instructional experiences include Brooklyn College, Pace University, and Touro College Graduate School of Technology. His high school teaching experiences include public, charter, Catholic, and Jewish schools.

His administrative experiences include serving as the Student Data Systems Manager at a charter management organization and high school and as the Academic Dean and Mathematics Chairperson at a Manhattan business college. For several years, he was responsible for student scheduling in a high school of about 1000 students.

He has taught courses in basic mathematics, logic, algebra, geometry, probability, discrete structures, computer programming, web design, data structures, switching theory, computer architecture, and application software. He has taught all three Regents level mathematics courses and AP Computer Science A and AP Computer Science Principles.

He owned and operated a successful part-time disc jockey business (Sound Sensation), performing personally at nearly 1000 weddings and private events over a 12 year span.

He has managed and coached high school baseball teams, after playing for over 40 years in organized baseball and softball leagues. He currently works as a high school baseball umpire.

He lives in Staten Island, NY, with his wife, Camille, and their cats.

COVER IMAGE

For me, the Next Generation moniker has always evoked a Star Trek theme; hence, the space-related image. Overlying the image are graphs, created in Desmos, from the three main families of functions covered in this course – linear ($y = 2x$ in green), quadratic ($y = x^2$ in red), and exponential ($y = 2^x$ in blue) – all meeting at their glowing point of intersection, (2,4). The yellow rocket ship completes the familiar four-color scheme associated with the CourseWorkBooks logo.

ABOUT THIS BOOK

The topics in this book are aligned to the national Common Core high school curriculum (C.C.S.S.). The book is intended for use in any state, but it specifically corresponds to the scope and sequence of topics established by New York State for the successful completion of a course leading to the "Next Generation" Regents examination.

Every topic section begins with an explanation of the key terms and concepts. I have intentionally limited the content here to the most essential ideas. The notes should supplement a fuller presentation of the concepts by the teacher through a more developmental approach.

Calculator Tips explain how to use the graphing calculator to solve problems or check solutions. Keystrokes include button names in rectangles, STO▶, alternate button features in brackets, [SIN⁻¹], and on-screen text within angle brackets, <NUM>. Directions for selecting on-screen text (arrow keys) are usually omitted. Screenshots from a TI-84 Plus CE are shown by default, but differences from the TI-83 or other TI-84 models are noted.

Topic sections include one or more Model Problems, each with a solution and an explanation of steps needed to solve the problem. Steps lettered (A), (B), etc., in the explanations refer to the corresponding lettered steps shown in the solutions. General wording is used in the explanations so that students may apply the steps directly to new but similar problems. However, for clarity, the text often refers to the specific model problem by using *[italicized text in brackets]*. To make the most sense of this writing style, insert the words "in this case" before reading any *[italicized text in brackets]*.

After each Model Problem are a number of Practice Problems in boxed work spaces. These numbered problems are generally arranged in order of increasing level of difficulty.

At the end of each section are all of the Algebra I Common Core Regents Questions related to that section, through the most recent January exam. As the upcoming Next Generation exams are administered, the annual updates of this book will include all of those exams' questions, as well. The Algebra I Reference Sheet is provided in Appendix I.

The Answer Key is available at CourseWorkBooks.com, both as a free PDF and as a paperback.

Also available at CourseWorkBooks.com are Digital Site Licenses (including Instructional Videos created by a NYS certified teacher), Mock Regents Exams, and a Standards and Pacing Guide.

Recently added to our catalog is the new *CWB's Course Workbook for AP® Statistics*.

My goal is to have an error-free book and answer key and I am willing to offer a monetary reward to the first person who contacts me about any error in mathematical content. Simply email me at donny@courseworkbooks.com and be as specific as possible about the error, including the title of the book, the edition (year), the section number, and the problem number (if applicable).

CHAPTER 1.   EQUATIONS AND INEQUALITIES

1.1   Properties of Real Numbers

Real numbers include all the rational and irrational numbers, and are represented by all the points that make up a number line or a coordinate axis.

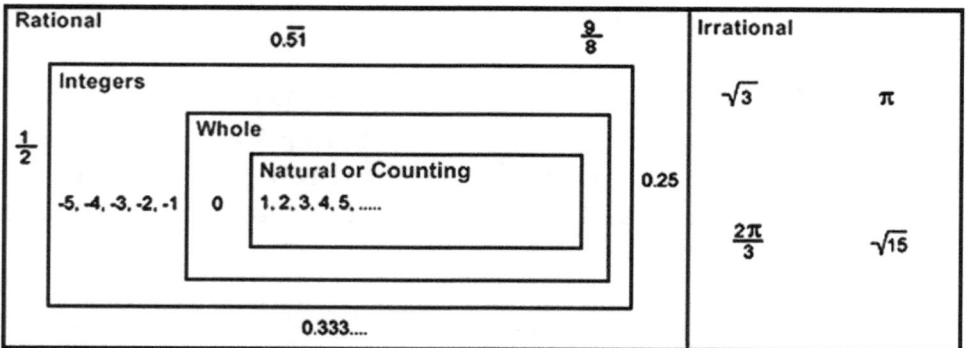

Rational numbers can be expressed as $\frac{a}{b}$ where a and b are integers and b ≠ 0.
Every rational number can be expressed as a **terminating or repeating decimal**.

Irrational numbers are all the real numbers that are *not* rational; that is, they cannot be expressed as a quotient of integers, and their decimals are non-repeating and non-terminating. π is an irrational number. 3.14 is an *approximation* of π, rounded to the nearest hundredth. *(We'll take a closer look at irrational numbers in Chapter 10.)*

Counting numbers (also called **Natural numbers**):   {1, 2, 3, 4, 5, ...}

Whole numbers include the counting numbers and zero: {0, 1, 2, 3, 4, 5, ...}

Integers include the whole numbers and their opposites: {... -3, -2, -1, 0, 1, 2, 3, ...}

Identities: The word "identity" comes from the Latin idem, meaning the same.
The **additive identity** is 0, because when you add 0, the value remains the same.
The **multiplicative identity** is 1, because when you multiply by 1, the value remains the same.
$$a + 0 = a \qquad\qquad a \times 1 = a$$

Inverses: When an operation is performed between a value and its inverse, the result is the identity for that operation.
So, the **additive inverse** of a number is its opposite, since adding a number and its opposite results in zero, the identity for addition: $\qquad\qquad a + (-a) = 0$
The **multiplicative inverse** of a number is its reciprocal, since multiplying a number and its reciprocal results in one, the identity for multiplication: $\qquad a \times \frac{1}{a} = 1$

Examples: The additive inverse of 2 is -2, and the multiplicative inverse of 2 is $\frac{1}{2}$.

The **Commutative Property** states that for an operation, the order of the operands doesn't matter. Addition and multiplication are commutative; subtraction and division are not. In other words, for addition, $a + b = b + a$, and for multiplication, $ab = ba$.

The **Associative Property** states that when performing the same operation on three operands, the order in which we perform the operations (grouping by parentheses) doesn't matter. Addition and multiplication are associative; subtraction and division are not. In other words, for addition, $a + (b + c) = (a + b) + c$, and for multiplication, $a(bc) = (ab)c$.

The **Distributive Property** states that when a value is multiplied by a sum, we get the same result if we were to multiply the value by each addend separately and then add the products. In other words, $x(a + b) = xa + xb$.

The **Reflexive Property of Equality** states that a value is equal to itself. That is, for all real values of a, $a = a$.

The **Symmetric Property of Equality** states that if we switch the two sides of an equation, the equation remains true. That is, if $a = b$, then $b = a$.

Note that if we switch the two sides of an *inequality*, we must also *reverse* the inequality symbol in order for it to remain true. This is called the **Reversal Property of Inequality**.
Examples: If $a > b$, then $b < a$.   If $a \geq b$, then $b \leq a$.

The **Transitive Property of Equality** states that if $a = b$ and $b = c$, then $a = c$.

The **Addition Property of Equality** states that if we add the same value to both sides of an equation, the equality remains. That is, if $a = b$, then $a + c = b + c$.

This extends to the other basic operations as well.
Subtraction Property of Equality: If $a = b$, then $a - c = b - c$.
Multiplication Property of Equality: If $a = b$, then $ac = bc$.

Division Property of Equality: If $a = b$, then $\dfrac{a}{c} = \dfrac{b}{c}$ (where $c \neq 0$).

To summarize the properties:

Commutative Property of Addition	$a + b = b + a$
Commutative Property of Multiplication	$ab = ba$
Associative Property of Addition	$a + (b + c) = (a + b) + c$
Associative Property of Multiplication	$a(bc) = (ab)c$
Distributive Property	$x(a + b) = xa + xb$
Reflexive Property of Equality	$a = a$
Symmetric Property of Equality	$(a = b) \leftrightarrow (b = a)$
Reversal Property of Inequality	$(a > b) \leftrightarrow (b < a)$ $(a \geq b) \leftrightarrow (b \leq a)$
Transitive Property of Equality	$(a = b)$ and $(b = c) \rightarrow (a = c)$
Addition Property of Equality	$(a = b) \rightarrow (a + c = b + c)$
Subtraction Property of Equality	$(a = b) \rightarrow (a - c = b - c)$
Multiplication Property of Equality	$(a = b) \rightarrow (ac = bc)$
Division Property of Equality	$(a = b) \rightarrow \left(\dfrac{a}{c} = \dfrac{b}{c} \right)$

Closure: A set is **closed** under an operation if, for <u>every pair</u> of elements, when the operation is performed on them, the result is an element of the same set.

Examples: (a) The set of integers is closed under addition because, whenever we add two integers, the result is always another integer.

(b) The set of integers is _not_ closed under division because when we divide two integers we _may_ get a result that is a fraction (decimal) and not an integer. For example, if we divide 1 by 2, the result is one half, which is not an integer.

The table below shows whether the given sets are closed under the specified operations. The set of whole numbers is often called the set of natural numbers. Note that division is _not_ closed for any set that includes zero, since division by zero is undefined.

	Addition	Subtraction	Multiplication	Division
WHOLE NUMBERS	Y	N	Y	N
INTEGERS	Y	Y	Y	N
RATIONAL NUMBERS	Y	Y	Y	Y*
REAL NUMBERS	Y	Y	Y	Y*

* The sets of <u>non-zero</u> rational or real numbers only.

We can show that the set of **rational numbers is closed under multiplication**:
If $\frac{a}{b}$ and $\frac{c}{d}$ are rational numbers and a, b, c, and d are integers, then $\frac{a}{b} \times \frac{c}{d} = \frac{ac}{bd}$ by the rules for multiplying fractions. Since the set of integers is closed under multiplication, then ac and bd are integers, so $\frac{ac}{bd}$ is rational.

Similarly, we can show that the set of **rational numbers is closed under addition**:
If $\frac{a}{b}$ and $\frac{c}{d}$ are rational numbers and a, b, c, and d are integers, then $\frac{a}{b} + \frac{c}{d} = \frac{ad+bc}{bd}$ by the rules for adding fractions. Since the set of integers is closed under addition and multiplication, then $ad + bc$ and bd are integers, so $\frac{ad+bc}{bd}$ is rational.

MODEL PROBLEM 1:   *IDENTITIES AND INVERSES*

Given the value $\frac{1}{4}$, what is the (a) additive inverse and (b) multiplicative inverse?

Solution:	**Explanation of steps:**
(A) $-\frac{1}{4}$	(A) The additive inverse is its opposite, which we get by negating.
(B) 4	(B) The multiplicative inverse is its reciprocal $\left[\frac{4}{1} = 4\right]$.

PRACTICE PROBLEMS

1. Which statement illustrates an additive identity? (1) $6 + 0 = 6$ (2) $-6 + 6 = 0$ (3) $4(6 + 3) = 4(6) + 4(3)$ (4) $(4 + 6) + 3 = 4 + (6 + 3)$	2. Which equation illustrates a multiplicative inverse? (1) $a \cdot 1 = a$ (2) $a \cdot 0 = 0$ (3) $a\left(\frac{1}{a}\right) = 1$ (4) $(-a)(-a) = a^2$
3. What is the additive inverse of $\frac{2}{3}$?	4. What is the multiplicative inverse of $\frac{2}{3}$?
5. What is the additive inverse of the expression $a - b$?	6. What is the multiplicative inverse of the expression $-\frac{1}{ab}$?

MODEL PROBLEM 2:   IDENTIFY PROPERTIES

Quenton determined that the equation $x - 7 = y$ is true. He decides to rewrite the equation as $y = x - 7$. Which property justifies this step?

Solution:

The Symmetric Property of Equality

Explanation of steps:

(A) Determine what changes are being made in the step.
   [Quenton is switching the two sides of the equation. If we let a represent the original left side (x − 7) and b represent the original right side (y), he is changing a = b into b = a.]

(B) Select the property that justifies the changes made in the step.
   [Consulting the table, (a = b) ↔ (b = a) is called the Symmetric Property of Equality.]

PRACTICE PROBLEMS

7. Which equation is an example of the use of the associative property of addition? (1) $x + 7 = 7 + x$ (2) $3(x + y) = 3x + 3y$ (3) $(x + y) + 3 = x + (y + 3)$ (4) $3 + (x + y) = (x + y) + 3$	8. The equation $3(4x) = (4x)3$ illustrates which property? (1) commutative (2) associative (3) distributive (4) multiplicative inverse
9. Which property is illustrated by the equation $(ab)c = a(bc)$?	10. Which property is illustrated by the equation $ax + ay = a(x + y)$?
11. To find the product of x and 5, which property allows us to perform the following step? $$x \cdot 5 = 5x$$	12. When adding 2 to the sum, $3 + x$, which property allows us to perform the first step below? $$2 + (3 + x) =$$ $$(2 + 3) + x =$$ $$5 + x$$

MODEL PROBLEM 3:   *APPLICATION OF THE DISTRIBUTIVE PROPERTY*

Distribute: $-4(1-x)$

Solution: $-4(1-x) = (-4)(1) + (-4)(-x) = -4 + 4x$

Explanation of steps:
The distributive property, $x(a+b) = xa + xb$, allows us to multiply the value outside the parentheses by each value inside the parentheses, and add the products.
[So, $-4(1-x) = (-4)(1) + (-4)(-x)$]

PRACTICE PROBLEMS

13. Distribute: $5(x+5)$	14. Distribute: $4(b-4)$
15. Distribute: $-2(x-1)$	16. Distribute: $-3(a-b)$
17. Simplify: $(1+y)(-1)$	18. Simplify: $-(-a-1)$
19. Apply the distributive property to: $rs + rt$	20. Apply the distributive property to: $2x + 10$

MODEL PROBLEM 4:   _JUSTIFYING ASSERTIONS_

Justify the statement, "Subtraction is _not_ associative."

Solution:
If subtraction were associative, then $a - (b - c)$ would equal $(a - b) - c$ for all real numbers a, b, and c. For example, $10 - (5 - 2)$ would equal $(10 - 5) - 2$. However, $10 - (5 - 2) = 7$ and $(10 - 5) - 2 = 3$. So, subtraction is not associative.

Explanation of steps:
When asked to justify that something is _not_ true, it is usually best to offer a counterexample.

PRACTICE PROBLEMS

21. Justify the statement, "Division is _not_ commutative."	22. Is the set of whole numbers closed under subtraction? Justify your answer.
23. Is the set of _non-zero_ integers closed under division? Justify your answer.	24. Show that the set of _non-zero_ rational numbers is closed under division.

Regents Questions

MULTIPLE CHOICE

1. When solving the equation $4(3x^2 + 2) - 9 = 8x^2 + 7$, Emily wrote $4(3x^2 + 2) = 8x^2 + 16$ as her first step. Which property justifies Emily's first step?
    - (1) addition property of equality
    - (2) commutative property of addition
    - (3) multiplication property of equality
    - (4) distributive property of multiplication over addition

2. A part of Jennifer's work to solve the equation $2(6x^2 - 3) = 11x^2 - x$ is shown below.
    Given: $2(6x^2 - 3) = 11x^2 - x$
    Step 1: $12x^2 - 6 = 11x^2 - x$
    Which property justifies her first step?
    - (1) identity property of multiplication
    - (2) multiplication property of equality
    - (3) commutative property of multiplication
    - (4) distributive property of multiplication over subtraction

3. When solving the equation $12x^2 - 7x = 6 - 2(x^2 - 1)$, Evan wrote $12x^2 - 7x = 6 - 2x^2 + 2$ as his first step. Which property justifies this step?
    - (1) subtraction property of equality
    - (2) multiplication property of equality
    - (3) associative property of multiplication
    - (4) distributive property of multiplication over subtraction

4. Britney is solving a quadratic equation. Her first step is shown below.
    Problem: $3x^2 - 8 - 10x = 3(2x + 3)$
    Step 1: $3x^2 - 10x - 8 = 6x + 9$
    Which two properties did Britney use to get to step 1?
    - I.   addition property of equality
    - II. commutative property of addition
    - III. multiplication property of equality
    - IV. distributive property of multiplication over addition

(1) I and III	(3) II and III
(2) I and IV	(4) II and IV

5. When solving $p^2 + 5 = 8p - 7$, Kate wrote $p^2 + 12 = 8p$. The property she used is

(1) the associative property	(3) the distributive property
(2) the commutative property	(4) the addition property of equality

6. In the process of solving the equation $10x^2 - 12x - 16x = 6$, George wrote $2(5x^2 - 14x) = 2(3)$, followed by $5x^2 - 14x = 3$. Which properties justify George's process?

    A. addition property of equality
    B. division property of equality
    C. commutative property of addition
    D. distributive property

    (1) A and C (3) D and C
    (2) A and B (4) D and B

CONSTRUCTED RESPONSE

7. A student is in the process of solving an equation. The original equation and the first step are shown below.

    Original: $3a + 6 = 2 - 5a + 7$
    Step one: $3a + 6 = 2 + 7 - 5a$

    Which property did the student use for the first step? Explain why this property is correct.

8. John was given the equation $4(2a + 3) = -3(a - 1) + 31 - 11a$ to solve. Some of the steps and their reasons have already been completed. State a property of numbers for each missing reason.

    $4(2a + 3) = -3(a - 1) + 31 - 11a$ Given
    $8a + 12 = -3a + 3 + 31 - 11a$ _____
    $8a + 12 = 34 - 14a$ Combining like terms
    $22a + 12 = 34$ _____

1.2   Solve Linear Equations in One Variable

An **equation** is a statement that one expression is equal to another. It contains an = sign.
Example: $3x - 1 = x + 5$

Domain (replacement set): the set of numbers that can replace a variable. Usually, you can assume the domain is {real numbers}. However, if, for example, the variable represents the length of the side of a polygon, the domain should be {positive real numbers}.
Solution set: set of values from the domain that make an equation or inequality true.
Solution: each element of the solution set.

A **variable term** in an equation includes the variable as a factor (or the variable by itself). A variable term may include a numerical factor called a **coefficient.**
A **constant term** in an equation does not include a variable factor.
Example: $3x + 5 = 35$ has a variable term [$3x$, which has a coefficient of 3] and a constant term [5] on the left side, and a constant term [35] on the right side.

A **linear equation** is an equation in which each term is either a *constant term* or a *variable term* that includes only a *single variable to the first power* (ie, the variable can only have an exponent of 1, which is usually not written at all).

The goal when solving an equation is to **isolate the variable** (that is, transform it into the form, $x = some\ value$).

In a previous course, you may have learned how to solve one- and two-step equations. These are equations that have only a variable term and optional constant term on one side and only a constant term on the other side. To isolate the variable in these equations:
  1. eliminate the constant term by subtracting it from (adding its opposite to) both sides.
  2. eliminate the coefficient of the variable by dividing both sides by the coefficient.
Example: $3x + 5 = 35$
                  $\underline{\quad -5 \quad -5\quad}$ eliminate the constant term 5 by subtracting 5 from both sides
                  $\dfrac{3x}{3} = \dfrac{30}{3}$ eliminate the coefficient 3 by dividing both sides by 3
                  $x = 10$

To check your solution:
Substitute your solution for the variable in the original equation. *It is usually best to use parentheses around the value when substituting.* Then, evaluate both sides of the equation to determine whether the solution makes the equation true.
Example: To check whether 10 is a solution for the above equation, $3x + 5 = 35$,
                  $3(10) + 5 = 35$
                  $35 = 35 \checkmark$

▦▢ CALCULATOR TIP

Checking a solution on the calculator:

1. Type your answer, then press $\boxed{\text{STO▸}}\boxed{\text{X,T,θ,}n}$ to store your answer into the variable x, or press $\boxed{\text{STO▸}}\boxed{\text{ALPHA}}$ and a letter key to store your answer in a different variable, and press $\boxed{\text{ENTER}}$.

2. Type in the equation, using $\boxed{\text{2nd}}\boxed{\text{TEST}}\boxed{1}$ to enter the equal sign. Use $\boxed{\text{X,T,θ,}n}$ for x or $\boxed{\text{ALPHA}}$ and a letter key for a different variable. Then, press $\boxed{\text{ENTER}}$.

3. The calculator will display 1 if the equation is true for the given value of the variable, or 0 if it is false.

Example: These screenshots show how to check whether $x = 10$ is a solution of $3x + 5 = 35$.

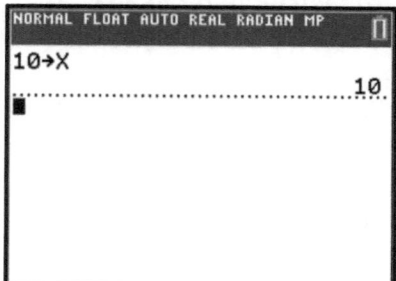 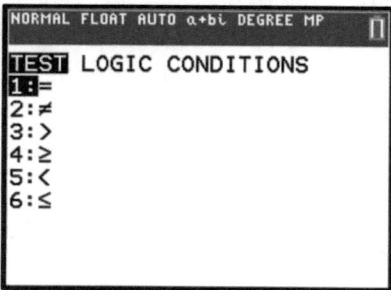 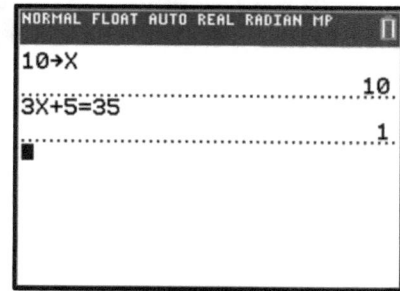

Note: In step 1 above, after pressing $\boxed{\text{ALPHA}}$, you can press any of the letters associated with the $\boxed{\text{ALPHA}}$ key to store the value in that variable. For examples, $\boxed{\text{MATH}}$ for [A], $\boxed{\text{APPS}}$ for [B], etc.

In this section, we will learn how to solve more complex equations by first transforming them into simpler one- or two-step equations like the one above.

In Section 1.3, we will learn how to solve equations involving fractions.

Like Terms: have the same exact variable parts (the numerical coefficients may differ). Like terms may be combined (added or subtracted); unlike terms may not.

Examples: Like terms: $2x$ and $3x$ $-4x^2y$ and x^2y
            <u>Not</u> like terms: $2x$ and $3x^2$ $-4x^2y$ and x^2

To combine like terms:
1. Add or subtract the coefficients
2. Keep the variable part the same

Example: $2x + 5x - 2$ is equivalent to $7x - 2$, since $2x$ and $5x$ are like terms.

Like terms may be combined only if they are on the <u>same side</u> of the equation. If there are variable terms on <u>opposite sides</u> of the equation, eliminate one by adding its opposite to both sides.

Examples: (a) $2x + 5x - 2 = 12$

$7x - 2 = 12$ _[combine like terms $2x + 5x$ on the same side of the =]_

$\underline{+2 \quad\quad +2}$ _[add 2 to both sides]_

$\underline{7x} = \underline{14}$

$7 \quad\quad 7$ _[divide both sides by 7]_

$x = 2$

        (b) $2x + 13 = 5x - 2$ _[we CANNOT add $2x + 5x$; they're on opposite sides!]_

$\underline{-2x \quad\quad\quad\quad -2x}$ _[eliminate $2x$ by subtracting it from both sides]_

$13 = 3x - 2$

$\underline{+2 \quad\quad\quad +2}$ _[eliminate –2 by adding 2 to both sides]_

$\underline{15} = \underline{3x}$

$3 \quad\quad 3$ _[eliminate the coefficient 3 by dividing both sides by 3]_

$5 = x$

When there are parentheses in an equation, you may be able to eliminate them by using the distributive property.

Example: $-2(3x - 5) = 28$

$-6x + 10 = 28$ [apply the distributive property to eliminate parentheses]

$\underline{-10 \quad -10}$ [eliminate 10 by subtracting 10 from both sides]

$\underline{-6x} = \underline{18}$

$-6 \quad -6$ [eliminate the coefficient –6 by dividing both sides by –6]

$x = -3$

Below is a summary of the steps needed to solve equations, which should be followed in order.

To solve a linear equation for one variable:
1. **SIMPLIFY:** Simplify each side of the equation to one or two terms.
   (a) Use the **distributive property** where possible to remove parentheses.
   (b) **Combine like terms** <u>on the same side of the equal sign</u> where possible.
2. **VARIABLE TERMS TO ONE SIDE:** If there are variable terms on both sides, eliminate a variable term from one side by adding its opposite to both sides.
   Eliminate from which side? Here's a good rule of thumb...
   (a) If one side has more terms, eliminate from that side.
   (b) Otherwise, eliminate the term with the smaller coefficient.
3. **ISOLATE THE VARIABLE:** Use the <u>reverse order of operations</u> to isolate the variable:
   (a) Add the opposite of the constant term to both sides.
   (b) Divide both sides by the variable term's coefficient.

MODEL PROBLEM 1:   ONE- AND TWO-STEP EQUATIONS

Solve for x: $-5x - 2 = 43$

Solution:
$$-5x - 2 = 43$$
(A) $\underline{+2 \quad +2}$
$$-5x = 45$$
(B) $-5 \quad -5$
$$x = -9$$

Explanation of steps:
(A)   Add the opposite of the constant term to both sides.
(B)   Divide both sides by the variable term's coefficient.

Check by substituting: $-5(-9) - 2 = 43$
Evaluating the left side gives us $43 = 43$ ✓

Checking on the calculator (1 means true):

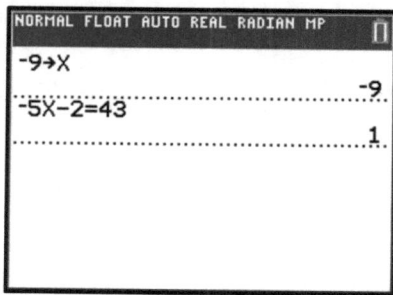

PRACTICE PROBLEMS

1. Solve: $-4x = -12$	2. Solve: $5 = x + 9$
3. Solve: $6x - 5 = -29$	4. Solve: $18 = -10 + 7x$

MODEL PROBLEM 2:   *LIKE TERMS ON THE SAME SIDE*

Solve for x: $18x - 6x - 8 = 28$

Solution: **Check:**

$$18x - 6x - 8 = 28$$
(A) $12x - 8 = 28$
(B) $12x = 36$
(C) $x = 3$

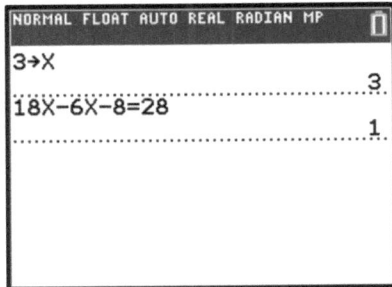

Explanation of steps:
  (A)   Combine like terms on the same side of the equal sign *[18x − 6x = 12x].*
  (B)   Add the opposite of the constant term to both sides *[add 8 to both sides].*
  (C)   Divide both sides by the variable term's coefficient *[divide both sides by 12].*

PRACTICE PROBLEMS

5. Solve: $4n - n = -12$	6. Solve: $25 = 3x - 10 - 8x$

MODEL PROBLEM 3:   *ELIMINATING PARENTHESES*

Solve: $5(x - 2) + 3x = 30$

Solution: **Explanation of steps:**

$$5(x - 2) + 3x = 30$$ (A)   Distribute *[5(x − 2) = 5x − 10]*
(A) $5x - 10 + 3x = 30$ (B)   Combine like terms *[5x + 3x = 8x]*
(B) $8x - 10 = 30$ (C)   Solve the two-step equation
(C) $8x = 40$ *[add 10 to both sides then divide both sides by 8].*
              $x = 5$

Check by substituting: $5(5 - 2) + 3(5) = 30$
Evaluating the left side gives us $30 = 30$ ✓

Check on the calculator: $\boxed{5}\,\boxed{\text{STO}\blacktriangleright}\,\boxed{\text{X,T,}\Theta\text{,}n}\,\boxed{\text{ENTER}}$

(1 means true.) $\boxed{5}\,\boxed{(}\,\boxed{\text{X,T,}\Theta\text{,}n}\,\boxed{-}\,\boxed{2}\,\boxed{)}\,\boxed{+}\,\boxed{3}\,\boxed{\text{X,T,}\Theta\text{,}n}\,\boxed{\text{2nd}}\,\boxed{\text{TEST}}\,\boxed{1}\,\boxed{3}\,\boxed{0}\,\boxed{\text{ENTER}}$

Practice Problems

7. Solve: $3(m - 2) = 18$

8. Solve: $28 = -4(x - 1)$

9. Solve: $2(x - 4) + 7 = 3$

10. Solve: $0.2(n - 6) = 2.8$

11. Solve: $-5 = -(y + 1) - y$

12. Solve: $15x - 3(3x + 4) = 6$

MODEL PROBLEM 4: EQUATIONS WITH VARIABLES ON BOTH SIDES

Solve for x: $2(x-3) - 3 = 8x + 3$

Solution:

$$2(x-3) - 3 = 8x + 3$$
(A) $2x - 6 - 3 = 8x + 3$
$$2x - 9 = 8x + 3$$
(B) $-2x \quad\quad -2x$
$$-9 = 6x + 3$$
(C) $-12 = 6x$
$$-2 = x$$

Explanation of steps:

(A) Simplify the left side by distributing [$2(x-3)$ *becomes* $2x - 6$] and then by combining like terms [$-6 - 3 = -9$].

(B) Since there are variables on both sides, eliminate the variable term with the smaller coefficient [$2x$, *since* $2 < 8$] by adding its opposite [$-2x$] to both sides [$8x - 2x = 6x$].

(C) Now solve the simpler equation [*by adding –3 to both sides, then dividing both sides by 6*].

Check on the calculator: (-)2 STO▶ X,T,Θ,n ENTER

(1 means true.) 2 (X,T,Θ,n − 3) − 3 2nd[TEST]1 8 X,T,Θ,n + 3 ENTER

PRACTICE PROBLEMS

13. Solve for *x*: $3x + 8 = 5x$	14. Solve for *g*: $3 + 2g = 5g - 9$
15. What is the value of *p* in the equation $8p + 2 = 4p - 10$?	16. What is the value of *p* in the equation $5p - 1 = 2p + 20$?

17. Solve for y: $0.06y + 200 = 0.03y + 350$	18. Solve for x: $5 - 2x = -4x - 7$
19. What is the value of x in the equation $5(2x - 7) = 15x - 10$?	20. Solve for x: $5(x - 2) = 2(10 + x)$

21. Solve for x:
$$2(x - 4) = 4(2x + 1)?$$

22. Solve for x:
$$3(x + 1) - 5x = 12 - (6x - 7)$$

23. Solve for y:
$$-4(y - 3) = 5(2y - 6)$$

24. Solve for x:
$$3(x - 2) - 2(x + 1) = 5(x - 4)$$

Regents Questions

MULTIPLE CHOICE

1. An equation is given below.
$$4(x - 7) = 0.3(x + 2) + 2.11$$
The solution to the equation is
   - (1) 8.3
   - (2) 8.7
   - (3) 3
   - (4) −3

2. The solution to $-2(1 - 4x) = 3x + 8$ is
   - (1) $\frac{6}{11}$
   - (2) 2
   - (3) $-\frac{10}{7}$
   - (4) −2

3. The expression $3(x + 4) - (2x + 7)$ is equivalent to
   - (1) $x + 5$
   - (2) $x - 10$
   - (3) $x - 3$
   - (4) $x + 11$

4. The solution to $3(x - 8) + 4x = 8x + 4$ is
   - (1) 12
   - (2) 28
   - (3) −12
   - (4) −28

5. What is the solution to $2 + 3(2a + 1) = 3(a + 2)$?
   - (1) $\frac{1}{7}$
   - (2) $\frac{1}{3}$
   - (3) $-\frac{3}{7}$
   - (4) $-\frac{1}{3}$

CONSTRUCTED RESPONSE

6. Solve the equation algebraically for x: $-2.4(x + 1.4) = 6.8x - 22.68$

1.3   Solve Equations with Fractions

When an equation involves fractions, we can solve the equation by **eliminating the fractions.** We can accomplish this by **multiplying each term** of the equation by the **least common multiple (LCM)** of the denominators, also called the **least common denominator (LCD)**. This will eliminate the denominator of each term since it will divide evenly into the LCD.
(For this course, equations of this type will have no variables in the denominators.)

Why can we multiply each term by the LCD? This is because we are actually multiplying both sides of the equation by the LCD (*multiplication property of equality*) and then distributing, without actually showing the distribution step.

Example: To solve $\dfrac{x}{4} - \dfrac{1}{2} = 2$,

$$\boxed{4\left(\frac{x}{4} - \frac{1}{2}\right) = 4(2)} \;\Rightarrow\; 4\left(\frac{x}{4}\right) - 4\left(\frac{1}{2}\right) = 4(2) \;\Rightarrow\; x - 2 = 8 \;\Rightarrow\; x = 10$$

you may skip this step

▦▦ CALCULATOR TIP

To enter a fraction on the calculator:

- On the TI-83 models, fractions are entered as division; for example, $\frac{1}{2}$ is entered as ⑴÷②. If there are multiple terms joined by addition or subtraction in the numerator or denominator, be sure to place the terms in parentheses.

- On the TI-84 models, fractions can be entered by pressing ALPHA F1 ①. Type the numerator, press ▶, type the denominator, and press ▶ again. Mixed fractions can also be entered using ALPHA F1 ②.

Example: The following TI-84 screenshots show how to check the solution above. Remember to press MATH TEST ① for the equal sign.

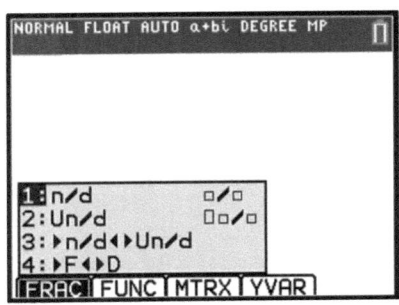 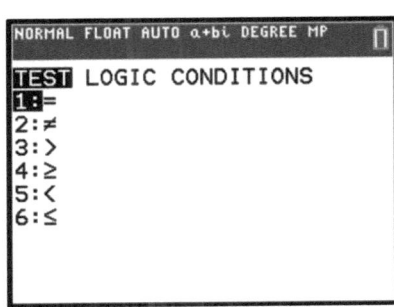 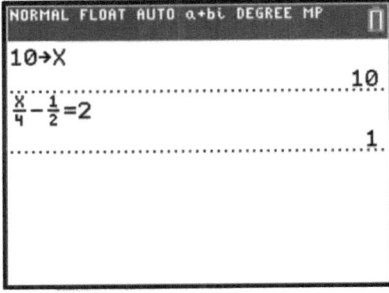

MODEL PROBLEM

Solve for x: $\dfrac{x+1}{4} - \dfrac{2x}{3} = \dfrac{x}{12}$

Solution:

(A) $12\left(\dfrac{x+1}{4}\right) - 12\left(\dfrac{2x}{3}\right) = 12\left(\dfrac{x}{12}\right)$

(B) $3(x+1) - 4(2x) = x$

(C)
$$3x + 3 - 8x = x$$
$$-5x + 3 = x$$
$$3 = 6x$$
$$\frac{1}{2} = x$$

Explanation of steps:

(A) Find the LCD *[12]*. Multiply each term by the LCD.

(B) Each denominator will divide evenly into the LCD, thereby eliminating the denominator.

(C) Solve the resulting equation.

PRACTICE PROBLEMS

1. Solve for x: $\dfrac{x}{16} + \dfrac{1}{4} = \dfrac{1}{2}$	2. Solve for x: $\dfrac{x}{2} + \dfrac{x}{6} = 2$
3. Solve for x: $\dfrac{2x}{3} + \dfrac{x}{6} = 5$	4. Solve for x: $\dfrac{3}{5}x + \dfrac{2}{5} = 4$
5. Solve for x: $\dfrac{3}{4}x + 2 = \dfrac{5}{4}x - 6$	6. Solve for x: $\dfrac{2}{3}x + \dfrac{1}{2} = \dfrac{5}{6}$
7. Solve for x: $\dfrac{3}{4}x = \dfrac{1}{3}x + 5$	8. Solve for x: $\dfrac{x}{3} + \dfrac{x+1}{2} = x$

9. Solve for x: $\dfrac{2x}{5} + \dfrac{1}{3} = \dfrac{7x-2}{15}$

10. Solve for x: $\dfrac{1}{7} + \dfrac{2x}{3} = \dfrac{15x-3}{21}$

11. Solve for x: $\dfrac{3}{4}(x+3) = 9$

12. Solve for x: $\dfrac{3}{5}(x+2) = x - 4$

13. Solve for x: $\dfrac{1}{2}(18 - 5x) = \dfrac{1}{3}(6 - 4x)$

14. Solve for x: $\dfrac{2}{3}\left(2x - \dfrac{1}{2}\right) = 13$

15. Solve for m: $\dfrac{m}{5} + \dfrac{3(m-1)}{2} = 2(m-3)$

Regents Questions

MULTIPLE CHOICE

1. Which value of x satisfies the equation $\frac{7}{3}\left(x + \frac{9}{28}\right) = 20$?

   (1) 8.25

   (2) 8.89

   (3) 19.25

   (4) 44.92

2. What is the value of x in the equation $\frac{x-2}{3} + \frac{1}{6} = \frac{5}{6}$?

   (1) 4

   (2) 6

   (3) 8

   (4) 11

3. Which value of x satisfies the equation $\frac{5}{6}\left(\frac{3}{8} - x\right) = 16$?

   (1) −19.575

   (2) −18.825

   (3) −16.3125

   (4) −15.6875

4. The value of x which makes $\frac{2}{3}\left(\frac{1}{4}x - 2\right) = \frac{1}{5}\left(\frac{4}{3}x - 1\right)$ true is

   (1) −10

   (2) −2

   (3) $-9.\overline{09}$

   (4) $-11.\overline{3}$

5. What is the solution to the equation $\frac{3}{5}\left(x + \frac{4}{3}\right) = 1.04$?

   (1) $3.0\overline{6}$

   (2) 0.4

   (3) $-0.4\overline{8}$

   (4) $-0.709\overline{3}$

6. The value of x that satisfies the equation $\frac{4}{3} = \frac{x+10}{15}$ is

   (1) −6

   (2) 5

   (3) 10

   (4) 30

7. Which value of x makes $\frac{x-3}{4} + \frac{2}{3} = \frac{17}{12}$ true?

   (1) 8

   (2) 6

   (3) 0

   (4) 4

8. Which of the equations below have the same solution?

   I. $10(x - 5) = -15$

   II. $4 + 2(x - 2) = 9$

   III. $\frac{1}{3}x = \frac{3}{2}$

   (1) I and II, only

   (2) I and III, only

   (3) II and III, only

   (4) I, II, and III

9.   What is the value of x in the equation $\dfrac{5(2x-4)}{3}+9=14$?

    (1) 1.9   (3) 5.3

    (2) 3.5   (4) 8.9

10.   The solution to $\dfrac{2}{3}(3-2x)=\dfrac{3}{4}$ is

    (1) $-\dfrac{11}{8}$ (3) $-\dfrac{33}{16}$

    (2) $\dfrac{5}{8}$ (4) $\dfrac{15}{16}$

CONSTRUCTED RESPONSE

11.   Solve the equation below algebraically for the exact value of x.
$$6-\frac{2}{3}(x+5)=4x$$

12.   Solve algebraically for x: $-\dfrac{2}{3}(x+12)+\dfrac{2}{3}x=-\dfrac{5}{4}x+2$

1.4   <u>Solve Linear Inequalities in One Variable</u>

An **inequality** is a statement that one expression is compared to, but not equal to, another. It contains one of the other relational symbols: $<, >, \leq, \geq,$ or $\neq$.
Example: $15 < 2x + 1$

Solving an inequality: Follow the same steps as in solving an equation EXCEPT when <u>multiplying or dividing</u> both sides by a <u>negative</u> number, <u>REVERSE</u> the inequality symbol.
Example: If $-3x \leq 15$, then dividing both sides by -3 gives us $x \geq -5$.

It may help to "flip" after solving: Once the variable is isolated, if it appears on the right side of the inequality, it is usually helpful to "flip" the entire inequality (switch the two sides and reverse the inequality symbol) so that the variable appears on the left side instead.
Example: The inequality $6 > y$ can be "flipped" to $y < 6$.

Graphing an inequality: Once a simple inequality is solved for a variable, it can be graphed on a number line using the following steps.
1. On the number line, find the value that the variable is being compared to.
2. If the inequality symbol is:
   a. **> or <**, draw an open circle ◯ at that value, which means it is *not* in the solution set.
   b. **≥ or ≤**, draw a closed circle ● at that value, which means it is in the solution set.
3. If the variable is > or ≥ the value, shade an arrow to the right; otherwise, if the variable is < or ≤ the value, shade an arrow to the left. All the values represented by the shaded arrow, extending infinitely in that direction, are in the solution set.

Examples: (a) The number line below represents $x > 2$.

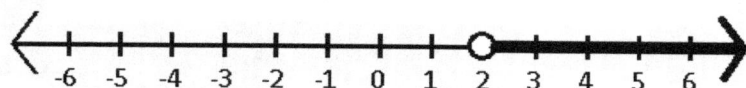

           (b) The number line below represents $x \leq 3$.

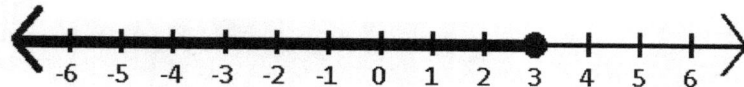

MODEL PROBLEM

Solve and graph the solution set: $7 > 5 - 2x$

Solution:

$7 > 5 - 2x$

(A) $\quad \underline{-5 \quad -5}$

$\quad \dfrac{2 > -2x}{-2 \quad -2}$

(B) $\quad -1 < x$

(C) $\quad x > -1$

Explanation of steps:

(A) Solve like an equation. *[Here, we subtract 5 from both sides, then divide both sides by −2.]*

(B) If, in the process of solving, we multiply or divide both sides by a negative number, reverse the inequality symbol. *[Since we divided by−2, we reverse the symbol from > to <.]*

(C) After solving, if the variable is on the right side, it is helpful to "flip" the entire inequality, including the inequality symbol, so that it appears on the left side.

(D) Graph the result *[open circle, arrow to the greater side].*

(D)

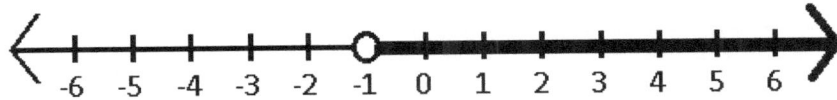

PRACTICE PROBLEMS

1. Write an inequality that represents the graph below. 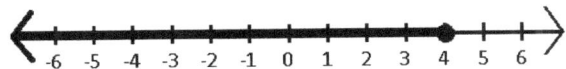	2. Write an inequality that represents the graph below. 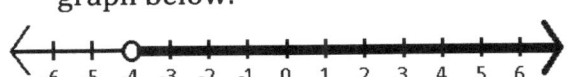
3. Solve and graph the solution set: $2x - 5 \leq 11$	4. Solve and graph the solution set: $-6y + 1 > 25$

5. Solve and graph the solution set: $-4 > 2(r - 3)$	6. Solve and graph the solution set: $-\frac{4}{3}(x - 3) \le 12$
7. Solve: $-6x - 17 \ge 8x + 25$	8. Solve: $-5(x - 7) < 15$
9. Which graph represents the solution set of $2x - 5 < 3$? (1) (2) (3) (4)	10. Solve: $3(2m - 1) \le 4m + 7$
11. Solve: $-4(2m - 6) + m > 3m + 4$	12. Solve: $-5(p + 1) \ge -p + 11$

Regents Questions

MULTIPLE CHOICE

1. The inequality $7 - \frac{2}{3}x < x - 8$ is equivalent to

    (1) $x > 9$

    (2) $x > -\frac{3}{5}$

    (3) $x < 9$

    (4) $x < -\frac{3}{5}$

2. When $3x + 2 \le 5(x - 4)$ is solved for x, the solution is

    (1) $x \le 3$

    (2) $x \ge 3$

    (3) $x \le -11$

    (4) $x \ge 11$

3. What is the solution to $2h + 8 > 3h - 6$?

    (1) $h < 14$

    (2) $h < \frac{14}{5}$

    (3) $h > 14$

    (4) $h > \frac{14}{5}$

4. Which value would be a solution for x in the inequality $47 - 4x < 7$?

    (1) -13

    (2) -10

    (3) 10

    (4) 11

5. What is the solution to the inequality $2 + \frac{4}{9}x \ge 4 + x$?

    (1) $x \le -\frac{18}{5}$

    (2) $x \ge -\frac{18}{5}$

    (3) $x \le \frac{54}{5}$

    (4) $x \ge \frac{54}{5}$

6. Given the set $\{x| -2 \le x \le 2$, where x is an integer$\}$, what is the solution of $-2(x - 5) < 10$?

    (1) $0, 1, 2$

    (2) $1, 2$

    (3) $-2, -1, 0$

    (4) $-2, -1$

7. The solution to $4p + 2 < 2(p + 5)$ is

    (1) $p > -6$

    (2) $p < -6$

    (3) $p > 4$

    (4) $p < 4$

8. Given $7x + 2 \ge 58$, which number is _not_ in the solution set?

    (1) 6

    (2) 8

    (3) 10

    (4) 12

9. What is the solution to $\frac{3}{2}b + 5 < 17$?

    (1) $b < 8$

    (2) $b > 8$

    (3) $b < 18$

    (4) $b > 18$

10. What is the solution to the inequality below?

$$4 - \frac{2}{5}x \geq \frac{1}{3}x + 15$$

   (1) $x \leq 11$ (3) $x \leq -15$

   (2) $x \geq 11$ (4) $x \geq -15$

11. What is the solution to $-3(x - 6) > 2x - 2$?

   (1) $x > 4$ (3) $x > -16$

   (2) $x < 4$ (4) $x < -16$

12. What is the solution to the inequality $2x - 7 > 2.5x + 3$?

   (1) $x > -5$ (3) $x > -20$

   (2) $x < -5$ (4) $x < -20$

CONSTRUCTED RESPONSE

13. Given $2x + ax - 7 > -12$, determine the largest integer value of a when $x = -1$.

14. Solve the inequality below to determine and state the smallest possible value for x in the solution set.

$$3(x + 3) \leq 5x - 3$$

15. Determine the smallest integer that makes $-3x + 7 - 5x < 15$ true.

16. Solve for x algebraically: $7x - 3(4x - 8) \leq 6x + 12 - 9x$

   If x is a number in the interval [4,8], state all integers that satisfy the given inequality. Explain how you determined these values.

17. Solve the inequality below:

$$1.8 - 0.4y \geq 2.2 - 2y$$

18. Solve algebraically for x: $3600 + 1.02x < 2000 + 1.04x$

19. Solve $\frac{3}{5}x + \frac{1}{3} < \frac{4}{5}x - \frac{1}{3}$ for x.

20. Solve algebraically for y: $4(y - 3) \leq 4(2y + 1)$

21. Solve the inequality $-\frac{2}{3}x + 6 > -12$ algebraically for x.

1.5   Solve Literal Equations and Inequalities

Literal equations and **literal inequalities** contain several variables or letters. To solve a literal equation or literal inequality for a specific variable, perform the same steps to isolate that variable as you would for any equation or inequality. In other words, solve for the specific variable by transforming the equation into one where the variable is written as equal to an expression **in terms of** the other variables.

Example: To solve $c = a + b$ for a, we can isolate a by subtracting b from both sides. This gives us $c - b = a$, an equation where a is written in terms of b and c. By the symmetric property of equality, we could also rewrite this as $a = c - b$.

If the specified variable appears in **more than one term** that cannot be combined, use the distributive property to factor it out before dividing both sides by the other factor.

Example: $mx + nx$ can be rewritten as $x(m + n)$ if you need to isolate x.

When solving a *literal inequality*, remember that the inequality symbol needs to switch whenever you multiply or divide both sides by a negative number.

Example: To solve $\frac{x}{a} > b$, we need to know whether a is positive or negative.

(We know $a \neq 0$ since a value of zero would make the fraction undefined.)

If $a > 0$, then $x > ab$, but if $a < 0$, then $x < ab$.

A formula is an equation made up of **dependent and independent variables**. The dependent variable's value is determined based on the freely chosen values of the independent variables. In a formula, the dependent variable is written *in terms of* the independent variables.

Example: In the formula for the area of a triangle, $A = \frac{1}{2}bh$, A is the dependent variable and b and h are the independent variables. A is dependent on b and h.

When we are asked to solve a formula for a different variable, the status of the variables in the resulting formula will change.

Example: If we solve the formula $A = \frac{1}{2}bh$ for h, the resulting formula is $h = \frac{2A}{b}$. In this new formula, h is the dependent variable and A and b are independent variables.

MODEL PROBLEM 1:   *MULTI-STEP LITERAL EQUATIONS*

Solve for x in terms of a, b, c, and y: $ax + by = c$

Solution:

$$ax + by = c$$

(A) $\dfrac{-by \quad -by}{}$

$$\dfrac{ax}{a} = \dfrac{c - by}{a}$$

(B)

$$x = \dfrac{c-by}{a}$$

Explanation of steps:

(A) Isolate the specified variable *[x]* by first eliminating terms that do not contain the variable. Add the opposite of any such term *[–by]* to both sides.

(B) Eliminate any coefficient (numerical or literal) of the specified variable *[a]* by dividing both sides by it.

PRACTICE PROBLEMS

1. Solve for p in terms of m: $2m + 2p = 16$	2. Solve for x in terms of b and K: $bx - 2 = K$
3. Solve for m in terms of c and d: $c > 2m + d$	4. Solve for x in terms of a, b, and c: $bx - 3a = c$
5. Solve for w in terms of V, l, and h: $V = lwh$	6. Solve for h in terms of A and b: $A = \dfrac{bh}{2}$

7. If $abx - 5 = 0$, then what is x in terms of a and b?	8. If $2y + 2w = x$, then what is w in terms of x and y?
9. Solve for x in terms of r, s, and t: $$s = \frac{2x + t}{r}$$	10. The formula for the volume of a pyramid is $V = \frac{1}{3}Bh$. What is h expressed in terms of B and V?
11. A formula used for calculating velocity is $v = \frac{1}{2}at^2$. What is a expressed in terms of v and t?	12. If $\frac{ey}{n} + k \geq t$, what is y in terms of e, n, k, and t, given that n and k are both positive and e is negative?

MODEL PROBLEM 2:   *USING THE DISTRIBUTIVE PROPERTY*

Solve for x in terms of a, b, and c: $ax + bx = c$

Solution:

$$ax + bx = c$$

(A) $x(a + b) = c$

(B) $\dfrac{x(a + b)}{a + b} = \dfrac{c}{a + b}$

$x = \dfrac{c}{a + b}$

Explanation of steps:

(A) If the variable we're trying to isolate *[x]* appears in more than one term on the same side of the equation, use the distributive property to factor it out.
       [ax + bx = x(a + b)]

(B) Then, divide both sides by the other factor *[a + b]*.

PRACTICE PROBLEMS

13. Solve for x in terms of a and b:
     $3x - ax = b$
	14. Solve for c in terms of a and b:
	     $bc + ac = ab$

15. If $k = am + 3mx$, what is m in terms of a, k, and x?	16. Solve for x in terms of a and b: $2ax = -bx + 1$
17. If $ax + 3 = 7 - bx$, what is x expressed in terms of a and b?	18. If $z + y = x + xy^2$, what is x expressed in terms of y and z?

Regents Questions

MULTIPLE CHOICE

1. Michael borrows money from his uncle, who is charging him simple interest using the formula $I = Prt$. To figure out what the interest rate, r, is, Michael rearranges the formula to find r. His new formula is r equals

    (1) $\dfrac{I - P}{t}$

    (2) $\dfrac{P - I}{t}$

    (3) $\dfrac{I}{Pt}$

    (4) $\dfrac{Pt}{I}$

2. Boyle's Law involves the pressure and volume of gas in a container. It can be represented by the formula $P_1V_1 = P_2V_2$. When the formula is solved for P_2, the result is

    (1) $P_1V_1V_2$

    (2) $\dfrac{V_2}{P_1V_1}$

    (3) $\dfrac{P_1V_1}{V_2}$

    (4) $\dfrac{P_1V_2}{V_1}$

3. The formula for the surface area of a right rectangular prism is $A = 2lw + 2hw + 2lh$, where l, w, and h represent the length, width, and height, respectively. Which term of this formula is *not* dependent on the height?

    (1) A

    (2) $2lw$

    (3) $2hw$

    (4) $2lh$

4. The formula for blood flow rate is given by $F = \dfrac{p_1 - p_2}{r}$, where F is the flow rate, p_1 the initial pressure, p_2 the final pressure, and r the resistance created by blood vessel size. Which formula can *not* be derived from the given formula?

    (1) $p_1 = Fr + p_2$

    (2) $p_2 = p_1 - Fr$

    (3) $r = F(p_2 - p_1)$

    (4) $r = \dfrac{p_1 - p_2}{F}$

5. Students were asked to write a formula for the length of a rectangle by using the formula for its perimeter, $p = 2l + 2w$. Three of their responses are shown below.

    I. $l = \dfrac{1}{2}p - w$

    II. $l = \dfrac{1}{2}(p - 2w)$

    III. $l = \dfrac{p - 2w}{2}$

    Which responses are correct?

    (1) I and II, only

    (2) II and III, only

    (3) I and III, only

    (4) I, II, and III

6. The formula for electrical power, P, is $P = I^2 R$, where I is current and R is resistance. The formula for I in terms of P and R is

    (1) $I = \left(\dfrac{P}{R}\right)^2$

    (2) $I = \sqrt{\dfrac{P}{R}}$

    (3) $I = (P - R)^2$

    (4) $I = \sqrt{P - R}$

7. When $3a + 7b > 2a - 8b$ is solved for a, the result is

    (1) $a > -b$

    (2) $a < -b$

    (3) $a < -15b$

    (4) $a > -15b$

8. The formula $Ax + By = C$ represents the equation of a line in standard form. Which expression represents y in terms of A, B, C, and x?

    (1) $\dfrac{C - Ax}{B}$

    (2) $\dfrac{C - A}{Bx}$

    (3) $\dfrac{C - A}{x + B}$

    (4) $\dfrac{C - B}{Ax}$

9. When the equation $\dfrac{x - 1}{2} - \dfrac{a}{4} = \dfrac{3a}{4}$ is solved for x in terms of a, the solution is

    (1) $\dfrac{3a}{2} + 1$

    (2) $a + 1$

    (3) $\dfrac{4a + 1}{2}$

    (4) $2a + 1$

10. The volume of a trapezoidal prism can be found using the formula $V = \frac{1}{2}a(b + c)h$. Which equation is correctly solved for b?

    (1) $b = \dfrac{V}{2ah} + c$

    (2) $b = \dfrac{V}{2ah} - c$

    (3) $b = \dfrac{2V}{ah} + c$

    (4) $b = \dfrac{2V}{ah} - c$

11. The amount of energy, Q, in joules, needed to raise the temperature of m grams of a substance is given by the formula $Q = mC(T_f - T_i)$, where C is the specific heat capacity of the substance. If its initial temperature is T_i, an equation to find its final temperature, T_f, is

    (1) $T_f = \dfrac{Q}{mC} - T_i$

    (2) $T_f = \dfrac{Q}{mC} + T_i$

    (3) $T_f = \dfrac{T_i + Q}{mC}$

    (4) $T_f = \dfrac{Q - mC}{T_i}$

12. The formula for the area of a trapezoid is $A = \frac{1}{2}(b_1 + b_2)h$. The height, h, of the trapezoid may be expressed as

    (1) $2A - b_1 - b_2$

    (2) $\dfrac{2A - b_1}{b_2}$

    (3) $\frac{1}{2}A - b_1 - b_2$

    (4) $\dfrac{2A}{b_1 + b_2}$

13. An equation used to find the velocity of an object is given as $v^2 = u^2 + 2as$, where u is the initial velocity, v is the final velocity, a is the acceleration of the object, and s is the distance traveled. When this equation is solved for a, the result is

(1) $a = \dfrac{v^2 u^2}{2s}$ (3) $a = v^2 - u^2 - 2s$

(2) $a = \dfrac{v^2 - u^2}{2s}$ (4) $a = 2s(v^2 - u^2)$

CONSTRUCTED RESPONSE

14. The formula for the area of a trapezoid is $A = \frac{1}{2}h(b_1 + b_2)$. Express b_1 in terms of A, h, and b_2.

The area of a trapezoid is 60 square feet, its height is 6 ft, and one base is 12 ft. Find the number of feet in the other base.

15. Given that $a > b$, solve for x in terms of a and b:
$$b(x - 3) \geq ax + 7b$$

16. The formula for the sum of the degree measures of the interior angles of a polygon is $S = 180(n - 2)$. Solve for n, the number of sides of the polygon, in terms of S.

17. Solve the equation below for x in terms of a.
$$4(ax + 3) - 3ax = 25 + 3a$$

18. The formula for converting degrees Fahrenheit (F) to degrees Kelvin (K) is:
$$K = \frac{5}{9}(F + 459.67)$$
Solve for F, in terms of K.

19. The formula for the volume of a cone is $V = \frac{1}{3}\pi r^2 h$. Solve the equation for h in terms of V, r, and π.

20. The formula $a = \dfrac{v_f - v_i}{t}$ is used to calculate acceleration as the change in velocity over the period of time. Solve the formula for the final velocity, v_f, in terms of initial velocity, v_i, acceleration, a, and time, t.

21. A formula for determining the finite sum, S, of an arithmetic sequence of numbers is $S = \frac{n}{2}(a + b)$, where n is the number of terms, a is the first term, and b is the last term. Express b in terms of a, S, and n.

22. The temperature inside a cooling unit is measured in degrees Celsius, C. Josh wants to find out how cold it is in degrees Fahrenheit, F. Solve the formula $C = \frac{5}{9}(F - 32)$ for F so that Josh can convert Celsius to Fahrenheit.

23. The formula $d = t\left(\dfrac{v_i + v_f}{2}\right)$ is used to calculate the distance, d, covered by an object in a given period of time, t. Solve the formula for v_f, the final velocity, in terms of d, t, and v_i, the initial velocity.

CHAPTER 2   VERBAL PROBLEMS

2.1   Translate Expressions

To translate verbal expressions into algebraic expressions, look for phrases commonly used to represent operations:

Addition	Subtraction	Multiplication	Division
increased by	decreased by	multiplied by	divided by
sum	difference	product	quotient
plus	minus	times	
more than	*less than*		
added to	*subtracted from*		

Of course, this is not a comprehensive list. For example, just as we can express subtraction using the word, "decreased," we can also use a synonym, such as "diminished" or "reduced."

Order of operands: Generally, operands are written in the order they appear in the verbal expression, with the important exception of the phrases written in *italics* above.

Example: The following are written as $a - b$,   but the following are written as $b - a$
                        a decreased by b a less than b
                        difference of a and b a subtracted from b
                        a minus b

Addition and multiplication are commutative, so the order of their operands shouldn't matter.

Note also: When we multiply a variable by a number, we generally write the number (called the numerical coefficient) first.

Example: "x times 5" is usually written $5x$ rather than $x \cdot 5$

Other common phrases:

Twice	means two times:	"twice x" is written $2x$
Fraction of	means fraction times:	"two-fifths of x" is written $\frac{2}{5}x$
The quantity	means parentheses:	"twice the quantity $x + 5$" is written $2(x + 5)$

Writing an expression "in terms of" a variable:
- Often in verbal expressions, a quantity is described by comparing it to another quantity.
- This latter quantity, to which it is being compared, will be represented by a variable.
- This variable quantity will often appear at the end of a verbal clause, and commonly *after* comparative words such as "than" or "as many as."
- Once the variable quantity is established, then the other quantities will be written as expressions containing, or in terms of, this variable.

Example: If "John's age is 5 less *than* Tom's age," let the variable t represent Tom's age.
                        John's age is represented by an expression in terms of t, namely: $t - 5$.

46

Ratios: if a ratio is stated between two (or more) quantities, then use a variable to represent their common factor.

Example: If the numbers of boys and girls in a class are in the ratio 3:2 (read as "three to two"), then express the number of boys as $3x$ and the number of girls as $2x$.

Consecutive integers are such that each integer is <u>1 larger than the previous</u>.

Examples: 4, 5, and 6 are three consecutive integers, as are –2, –1, and 0.

If x is the smallest of three consecutive integers, then the three numbers can be expressed as x, $x + 1$, and $x + 2$. The next consecutive integer after $x + 2$ is $x + 3$.

Consecutive even integers (for example, 6, 8, 10, and 12) and **consecutive odd integers** (for example, 5, 7, 9, and 11) are each <u>2 larger than the previous</u>. So, If x is the smallest of four consecutive even integers, then the four numbers can be expressed as x, $x + 2$, $x + 4$, and $x + 6$. (The same expressions are used if x is the smallest of four consecutive *odd* integers.)

Monetary values: to find the value of coins or bills, multiply the number of each coin or bill by its denomination, and add the products.

Example: x dimes and $3x$ nickels have a total value of $(0.10)x + (0.05)3x$, or $0.25x$.

Total costs: to find the total cost of items purchased, multiply the number of each item by its price, and add the products.

Example: x apples at \$0.25 each and $(10 - x)$ oranges at \$0.40 each costs a total of $(0.25)x + (0.40)(10 - x)$, or after simplifying, $4.00 - 0.15x$.

Chain comparisons: In some cases, a quantity is written as an expression in terms of a variable, and then another quantity is written in terms of *this expression*. Use a variable for the quantity at the *end of the chain*, and then work backwards to express the others.

Example: Mark is 4 years younger than Samuel and Charles is three times as old as Mark.
What is the sum of their ages written as an expression?
Mark is described in terms of Samuel and Charles is described in terms of Mark, so the chain is: Charles → Mark → Samuel.
Let s be Samuel's age.
If s is Samuel's age, Mark's age is $s - 4$, and Charles' age is $3(s - 4)$.
The sum of their ages is $s + (s - 4) + 3(s - 4)$, which simplifies to $5s - 16$.

MODEL PROBLEM 1:   *EXPRESSIONS IN TERMS OF ONE VARIABLE*

Abby, Barbie, and Carol are sisters. Abby's age is four times Carol's age and Barbie's age if 5 less than twice Carol's age. Write an expression, in terms of Carol's age, *c*, for the sum of their ages.

Solution:

Carol's age is *c*.   Abby's age is $4c$.   Barbie's age is $2c - 5$.

So the sum of their ages can be represented by $c + 4c + 2c - 5$, which simplifies to $7c - 5$.

Explanation of steps:

A common temptation is to create a separate variable for each quantity *[the three ages]*. But, since all other quantities *[Abby's and Barbie's ages]* are expressed using comparisons to a single value *[Carol's age]*, it is much better to use only one variable *[c]* and then write expressions for the others in terms of this variable.

PRACTICE PROBLEMS

1. Which verbal expression can be represented by $2(x - 5)$? (1) 5 less than 2 times *x* (2) 2 multiplied by x less than 5 (3) twice the difference of *x* and 5 (4) the product of 2 and *x*, decreased by 5	2. Which verbal expression is represented by $\frac{1}{2}(n - 3)$? (1) one-half *n* decreased by 3 (2) one-half *n* subtracted from 3 (3) the difference of one-half *n* and 3 (4) one-half the difference of *n* and 3
3. Write an expression for 5 less than the product of 7 and *x*.	4. Write an expression for twice the difference of *x* and 8.
5. The sum of Scott's age and Greg's age is 33 years. If Greg's age is represented by *g*, what is Scott's age in terms of *g*?	6. Tara buys two items that cost *d* dollars each. She gives the cashier $20, which is more than the total cost. Write an expression to represent the change she should receive.

7. John is four times as old as Ashley. If x represents Ashley's age, write an expression to represent how old John will be in 10 years.

8. If n represents the height of an object in inches, write an expression in terms of n to represent the height of the object in feet. (12 inches = 1 foot.)

9. If Jose's weekly allowance is d dollars, write an expression for his allowance, in dollars, for x weeks?

10. A skateboard and two helmets cost a total of d dollars. If each helmet cost h dollars, write an expression for the cost of the skateboard.

11. What is the perimeter of a regular pentagon with a side whose length is $x + 4$?

12. Angelina determined that her father's age is four less than three times her age. If x represents Angelina's age, write an expression for her father's age?

13. Marcy bought d dollars worth of stock. During the first year, the value of the stock tripled. The next year, the value of the stock decreased by $1200. Write an expression in terms of d to represent the value of the stock after two years.

14. Charles gets paid $280 per week plus 5% commission on all sales for selling gym memberships. If he sells n dollars worth of gym memberships in one week, write an expression for the amount of money he will earn that week.

MODEL PROBLEM 2:   *CONSECUTIVE INTEGERS*

Write an expression, in simplest terms, for the sum of three consecutive odd integers.

Solution:
$x + x + 2 + x + 4 = 3x + 6$

Explanation of steps:
Let a variable *[x]* represent the smallest of the integers. Write expressions involving this variable for each of the other integers *[x+2 and x+4]*. Express the sum and simplify the result by combining like terms.

PRACTICE PROBLEMS

15. If *y* is the smallest of four consecutive integers, write an expression, in simplest terms, for the sum of the four integers.	16. If the smallest of three consecutive even integers is $x + 3$, write the sum of the three integers in simplest terms.
17. The ages of three children are consecutive odd integers. If *x* represents the youngest child's age, write an expression, in simplest terms, for the sum of the children's ages.	18. Two numbers are consecutive integers. If *x* represents the smaller number, write an expression, in simplest terms, for the product of the numbers.

MODEL PROBLEM 3: *CHAIN COMPARISONS*

Ashanti and Maria went to the store to buy snacks for their back-to-school party. They bought bags of chips, pretzels, and nachos. They bought three times as many bags of pretzels as bags of chips, and two fewer bags of nachos than bags of pretzels. If x represents the number of bags of chips they bought, express, in terms of x, how many bags of snacks they bought in all.

Solution:
- (A) Let x be the number of bags of chips bought.
- (B) Then, $3x$ represents the bags of pretzels, and $3x - 2$ represents the bags of nachos.
- (C) Altogether, they bought $x + 3x + 3x - 2$, or $7x - 2$, bags of snacks.

Explanation of steps:
- (A) Create the comparison chain and let the variable represent the quantity at the end of the chain. *[Pretzels are expressed in terms of chips and nachos are expressed in terms of pretzels. So, the chain is Nachos → Pretzels → Chips. Let x represent the chips.]*
- (B) By working backwards in the chain, express the other quantities in terms of this variable.
- (C) Combine the expressions using the operation specified.

PRACTICE PROBLEMS

19. Camille is 7 years older than Donny, and Donny is 4 years younger than Tommy. Write an expression for the total ages of the three people, in simplest form.	20. The life span of a whale is 4 times than of a stork, which lives 70 years longer than a horse. Write an expression, in simplest form, for the total life spans of the three creatures.

21. Carl ate four more cookies than Alice. Bob ate twice as many cookies as Carl. Write an expression that represents the number of cookies Bob ate.

22. Two friends went to the store to buy snacks for a party. They bought bags of chips, pretzels, and nachos. They bought three times as many bags of pretzels as bags of chips, and two fewer bags of nachos than bags of pretzels. If x represents the number of bags of chips they bought, express, in terms of x, how many bags of snacks they bought in all.

Regents Questions

MULTIPLE CHOICE

1. To watch a varsity basketball game, spectators must buy a ticket at the door. The cost of an adult ticket is $3.00 and the cost of a student ticket is $1.50. If the number of adult tickets sold is represented by a and student tickets sold by s, which expression represents the amount of money collected at the door from the ticket sales?

    (1) $4.50as$ (3) $(3.00a)(1.50s)$

    (2) $4.50(a + s)$ (4) $3.00a + 1.50s$

2. Andy has $310 in his account. Each week, w, he withdraws $30 for his expenses. Which expression could be used if he wanted to find out how much money he had left after 8 weeks?

    (1) $310 - 8w$ (3) $310w - 30$

    (2) $280 + 30(w - 1)$ (4) $280 - 30(w - 1)$

3. Konnor wants to burn 250 Calories while exercising for 45 minutes at the gym. On the treadmill, he can burn 6 Cal/min. On the stationary bike, he can burn 5 Cal/min.

    If t represents the number of minutes on the treadmill and b represents the number of minutes on the stationary bike, which expression represents the number of Calories that Konnor can burn on the stationary bike?

    (1) b (3) $45 - b$

    (2) $5b$ (4) $250 - 5b$

4. Bryan's hockey team is purchasing jerseys. The company charges $250 for a onetime set-up fee and $23 for each printed jersey. Which expression represents the total cost of x number of jerseys for the team?

    (1) $23x$ (3) $23x + 250$

    (2) $23 + 250x$ (4) $23(x + 250)$

2.2 Translate Equations

An **equation** is simply a statement in which two expressions are set equal to each other. So, to translate a verbal problem into an equation:
1. translate each quantity into an expression in terms of a variable
2. determine from the problem how the expressions form an equation

Examples: the sum of x and $2x$ is 42   becomes $x + 2x = 42$
                 $4s$ is 15 more than s becomes $4s = s + 15$

Sometimes a **known formula** needs to be applied in order to write the equation.

Example: If a rectangle's length is $4x$, its width is x, and its area is 100, use the formula $A = lw$ to produce the equation, $100 = (4x)(x)$, or $100 = 4x^2$

MODEL PROBLEM

Write an equation to find three consecutive even integers whose sum is 84.

Solution:
(A)   Let x be the first even integer.
(B)   Therefore, $x + 2$ is the second even integer and $x + 4$ is the third even integer.
(C)   So, $x + (x + 2) + (x + 4) = 84$, or, simplifying the left side of the equation, $3x + 6 = 84$.

Explanation of steps:
(A)   Determine which quantity will be represented by the variable *[in a consecutive integers problem, x can represent the first in the sequence]*.
(B)   Express the other quantities as expressions in terms of this variable *[each consecutive even integer is two larger than the previous]*.
(C)   Arrange the expressions to form a correct equation *[the sum is 84]*.

PRACTICE PROBLEMS

1. If h represents a number, write an equation for the statement, "Sixty more than 9 times the number is 375"	2. Three times the sum of a number and four is equal to five times the number, decreased by two. If x represents the number, write an equation that could be used to find x.
3. The width of a rectangle is 4 less than half the length. If l represents the length, write an equation that could be used to find the width, w.	4. The width of a rectangle is 3 less than twice the length, x. If the area of the rectangle is 43 square feet, write an equation that could be used to find the length, in feet.
5. The radius of a circle is represented by $3x + 2$, and the length of the diameter is 22 centimeters. Write an equation to find the value of x, in centimeters.	6. Jerome purchased four more apples than oranges. Apples cost 30 cents each and oranges cost 50 cents each. Jerome spent a total of $3.60. Write an equation to find how many oranges he purchased.

7. Byron has $1.35 in nickels and dimes in his piggy bank. If he has six more dimes than nickels, write an equation that could be used to determine n, the number of nickels he has.	8. Rhonda has 72 coins in a jar. The jar contains only dimes and quarters. If the jar contains $14.70, write an equation that could be used to determine q, the number of quarters in the jar.
9. Write an equation that could be used to find two consecutive integers whose product is 20.	10. Write an equation that could be used to find three consecutive odd integers whose sum is –3.

Regents Questions

MULTIPLE CHOICE

1. The length of the shortest side of a right triangle is 8 inches. The lengths of the other two sides are represented by consecutive odd integers. Which equation could be used to find the lengths of the other sides of the triangle?

 (1) $8^2 + (x + 1)^2 = x^2$ (3) $8^2 + (x + 2)^2 = x^2$

 (2) $x^2 + 8^2 = (x + 1)^2$ (4) $x^2 + 8^2 = (x + 2)^2$

2. John has four more nickels than dimes in his pocket, for a total of $1.25. Which equation could be used to determine the number of dimes, x, in his pocket?

   (1) $0.10(x + 4) + 0.05(x) = \$1.25$ (3) $0.10(4x) + 0.05(x) = \$1.25$
   (2) $0.05(x + 4) + 0.10(x) = \$1.25$ (4) $0.05(4x) + 0.10(x) = \$1.25$

3. Joe has a rectangular patio that measures 10 feet by 12 feet. He wants to increase the area by 50% and plans to increase each dimension by equal lengths, x. Which equation could be used to determine x?

   (1) $(10 + x)(12 + x) = 120$ (3) $(15 + x)(18 + x) = 180$
   (2) $(10 + x)(12 + x) = 180$ (4) $(15)(18) = 120 + x^2$

4. Kendal bought x boxes of cookies to bring to a party. Each box contains 12 cookies. She decides to keep two boxes for herself. She brings 60 cookies to the party. Which equation can be used to find the number of boxes, x, Kendal bought?

   (1) $12x - 12 = 60$ (3) $12x - 24 = 60$
   (2) $12x - 2 = 60$ (4) $24 - 12x = 60$

5. Nicci's sister is 7 years less than twice Nicci's age, a. The sum of Nicci's age and her sister's age is 41. Which equation represents this relationship?

   (1) $a + (7 - 2a) = 41$ (3) $2a - 7 = 41$
   (2) $a + (2a - 7) = 41$ (4) $a = 2a - 7$

6. The length of a rectangular patio is 7 feet more than its width, w. The area of a patio, $A(w)$, can be represented by the function

   (1) $A(w) = w + 7$ (3) $A(w) = 4w + 14$
   (2) $A(w) = w^2 + 7w$ (4) $A(w) = 4w^2 + 28w$

7. Joe has dimes and nickels in his piggy bank totaling $1.45. The number of nickels he has is 5 more than twice the number of dimes, d. Which equation could be used to find the number of dimes he has?

   (1) $0.10d + 0.05(2d + 5) = 1.45$ (3) $d + (2d + 5) = 1.45$
   (2) $0.10(2d + 5) + 0.05d = 1.45$ (4) $(d - 5) + 2d = 1.45$

CONSTRUCTED RESPONSE

8. The cost of one pound of grapes, g, is 15 cents more than one pound of apples, a. The cost of one pound of bananas, b, is twice as much as one pound of grapes. Write an equation that represents the cost of one pound of bananas in terms of the cost of one pound of apples.

2.3   Linear Model in Two Variables

As we have seen, a *linear equation* is an equation in which each term is either a *constant term* or a *variable term*, and where each variable term includes only a *single variable to the first power* and an optional *numerical coefficient*. A linear equation can have any number of variables, but to be graphed on a two-dimensional coordinate grid (as we'll see in the next unit), it can have at most two variables.

If we use x and y as the two variables, any linear equation can be simplified into the form:
$$y = \Box x + \Box$$
where the boxes represent numerical values. On the right side of the equation above, the first term is the *variable term* (it includes the variable x) and the second term is the *constant term*.
Example: $y = 16x - 12$ (note that either numerical value can be negative)

We have already seen the word "each" used in a number of verbal expressions. Phrases such as "for each," "for every," or "per" are frequently used to represent the **variable term** in a linear expression. Often, the situation will also involve a "starting" or "one-time" value, which usually represents a **constant term**.
Example: Taxis in a certain city charge $2.50 just to enter the taxi and $1.50 <u>for each</u> mile driven. If we let m represent the number of miles, the cost is 2.50 + 1.50m.

The **isolated variable** in the equation is the value that the linear model is used to calculate.
Example: If c represents the cost of the taxi ride in the previous example, we can write the linear equation $c = 1.50m + 2.50$ to calculate the cost. (Note that the commutative property of addition allows us to switch the order of the two terms.)

A linear model may be written in **function notation**. For now, this only means that the isolated variable is written as a "function of" the other variable, using the name of the function (usually an upper or lower case letter) followed by the other variable in parentheses.
Example: The previous equation, $c = 1.50m + 2.50$, could have been written as $c(m) = 1.50m + 2.50$, meaning that c (the cost) is a "function of" m.

Functions will be more formally introduced later, in Chapter 6.

If a problem defines the rate "for each additional" unit, then one or more of the units are already included in the constant term and need to be subtracted from the variable in the variable term.
Example: Suppose the taxi charges $2.50 for the <u>first</u> mile and $1.50 <u>for each additional</u> mile. The equation is $c = 1.50(m - 1) + 2.50$ for whole numbers $m \geq 1$.

MODEL PROBLEM

Oberon Cell Phone Company advertises service for 3 cents per minute plus a monthly fee of $29.95. Write an equation to calculate the monthly cost, *c*, if *n* call minutes are used.

Solution:

$c = 0.03n + 29.95$

Explanation:

The monthly fee is a one-time starting cost for the month, so it is the constant term. The cost "per" minute *[3 cents times n]* is the variable term. We are calculating the cost *[c]*.

PRACTICE PROBLEMS

1. Kim and Cyndi are starting a business tutoring students in math. They rent an office for $400 per month and charge their students $40 per hour. They write an equation for their profit (or loss) as $y = 40x - 400$. In this equation, what does the variable *x* represent?	2. A car rental company charges its customers a fixed rental fee plus 30 cents per mile driven. If a customer's cost is expressed as the function $C(x) = 0.30x + 100$, what does the variable *x* represent?
3. Essence of Yoga charges $80 per month with a $75 registration fee. Write an equation for the cost, *c*, of an *x*-month membership.	4. Andy deposits $100 in a bank account that earns $5 interest annually. Write a function, *P(y)*, for the balance in the account after *y* years.

5. Abbey starts with $20 and plays an arcade game that costs 50 cents per game. Write an equation for the amount of money, m, remaining after g games are played.

6. An airplane 30,000 feet above the ground begins descending at the constant rate of 2000 feet per minute. Write an function for the plane's altitude, $h(m)$, after m minutes.

7. A video rental company charges its customers $5 for the first day's rental and $2 for each additional day. Write an equation for the cost, c, of renting a video for n days, where $n \geq 1$.

8. An employee is paid $800 per week for the first 40 hours of work, plus overtime pay of $30 per hour for each hour over 40. Write a function for the employee's total wages, $w(h)$, for working h hours in a week ($h \geq 40$).

Regents Questions

MULTIPLE CHOICE

1. A company that manufactures radios first pays a start-up cost, and then spends a certain amount of money to manufacture each radio. If the cost of manufacturing r radios is given by the function $c(r) = 5.25r + 125$, then the value 5.25 best represents
    - (1) the start-up cost
    - (2) the profit earned from the sale of one radio
    - (3) the amount spent to manufacture each radio
    - (4) the average number of radios manufactured

2. A cell phone company charges $60.00 a month for up to 1 gigabyte of data. The cost of additional data is $0.05 per megabyte. If d represents the number of additional megabytes used and c represents the total charges at the end of the month, which linear equation can be used to determine a user's monthly bill?
    - (1) $c = 60 - 0.05d$
    - (2) $c = 60.05d$
    - (3) $c = 60d - 0.05$
    - (4) $c = 60 + 0.05d$

3. A satellite television company charges a one-time installation fee and a monthly service charge. The total cost is modeled by the function $y = 40 + 90x$. Which statement represents the meaning of each part of the function?
    - (1) y is the total cost, x is the number of months of service, $90 is the installation fee, and $40 is the service charge per month.
    - (2) y is the total cost, x is the number of months of service, $40 is the installation fee, and $90 is the service charge per month.
    - (3) x is the total cost, y is the number of months of service, $40 is the installation fee, and $90 is the service charge per month.
    - (4) x is the total cost, y is the number of months of service, $90 is the installation fee, and $40 is the service charge per month.

4. The owner of a small computer repair business has one employee, who is paid an hourly rate of $22. The owner estimates his weekly profit using the function $P(x) = 8600 - 22x$. In this function, x represents the number of
    - (1) computers repaired per week
    - (2) hours worked per week
    - (3) customers served per week
    - (4) days worked per week

5. In 2013, the United States Postal Service charged $0.46 to mail a letter weighing up to 1 oz. and $0.20 per ounce for each additional ounce. Which function would determine the cost, in dollars, $c(z)$, of mailing a letter weighing z ounces where z is an integer greater than 1?
    - (1) $c(z) = 0.46z + 0.20$
    - (2) $c(z) = 0.20z + 0.46$
    - (3) $c(z) = 0.46(z - 1) + 0.20$
    - (4) $c(z) = 0.20(z - 1) + 0.46$

6. The cost of airing a commercial on television is modeled by the function $C(n) = 110n + 900$, where n is the number of times the commercial is aired. Based on this model, which statement is true?

    (1) The commercial costs $0 to produce and $110 per airing up to $900.

    (2) The commercial costs $110 to produce and $900 each time it is aired.

    (3) The commercial costs $900 to produce and $110 each time it is aired.

    (4) The commercial costs $1010 to produce and can air an unlimited number of times.

7. A typical cell phone plan has a fixed base fee that includes a certain amount of data and an overage charge for data use beyond the plan. A cell phone plan charges a base fee of $62 and an overage charge of $30 per gigabyte of data that exceed 2 gigabytes. If C represents the cost and g represents the total number of gigabytes of data, which equation could represent this plan when more than 2 gigabytes are used?

    (1) $C = 30 + 62(2 - g)$ (3) $C = 62 + 30(2 - g)$

    (2) $C = 30 + 62(g - 2)$ (4) $C = 62 + 30(g - 2)$

8. A parking garage charges a base rate of $3.50 for up to 2 hours, and an hourly rate for each additional hour. The sign below gives the prices for up to 5 hours of parking.

Parking Rates	
2 hours	$3.50
3 hours	$9.00
4 hours	$14.50
5 hours	$20.00

Which linear equation can be used to find x, the additional hourly parking rate?

    (1) $9.00 + 3x = 20.00$ (3) $2x + 3.50 = 14.50$

    (2) $9.00 + 3.50x = 20.00$ (4) $2x + 9.00 = 14.50$

9. A car leaves Albany, NY, and travels west toward Buffalo, NY. The equation $D = 280 - 59t$ can be used to represent the distance, D, from Buffalo after t hours. In this equation, the 59 represents the

    (1) car's distance from Albany   (3) distance between Buffalo and Albany

    (2) speed of the car   (4) number of hours driving

10. A plumber has a set fee for a house call and charges by the hour for repairs. The total cost of her services can be modeled by $c(t) = 125t + 95$. Which statements about this function are true?

    I. A house call fee costs $95.

    II. The plumber charges $125 per hour.

    III. The number of hours the job takes is represented by t.

    (1) I and II, only   (3) II and III, only

    (2) I and III, only   (4) I, II, and III

11. Last weekend, Emma sold lemonade at a yard sale. The function $P(c) = .50c - 9.96$ represented the profit, $P(c)$, Emma earned selling c cups of lemonade. Sales were strong, so she raised the price for this weekend by 25 cents per cup. Which function represents her profit for this weekend?

    (1) $P(c) = .25c - 9.96$ (3) $P(c) = .50c - 10.21$

    (2) $P(c) = .50c - 9.71$ (4) $P(c) = .75c - 9.96$

12. The amount Mike gets paid weekly can be represented by the expression $2.50a + 290$, where a is the number of cell phone accessories he sells that week. What is the constant term in this expression and what does it represent?

    (1) $2.50a$, the amount he is guaranteed to be paid each week

    (2) $2.50a$, the amount he earns when he sells a accessories

    (3) 290, the amount he is guaranteed to be paid each week

    (4) 290, the amount he earns when he sells a accessories

13. Each day, a local dog shelter spends an average of $2.40 on food per dog. The manager estimates the shelter's daily expenses, assuming there is at least one dog in the shelter, using the function $E(x) = 30 + 2.40x$. Which statements regarding the function $E(x)$ are correct?

        I. x represents the number of dogs at the shelter per day.
        II. x represents the number of volunteers at the shelter per day.
        III. 30 represents the shelter's total expenses per day.
        IV. 30 represents the shelter's nonfood expenses per day.

    (1) I and III   (3) II and III

    (2) I and IV   (4) II and IV

14. A high school club is researching a tour package offered by the Island Kayak Company. The company charges $35 per person and $245 for the tour guide. Which function represents the total cost, $C(x)$, of this kayak tour package for x club members?

    (1) $C(x) = 35x$ (3) $C(x) = 35(x + 245)$

    (2) $C(x) = 35x + 245$ (4) $C(x) = 35 + (x + 245)$

15. At Benny's Cafe, a mixed-greens salad costs $5.75. Additional toppings can be added for $0.75 each. Which function could be used to determine the cost, $c(s)$, in dollars, of a salad with s additional toppings?

    (1) $c(s) = 5.75s + 0.75$ (3) $c(s) = 5.00s + 0.75$

    (2) $c(s) = 0.75s + 5.75$ (4) $c(s) = 0.75s + 5.00$

16. The Speedy Jet Ski Rental Company charges an insurance fee and an hourly rental rate. The total cost is modeled by the function $R(x) = 30 + 40x$. Based on this model, which statements are true?

      I. $R(x)$ represents the total cost.
      II. x is the number of hours rented.
      III. $40 is the insurance fee.
      IV. $30 is the hourly rental rate.

(1) I, only

(2) I and II, only

(3) I, III, and IV, only

(4) I, II, III, and IV

CONSTRUCTED RESPONSE

17. Alex is selling tickets to a school play. An adult ticket costs $6.50 and a student ticket costs $4.00. Alex sells x adult tickets and 12 student tickets. Write a function, $f(x)$, to represent how much money Alex collected from selling tickets.

18. Sandy programmed a website's checkout process with an equation to calculate the amount customers will be charged when they download songs. The website offers a discount. If one song is bought at the full price of $1.29, then each additional song is $.99. State an equation that represents the cost, C, when s songs are downloaded.

Sandy figured she would be charged $52.77 for 52 songs. Is this the correct amount? Justify your answer.

19. Jim is a furniture salesman. His weekly pay is $300 plus 3.5% of his total sales for the week. Jim sells x dollars' worth of furniture during the week. Write a function, $p(x)$, which can be used to determine his pay for the week.

Use this function to determine Jim's pay to the *nearest cent* for a week when his sales total is $8250.

2.4   Word Problems – Linear Equations

To solve a word problem using an equation:
1. Translate each quantity into an expression in terms of a variable
2. Determine from the problem how the expressions form an equation
3. Write the equation
4. Solve for the variable, and check your solution
5. Be sure to answer the question asked by the problem, and make sure your answer is reasonable!

How to check your solution: Substitute the solution for the variable into the original equation and evaluate both sides of the equation to make sure the solution makes the equation true.

When is an answer unreasonable?
For examples, if your answer is that the length of a side of a triangle is –4 inches, or that the number of people on a bus is 10.75, dismiss these as unreasonable and retrace your steps.

MODEL PROBLEM 1: *LINEAR MODELS*

Jack's streaming service charges a $10 membership fee to join plus $2 for each movie purchased. Jill's streaming service does not charge a membership fee, but charges $5 for the first movie purchased and $3 for each additional movie. If Jack and Jill each paid $20 this month, how many movies did each of them purchase?

Solution:

(A)   Jack   Jill

(B) $2x + 10 = 20$ $3(y - 1) + 5 = 20$

(C) $2x = 10$ $3y - 3 + 5 = 20$

        $x = 5$ $3y + 2 = 20$

                              $3y = 18$

                              $y = 6$

(D)   Jack purchased 5 movies and Jill purchased 6

Explanation of steps:
(A)   With linear model problems, be sure to look for word clues such as "each additional," which means that one or more units are already accounted for in the constant term.
     [Jack's $10 fee does not include any movies, so $2x$ represents the cost of x movies at $2 each. On the other hand, Jill's initial $5 purchase covered the cost of her first movie, so $3(y - 1)$ represents the cost of $y - 1$ movies at $3 each.]
(B)   Write an equation to model the situation.
(C)   Solve.
(D)   Be sure to answer the question.

PRACTICE PROBLEMS

1. Marilyn spent $17 at an amusement park for admission and rides. If she paid $5 for admission, and rides cost $3 each, what is the total number of rides that she went on?	2. A passenger pays $44.25 in taxi fare from the hotel to the airport. The taxi charged $2.25 for the first mile plus $3.50 for each additional mile. How many miles did the taxi travel?

MODEL PROBLEM 2: *USING EXPRESSIONS IN TERMS OF A VARIABLE*

Tamara has five less than four times as many friendship bracelets as Allison. If they have a total of 55 bracelets, how many bracelets does Tamara have?

Solution:

(A) Let a = how many bracelets Allison has;

(B) Tamara has $4a - 5$ bracelets.
$$a + 4a - 5 = 55$$

(C) $$5a - 5 = 55$$
$$5a = 60$$
$$a = 12$$

(D) To check the solution:
$$a + 4a - 5 = 55$$
$$(12) + 4(12) - 5 = 55$$
$$12 + 48 - 5 = 55$$
$$55 = 55 \checkmark$$

(E) If Allison has 12 bracelets, then Tamara has $4a - 5 = 4(12) - 5 = 43$ bracelets. This is reasonable: $12 + 43 = 55$. Tamara has 43 bracelets.

Explanation of steps:

(A) Determine what the variable is.

(B) Express other quantities *[Tamara's bracelets]* in terms of the variable, and write an appropriate equation *[the sum of a and 4a – 5 is 55]*.

(C) Solve for the variable.

(D) Check the solution *[a = 12]* by substituting it into the original equation *[substitute (12) for a]* and then evaluate each side to see if the equation is true.

(E) Answer the question asked by the problem; sometimes the value of the variable does **not** answer the question *[a is the number that Allison has, but we need to say how many Tamara has, which is 4a – 5]*. Check that the answer is reasonable.

PRACTICE PROBLEMS

3. Jamie is 5 years older than her sister Amy. If the sum of their ages is 19, how old is Jamie?

4. Arielle has a collection of grasshoppers and crickets. She has 561 insects in all. The number of grasshoppers is twice the number of crickets. Find the number of *each* type of insect that she has.

5. Three times as many robins as cardinals visited a bird feeder. If a total of 20 robins and cardinals visited the feeder, how many were robins?

6. On the JV baseball team, the number of sophomores is four more than twice the number of freshmen. If there are 16 players combined on the team, how many of each grade level are there?

7. Every year, Jack buys pizzas to serve at a Super Bowl party for his friends. This year, he bought three more than twice the number of pizzas he bought last year. If he bought 15 pizzas this year, how many pizzas did he buy last year?	8. A DVD costs twice as much as a music CD. Omar buys 2 DVDs and 2 CDs and spends $45. How much does one CD cost?
9. Keisha has 28 video discs, which is 8 less than 4 times the number of video discs in Minnie's collection. How many video discs does Minnie own?	10. A jar contains red and black marbles. The number of red marbles in the jar is three more than twice the number of black marbles. There are 42 marbles in all. How many red marbles are there?

MODEL PROBLEM 3:   *RATIO PROBLEMS*

In a school, the student-teacher ratio is 25:2. If there are 216 total students and teachers in the school, how many teachers are there?

Solution:

(A) $25x + 2x = 216$
(B) $27x = 216$
(C) $x = 8$
(D)   The number of teachers is $2x = 16$

Explanation of steps:

(A)   In a ratio, the numbers represent the coefficients of the variable. *[25:2 means there are 25x students and 2x teachers.]*
(B)   Write an equation to model the problem.
(C)   Solve.
(D)   Be sure to answer the question.

PRACTICE PROBLEMS

11. During a recent winter, the ratio of deer to foxes was 7 to 3 in one county of New York State. If there were 210 foxes in the county, what was the number of deer in the county?	12. There are 357 seniors in Harris High School. The ratio of boys to girls is 7:10. How many boys are in the senior class?

MODEL PROBLEM 4:   *CONSECUTIVE INTEGER PROBLEMS*

The sum of three consecutive odd integers is 75. What are the integers?

Solution:

(A)   Let $x, x + 2$, and $x + 4$ represent the integers.
(B) $x + (x + 2) + (x + 4) = 75$
(C) $3x + 6 = 75$
       $3x = 69$
       $x = 23$
(D)   Integers are 23, 25, and 27

Explanation of steps:

(A)   Remember that *consecutive integers* are such that each integer is 1 more than the previous $(x, x + 1, x + 2$, etc.), but *consecutive odd integers* or *consecutive even integers* are each 2 more than the previous $(x, x + 2, x + 4$, etc.).
(B)   Write an equation to model the problem.
(C)   Solve.
(D)   Be sure to answer the question.

69

PRACTICE PROBLEMS

13. The sum of three consecutive integers is 39. What are the three integers?	14. The ages of three brothers are consecutive even integers. Three times the age of the youngest brother exceeds the oldest brother's age by 48 years. What is the age of the youngest brother?

MODEL PROBLEM 5: *TOTAL VALUE PROBLEMS*

Carnival tickets cost $5 per child, $10 per adult, and $8 per senior. Last weekend, the carnival sold twice as many children's tickets as seniors' tickets and 100 more adults' tickets than seniors' tickets. If the total revenue was $5200, how many of each type of ticket were sold?

Solution:
- (A) Let s = the number of seniors' tickets sold
     $2s$ children's tickets and $s + 100$ adults' tickets were sold
- (B) $5(2s) + 10(s + 100) + 8(s) = 5200$
- (C) $10s + 10s + 1000 + 8s = 5200$
     $28s + 1000 = 5200$
     $28s = 4200$
- (D) $s = 150$ senior tickets sold
     $2s = 2(150) = 300$ children's tickets sold
     $s + 100 = 150 + 100 = 250$ adults' tickets sold

Explanation of steps:
- (A) Determine what the variable is, and represent the other quantities as expressions in terms of that variable.
- (B) Write an equation to model the problem. For problems such as these, this often looks like:
     $price_1(quantity_1) + price_2(quantity_2) + \cdots + price_n(quantity_n) = total\ value$
- (C) Solve.
- (D) Be sure to answer the question.

PRACTICE PROBLEMS

15. In his piggy bank, Neil has three times as many dimes as nickels and he has four more quarters than nickels. The coins total $4.60. How many of each coin does he have?

16. There were 100 more balcony tickets than main-floor tickets sold for a concert. The balcony tickets sold for $4 and the main-floor tickets sold for $12. The total amount of sales for both types of tickets was $3,056. Find the number of balcony tickets that were sold.

17. A craft shop sold 150 pillows. Small pillows were $6.50 each and large pillows were $9.00 each. If the total amount collected from the sale of these items was $1180.00, what is the total number of each size pillow that was sold?

Regents Questions

MULTIPLE CHOICE

1. Last week, a candle store received $355.60 for selling 20 candles. Small candles sell for $10.98 and large candles sell for $27.98. How many large candles did the store sell?

    (1) 6   (3) 10

    (2) 8   (4) 12

CONSTRUCTED RESPONSE

2. Donna wants to make trail mix made up of almonds, walnuts and raisins. She wants to mix one part almonds, two parts walnuts, and three parts raisins. Almonds cost $12 per pound, walnuts cost $9 per pound, and raisins cost $5 per pound. Donna has $15 to spend on the trail mix. Determine how many pounds of trail mix she can make.

3. Hannah went to the school store to buy supplies and spent $16. She bought four more pencils than pens and two fewer erasers than pens. Pens cost $1.25 each, pencils cost $0.55 each, and erasers cost $0.75 each. If x represents the number of pens Hannah bought, write an equation in terms of x that can be used to find how many of each item she bought. Use your equation to determine algebraically how many pens Hannah bought.

2.5   Translate Inequalities

Translate a verbal problem into an **inequality** the same way as you do for an equation, but instead of using an = sign for equality, use the appropriate **inequality symbol**:

>	"is more than"	"is greater than"	
<	"is less than"		
≥	"greater than or equal to"	"at least"	"not less than"
≤	"less than or equal to"	"at most"	"not more than"

Caution: recognize the distinction in verbal problems between the phrases "more than" and "less than" for addition and subtraction and the phrases "is more than" and "is less than" for an inequality.

Example: 5 more than x → $x + 5$ 5 is more than x → $5 > x$
                     5 less than x → $x - 5$ 5 is less than x → $5 < x$

MODEL PROBLEM

Barbara has $10 to buy milk and cookies for her child's party. If each pint of milk costs $0.55 and each cookie costs $0.15, and she wants to buy 6 times as many cookies as pints of milk, write an inequality that shows how much she can buy for at most $10.

Solution:
(A)   Let m = the number of pints of milk she buys; $6m$ = the number of cookies she buys.
(B) $0.55m + 0.15(6m) \leq 10.00$

Explanation of steps:
(A)   Write expressions in terms of a variable.
(B)   Create an appropriate inequality using the expressions.

PRACTICE PROBLEMS

1. If x represents a number, write an inequality to represent the statement, "Eight less than three times a number is greater than fifteen."	2. A sign in front of a roller coaster ride says that all riders must be at least 48 inches tall. If h represents the height of a rider in inches, write an inequality for the statement on this sign.
3. Allison is nine inches taller than Ben. The sum of their heights is less than 144 inches. Write an appropriate inequality to describe this situation.	4. Cyril is going to buy a coat and a hat. The coat costs 3 times as much as the hat. He cannot spend more than $120. Write an inequality to describe this situation.
5. Abe and Betty need to sell at least 90 magazine subscriptions between them. If Abe sells twice as many subscriptions as Betty, write an inequality that could be used to determine how many subscriptions, x, Betty needs to sell.	6. You need to purchase a apples and b bananas, and you can spend no more than $25. Apples cost 75 cents each and bananas cost $1.25 each. Write an inequality, in terms of a and b, to describe this situation.

7. The length of a rectangle is 15 and its width is w. The perimeter of the rectangle is, *at most*, 50. Write an inequality that could be used to find the longest possible width.

8. The length of a rectangle is three feet less than twice its width. The area of the rectangle is at most 30 square feet. If w represents the width of the rectangle, in feet, write an inequality that could be used to find the width.

9. A local high school needs to pay $250.00 for a hall where a school dance will be held. Each student attending the dance pays $0.75 and each guest pays $1.25. If 200 students attend the dance, write an inequality that could be used to determine the number of guests, x, needed to cover the cost of the hall.

Regents Questions

MULTIPLE CHOICE

1. The cost of a pack of chewing gum in a vending machine is $0.75. The cost of a bottle of juice in the same machine is $1.25. Julia has $22.00 to spend on chewing gum and bottles of juice for her team and she must buy seven packs of chewing gum. If b represents the number of bottles of juice, which inequality represents the maximum number of bottles she can buy?

    (1) $0.75b + 1.25(7) \geq 22$ (3) $0.75(7) + 1.25b \geq 22$

    (2) $0.75b + 1.25(7) \leq 22$ (4) $0.75(7) + 1.25b \leq 22$

2. The acidity in a swimming pool is considered normal if the average of three pH readings, p, is defined such that $7.0 < p < 7.8$. If the first two readings are 7.2 and 7.6, which value for the third reading will result in an overall rating of normal?

    (1) 6.2   (3) 8.6

    (2) 7.3   (4) 8.8

3. Joy wants to buy strawberries and raspberries to bring to a party. Strawberries cost $1.60 per pound and raspberries cost $1.75 per pound. If she only has $10 to spend on berries, which inequality represents the situation where she buys x pounds of strawberries and y pounds of raspberries?

    (1) $1.60x + 1.75y \leq 10$ (3) $1.75x + 1.60y \leq 10$

    (2) $1.60x + 1.75y \geq 10$ (4) $1.75x + 1.60y \geq 10$

4. David wanted to go on an amusement park ride. A sign posted at the entrance read "You must be greater than 42 inches tall and no more than 57 inches tall for this ride." Which inequality would model the height, x, required for this amusement park ride?

    (1) $42 < x \leq 57$ (3) $42 < x$ or $x \leq 57$

    (2) $42 > x \geq 57$ (4) $42 > x$ or $x \geq 57$

5. An ice cream shop sells ice cream cones, c, and milkshakes, m. Each ice cream cone costs $1.50 and each milkshake costs $2.00. Donna has $19.00 to spend on ice cream cones and milkshakes. If she must buy 5 ice cream cones, which inequality could be used to determine the maximum number of milkshakes she can buy?

    (1) $1.50(5) + 2.00m \geq 19.00$ (3) $1.50c + 2.00(5) \geq 19.00$

    (2) $1.50(5) + 2.00m \leq 19.00$ (4) $1.50c + 2.00(5) \leq 19.00$

6. Peter has $100 to spend on drinks for his party. Bottles of lemonade cost $2 each, and juice boxes cost $0.50 each. If x is the number of bottles of lemonade and y is the number of juice boxes, which inequality models this situation?

    (1) $0.50x + 2y \leq 100$ (3) $2x + 0.50y \leq 100$

    (2) $0.50x + 2y \geq 100$ (4) $2x + 0.50y \geq 100$

7. Two texting plans are advertised. Plan A has a monthly fee of $15 with a charge of $0.08 per text. Plan B has a monthly fee of $3 with a charge of $0.12 per text. If t represents the number of text messages in a month, which inequality should be used to show that the cost of Plan A is *less* than the cost of Plan B?

    (1) $15 + 0.08t < 3 + 0.12t$ (3) $15t + 0.08 < 3t + 0.12$

    (2) $15 + 0.08t > 3 + 0.12t$ (4) $15t + 0.08 > 3t + 0.12$

2.6   Word Problems – Inequalities

To solve a word problem using an inequality:
1. Translate each quantity into an expression in terms of a variable
2. Determine from the problem how the expressions form an inequality
3. Write and solve the inequality
4. Be sure to answer the question asked by the problem

MODEL PROBLEM

Thelma and Laura start a lawn-mowing business and buy a lawnmower for $225. They plan to charge $15 to mow one lawn. What is the *minimum* number of lawns they need to mow if they wish to earn a profit of *at least* $750?

Solution:
(A) Let x represent the number of lawns they need to mow. Total profit is $15x - 225$.
(B) $15x - 225 \geq 750$
(C)
$$\underline{\quad +225 \quad +225\quad}$$
$$\frac{15x}{15} \geq \frac{975}{15}$$
$$x \geq 65$$
(D) They need to mow a minimum of 65 lawns.

Explanation of steps:
(A) Translate each quantity into an expression in terms of a variable *[profit would be $15 per mowed lawn, less the $225 cost of the lawnmower, or 15x – 225]*.
(B) Determine from the problem how the expressions form an inequality *[the profit must be at least – greater than or equal to – $750]*.
(C) Write and solve the inequality.
(D) Be sure to answer the question asked by the problem.

PRACTICE PROBLEMS

1. Find the smallest integer such that five less than twice the integer is greater than 23.	2. If five times a number is less than 55, what is the greatest possible integer value of the number?
3. The larger of two integers is 7 times the smaller. The sum of the integers is at most 60. What are the largest two integers that can make these statements true?	4. Tony's job pays him $155 a week. If he has already saved $375, what is the minimum number of weeks he needs to work in order to have enough money to buy a drone for $900?
5. Andy earns $5.95 per hour working after school. He needs at least $215 for his holiday shopping. How many hours must he work to reach his goal?	6. The cost per month of making n number of wooden toys is $3n + 30$. The income from selling n toys is $6n$. How many toys must the company make to make a profit (the income is greater than the cost)?

7. A music club has a registration fee of $13.95 and charges $0.49 to buy each song. Nelly has $50.00 to join the club and buy songs. What is the maximum number of songs she can buy?

8. A cell phone plan charges $0.07 per minute plus a monthly fee of $19.00. Tamara budgets $29.50 per month for total cell phone expenses. What is the maximum number of minutes she could use each month in order to stay within her budget?

9. Parking charges at Superior Parking Garage are $5.00 for the first hour and $1.50 for each additional 30 minutes. If Margo has $12.50, what is the maximum amount of time she will be able to park her car at the garage?

10. Members of the band boosters are planning to sell programs at football games. The cost to print the programs is $150 plus $0.50 per program. They plan to sell each program for $2. How many programs must they sell to make a profit of at least $500?

Regents Questions

MULTIPLE CHOICE

1. Connor wants to attend the town carnival. The price of admission to the carnival is $4.50, and each ride costs an additional 79 cents. If he can spend at most $16.00 at the carnival, which inequality can be used to solve for r, the number of rides Connor can go on, and what is the maximum number of rides he can go on?

   (1) $0.79 + 4.50r \leq 16.00$; 3 rides   (3) $4.50 + 0.79r \leq 16.00$; 14 rides

   (2) $0.79 + 4.50r \leq 16.00$; 4 rides   (4) $4.50 + 0.79r \leq 16.00$; 15 rides

2. Natasha is planning a school celebration and wants to have live music and food for everyone who attends. She has found a band that will charge her $750 and a caterer who will provide snacks and drinks for $2.25 per person. If her goal is to keep the average cost per person between $2.75 and $3.25, how many people, p, must attend?

   (1) $225 < p < 325$ (3) $500 < p < 1000$

   (2) $325 < p < 750$ (4) $750 < p < 1500$

3. The math department needs to buy new textbooks and laptops for the computer science classroom. The textbooks cost $116.00 each, and the laptops cost $439.00 each. If the math department has $6500 to spend and purchases 30 textbooks, how many laptops can they buy?

   (1) 6   (3) 11

   (2) 7   (4) 12

4. Ashley only has 7 quarters and some dimes in her purse. She needs at least $3.00 to pay for lunch. Which inequality could be used to determine the number of dimes, d, she needs in her purse to be able to pay for lunch?

   (1) $1.75 + d \geq 3.00$ (3) $1.75 + d \leq 3.00$

   (2) $1.75 + 0.10d \geq 3.00$ (4) $1.75 + 0.10d \leq 3.00$

CONSTRUCTED RESPONSE

5. David has two jobs. He earns $8 per hour babysitting his neighbor's children and he earns $11 per hour working at the coffee shop.

   Write an inequality to represent the number of hours, x, babysitting and the number of hours, y, working at the coffee shop that David will need to work to earn a minimum of $200.

   David worked 15 hours at the coffee shop. Use the inequality to find the number of full hours he must babysit to reach his goal of $200.

6. Sarah wants to buy a snowboard that has a total cost of $580, including tax. She has already saved $135 for it. At the end of each week, she is paid $96 for babysitting and is going to save three-quarters of that for the snowboard.

   Write an inequality that can be used to determine the *minimum* number of weeks Sarah needs to babysit to have enough money to purchase the snowboard.

   Determine and state the *minimum* number of full weeks Sarah needs to babysit to have enough money to purchase this snowboard.

7. A school plans to have a fundraiser before basketball games selling shirts with their school logo. The school contacted two companies to find out how much it would cost to have the shirts made. Company *A* charges a $50 set-up fee and $5 per shirt. Company *B* charges a $25 set-up fee and $6 per shirt.

   Write an equation for Company *A* that could be used to determine the total cost, *A*, when *x* shirts are ordered.

   Write a second equation for Company *B* that could be used to determine the total cost, *B*, when *x* shirts are ordered.

   Determine algebraically and state the minimum number of shirts that must be ordered for it to be cheaper to use Company *A*.

8. Maria orders T-shirts for her volleyball camp. Adult-sized T-shirts cost $6.25 each and youth-sized T-shirts cost $4.50 each. Maria has $550 to purchase both adult-sized and youth-sized T-shirts. If she purchases 45 youth-sized T-shirts, determine algebraically the maximum number of adult-sized T-shirts she can purchase.

9. A store sells grapes for $1.99 per pound, strawberries for $2.50 per pound, and pineapples for $2.99 each. Jonathan has $25 to buy fruit. He plans to buy 2 more pounds of strawberries than grapes. He also plans to buy 2 pineapples. If *x* represents the number of pounds of grapes, write an inequality in one variable that models this scenario. Determine algebraically the maximum number of whole pounds of grapes he can buy.

10. The senior class at Hills High School is purchasing sports drinks and bottled water to sell at the school field day. At the local discount store, a case of sports drinks costs $15.79, and a case of bottled water costs $5.69. The senior class has $125 to spend on the drinks. If *x* represents the number of cases of sports drinks and *y* represents the number of cases of bottled water purchased, write an inequality that models this situation.

    Nine cases of bottled water are purchased for this year's field day. Use your inequality to determine algebraically the maximum number of full cases of sports drinks that can be purchased. Explain your answer.

2.7   Conversions

At times, the solution of a problem may require a conversion between units of measure, or between unit rates.

A **measurement equivalent** is an equation or approximation expressing the relationship between two different units of measure, and is used to convert between the two units.

Examples: 60 minutes = 1 hour

 1.61 kilometers ≈ 1 mile

A **conversion fraction** is a fraction (*ratio*) which derives from a measurement equivalent. The numerator and denominator are equivalent measures, so the fraction is equal to 1. Therefore, you can multiply a value by a conversion fraction without changing its value.

Examples: $\dfrac{60\ mins}{1\ hr}$ $\dfrac{1\ hr}{60\ mins}$ $\dfrac{1.61\ km}{1\ mi}$ $\dfrac{1\ mi}{1.61\ km}$

To convert units of measure using measurement equivalents:
1. Write the given measurement.
2. Create a *conversion fraction* from one of the *measurement equivalents*, placing the unit of measure from which we need to change in the denominator. Write as a product and cancel the common unit of measure.
3. Repeat step 2 with any additional conversion fractions that are needed until only the desired unit of measure remains.
4. Multiply.
5. Simplify and write the answer, rounding if necessary.

Example: To convert 6 feet into inches, start by writing 6 feet. Then, use the conversion fraction $\dfrac{12\ inches}{1\ foot}$, making sure to keep the number of feet in the denominator.
Cancel the common unit, feet, and multiply. This gives us $6 \times 12 = 72$ inches.

$$6\ \cancel{feet} \cdot \frac{12\ inches}{1\ \cancel{foot}} = 72\ inches$$

To convert unit rates using measurement equivalents:
1. Write the given rate in fraction form.
2. To change the unit in the numerator, create a *conversion fraction* from one of the *measurement equivalents*, placing this unit in the denominator. Write as a product and cancel out the common unit.
3. Similarly, to change the unit in the denominator, create another conversion fraction, but place this unit in the numerator. Write as a product and cancel out the common unit.
4. Multiply the fractions.
5. Simplify and write the answer, rounding if necessary.

Example: To convert 300 miles per hour into miles per minute,

$$\frac{300\ miles}{1\ \cancel{hour}} \cdot \frac{1\ \cancel{hour}}{60\ minutes} = \frac{300}{60} = 5\ miles\ per\ minute$$

MODEL PROBLEM 1:   *CONVERTING UNITS OF MEASURE*

How many feet are in 3 furlongs?

$$8 \text{ furlongs} = 1 \text{ mile}$$
$$5280 \text{ feet} = 1 \text{ mile}$$

Solution:

(A)   (B)   (C)   (D)   (E)

$$3 \text{ furlongs} \times \frac{1 \text{ mile}}{8 \text{ furlongs}} \times \frac{5280 \text{ feet}}{1 \text{ mile}} = \frac{(3)(1)(5280) \text{ feet}}{(8)(1)} = 1{,}980 \text{ feet}$$

Explanation of steps:

(A)   Write the given measurement *[3 furlongs]*.

(B)   Create a conversion fraction from one of the measurement equivalents, placing the unit of measure from which we need to change *[furlongs]* in the denominator. Write as a product and cancel the common unit of measure *[furlongs]*.

(C)   Repeat step B with any additional conversion fractions that are needed *[we still need to change miles to feet]*, until only the desired unit of measure *[feet]* remains.

(D)   Multiply.

(E)   Simplify and write the answer, rounding if necessary.

PRACTICE PROBLEMS

1.   Convert 20 inches into centimeters (*cm*), rounded to the *nearest centimeter*.
 	
     2.54 cm = 1 inch
	2. Saul walked 8,900 feet from home to school. How far, to the *nearest tenth of a mile*, did he walk?
	
	     1 mile = 5,280 feet

3. A pet store manager needs to pack 1680 ounces of pet food into 5-pound bags. How many 5-pound bags of pet food can be packed?

1 pound = 16 ounces

4. Liz is baking cookies. A single batch uses ¾ teaspoon of vanilla. If Liz is mixing the ingredients for five batches at the same time, how many tablespoons of vanilla will she use?

3 teaspoons = 1 tablespoon

5. Little league baseball bats have barrels that measure 2.625 inches in diameter. What is the diameter of a bat in millimeters (*mm*), to the *nearest millimeter*?

2.54 cm = 1 inch
10 mm = 1 cm

6. The most common distance for a thoroughbred horse race is 6 furlongs. What is this distance in kilometers (*km*), to the *nearest hundredth of a kilometer*?

8 furlongs = 1 mile
1.61 km = 1 mile

7. Ribbon sells for $3.75 per yard. Find the cost, in dollars, for 48 inches of ribbon.
 	
        12 inches = 1 foot
        3 feet = 1 yard
	8. Convert 60 feet into meters, rounded to the *nearest tenth of a meter*.
	
	        12 inches = 1 foot
	        2.54 cm = 1 inch
	        100 cm = 1 meter

MODEL PROBLEM 2:   CONVERTING FROM ONE UNIT RATE TO ANOTHER

A cheetah runs at a rate of 70 miles per hour. What is the cheetah's speed in feet per minute?

$$5280 \text{ feet} = 1 \text{ mile}$$
$$60 \text{ minutes} = 1 \text{ hour}$$

Solution:

$$\underset{(A)}{\frac{70 \ \cancel{miles}}{1 \ \cancel{hour}}} \times \underset{(B)}{\frac{5280 \ ft}{1 \ \cancel{mile}}} \times \underset{(C)}{\frac{1 \ \cancel{hour}}{60 \ mins}} = \underset{(D)}{\frac{(70)(5280)(1) \ ft}{(1)(1)(60) \ mins}} = \underset{(E)}{6{,}160 \ ft/min}$$

Explanation of steps:

(A) Write the given rate *[70 miles per hour]* in fraction form.

(B) To change the unit in the numerator *[miles]*, create a conversion fraction from one of the measurement equivalents, placing this unit *[miles]* in the denominator. Write as a product and cancel out the common unit *[miles]*, leaving the desired unit *[feet, or ft]* in the numerator of the new fraction.

(C) Similarly, to change the unit in the denominator *[hours]*, create another conversion fraction, but place this unit *[hours]* in the numerator. Write as a product and cancel out the common unit *[hours]*, leaving the desired unit *[minutes]* in the denominator.

(D) Multiply the fractions.

(E) Simplify and write the answer, rounding if necessary.

PRACTICE PROBLEMS

9. Roger ran a distance of 150 meters in 1½ minutes. What is his speed in meters per hour?	10. If the speed of sound is 344 meters per second, what is the approximate speed of sound, in meters per hour?
11. A Toyota Camry Hybrid automobile boasts a gas mileage of 43 miles per gallon. What is its gas mileage in kilometers (*km*) per liter, to the *nearest tenth*. 1.61 km = 1 mile 3.79 liters = 1 gallon	12. At a grocery store, a 2-liter bottle of soda costs $1.50 and a 1-gallon bottle costs $2.50. Which is a better buy? Justify your answer. 3.79 liters = 1 gallon

13. Nikita rode her bicycle a total of 8000 miles last year. To the *nearest yard*, Nikita rode an average of how many yards per day?

    1 mile = 1760 yards
    1 year = 365 days

14. If you travel 30 miles per hour, how many feet per second are you traveling?

    5280 feet = 1 mile
    60 minutes = 1 hour
    60 seconds = 1 minute

15. A football player runs the length of a 100 yard football field in 11 seconds. How fast did he run in miles per hour, rounded to the *nearest tenth*?

    3 feet = 1 yard
    5280 feet = 1 mile
    60 minutes = 1 hour
    60 seconds = 1 minute

Regents Questions

MULTIPLE CHOICE

1. Peyton is a sprinter who can run the 40-yard dash in 4.5 seconds. He converts his speed into miles per hour, as shown below.

$$\frac{40\ yd}{4.5\ sec} \cdot \frac{3\ ft}{1\ yd} \cdot \frac{5280\ ft}{1\ mi} \cdot \frac{60\ sec}{1\ min} \cdot \frac{60\ min}{1\ hr}$$

   Which ratio is *incorrectly* written to convert his speed?

   (1) $\dfrac{3\ ft}{1\ yd}$ (3) $\dfrac{60\ sec}{1\ min}$

   (2) $\dfrac{5280\ ft}{1\ mi}$ (4) $\dfrac{60\ min}{1\ hr}$

2. Dan took 12.5 seconds to run the 100-meter dash. He calculated the time to be approximately

   (1) 0.2083 minute   (3) 0.2083 hour

   (2) 750 minutes   (4) 0.52083 hour

3. Patricia is trying to compare the average rainfall of New York to that of Arizona. A comparison between these two states for the months of July through September would be best measured in

   (1) feet per hour   (3) inches per month

   (2) inches per hour   (4) feet per month

4. A construction worker needs to move 120 ft³ of dirt by using a wheelbarrow. One wheelbarrow load holds 8 ft³ of dirt and each load takes him 10 minutes to complete. One correct way to figure out the number of hours he would need to complete this job is

   (1) $\dfrac{120\ ft^3}{1} \cdot \dfrac{10\ min}{1\ load} \cdot \dfrac{60\ min}{1\ hr} \cdot \dfrac{1\ load}{8\ ft^3}$ (3) $\dfrac{120\ ft^3}{1} \cdot \dfrac{1\ load}{10\ min} \cdot \dfrac{8\ ft^3}{1\ load} \cdot \dfrac{1\ hr}{60\ min}$

   (2) $\dfrac{120\ ft^3}{1} \cdot \dfrac{60\ min}{1\ hr} \cdot \dfrac{8\ ft^3}{10\ min} \cdot \dfrac{1}{1\ load}$ (4) $\dfrac{120\ ft^3}{1} \cdot \dfrac{1\ load}{8\ ft^3} \cdot \dfrac{10\ min}{1\ load} \cdot \dfrac{1\ hr}{60\ min}$

5. The Utica Boilermaker is a 15-kilometer road race. Sara is signed up to run this race and has done the following training runs:

         I.   10 miles
         II.   44,880 feet
         III.   15,560 yards

   Which run(s) are at least 15 kilometers?

   (1) I, only   (3) I and III

   (2) II, only   (4) II and III

6. Olivia entered a baking contest. As part of the contest, she needs to demonstrate how to measure a gallon of milk if she only has a teaspoon measure. She converts the measurement using the ratios below:

$$\frac{4\ quarts}{1\ gallon} \cdot \frac{2\ pints}{1\ quart} \cdot \frac{2\ cups}{1\ pint} \cdot \frac{\frac{1}{4}\ cup}{4\ tablespoons} \cdot \frac{3\ teaspoons}{1\ tablespoon}$$

Which ratio is *incorrectly* written in Olivia's conversion?

(1) $\dfrac{4\ quarts}{1\ gallon}$ (3) $\dfrac{\frac{1}{4}\ cup}{4\ tablespoons}$

(2) $\dfrac{2\ pints}{1\ quart}$ (4) $\dfrac{3\ teaspoons}{1\ tablespoon}$

7. The following conversion was done correctly:

$$\frac{3\ miles}{1\ hour} \cdot \frac{1\ hour}{60\ minutes} \cdot \frac{5280\ feet}{1\ mile} \cdot \frac{12\ inches}{1\ foot}$$

What were the final units for this conversion?

(1) minutes per foot   (3) feet per minute

(2) minutes per inch   (4) inches per minute

8. Bamboo plants can grow 91 centimeters per day. What is the approximate growth of the plant, in inches per hour?

(1) 1.49   (3) 9.63

(2) 3.79   (4) 35.83

9. The owner of a landscaping business wants to know how much time, on average, his workers spend mowing one lawn. Which is the most appropriate rate with which to calculate an answer to his question?

(1) lawns per employee   (3) employee per lawns

(2) lawns per day   (4) hours per lawn

10. Sarah travels on her bicycle at a speed of 22.7 miles per hour. What is Sarah's approximate speed, in kilometers per minute?

(1) 0.2   (3) 36.5

(2) 0.6   (4) 36.6

11. It takes Tim 4.5 hours to run 50 kilometers. Which expression will allow him to change this rate to minutes per mile?

(1) $\dfrac{4.5\ hr}{50\ km} \cdot \dfrac{1.609\ km}{1\ mi} \cdot \dfrac{60\ min}{1\ hr}$ (3) $\dfrac{50\ km}{4.5\ hr} \cdot \dfrac{1\ mi}{1.609\ km} \cdot \dfrac{1\ hr}{60\ min}$

(2) $\dfrac{50\ km}{4.5\ hr} \cdot \dfrac{1\ mi}{1.609\ km} \cdot \dfrac{60\ min}{1\ hr}$ (4) $\dfrac{4.5\ hr}{50\ km} \cdot \dfrac{1\ mi}{1.609\ km} \cdot \dfrac{60\ min}{1\ hr}$

12. A swimmer set a world record in the women's 1500-meter freestyle, finishing the race in 15.42 minutes. If 1 meter is approximately 3.281 feet, which set of calculations could be used to convert her speed to miles per hour?

(1) $\dfrac{1500\ meters}{15.42\ min} \cdot \dfrac{60\ min}{1\ hour} \cdot \dfrac{1\ meter}{3.281\ feet} \cdot \dfrac{1\ mile}{5280\ feet}$

(2) $\dfrac{1500\ meters}{15.42\ min} \cdot \dfrac{60\ min}{1\ hour} \cdot \dfrac{3.281\ feet}{1\ meter} \cdot \dfrac{1\ mile}{5280\ feet}$

(3) $\dfrac{1500\ meters}{15.42\ min} \cdot \dfrac{3.281\ feet}{1\ meter} \cdot \dfrac{1\ mile}{5280\ feet}$

(4) $\dfrac{1500\ meters}{15.42\ min} \cdot \dfrac{60\ min}{1\ hour} \cdot \dfrac{1\ mile}{5280\ feet}$

13. Morgan read that a snail moves about 72 feet per day. He performs the calculation $\dfrac{72\ feet}{1\ day} \cdot \dfrac{1\ day}{24\ hours} \cdot \dfrac{1\ hour}{60\ minutes} \cdot \dfrac{12\ inches}{1\ foot}$ to convert this rate to different units. What are the units for the converted rate?

(1) hours/inch
(2) minutes/inch
(3) inches/hour
(4) inches/minute

14. A company ships an average of 30,000 items each week. The approximate number of items shipped each minute is calculated using the conversion

(1) $\dfrac{30{,}000\ items}{1\ week} \cdot \dfrac{7\ days}{1\ week} \cdot \dfrac{60\ min}{1\ hr} \cdot \dfrac{1\ day}{24\ hrs}$

(2) $\dfrac{30{,}000\ items}{1\ week} \cdot \dfrac{1\ week}{7\ days} \cdot \dfrac{1\ day}{24\ hrs} \cdot \dfrac{1\ hr}{60\ min}$

(3) $\dfrac{1\ week}{30{,}000\ items} \cdot \dfrac{1\ week}{7\ days} \cdot \dfrac{1\ day}{24\ hrs} \cdot \dfrac{1\ hr}{60\ min}$

(4) $\dfrac{1\ week}{30{,}000\ items} \cdot \dfrac{7\ days}{1\ week} \cdot \dfrac{24\ hrs}{1\ day} \cdot \dfrac{60\ min}{1\ hr}$

15. Joe compared gas prices in England and New York State one day. In England, gas sold for 1.35 euros per liter, and one dollar equaled 0.622 euros. A correct way to figure out this cost, in dollars per gallon, is

(1) $\dfrac{1.35\ euros}{1\ L} \cdot \dfrac{1\ L}{0.264\ gal} \cdot \dfrac{\$1.00}{0.622\ euros}$

(2) $\dfrac{1.35\ euros}{1\ L} \cdot \dfrac{\$1.00}{0.622\ euros} \cdot \dfrac{0.264\ gal}{1\ L}$

(3) $\dfrac{1.35\ euros}{1\ L} \cdot \dfrac{1\ L}{0.264\ gal} \cdot \dfrac{0.622\ euros}{\$1.00}$

(4) $\dfrac{1.35\ euros}{1\ L} \cdot \dfrac{0.622\ euros}{\$1.00} \cdot \dfrac{0.264\ gal}{1\ L}$

16. When the temperature is 59°F, the speed of sound at sea level is 1225 kilometers per hour. Which process could be used to convert this speed into feet per second?

(1) $\dfrac{1225\ km}{1\ hr} \cdot \dfrac{0.62\ mi}{1\ km} \cdot \dfrac{1\ hr}{60\ min} \cdot \dfrac{1\ mi}{5280\ ft} \cdot \dfrac{1\ min}{60\ sec}$

(2) $\dfrac{1225\ km}{1\ hr} \cdot \dfrac{0.62\ mi}{1\ km} \cdot \dfrac{5280\ ft}{1\ mi} \cdot \dfrac{1\ hr}{60\ min} \cdot \dfrac{1\ min}{60\ sec}$

(3) $\dfrac{1225\ km}{1\ hr} \cdot \dfrac{1\ km}{0.62\ mi} \cdot \dfrac{5280\ ft}{1\ mi} \cdot \dfrac{1\ hr}{60\ min} \cdot \dfrac{1\ min}{60\ sec}$

(4) $\dfrac{1225\ km}{1\ hr} \cdot \dfrac{0.62\ mi}{1\ km} \cdot \dfrac{5280\ ft}{1\ mi} \cdot \dfrac{60\ min}{1\ hr} \cdot \dfrac{1\ min}{60\ sec}$

CONSTRUCTED RESPONSE

17. A typical marathon is 26.2 miles. Allan averages 12 kilometers per hour when running in marathons. Determine how long it would take Allan to complete a marathon, to the _nearest tenth of an hour_. Justify your answer.

18. A two-inch-long grasshopper can jump a horizontal distance of 40 inches. An athlete, who is five feet nine, wants to cover a distance of one mile by jumping. If this person could jump at the same ratio of body-length to jump-length as the grasshopper, determine, to the _nearest jump_, how many jumps it would take this athlete to jump one mile.

19. The distance traveled is equal to the rate of speed multiplied by the time traveled. If the distance is measured in feet and the time is measured in minutes, then the rate of speed is expressed in which units? Explain how you arrived at your answer.

20. A news report suggested that an adult should drink a minimum of 4 pints of water per day. Based on this report, determine the minimum amount of water an adult should drink, in fluid ounces, per week.

CHAPTER 3.   LINEAR GRAPHS

3.1   Determine Whether a Point is on a Line

A **line** consists of a set of **points**, each of which can be represented by an **ordered pair** stating its x value and y value as (x, y). This set of points represents the **solution set** of a **linear equation** involving the two variables, x and y.

One way to graph a line is to by **creating a table**. The first column will contain some sample x values; I usually prefer to choose –2, –1, 0, 1, and 2. The second column is used to substitute the x value into the equation in order to solve for y. The resulting y values are written in the third column. The last column gives the corresponding ordered pairs of x and y values.

Example: We can graph $y = 3x + 1$ by using a table and drawing a line through the ordered pairs on a coordinate graph.

x	$y = 3x + 1$	y	(x, y)
-2	$y = 3(-2) + 1$	-5	$(-2, -5)$
-1	$y = 3(-1) + 1$	-2	$(-1, -2)$
0	$y = 3(0) + 1$	1	$(0,1)$
1	$y = 3(1) + 1$	4	$(1,4)$
2	$y = 3(2) + 1$	7	$(2,7)$

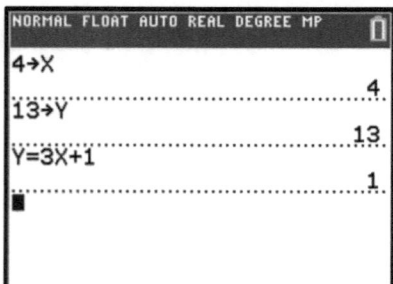

You can **determine whether a point is on a line** by substituting the x value and y value for the variables x and y in the equation and then checking if these values make the equation true.

Example: $(4,13)$ is on the line $y = 3x + 1$ because substituting 4 for x and 13 for y, we get $13 = 3(4) + 1$, which is true.

 CALCULATOR TIP

This can also be done using the calculator.

Example: For the example above, enter ④ⓈⓉⓄ▶Ⓧ,Ⓣ,Θ,ⓝ and ①③ⓈⓉⓄ▶ⒶⓁⓅⒽⒶ[Y].
            Then test by typing ⒶⓁⓅⒽⒶ[Y] ②ⁿᵈ[TEST]① ③Ⓧ,Ⓣ,Θ,ⓝ➕①.

```
NORMAL FLOAT AUTO REAL DEGREE MP      ▯

4→X
                                      4
13→Y
                                     13
Y=3X+1
                                      1
■
```

MODEL PROBLEM

Does the line whose equation is $2y + 6 = 4x$ contain the point $(1, -1)$?

Solution:

(A) $2y + 6 = 4x$ for $x = 1, y = -1$

(B) $2(-1) + 6 = 4(1)$?

(C) $4 = 4$, so yes, $(1, -1)$ is in the solution set.

Explanation of steps:

(A)   The first value in the ordered pair represents the value of x, the second is the value of y.

(B)   Substitute for x and y.

(C)   Evaluate both sides of the equation to determine if the equation is true.

PRACTICE PROBLEMS

1. Does the point $(3,7)$ lie on the line whose equation is $y = 3x - 2$?	2. Does the line whose equation is $y = \frac{1}{2}x + 5$ contain the point $(4,9)$?
3. Does the line whose equation is $y = 4x$ pass through the origin, $(0,0)$?	4. Does the point $(-2, -4)$ lie on the line whose equation is $2y - 3x = -2$?
5. Determine if the ordered pair $(-4, 3)$ is a solution of $4x - y = -13$.	6. Determine if the ordered pair $(-2, -4)$ is a solution of $5x - 2y = -2$.

7. Determine if the ordered pair $(-5, -1)$ is a solution of $2x - y = -11$.	8. Determine if the ordered pair $(3, -2)$ is a solution of $4x = 3y + 18$.
9. The graph of the equation $2x + 6y = 4$ passes through point $(x, -2)$. What is the value of x?	10. If $(k, 3)$ is a point on the line whose equation is $4x + y = -9$, what is the value of k?
11. Point $(k, -3)$ lies on the line whose equation is $x - 2y = -2$. What is the value of k?	12. Point $(5, k)$ lies on the line represented by the equation $2x + y = 9$. What is the value of k?

Regents Questions

MULTIPLE CHOICE

1. The solution of an equation with two variables, x and y, is
   - (1) the set of all x values that make $y = 0$
   - (2) the set of all y values that make $x = 0$
   - (3) the set of all ordered pairs, (x, y), that make the equation true
   - (4) the set of all ordered pairs, (x, y), where the graph of the equation crosses the y-axis

2. How many of the equations listed below represent the line passing through the points $(2,3)$ and $(4,-7)$?

$$5x + y = 13$$
$$y + 7 = -5(x - 4)$$
$$y = -5x + 13$$
$$y - 7 = 5(x - 4)$$

   (1) 1   (3) 3

   (2) 2   (4) 4

3. Which statement best describes the solutions of a two-variable equation?

   (1)   The ordered pairs must lie on the graphed equation.

   (2)   The ordered pairs must lie near the graphed equation.

   (3)   The ordered pairs must have $x = 0$ for one coordinate.

   (4)   The ordered pairs must have $y = 0$ for one coordinate.

4. Mrs. Rossano asked her students to explain why $(3, -4)$ is a solution to $2y + 3x = 1$. Three student responses are given below.

Andrea:
  "When the equation is graphed on a calculator, the point can be found within its table."
Bill:
  "Substituting $x = 3$ and $y = -4$ into the equation makes it true."
Christine:
  "The graph of the line passes through the point $(3, -4)$."

Which students are correct?

   (1)   Andrea and Bill, only   (3)   Andrea and Christine, only

   (2)   Bill and Christine, only   (4)   Andrea, Bill, and Christine

5. If point $(K, -5)$ lies on the line whose equation is $3x + y = 7$, then the value of K is

   (1)   –8   (3)   22

   (2)   –4   (4)   4

6. The point $(3, w)$ is on the graph of $y = 2x + 7$. What is the value of w?

   (1)   –2   (3)   10

   (2)   –4   (4)   13

7. Which linear equation represents a line that passes through the point $(-3, -8)$?

   (1) $y = 2x - 2$ (3) $y = 2x + 13$

   (2) $y = 2x - 8$ (4) $y = 2x - 14$

3.2   Lines Parallel to Axes

If a linear equation has only one variable (that is, x or y is equal to a constant), then it represents a line that is parallel to one of the axes.

If y is equal to a constant, the line is parallel to the x-axis and crosses the y-axis at that constant.

Example: $y = 3$ represents a line parallel to the x-axis but 3 units above it. No matter what values we choose for x, the y value is always 3, as shown below.

x	$y = 3$	y	(x, y)
-2	$y = 3$	3	$(-2,3)$
-1	$y = 3$	3	$(-1,3)$
0	$y = 3$	3	$(0,3)$
1	$y = 3$	3	$(1,3)$
2	$y = 3$	3	$(2,3)$

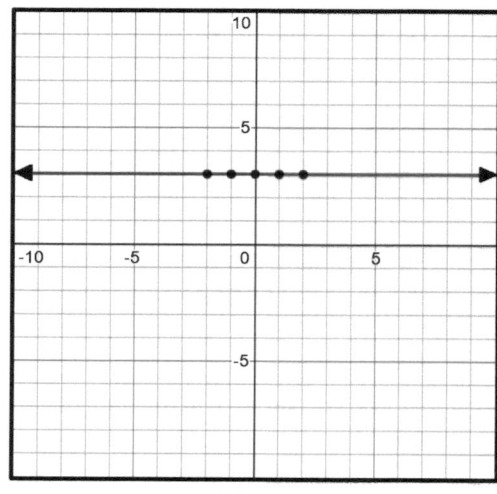

If x is equal to a constant, the line is parallel to the y-axis and crosses the x-axis at that constant.

Example: $x = -5$ represents a line parallel to the y-axis but 5 units to the left of it. The line contains the points $(-5, y)$ where y is any real number.

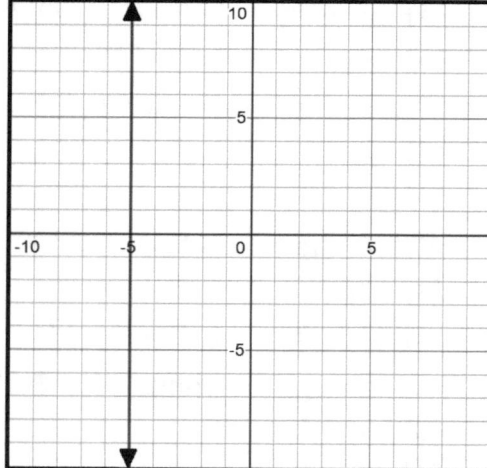

MODEL PROBLEM

Write the equation of a line parallel to the *x*-axis but 10 units below it.

Solution:
$y = -10$

Explanation of steps:
If a line is parallel to an axis, then the other variable is equal to a constant. *[A line parallel to the x-axis has y = constant; since it is 10 units below the axis, $y = -10$.]*

PRACTICE PROBLEMS

1. The graph of $y = -2$ is a line
 	
    (1)   parallel to the *x*-axis
    (2)   parallel to the *y*-axis
    (3)   passing through the origin
    (4)   passing through the point $(-2,0)$
	2. Which equation represents a line that is parallel to the *y*-axis and passes through the point $(4,3)$?
	
	    (1) $x = 3$ $\qquad$ (3) $y = 3$
	    (2) $x = 4$ $\qquad$ (4) $y = 4$

3. Write the equation of the line parallel to the *y*-axis but 9 units to the right of it.	4. Write the equation of the line parallel to the *x*-axis but 1 unit above it.
5. Write the equation of the line that lies on the *y*-axis.	6. Write the equation of the line that lies on the *x*-axis.

7. At what point does the line whose equation is $x = 5$ intersect the x-axis, written as an ordered pair?

8. Graph the line whose equation is $x = 7$.

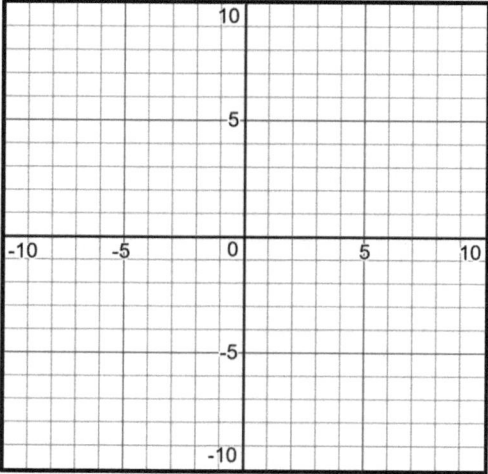

9. Graph the line whose equation is $y = -4$.

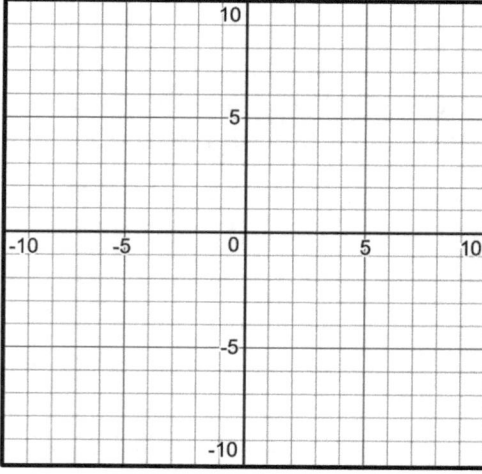

10. Graph the line whose equation is $y - 4 = -1$.

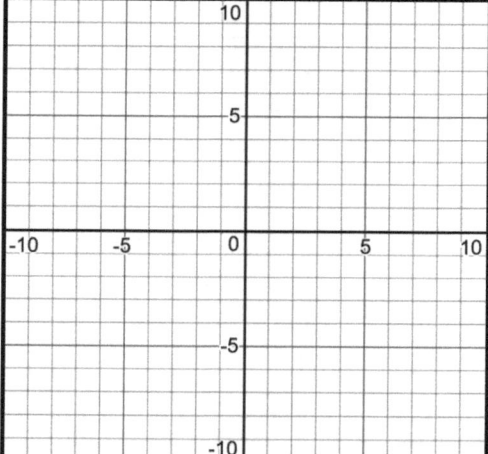

Regents Questions

There are no Regents exam questions on this topic.

3.3   Find Intercepts

The **y-intercept** is the value of y at the point where a graph
        intersects the y-axis and the **x-intercept** is the
        value of x at the point where the graph intersects
        the x-axis.

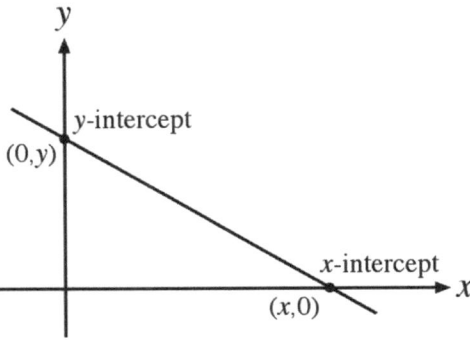

Algebraically, we can find the y-intercept by substituting 0
for x in the equation, and we can find the x-intercept by
substituting 0 for y in the equation. This works because
$x = 0$ all along the y-axis, and $y = 0$ all along the x-axis.

Example: For the equation $y = 2x - 1$, the y-intercept is
                $y = 2(0) - 1 = -1$. We can find the x-intercept by solving for $0 = 2x - 1$, which
                gives us $x = \frac{1}{2}$.

Of course, if a line is parallel to an axis, it will never intersect that axis, so an intercept cannot be
found for that axis' variable.

MODEL PROBLEM

What are the y-intercept and x-intercept of the line whose equation is $2x + 3y = 12$?

Solution:

    (A)   (B)

    $2(0) + 3y = 12$ $2x + 3(0) = 12$

             $3y = 12$ $2x = 12$

               $y = 4$ $x = 6$

    (C) The y-intercept is 4 and the x-intercept is 6.

Explanation of steps:
    (A)   To find the y-intercept, substitute 0 for x in the equation and solve.
    (B)   To find the x-intercept, substitute 0 for y in the equation and solve.
    (C)   The resulting values of y and x are the intercepts.

PRACTICE PROBLEMS

1. State the x and y intercepts, both integers, of the graph below. 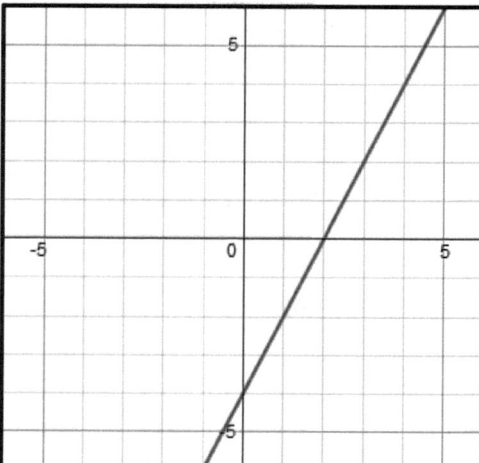	2. State the x and y intercepts, both integers, of the graph below. 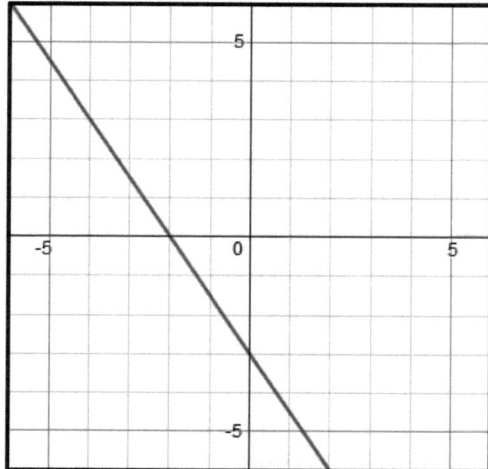
3. Find the x and y intercepts of the equation $3y + 2x = 6$.	4. Find the x and y intercepts of the equation $3x - 4y = 12$.
5. Find the x and y intercepts of the equation $y = -2x + 5$.	6. Find the x and y intercepts of the equation $9x - 6y + 5 = 0$.

Regents Questions

MULTIPLE CHOICE

1. The value of the x-intercept for the graph of $4x - 5y = 40$ is

   (1) 10

   (2) $\frac{4}{5}$

   (3) $-\frac{4}{5}$

   (4) -8

2. Which function has the same y-intercept as the graph below?

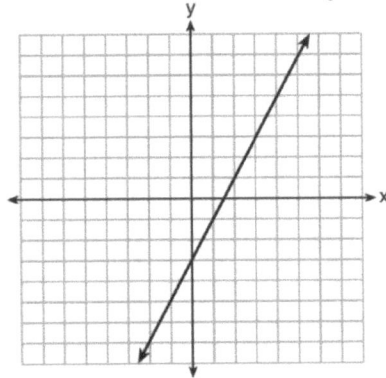

   (1) $y = \dfrac{12 - 6x}{4}$

   (2) $27 + 3y = 6x$

   (3) $6y + x = 18$

   (4) $y + 3 = 6x$

3. Which function has the largest y-intercept?

$$f(x) = -4x - 1$$

(1)

$$g(x) = |x| + 3$$

(3)

x	h(x)
−1	1.5
0	2
1	3
2	5

(2)

k(x)

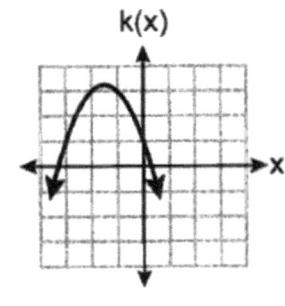

(4)

101

3.4   Find Slope Given Two Points

If you can imagine a person walking along a line *from left to right*, the **slope of the line** represents how steep the road is.

A **positive slope** would represent walking uphill and a **negative slope** would represent walking downhill. A **slope of zero** would mean the road is horizontal (parallel to the *x*-axis); it is neither uphill nor downhill. The person cannot walk from left to right along a vertical line (parallel to the *y*-axis), so we say that a vertical line has **no slope**. The larger the absolute value of the slope, the steeper the road: a slope of 3 is a steeper uphill climb than a slope of 1/3, and a slope of –3 is a steeper downhill descent than a slope of –1/3.

The slope is usually represented by the letter **m**. Given two points on a line, we can determine, either graphically or algebraically, the slope of the line.

Finding the slope graphically:

If we move from the left point to the right point, the slope $m = \dfrac{rise}{run}$.

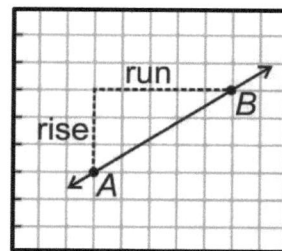

The **rise** is how many units we travel up (positive) or down (negative), and the **run** is how many units we travel to the right.

Example: From point A to point B on the graph at right, the rise is 3 and the run is 5, so the slope is $\dfrac{3}{5}$.

Finding the slope algebraically:

We can think of the rise, or the number of *y*-value units we need to travel, as the **difference in the y-values**, and the run, or the number of *x*-value units we need to travel, as the **difference in the x-values**. If we name the coordinates of the two points (x_1, y_1) and (x_2, y_2), the slope formula can be written as: $m = \dfrac{y_2 - y_1}{x_2 - x_1}$

Example: The slope of the line through (1,3) and (2,6) is $m = \dfrac{y_2 - y_1}{x_2 - x_1} = \dfrac{6 - 3}{2 - 1} = \dfrac{3}{1} = 3.$

If the two points lie on a horizontal line (parallel to the *x*-axis), the slope is zero.

Example: (2,5) and (3,5) lie on the horizontal line $y = 5$. $m = \dfrac{y_2 - y_1}{x_2 - x_1} = \dfrac{5 - 5}{3 - 2} = \dfrac{0}{1} = 0.$

If the two points lie on a vertical line (parallel to the *y*-axis), there is no slope.

Example: (4,1) and (4,3) lie on the vertical line $x = 4$. To use the slope formula would result in a *denominator of zero*, which means that the slope is *undefined*.

MODEL PROBLEM 1:   *GRAPHICALLY*

What is the slope of the line passing through the points (2,4) and (6,6)?

Solution:

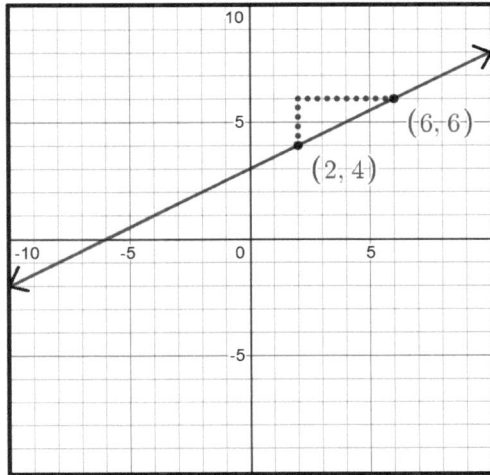

$$m = \frac{rise}{run} = \frac{2}{4} = \frac{1}{2}$$

Explanation of steps:
- (A) Using the graph, first trace how many units you need to travel up (+) or down (–), and call this the rise *[from y = 4 to y = 6 is 2 units]*.
- (B) Next, trace how many units you need to travel to the right, and call this the run *[from x = 2 to x = 6 is 4 units]*.
- (C) Write $\frac{rise}{run}$ as a fraction, and reduce.

PRACTICE PROBLEMS

1. What is the slope of the line passing through the points *A* and *B*, as shown on the graph below?	2. What is the slope of the line passing through the points *A* and *B*, as shown on the graph below?
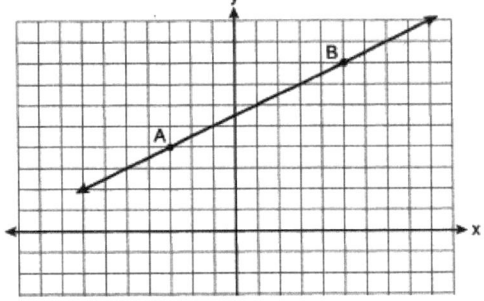	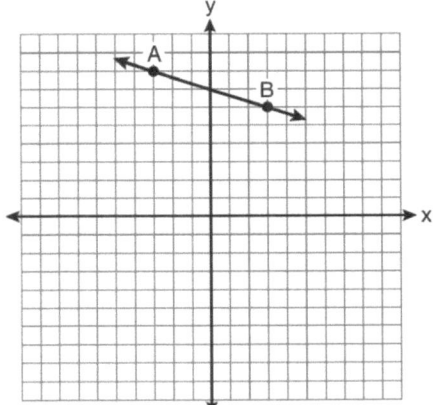

3. What is the slope of line ℓ in the accompanying diagram?

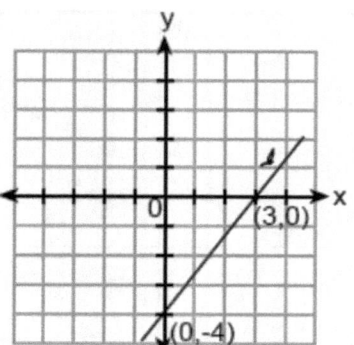

4. What is the slope of line ℓ in the accompanying diagram?

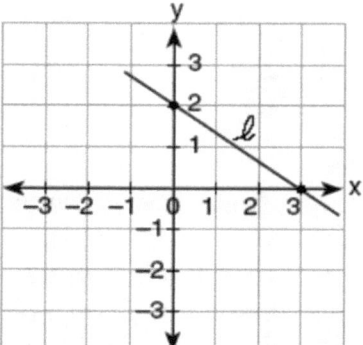

5. What is the slope of the line shown below?

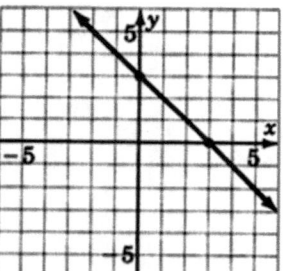

6. On a separate coordinate graph, draw a line through the points $(-4, -3)$ and $(1,2)$ and find the slope of the line.

MODEL PROBLEM 2:   ALGEBRAICALLY

What is the slope of the line passing through the points $(-3,4)$ and $(1,2)$?

Solution:

(A) $(x_1, y_1) = (-3,4)$ $(x_2, y_2) = (1,2)$
(B)   (C)   (D)   (E)

$$m = \frac{y_2 - y_1}{x_2 - x_1} = \frac{2-4}{1-(-3)} = \frac{-2}{4} = -\frac{1}{2}$$

Explanation of steps:

(A) Label the coordinates of the two points. Each point has an x and y value, but the first point has subscripts of 1 and the second point has subscripts of 2.
(B) Write the slope formula.
(C) Substitute the coordinates as labeled.
(D) Evaluate the numerator and denominator.
(E) Simplify, if possible.

PRACTICE PROBLEMS

7. What is the slope of the line that passes through points $(1,3)$ and $(5,13)$?	8. What is the slope of the line that passes through points $(3,-6)$ and $(1,8)$?

9. What is the slope of the line that passes through points $(4,5)$ and $(0,-3)$?

10. What is the slope of the line that passes through $(-4,-2)$ and $(2,-2)$?

11. What is the slope of the line that passes through the points $(2,5)$ and $(7,3)$?

12. What is the slope of the line that passes through the points $(3,5)$ and $(-2,2)$?

Regents Questions

There are no Regents exam questions on this topic.

3.5   Find Slope Given an Equation

The equation of a line is most commonly written in **slope-intercept form**: $y = mx + b$.
In this form, m (the coefficient of x) is the **slope** of the line and b is the **y-intercept**. The y-intercept is the value of y at the point where the line intersects the y-axis.

Transforming an equation into slope-intercept form
If the equation is not already in slope-intercept form, you will need to transform the equation by solving for y in terms of x.

Examples: To transform the equations $y + 2x = 4$ and $3y = 2x - 9$,

$$y + 2x = 4$$
$$\underline{-2x \quad - 2x}$$
$$y = -2x + 4$$

$$\frac{3y}{3} = \frac{2x - 9}{3}$$
$$y = \frac{2}{3}x - 3$$

If two distinct lines have the **same slope**, they are **parallel**. So, to determine whether two lines are parallel, write each equation in *slope-intercept form* to determine whether the slopes are the same for both equations.

MODEL PROBLEM 1:   *TRANSFORMING AN EQUATION*

What is the slope of the line whose equation is $2y - 3x = x + 2$?

Solution:

(A) $2y - 3x = x + 2$
$$\underline{+3x \quad + 3x}$$
$$\frac{2y}{2} = \frac{4x + 2}{2}$$
$$y = 2x + 1$$
(B) Slope is 2

Explanation of steps:

(A) If the equation is not already in slope-intercept form, transform it by solving for y in terms of x. *[Add 3x to both sides, then divide each by 2.]*

(B) For an equation in slope-intercept form, the slope is the coefficient of x *[the slope is 2].*

PRACTICE PROBLEMS

1. What is the slope of a line whose equation is $y = \frac{2}{5}x - 5$?	2. What is the slope of the line whose equation is $y - 3x = 1$?
3. What is the slope of the line whose equation is $2y = 5x + 4$?	4. What is the slope of the linear equation $5y - 10x = -15$?
5. What is the slope of the line represented by the equation $4x + 3y = 12$?	6. What is the slope of a line represented by the equation $2y = x - 4$?
7. What is the slope of the line whose equation is $3x - 2y = 12$?	8. What is the slope of the line whose equation is $3x - 4y - 16 = 0$?

MODEL PROBLEM 2:   *EQUATIONS OF PARALLEL LINES*

The equations of two distinct lines are $y = 3x - 6$ and $2y = 3x + 6$. Are the lines parallel?

Solution:

(A) For $y = 3x - 6$, the slope $m = 3$.
     Solving $2y = 3x + 6$ for y:
     $$\frac{2y}{2} = \frac{3x + 6}{2}$$
     $y = \frac{3}{2}x + 3$, so the slope $m = \frac{3}{2}$.

(B) The lines are *not* parallel because the slopes are not equal.

Explanation of steps:

(A) Write each equation in slope-intercept form to determine the slope of each line.
     [The first equation is already in slope-intercept form, $y = mx + b$, so the slope m = 3. The second equation needed to be transformed, resulting in a slope of $\frac{3}{2}$.]

(B) If the slopes are equal, the lines are parallel.
     [These slopes are 3 and $\frac{3}{2}$, so they are not parallel.]

PRACTICE PROBLEMS

9. Line ℓ has an equation of $y = -2x - 5$. Write the equation of a line that is parallel to line ℓ but has a y-intercept of 2.	10. Line ℓ has an equation of $y = \frac{1}{2}x + 2$. Write the equation of a line that is parallel to line ℓ but passes through the origin.

11. Which equation represents a line that is parallel to the line whose equation is $y = -3x - 7$?

(1) $y = -3x + 4$ (3) $y = \frac{1}{3}x + 5$

(2) $y = -\frac{1}{3}x - 7$ (4) $y = 3x - 2$

12. Which equation represents a line that is parallel to the line whose equation is $2x - 3y = 9$?

(1) $y = \frac{2}{3}x - 4$ (3) $y = \frac{3}{2}x - 4$

(2) $y = -\frac{2}{3}x + 4$ (4) $y = -\frac{3}{2}x + 4$

13. Which equation below represents a line that is parallel to the line, $y = -x + 4$?

$$2y + 2x = 6$$
$$2y - x = 6$$

14. Which equation below represents a line that is parallel to the line, $4x + 6y = 5$?

$$-3y = 2x + 5$$
$$-6y + 4x = 5$$

Regents Questions

There are no Regents exam questions on this topic.

3.6   Graph Linear Equations

Graphing an equation in slope-intercept form

Given an equation in the form $y = mx + b$, you can follow these steps to graph the line:

1. Use the y-intercept to plot the point $(0, b)$ on the y-axis.
2. Use the slope to determine at least two more points.
3. Draw a line through the points and label the line with the equation.

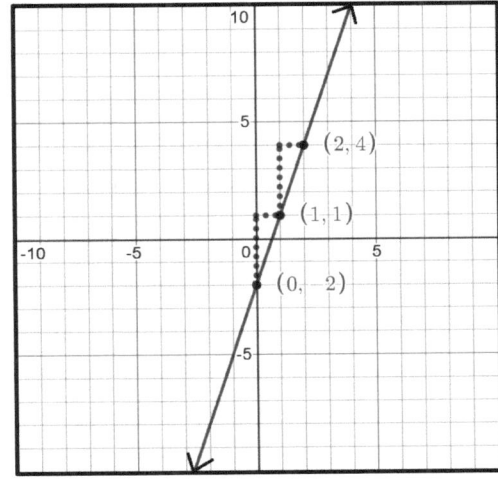

Example: For the equation $y = 3x - 2$, the slope $m = 3$ and the y-intercept $b = -2$.

The y-intercept of -2 gives the starting point, $(0, -2)$. Since the slope $m = 3 = \frac{3}{1} = \frac{rise}{run}$, use the rise of 3 and run of 1 to get two more points, $(1,1)$ and $(2,4)$.

It may help to remember that b tells us a point to *begin* graphing the line and the m tells us how to *move* to find other points on the line.

CALCULATOR TIP

We can also **graph the line using the calculator**:

1. Write the equation in slope-intercept form.
2. Enter the equation by pressing the Y= button, then type the right side of the equation. Use X,T,Θ,n to enter the variable x.
3. Press GRAPH.

Example: For the equation $y = 3x - 2$, enter Y= 3 X,T,Θ,n − 2 ENTER GRAPH ENTER.

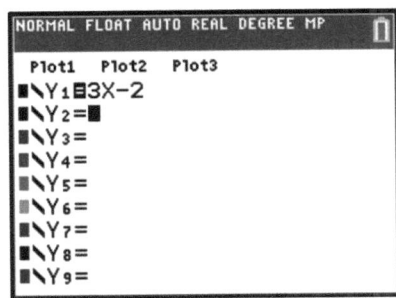   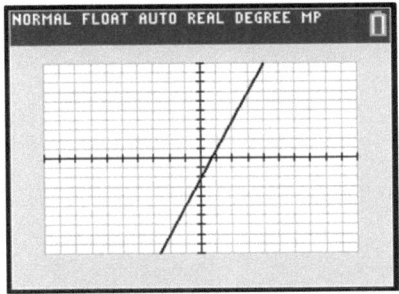

Note: After pressing the Y= button, if any of the Plots at the top of the screen are highlighted, use the arrow keys to move to them and press ENTER to turn them off.

▦▯ CALCULATOR TIP

To adjust the calculator's grid settings:

The calculator's default zoom size, **ZStandard**, sets the grid size to 20 by 20 units centered at the origin. For graphs that may not display well in the standard grid size, you may need to adjust the **Window** size. Press [WINDOW] and enter values for Xmin and Xmax, and also values for Ymin and Ymax. The values of Xscl and Yscl represent the difference between tick marks on each axis and can usually be left at 1. Then press [GRAPH].

If the grid is not shown on the graph screen, you can turn it on by pressing [2nd][FORMAT]. For calculators that will only show the grid as dots, select GridOn. On other calculators that allow gridlines, select GridLine.

Because the calculator screen is wider than it is tall, the cells in the grid are not squares in the ZStandard zoom size. Most of the time, you may want the zoom set to **ZSquare** by pressing [ZOOM][5]. This will graph your equation using square boxes on the grid, which will display the slopes of your lines more accurately, as shown below.

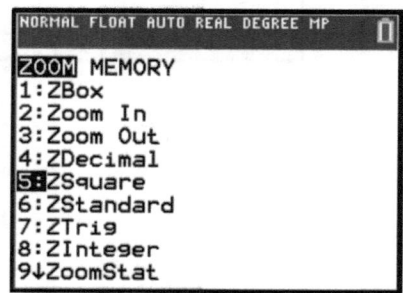 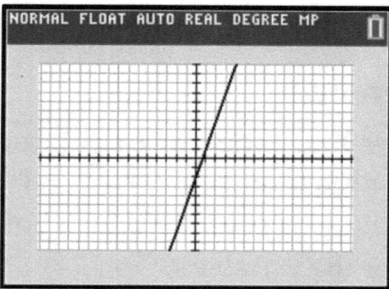

▥▤ CALCULATOR TIP

To view a table of points on the line with the calculator:

Press [2nd][TABLE]. You can then scroll with the arrow keys [▲][▼] to see more points.

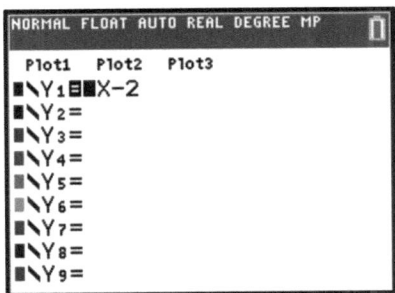

If you'd like to see both the graph and the table on the same screen, press [MODE], then scroll down and change from Full to G-T (or Graph-Table) mode.

Press [TRACE] and the [◄][►] keys to move the cursor along the line while the calculator scrolls to the corresponding points in the table. Or, press [2nd][TABLE] to switch to the table, then use the [▲][▼] keys to scroll through the table rows while the calculator moves to the corresponding points on the graph.

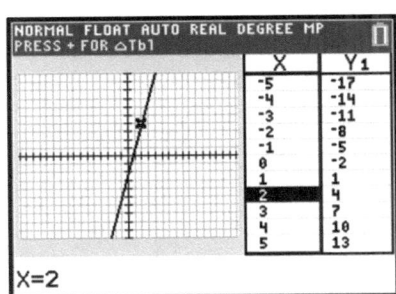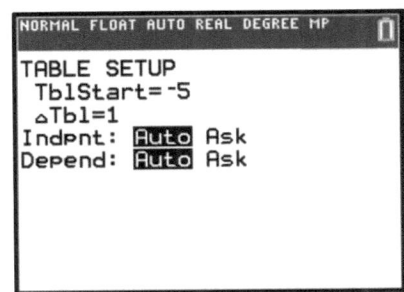

You can specify which values of x the calculator will use for its table. Press [2nd] [TBLSET], then enter values for TblStart and ΔTbl. For example, if you want the table to start at $x = -5$ with increments of 1, enter TblStart = –5 and ΔTbl = 1, as shown above right.

As a shortcut, you can change the ΔTbl value by pressing [+] while the cursor is in the table.

MODEL PROBLEM

Graph the equation $y + 4x = 6$.

(D)

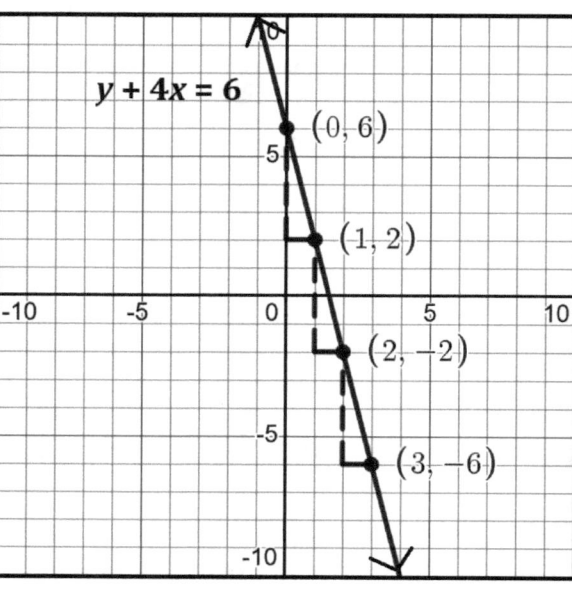

Solution:

(A) $y + 4x = 6$
$$\underline{-4x \qquad -4x}$$
$$y = -4x + 6$$

(B)

Explanation of steps:

(C)

(A) Solve for y if necessary.
[Isolate y by subtracting $4x$ from both sides.]

(B) Plot the y-intercept.
[The y-intercept is 6, so plot the point (0,6).]

(C) Use the slope to plot additional points.
[The slope is -4, so starting at (0,6), go down 4 and to the right 1 to get to the next point. This takes us to (1,2), (2,−2), (3,−6), etc.]

(D) Connect the points with a straight line and add arrowheads at both ends. Label the line with the original equation.

PRACTICE PROBLEMS

1. Graph the equation $y = x - 5$.

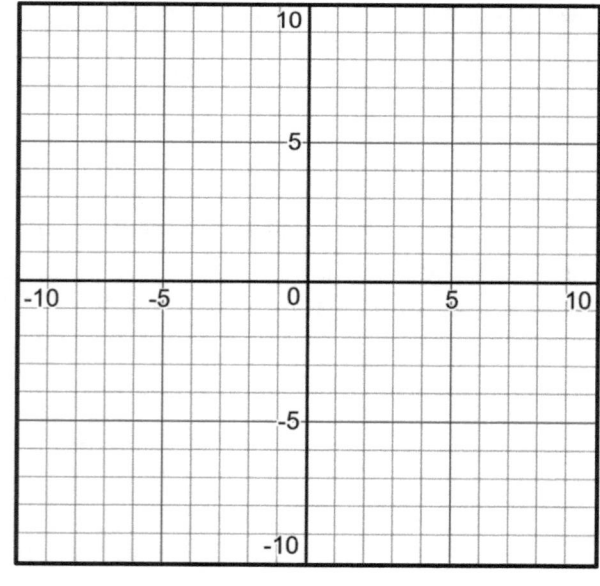

2. Graph the equation $y = -2x + 4$.

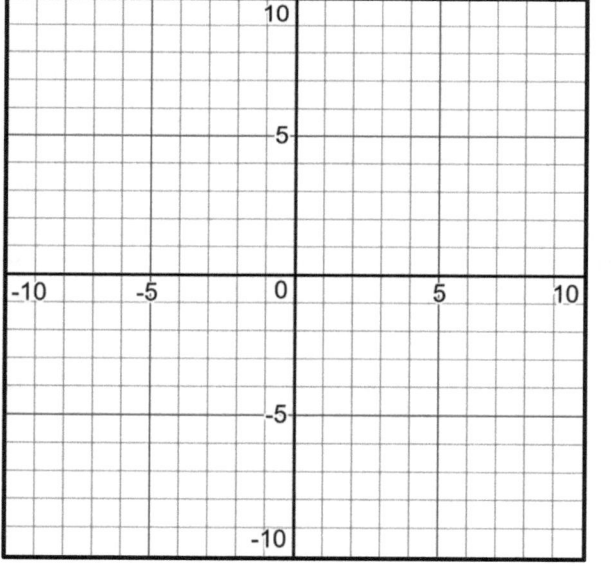

3.   Graph the equation $y - 3x = 4$.

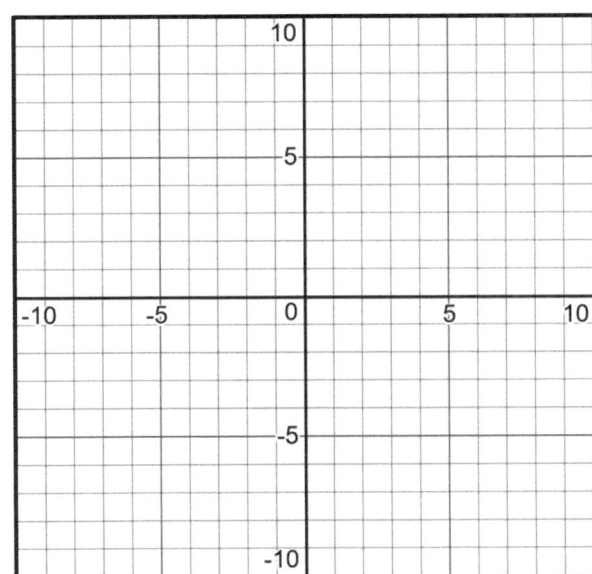

4.   Graph the equation $2y + 2x = x - 2$.

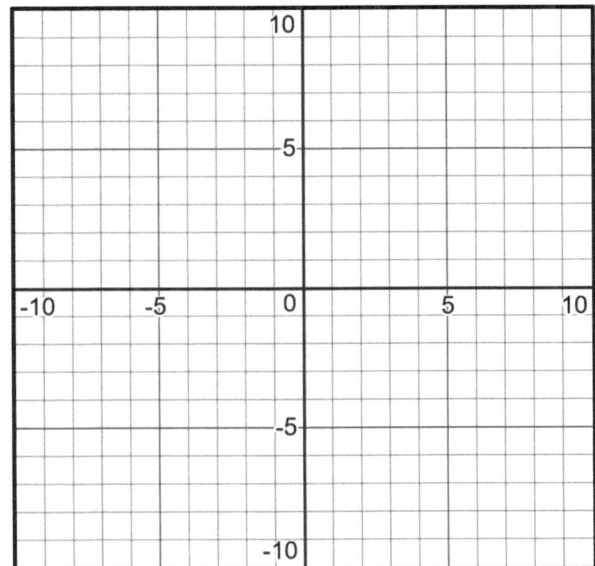

5.   Which equation is represented by the graph?

   (1) $2y + x = 10$

   (2) $y - 2x = -5$

   (3) $-2y = 10x - 4$

   (4) $2y = -4x - 10$

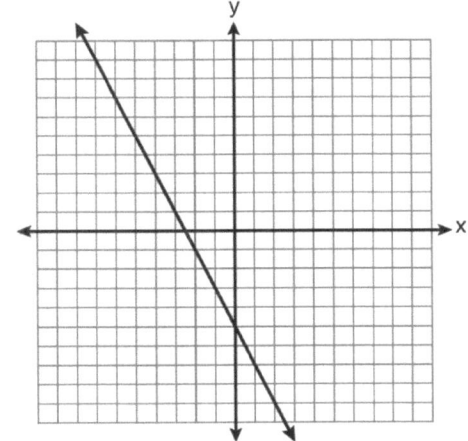

Regents Questions

MULTIPLE CHOICE

1. Which graph shows a line where each value of y is three more than half of x?

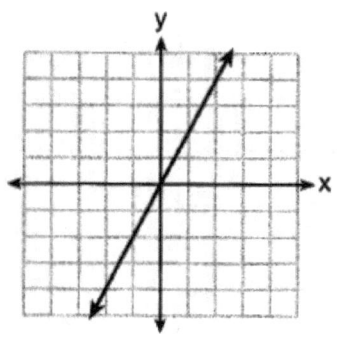

(1)

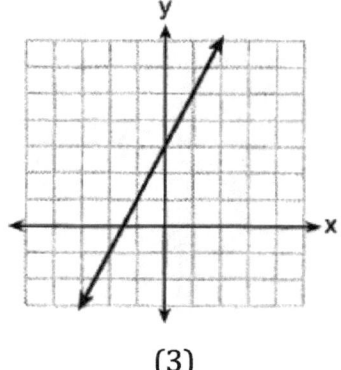

(3)

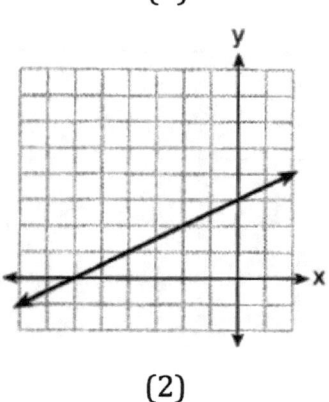

(2)

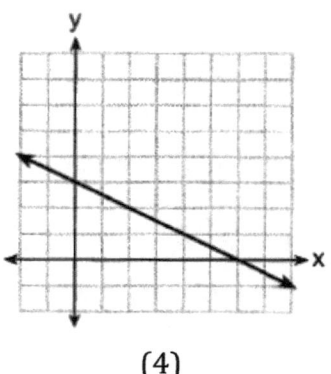

(4)

CONSTRUCTED RESPONSE

2. On the set of axes below, draw the graph of the equation $y = -\frac{3}{4}x + 3$.

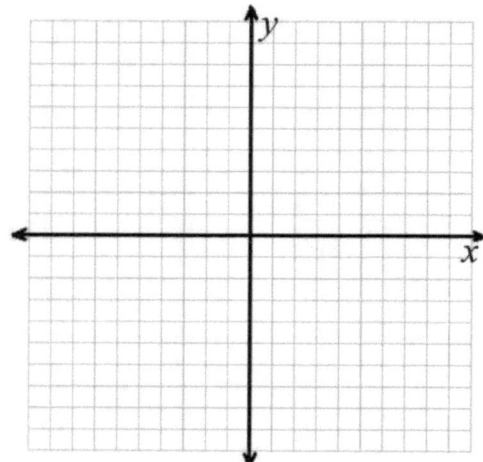

Is the point (3,2) a solution to the equation? Explain your answer based on the graph drawn.

3.   On the set of axes below, graph the line whose equation is $2y = -3x - 2$.

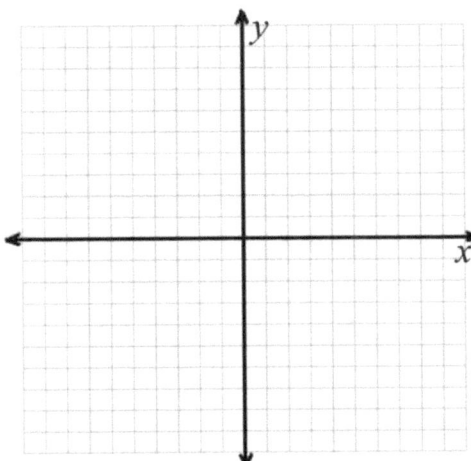

      This linear equation contains the point $(2, k)$. State the value of k.

4.   The table below shows the height in feet, $h(t)$, of a hot-air balloon and the number of minutes, t, the balloon is in the air.

Time (min)	2	5	7	10	12
Height (ft)	64	168	222	318	369

The function $h(t) = 30.5t + 8.7$ can be used to model this data table. Explain the meaning of the slope in the context of the problem.

Explain the meaning of the y-intercept in the context of the problem.

5.   On the set of axes below, graph the equation $3y + 2x = 15$.

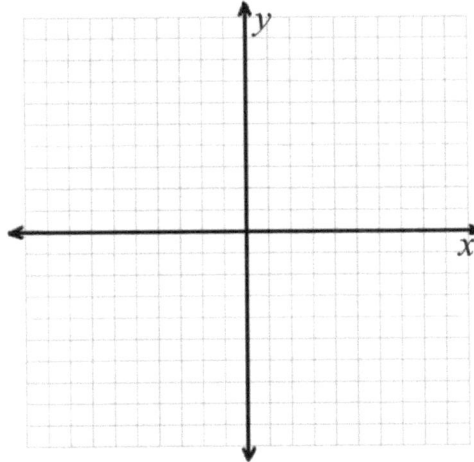

      Explain why $(-6, 9)$ is a solution to the equation.

3.7   Write an Equation Given a Point and Slope

If given the **slope** of a line and the coordinates of **one point**, substitute the slope for m and the coordinates of the point for x and y in the general **slope-intercept form**, $y = mx + b$. This will allow us to solve for b. Once m and b are known, the equation of the line may be written.

An alternative method uses the general **point-slope form** of a linear equation:
$$y - y_1 = m(x - x_1)$$ where m is the slope and (x_1, y_1) is a point on the line.

Using this form, we can substitute the given slope for m and the coordinates of the point for x_1 and y_1. If we want to transform the resulting equation into slope-intercept form, we need to solve for y in terms of x.

MODEL PROBLEM

Write the equation of the line through $(1, -3)$ with a slope of 2.

Solution:

(A) $y = mx + b$
(B) $-3 = 2(1) + b$
(C) $-3 = 2 + b$
      $-5 = b$
(D) $y = 2x - 5$

Explanation of steps:

(A) Write the slope-intercept form of an equation.
(B) Substitute the given values *[x = 1, y = −3, m = 2]*.
(C) Solve for b.
(D) Write the equation using the given slope m and the value of b.

PRACTICE PROBLEMS

1. Write an equation of a line through point $(1, 4)$ with a slope of 2.	2. Write an equation of the line passing through point $(-6, 5)$ with a slope of 5.

3. Write an equation of a line that passes through $(-3,2)$ and has a slope of $\frac{1}{3}$.	4. Write an equation of the line passing through point $(8,-3)$ with a slope of $\frac{3}{4}$.
5. A line having a slope of $\frac{3}{4}$ passes through the point $(-8,4)$. Write the equation of this line in slope-intercept form.	6. Write an equation of the line that passes through the point $(3,-7)$ and has a slope of $-\frac{4}{3}$.

Regents Questions

MULTIPLE CHOICE

1. What is the equation of the line that passes through the point $(6,-3)$ and has a slope of $-\frac{4}{3}$?

    (1) $3y = -4x + 15$ (3) $-3y = 4x + 15$

    (2) $3y = -4x + 6$ (4) $-3y = 4x + 6$

3.8   Write an Equation Given Two Points

Given two points on a line, we can write the equation of the line in **slope-intercept form**. First, find the slope m. Then, substitute the slope and one of the point's coordinates into the general equation, $y = mx + b$, in order to solve for b. Once you have m and b, you can write the equation in slope-intercept form.

We can also have the calculator find the equation for us using the LinReg function.

 CALCULATOR TIP

To find the equation of a line given two points:

1. Press [STAT][1] to select Edit.
2. If any values already appear in the L1 or L2 columns, select the column heading and press [CLEAR][ENTER].
3. Enter the x values into the L1 column and the corresponding y values into the L2 column.
4. Press [STAT]<CALC>[4] for LinReg(ax+b).
   [This function represents the slope as "a" rather than "m", as in $y = ax + b$.]
5. On the next screen prompt, make sure L1 and L2 are selected for Xlist and Ylist. Next to Store RegEQ, enter [ALPHA][F4][1] to store the equation in Y1.
   [On the TI-83, you'll see a LinReg prompt instead. Enter [VARS]<Y-VARS>[1][1].]
6. The screen will show the equation y=ax+b along with the values of a (the slope, m) and b.
7. To view the graph, press [GRAPH]. To see the equation, press [Y=].

Example: The following screenshots show how to find the equation of the line passing through points $(3, -2)$ and $(6, 4)$. The result is $a = 2$ and $b = -8$, representing $y = 2x - 8$.

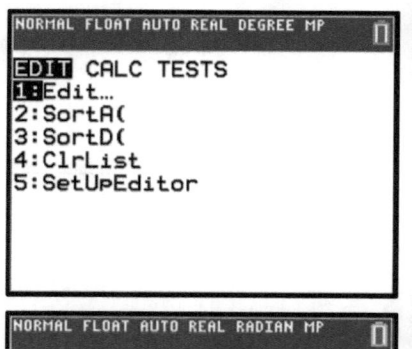
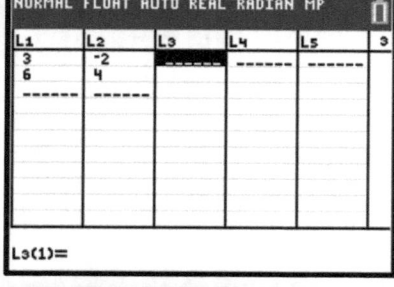
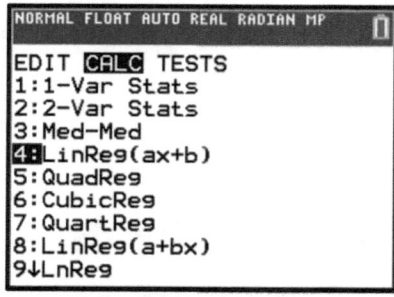

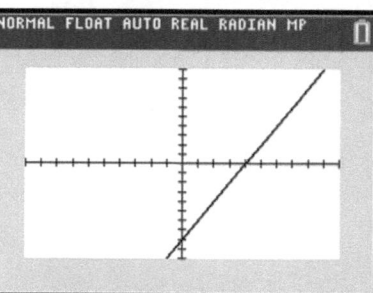

The calculator's LinReg function name is short for "linear regression." We'll learn more about linear regression in Chapter 19.

[Caution: If an exam question asks you to find the equation <u>algebraically</u>, then don't use the calculator method except to check your answer.]

An alternative method is to write the linear equation in point-slope form. The general point-slope form of an equation is $y - y_1 = m(x - x_1)$. Find the slope m. Then substitute one of the point's coordinates for x_1 and y_1.

MODEL PROBLEM

Write an equation of the line that passes through $(3, -2)$ and $(6, 4)$.

Solution:

(A) $m = \dfrac{y_2 - y_1}{x_2 - x_1} = \dfrac{4 - (-2)}{6 - 3} = \dfrac{6}{3} = 2$

(B) $y = mx + b$

(C) $4 = 2(6) + b$

(D) $4 = 12 + b$

$\quad\ -8 = b$

(E) $y = 2x - 8$

Explanation of steps:

(A) find the slope of the line

(B) write the general slope-intercept form of an equation

(C) substitute one point's coordinates *[(6,4)]* for x and y, and the slope *[2]* for m

(D) solve for b

(E) write the resulting equation using the calculated values of m and b

PRACTICE PROBLEMS

1. Write an equation of the line that passes through the points $(1,2)$ and $(5,6)$.	2. Write an equation of the line that passes through the points $(2, -1)$ and $(3,4)$.

3. Write an equation of the line that passes through the points $(-3, 0)$ and $(3, -2)$.

4. Write an equation of the line that passes through the points $(-2, 4)$ and $(2, 4)$.

5. Write an equation, in point-slope form, of the line that passes through the points $(1, 3)$ and $(8, 5)$.

6. A line passes through the points $(5, 4)$ and $(-5, 0)$.

    (a) Write an equation of the line in slope-intercept form.

    (b) Write an equation of the line in point-slope form.

Regents Questions

MULTIPLE CHOICE

1. The graph of a linear equation contains the points $(3,11)$ and $(-2,1)$. Which point also lies on the graph?
    - (1) $(2,1)$
    - (2) $(2,4)$
    - (3) $(2,6)$
    - (4) $(2,9)$

2. Which ordered pair does *not* fall on the line formed by the other three?
    - (1) $(16,18)$
    - (2) $(12,12)$
    - (3) $(9,10)$
    - (4) $(3,6)$

3. What is an equation of the line that passes through the points $(2,7)$ and $(-1,3)$?
    - (1) $y - 2 = \frac{3}{4}(x - 7)$
    - (2) $y - 2 = \frac{4}{3}(x - 7)$
    - (3) $y - 7 = \frac{3}{4}(x - 2)$
    - (4) $y - 7 = \frac{4}{3}(x - 2)$

CONSTRUCTED RESPONSE

4. Sue and Kathy were doing their algebra homework. They were asked to write the equation of the line that passes through the points $(-3,4)$ and $(6,1)$. Sue wrote $y - 4 = -\frac{1}{3}(x + 3)$ and Kathy wrote $y = -\frac{1}{3}x + 3$. Justify why both students are correct.

3.9   Graph Inequalities

To graph a linear inequality, start by isolating (solving for) the variable y. Then, consider the graph of the equation that would result if the inequality symbol was replaced by an equal sign.

The points on this line are *included in the solution set* if the inequality symbol is ≤ or ≥, but *not included in the solution set* if the inequality symbol is < or >. To show inclusion (≤ or ≥), draw a **solid line**; otherwise, draw a **dashed line**.

Examples: Use a solid line to graph $y \geq 3x + 1$ but a dashed line to graph $y < 3x + 1$.

The line divides the plane into two parts. The part above the line includes all points where $y > mx + b$ and the part below the line includes all points where $y < mx + b$. So, if the inequality starts with $y >$ or $y \geq$, shade **above the line**; if it starts with $y <$ or $y \leq$, then shade **below the line**. Points in the shaded area are *included in the solution set*.

Examples: The graph of $y \geq 3x + 1$ is shaded above the line, but the graph of $y < 3x + 1$ is shaded below the line.

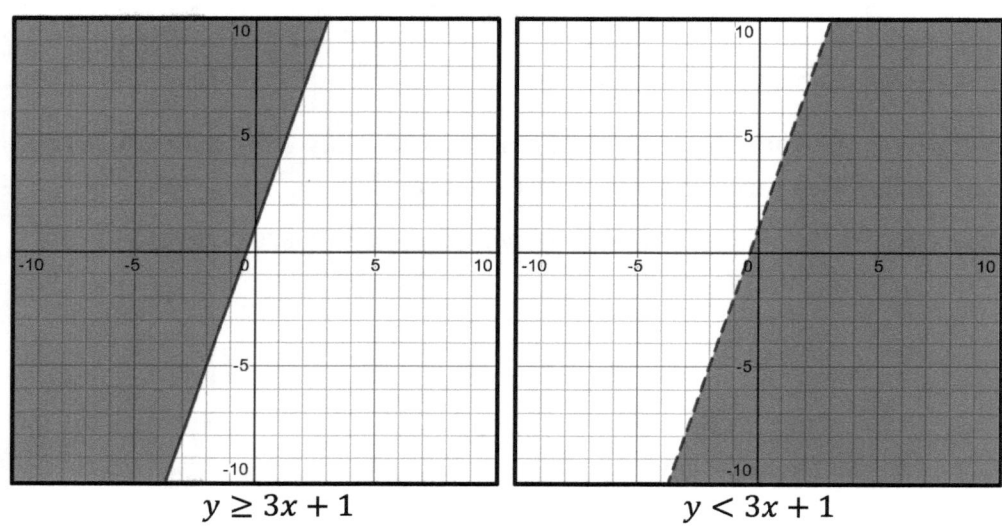

$$y \geq 3x + 1 \qquad\qquad\qquad y < 3x + 1$$

Special case: vertical lines
If the only variable in the inequality is x, solve for x and graph the corresponding vertical line (either solid or dashed according to the same rules). Then shade to the **right of the line** for $x >$ or $x \geq$, or to the **left of the line** for $x <$ or $x \leq$.

Example: The graph of $x \geq 5$ will have a solid vertical line at $x = 5$ and shading to the right of the line.

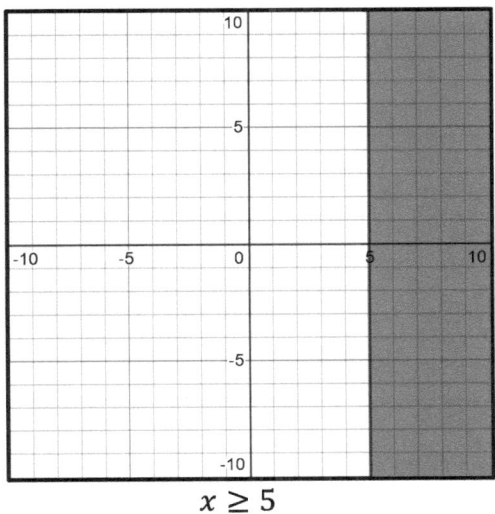

$$x \geq 5$$

CALCULATOR TIP

Inequalities can also be graphed on the calculator. Follow the same steps as entering an equation, but then change the symbol to the left of Y_1 from a line $\boxed{\ }$ to a "shade above" $\boxed{\ }$ or "shade below" $\boxed{\ }$ symbol by moving to the line symbol and pressing $\boxed{\text{ENTER}}$ repeatedly. (On some models, a popup will prompt you for the line type instead, as shown below.) You will still need to know whether to use a solid or dashed line when drawing your graph on paper, depending on the inequality symbol.

Example: The graph of $y \geq 3x - 2$ is graphed below.

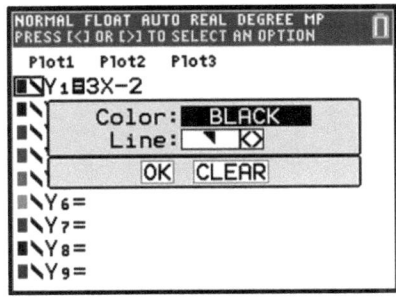
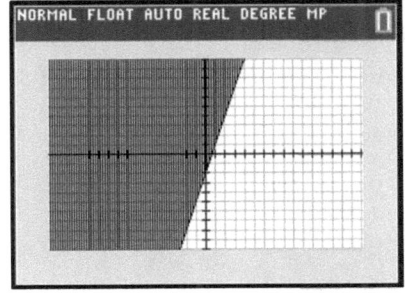

MODEL PROBLEM

Graph the inequality $-4y < 5x - 20$. Is the point $(1,2)$ in the solution set?

Solution:

(A) Solving for y, (B)
$$\frac{-4y}{-4} < \frac{5x - 20}{-4}$$
$$y > -\frac{5}{4}x + 5$$

(C) $(1,2)$ is not in the solution set.

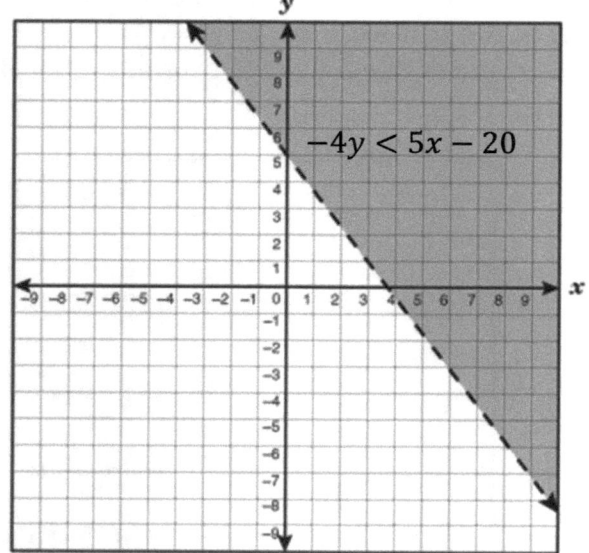

Explanation of steps:

(A) Solve the inequality for y. *[Remember that multiplying or dividing both sides of an inequality by a negative value requires that you reverse the inequality symbol.]*

(B) Use the y-intercept and slope to graph the line. Use a solid line for ≤ or ≥, or a dashed line for < or >. Shade above the line if y is > or ≥, or below the line is y is < or ≤. *[The > means dashed and shaded above.]*

(C) If a point lies on a solid line or in a shaded area, it is in the solution set; otherwise, it is not. *[(1,2) lies in the unshaded area below the line, so it is not a solution.]*

You could also graph $y > -\frac{5}{4}x + 5$ on the calculator:

1. Press $\boxed{Y=}$ then the left arrow $\boxed{\triangleleft}$ twice until the cursor is over the $\boxed{\diagdown}$ symbol.

2. Since the inequality starts with "$y >$" press the $\boxed{\text{ENTER}}$ two times until the shade above $\boxed{\blacksquare}$ symbol appears. Move the cursor back to the right, after the = sign.

3. Enter the right side of the inequality, $\boxed{(\text{-})}\boxed{5}\boxed{\div}\boxed{4}\boxed{X,T,\Theta,n}\boxed{+}\boxed{5}\boxed{\text{ENTER}}$. *[On the TI-84, you can enter the fraction using $\boxed{\text{ALPHA}}\boxed{\text{F1}}\boxed{1}\boxed{\triangleright}\boxed{4}\boxed{\triangleright}$ instead.]*

4. Press $\boxed{\text{GRAPH}}$.

PRACTICE PROBLEMS

1. Which graph represents the inequality $y > 3$?

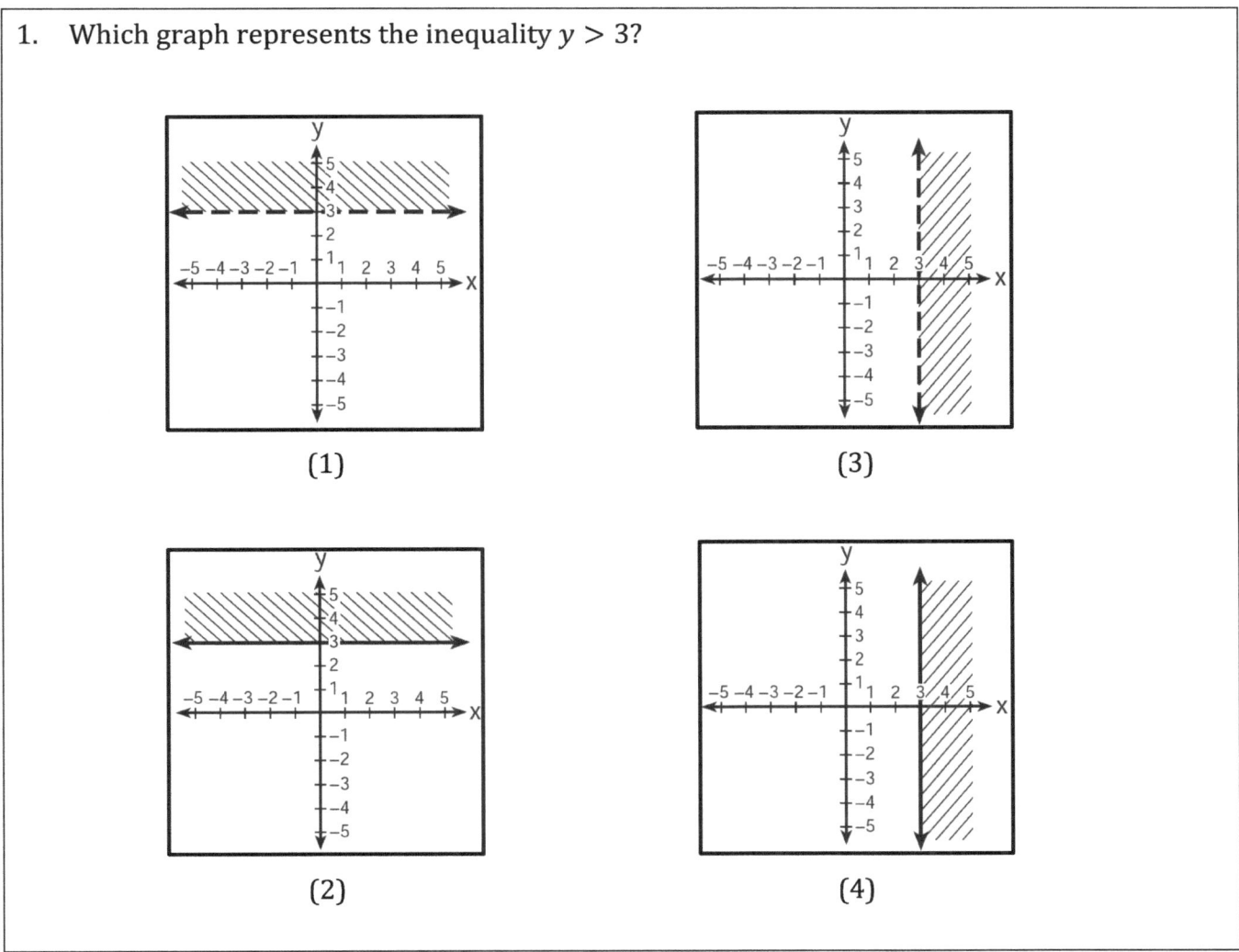

(1)   (3)

(2)   (4)

127

2.   Which graph represents the inequality $y \geq x + 3$?

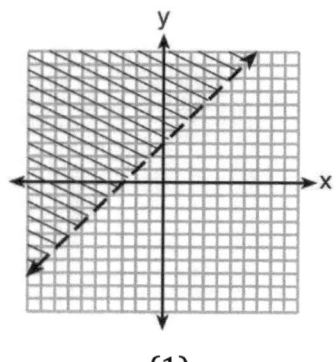

(1)

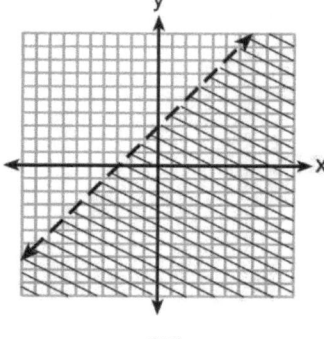

(3)

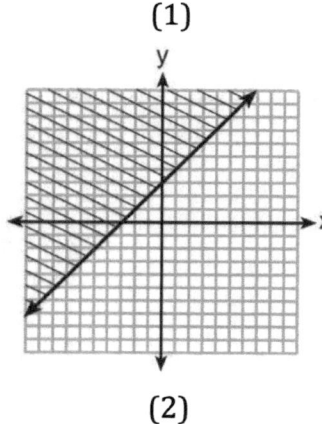

(2)

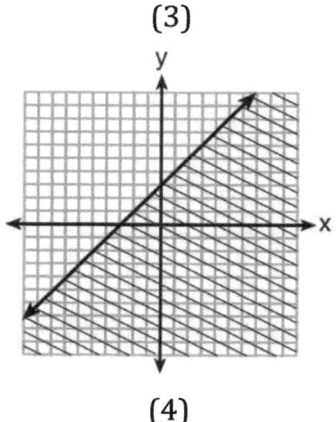

(4)

3.   Which graph represents the solution of $2y + 6 > 4x$?

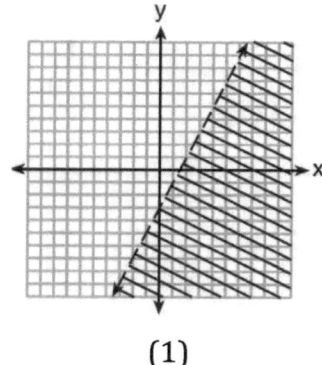

(1)

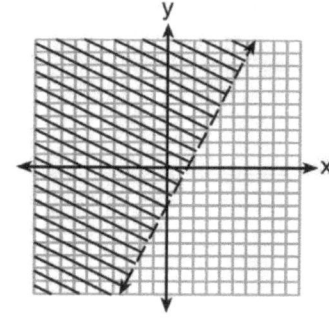

(3)

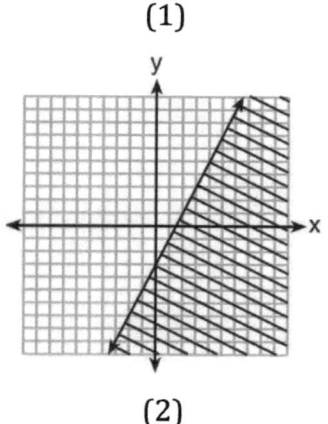

(2)

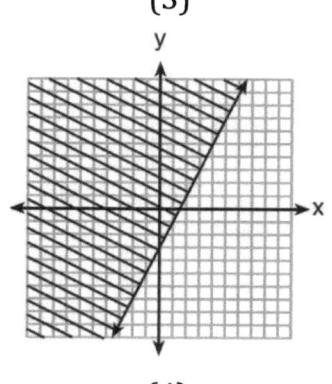

(4)

4. Write an inequality that is represented by the graph below.

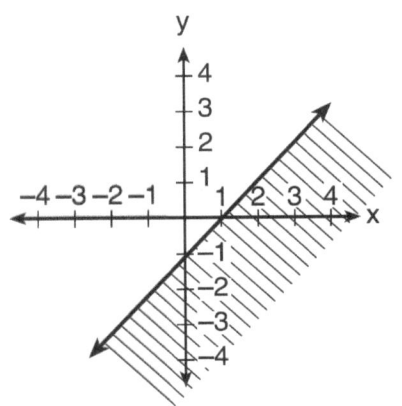

5. Write an inequality that is represented by the graph below.

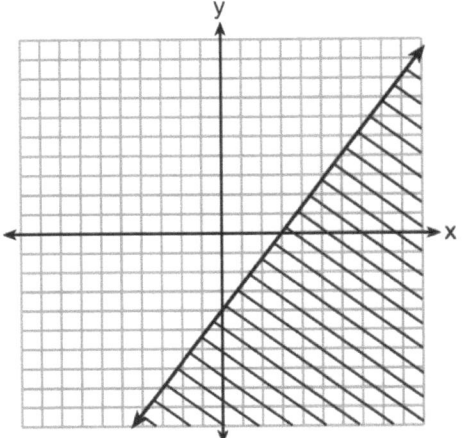

6. Write an inequality that is represented by the graph below.

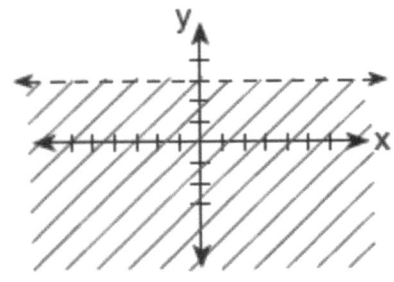

7. Write an inequality that is represented by the graph below.

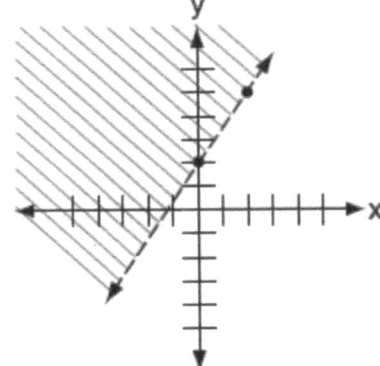

8. Graph the inequality $y \geq 4$.

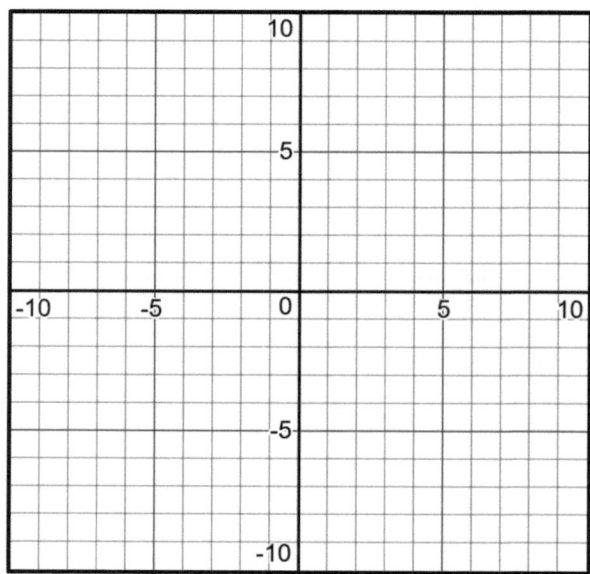

9. Graph the inequality $x < -1$.

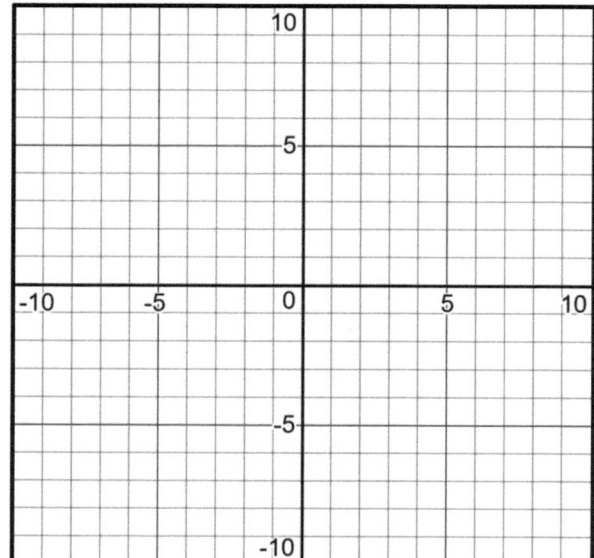

10. Graph the inequality $y > x - 2$.

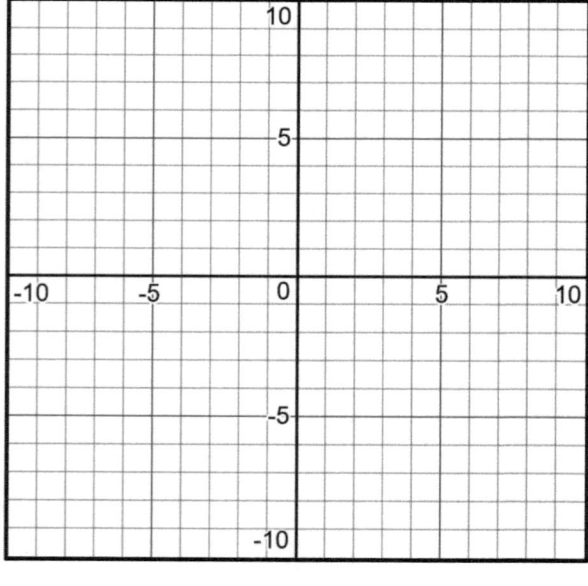

11. Graph the inequality $y \leq -\frac{2}{3}x + 5$.

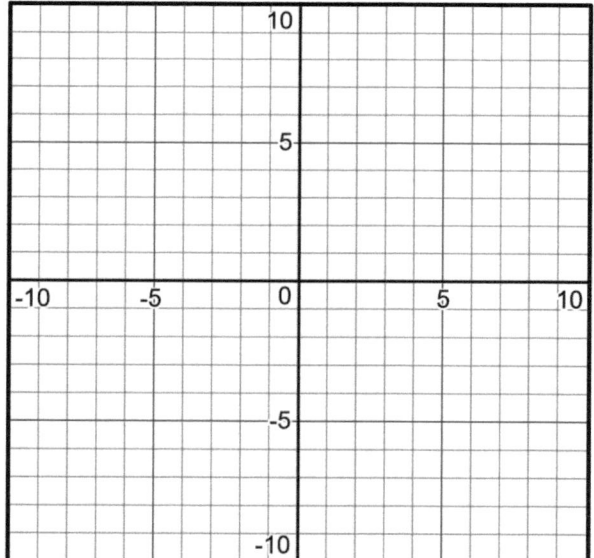

12. Graph the inequality $x + y \leq -3$.

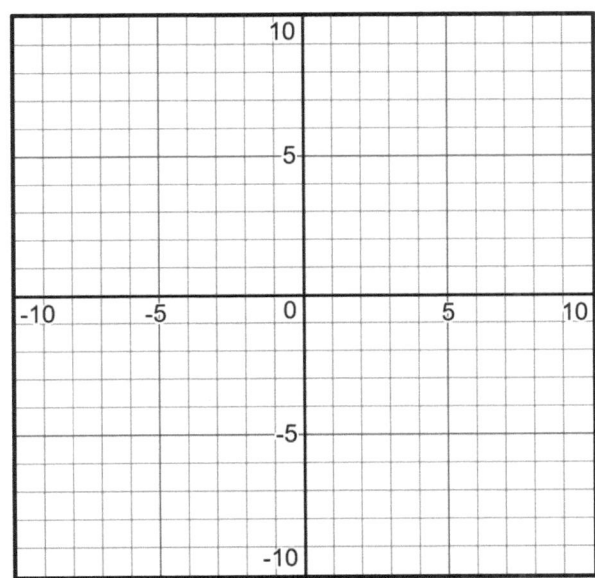

13. Graph the inequality $x - y \leq -1$.

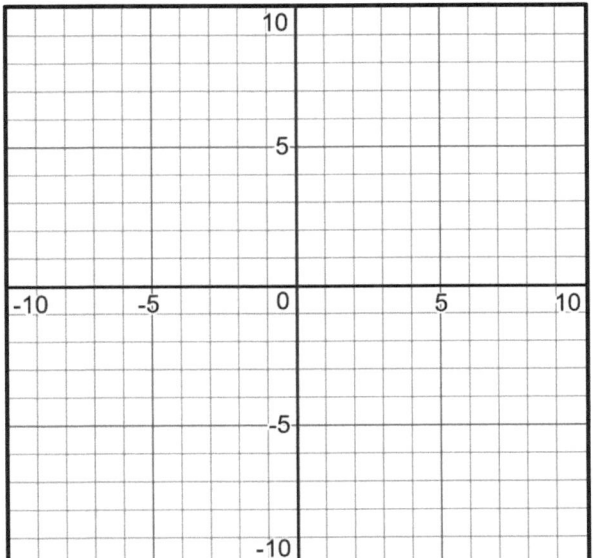

14. Graph the inequality $2y - 6x > 10$.

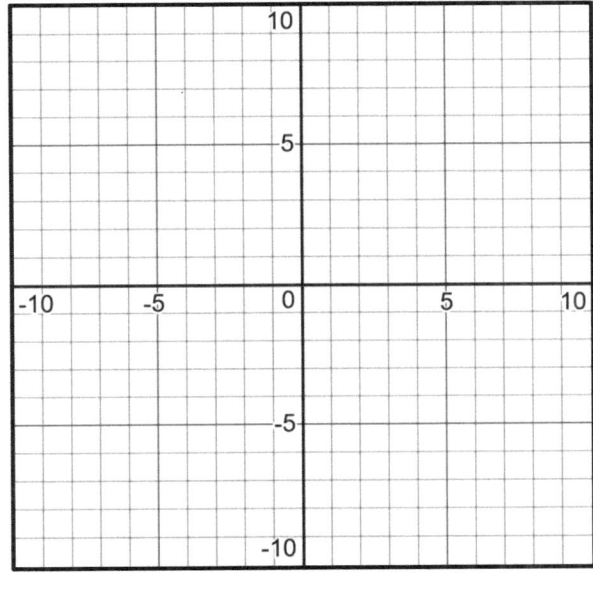

15. Graph the inequality $9 - x \geq 3y$.

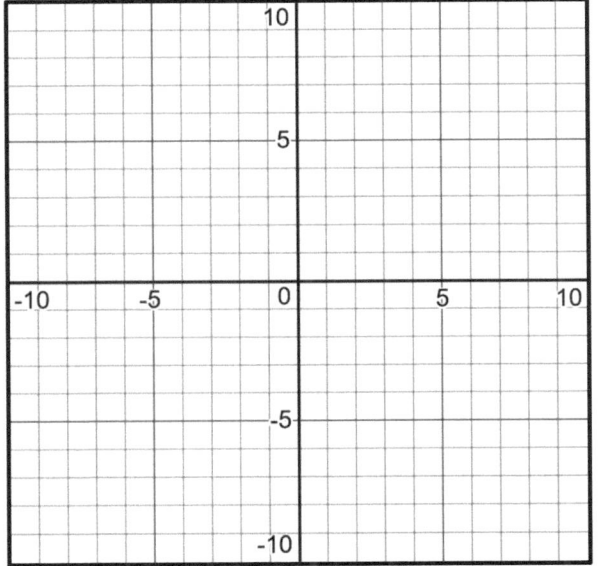

Regents Questions

MULTIPLE CHOICE

1. Which inequality is represented in the graph below?

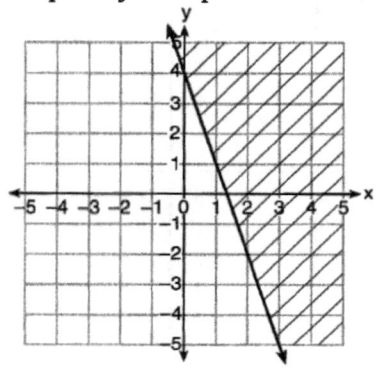

   (1) $y \geq -3x + 4$ (3) $y \geq -4x - 3$
   (2) $y \leq -3x + 4$ (4) $y \leq -4x - 3$

2. Which inequality is represented by the graph below?

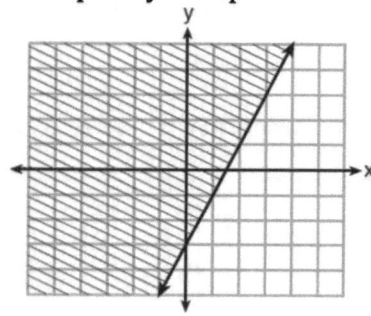

   (1) $y \leq 2x - 3$ (3) $y \leq -3x + 2$
   (2) $y \geq 2x - 3$ (4) $y \geq -3x + 2$

CONSTRUCTED RESPONSE

3.   On the set of axes below, graph the inequality $2x + y > 1$.

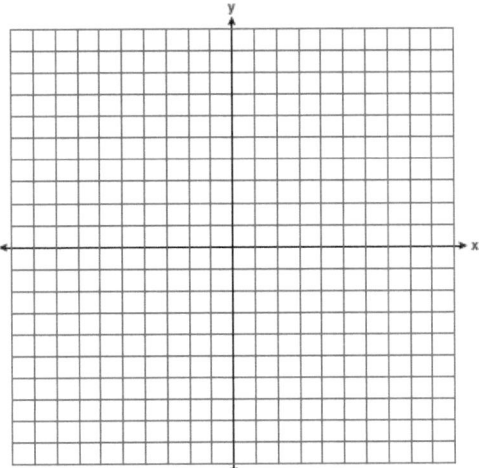

4.   Shawn incorrectly graphed the inequality $-x - 2y \geq 8$ as shown below.

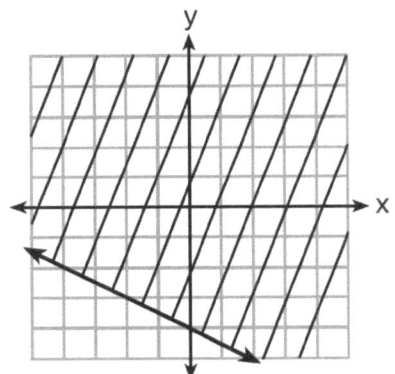

Explain Shawn's mistake. Graph the inequality correctly on the set of axes below.

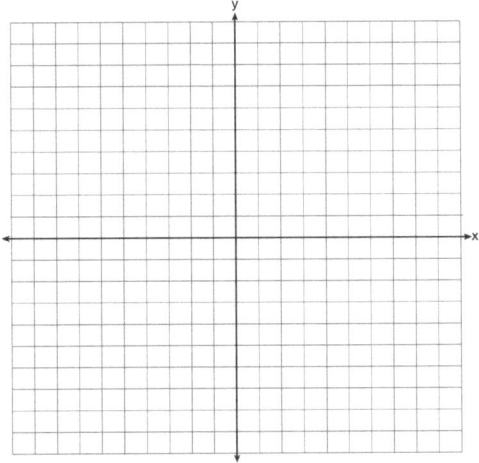

133

5.   Graph the inequality $y > 2x - 5$ on the set of axes below. State the coordinates of a point in its solution.

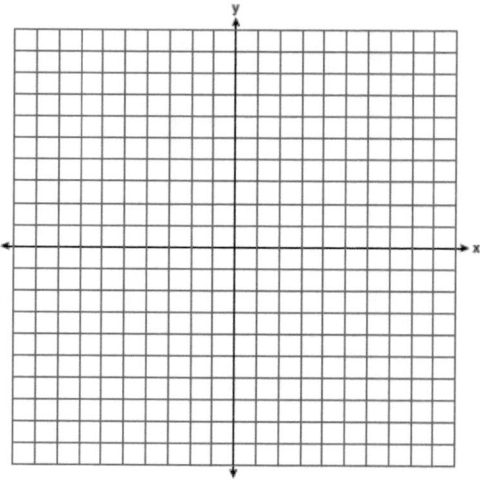

6.   Graph the inequality $y + 4 < -2(x - 4)$ on the set of axes below.

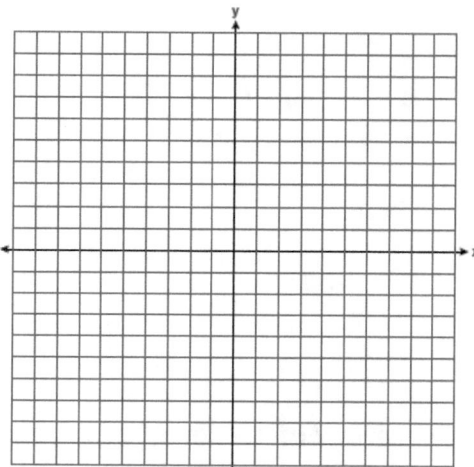

7.   Myranda received a movie gift card for $100 to her local theater. Matinee tickets cost $7.50 each and evening tickets cost $12.50 each.

If x represents the number of matinee tickets she could purchase, and y represents the number of evening tickets she could purchase, write an inequality that represents all the possible ways Myranda could spend her gift card on movies at the theater.

On the set of axes below, graph this inequality.

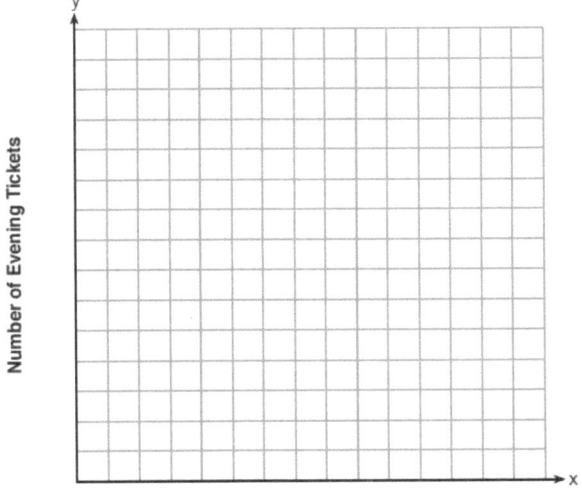

Number of Matinee Tickets

What is the maximum number of matinee tickets Myranda could purchase with her gift card? Explain your answer.

CHAPTER 4.   LINEAR SYSTEMS

4.1   Solve Linear Systems Algebraically

A **linear system of equations** consists of two equations in two variables. A solution for a linear system is a set of one value for each variable that solves both equations at the same time. If the variables are x and y, the solution is an ordered pair which represents the point where the two equations' lines intersect on a coordinate plane.

There are two common methods used to solve linear systems algebraically: the addition method and the substitution method. In both methods, we aim to develop a new equation in which one of the two variables has been eliminated.

The **addition method (elimination method)** depends on the fact that if two equations are true, both the left and right sides of each equation can be added to create a new equation:

    If $a = b$ and $c = d$, then $a + c = b + d$ or, written vertically $\searrow$

$$\begin{array}{l} a = b \\ c = d \\ \hline a + c = b + d \end{array}$$

It also depends on the fact that we can multiply both sides of an equation by the same value and the equation remains true:

    If $a = b$, then $ma = mb$

The goal in the addition method is to eliminate a variable by adding terms whose coefficients for that variable are additive inverses.

Example:
$$\begin{array}{l} 2x - y = 2 \\ x + y = 4 \\ \hline 3x \quad = 6 \end{array}$$

Adding $-y$ and $+y$ eliminates the variable y, allowing us to solve a simpler equation in one variable, $3x = 6$, or $x = 2$.

In the addition method, if two variable terms (in the same variable) are not already additive inverses, they can be made into additive inverses by multiplying one or both equations by values that will change the coefficients into inverses.

Example:
$$\begin{array}{ll} 5a + b = 13 & \times 3 \\ 4a - 3b = 18 & \rightarrow \end{array} \quad \begin{array}{l} 15a + 3b = 39 \\ 4a - 3b = 18 \\ \hline 19a \quad = 57 \end{array}$$

    $+3b$ and $-3b$ are inverses

    Therefore, $a = \dfrac{57}{19} = 3$.

Once we know the value of one variable, we can substitute that value for the variable in either of the two original equations to solve for the other variable.

Example: In the above example, since we know $a = 3$, we can substitute 3 for a in either of the original equations to find b.

$$5(3) + b = 13$$
$$15 + b = 13$$
$$b = -2$$

In the **substitution method**, an equation needs to have one of the variables expressed in terms of the other, or we will need to solve for one of the variables. Once we have a variable equal to an expression, we can substitute that expression for that variable in the other equation.

Example:

$$y = x + 1$$
$$x + 2y = 17 \;\rightarrow\; x + 2(x + 1) = 17$$

Now, solving the new equation for x gives us
$$x + 2(x + 1) = 17$$
$$x + 2x + 2 = 17$$
$$3x + 2 = 17$$
$$3x = 15$$
$$x = 5$$

As with the addition method, once we know the value of one variable, we can substitute the known value into either original equation to solve for the other variable.

Example: For the above example, $y = x + 1$, so $y = 5 + 1 = 6$.

CALCULATOR TIP

We can check our algebraic solutions using the calculator's matrix feature:

1. Press [2nd][MATRIX]<MATH>[ALPHA][B] to select the **rref** function (which stands for "reduced row echelon form").
2. Create a matrix with a size of 3 rows and 3 columns.
   - On the TI-84, press [ALPHA][F3], select 2 and 3 with the arrow keys, then select OK.
   - On the TI-83, press [2nd][MATRIX][1] and type 2 × 3 for 2 rows and 3 columns.
3. Each row of the matrix represents an equation. Enter the coefficients of the first variable in column 1, coefficients of the second variable in column 2, and constants in column 3.
   - On the TI-84, exit the matrix by pressing [▶]. Then, press [)][ENTER].
   - On the TI-83, press [2nd][QUIT][2nd][MATRIX][1][)][ENTER].
4. The resulting matrix will show the values of the variables, in order, in column 3.

Example: The screenshots below show how the calculator solves the system,
$$5a + b = 13$$
$$4a - 3b = 18$$
The solutions matrix shows that $a = 3$ and $b = -2$.

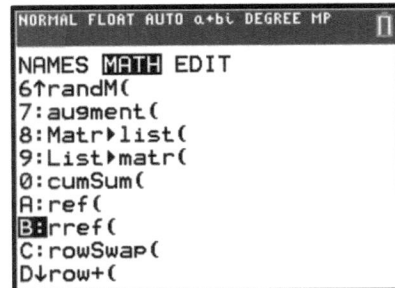
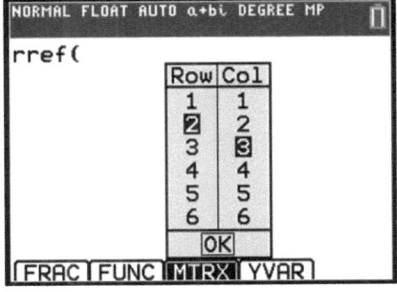
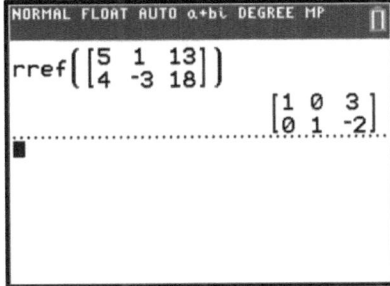

MODEL PROBLEM 1: *ADDITION METHOD*

Solve the following system of equations using the addition method:

$$4a + 2b = 22$$
$$-4a + 3b = 3$$

Solution:

$$4a + 2b = 22$$
$$\underline{-4a + 3b = 3}$$
(A) $\quad\quad 5b = 25$
(B) $\quad\quad\quad b = 5$

(C) $\quad 4a + 2(5) = 22$
(D) $\quad\quad 4a + 10 = 22$
$$\quad\quad\quad 4a = 12$$
$$\quad\quad\quad\quad a = 3$$
(E) $\quad$ Solution: *a = 3, b = 5*

Explanation of steps:

(A) If the coefficients for one of the variables are additive inverses *[4a and −4a]*, add the equations to derive a new equation without that variable.

(B) Now that we have an equation with only one variable *[b]*, solve for that one variable.

(C) Substitute this solution *[b=5]* into either one of the original equations.

(D) Solve for the other variable *[a]*.

(E) Write the solution by stating the values of both variables.

PRACTICE PROBLEMS

1. Solve the following system of equations for *x* and *y* algebraically and check your solutions on the calculator: $\quad\quad 3x - y = 8$ $\quad\quad x + y = 4$	2. Solve the following system of equations for *x* and *y* algebraically and check your solutions on the calculator: $\quad\quad 2x - 3y = 19$ $\quad\quad 3x + 3y = 21$
3. Solve the following system of equations for *x* and *y* algebraically and check your solutions on the calculator: $\quad\quad 3x + 2y = 12$ $\quad\quad 5x - 2y = 4$	4. Solve the following system of equations for *x* and *y* algebraically and check your solutions on the calculator: $\quad\quad 2x - 5y = 11$ $\quad\quad -2x + 3y = -9$

5. Solve the following system of equations for x and y algebraically and check your solutions on the calculator:
$$2x - 4y = 12$$
$$-2x + y = -9$$

6. Solve the following system of equations for x and y algebraically and check your solutions on the calculator:
$$3x + y = 0$$
$$-x - y = -4$$

MODEL PROBLEM 2: *ADDITION METHOD WITH MULTIPLIERS*

Solve the following system of equations using the addition method:
$$5x + 8y = 1$$
$$3x + 4y = -1$$

Solution:

	(A)	(B)	(C)
$5x + 8y = 1$	$\rightarrow$	$5x + 8y = 1$	$5(-3) + 8y = 1$
$3x + 4y = -1$	$\times (-2)$	$\underline{-6x - 8y = 2}$	$-15 + 8y = 1$
		$-x \quad\quad = 3$	$8y = 16$
		$x = -3$	$y = 2$

(D) Solution: $(-3, 2)$

Explanation of steps:

(A) If neither the x terms nor the y terms are additive inverses of each other, multiply one (or each) of the equations by a value to turn them into inverses *[to change the "y" terms into inverses, 8y and –8y, we can multiply the second equation by –2]*.

(B) Adding the equations eliminates the inverses and will now give us a new equation in one variable. Solve for that variable.

(C) Then, substitute the solution *[x = –3]* into either one of the original equations *[the first equation was used here]*, allowing you to solve for the other variable *[y = 2]*.

(D) The solution for variables x and y may be written as an ordered pair, (x,y).

PRACTICE PROBLEMS

7. Solve the following system of equations for x and y algebraically and check your solutions on the calculator:
$$3x + 2y = 4$$
$$-2x + 2y = 24$$

8. What is the value of y in the following system of equations?
$$2x + 3y = 6$$
$$2x + y = -2$$

9. Solve the following system of equations for x and y algebraically and check your solutions on the calculator:
$$-3x + 4y = 11$$
$$6x - 5y = -16$$

10. Solve the following system of equations for x and y algebraically and check your solutions on the calculator:
$$2x + 3y = 7$$
$$x + y = 3$$

11. Solve the following system of equations for x and y algebraically and check your solutions on the calculator:
$$2x + y = 8$$
$$x - 3y = -3$$

12. Solve the following system of equations for x and y algebraically and check your solutions on the calculator:
$$x + 2y = 9$$
$$x - y = 3$$

13. Solve the following system of equations for x and y algebraically and check your solutions on the calculator:
$$3x + 2y = 4$$
$$4x + 3y = 7$$

14. Solve the following system of equations for x and y algebraically and check your solutions on the calculator:
$$3x + 4y = 9$$
$$5x + 6y = 21$$

MODEL PROBLEM 3:   SUBSTITUTION METHOD

Solve the following system of equations using the substitution method:

$$3x - y = 16$$
$$y = x - 8$$

Solution:

(A) $3x - (x - 8) = 16$

(B) $3x - x + 8 = 16$

       $2x + 8 = 16$

         $2x = 8$

          $x = 4$

(C) $y = (4) - 8$

     $y = -4$

(D) Solution: $(4, -4)$

Explanation of steps:

(A) If one equation already has one of the variables isolated, as in *variable = expression*, substitute this expression for this variable in the other equation. *[y = x − 8 already has y expressed in terms of x, so substitute the expression x − 8 for the y in the first equation: 3x − y = 16 becomes 3x − (x − 8) = 16.]* It is always safest to use parentheses around the expression whenever you perform a substitution.

(B) Solve the equation for one variable.

(C) Substitute the solution found in step (B) into either original equation *[substitute 4 for x]*.

(D) Solve the equation for the other variable.

(E) State the solution.

PRACTICE PROBLEMS

15. Solve the following system of equations for x and y: $$y = 4x - 10$$ $$y = 5 - x$$	16. Solve the following system of equations for x and y: $$x = y - 2$$ $$y = 10 - 3x$$

17. Solve the following system of equations for x and y: $$y = 9 - 2x$$ $$3y - 2x = 11$$	18. Using the substitution method, solve the following system of equations for x and y: $$7x + 3y = 68$$ $$x - 4y = -8$$
19. Solve the following system of equations algebraically: $$2a + 3b = 12$$ $$a = \frac{1}{2}b - 6$$	20. Solve the following system of equations algebraically: $$c + 3d = 8$$ $$c = 4d - 6$$

21. To solve the following system of equations by the substitution method, which equivalent equation could be used?

$$2x - y = 5$$
$$3x + 2y = -3$$

    (1) $3x + 2(2x - 5) = -3$ (3) $3\left(y + \frac{5}{2}\right) + 2y = -3$

    (2) $3x + 2(5 - 2x) = -3$ (4) $3\left(\frac{5}{2} - y\right) + 2y = -3$

Regents Questions

MULTIPLE CHOICE

1. Which system of equations has the same solution as the system below?

$$2x + 2y = 16$$
$$3x - y = 4$$

   (1) $2x + 2y = 16$
        $6x - 2y = 4$

   (2) $2x + 2y = 16$
        $6x - 2y = 8$

   (3) $x + y = 16$
        $3x - y = 4$

   (4) $6x + 6y = 48$
        $6x + 2y = 8$

2. Which pair of equations could *not* be used to solve the following equations for x and y?

$$4x + 2y = 22$$
$$-2x + 2y = -8$$

   (1) $4x + 2y = 22$
        $2x - 2y = 8$

   (2) $4x + 2y = 22$
        $-4x + 4y = -16$

   (3) $12x + 6y = 66$
        $6x - 6y = 24$

   (4) $8x + 4y = 44$
        $-8x + 8y = -8$

3. A system of equations is given below.

$$x + 2y = 5$$
$$2x + y = 4$$

   Which system of equations does *not* have the same solution?

   (1) $3x + 6y = 15$
        $2x + y = 4$

   (2) $4x + 8y = 20$
        $2x + y = 4$

   (3) $x + 2y = 5$
        $6x + 3y = 12$

   (4) $x + 2y = 5$
        $4x + 2y = 12$

4. Which system of equations does *not* have the same solution as the system below?

$$4x + 3y = 10$$
$$-6x - 5y = -16$$

   (1) $-12x - 9y = -30$
        $12x + 10y = 32$

   (2) $20x + 15y = 50$
        $-18x - 15y = -48$

   (3) $24x + 18y = 60$
        $-24x - 20y = -64$

   (4) $40x + 30y = 100$
        $36x + 30y = -96$

5. A system of equations is shown below.

   Equation A: $5x + 9y = 12$
   Equation B: $4x - 3y = 8$

   Which method eliminates one of the variables?

   (1) Multiply equation A by $-\frac{1}{3}$ and add the result to equation B.

   (2) Multiply equation B by 3 and add the result to equation A.

   (3) Multiply equation A by 2 and equation B by -6 and add the results together.

   (4) Multiply equation B by 5 and equation A by 4 and add the results together.

6.   Which system of equations will yield the same solution as the system below?

$$x - y = 3$$
$$2x - 3y = -1$$

(1) $-2x - 2y = -6$
      $2x - 3y = -1$

(3) $2x - 2y = 6$
      $2x - 3y = -1$

(2) $-2x + 2y = 3$
      $2x - 3y = -1$

(4) $3x + 3y = 9$
      $2x - 3y = -1$

7.   Using the substitution method, Vito is solving the following system of equations algebraically:

$$y + 3x = -4$$
$$2x - 3y = -21$$

Which equivalent equation could Vito use?

(1) $2(-3x - 4) + 3x = -21$

(3) $2x - 3(-3x - 4) = -21$

(2) $2(3x - 4) + 3x = -21$

(4) $2x - 3(3x - 4) = -21$

8.   Which system of linear equations has the same solution as the one shown below?

$$x - 4y = -10$$
$$x + y = 5$$

(1) $5x = 10$
      $x + y = 5$

(3) $-3x = -30$
      $x + y = 5$

(2) $-5y = -5$
      $x + y = 5$

(4) $-5y = -5$
      $x - 4y = -10$

9.   Which system of equations has the same solutions as the one shown below?

$$3x - y = 7$$
$$2x + 3y = 12$$

(1) $6x - 2y = 14$
      $-6x + 9y = 36$

(3) $-9x - 3y = -21$
      $2x + 3y = 12$

(2) $18x - 6y = 42$
      $4x + 6y = 24$

(4) $3x - y = 7$
      $x + y = 2$

10.   Which system has the same solution as the system below?

$$x + 3y = 10$$
$$-2x - 2y = 4$$

(1) $-x + y = 6$
      $2x + 6y = 20$

(3) $x + y = 6$
      $2x + 6y = 20$

(2) $-x + y = 14$
      $2x + 6y = 20$

(4) $x + y = 14$
      $2x + 6y = 20$

CONSTRUCTED RESPONSE

11. Albert says that the two systems of equations shown below have the same solutions.

First System	Second System
$8x + 9y = 48$ $12x + 5y = 21$	$8x + 9y = 48$ $-8.5y = -51$

Determine and state whether you agree with Albert. Justify your answer.

4.2   Solve Linear Systems Graphically

To solve a system of linear equations **graphically**, simply graph the two equations as lines in the same coordinate plane. The point (if any) where the two lines **intersect** is the solution. This is because the point of intersection is the only point that satisfies *both* equations.

Example: The lines $y = -x + 5$ and $y = 2x - 4$ intersect at (3,2).
Therefore, the solution for the system of equations is $x = 3$ and $y = 2$.

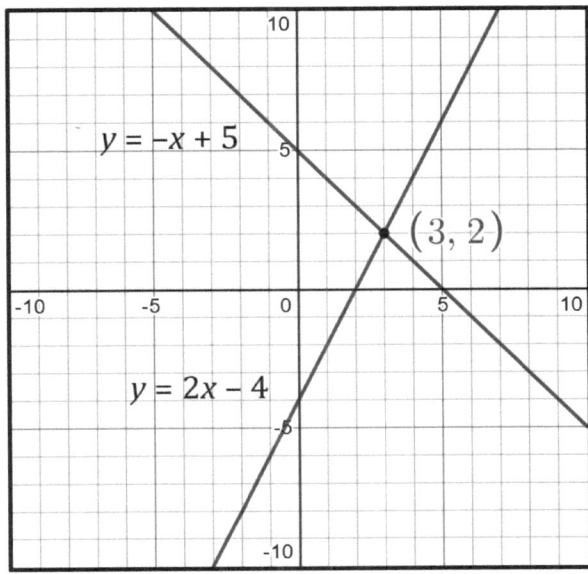

If the two lines are parallel, they never intersect, and so there is no solution. If the two lines are identical (coincide), then there are infinitely many solutions.

Whenever you graph two lines on the same set of axes, you should label both of them with their equations, as shown above.

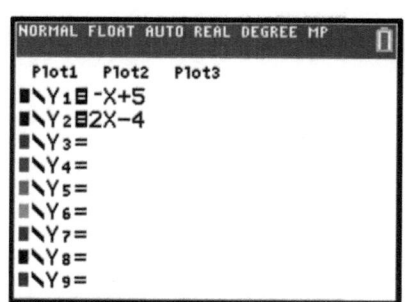 CALCULATOR TIP

Using a calculator, you can solve a system of equations by graphing both equations and then using the intersect feature.

Example: To solve the system, $y = -x + 5$ and $y = 2x - 4$, graphically,

Press [Y=] [(-)][X,T,Θ,n][+][5][ENTER]

[2][X,T,Θ,n][-][4][ENTER]

[2nd][CALC][5] to select intersect

First curve? [ENTER] Second curve? [ENTER] Guess? [ENTER]

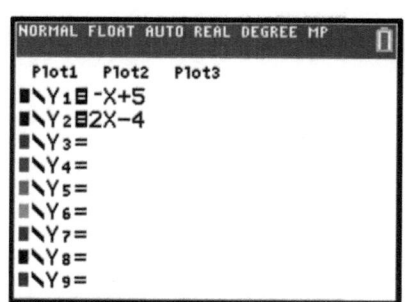 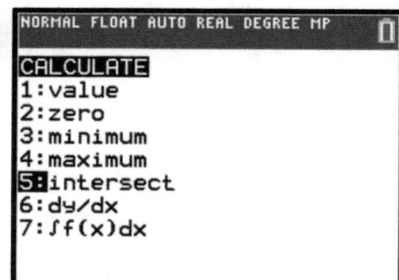 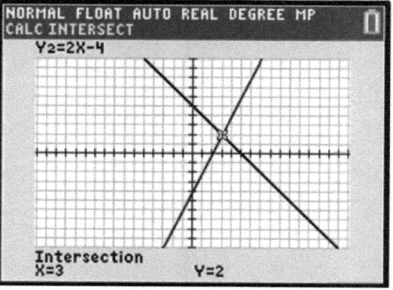

MODEL PROBLEM

Solve the following system of equations graphically:

$$x - y = -1$$
$$y = -3x + 9$$

Solution:

(A) $x - y = -1$
$$\underline{-x \qquad - x}$$
$$-y = -x - 1$$
$$\underline{-1 \qquad - 1}$$
$$y = x + 1$$ (B)

(C)   Solution: $(2,3)$

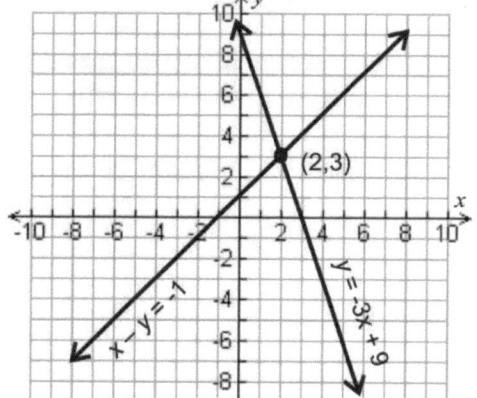

Explanation of steps:

(A) If either equation is not already in slope-intercept form, transform the equation by solving for y in terms of x.

(B) Graph both equations on the same coordinate plane, labeling each line.

(C) The point of intersection is the solution to the system of equations.
    [A solution of $(2,3)$ means $x = 2$, $y = 3$ solves both equations simultaneously.]

PRACTICE PROBLEMS

1. A system of equations is graphed below.

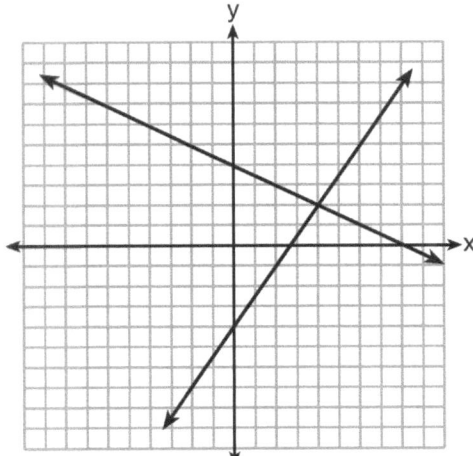

The solution of this system is

(1) (0,4)   (3) (4,2)
(2) (2,4)   (4) (8,0)

2. A system of equations is graphed below.

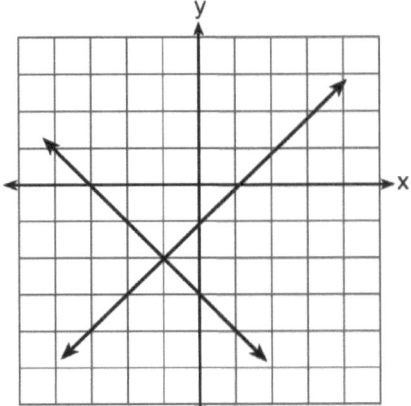

The solution of this system is

(1) (1,0) and (−3,0)
(2) (0,−3) and (0,−1)
(3) (−1,−2)
(4) (−2,−1)

3. If the lines whose equations are $x = -2$ and $y = 3$ were graphed on the same set of coordinate axes, what would be their point of intersection?

4. Solve the system of equations graphically:
$$y = 3x - 2$$
$$y = -x - 6$$

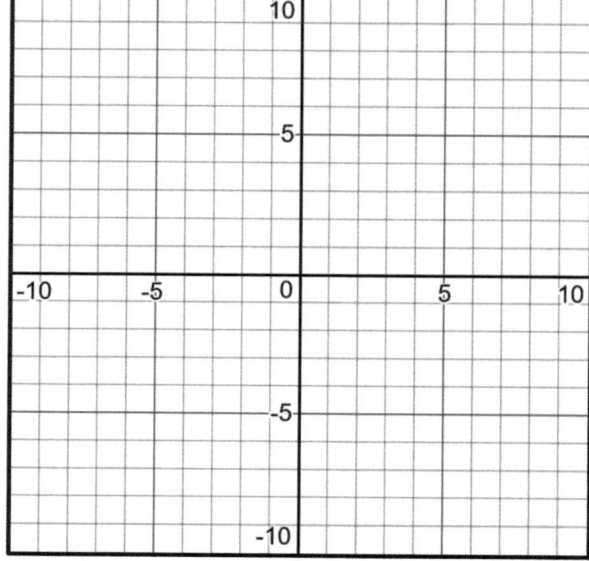

5. Solve the system of equations graphically:
$$x + y = 2$$
$$x - y = 4$$

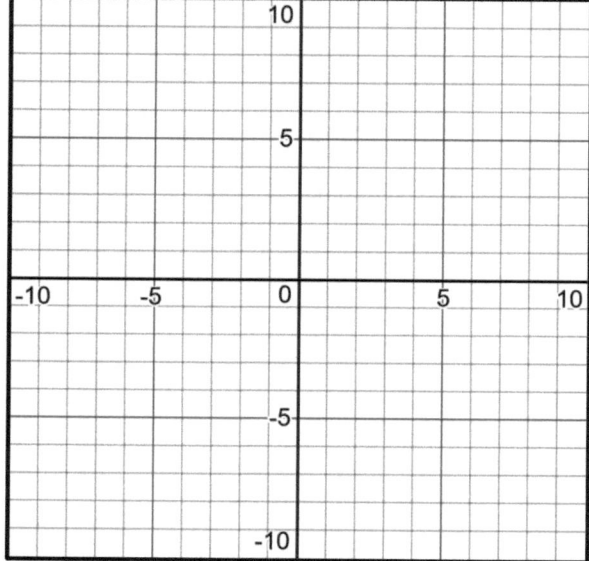

6. Solve the system of equations graphically:
$$3x - 5y = 15$$
$$y = 2x + 4$$

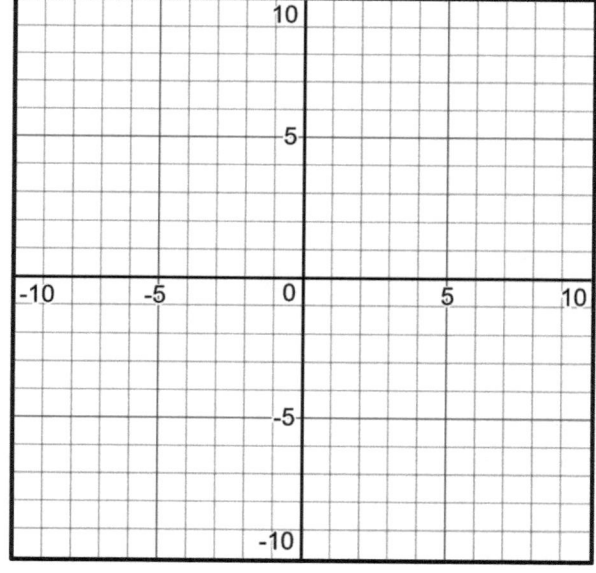

7. Solve the system of equations graphically:
$$y = \frac{2}{3}x + 5$$
$$x + 3y = -3$$

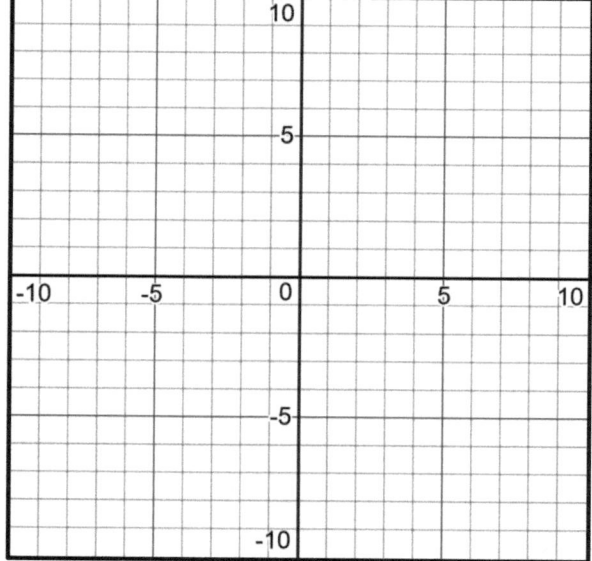

8. Solve the system of equations graphically:

$$4x - 2y = 10$$
$$y = -2x - 1$$

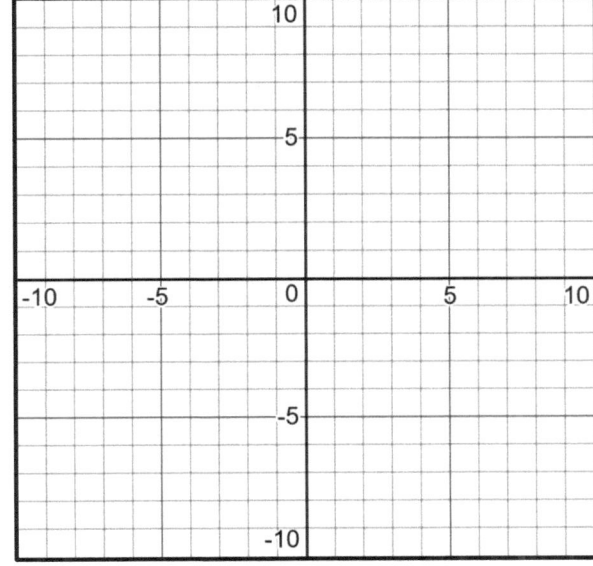

9. Solve the system of equations graphically:

$$y = 4x - 1$$
$$2x + y = 5$$

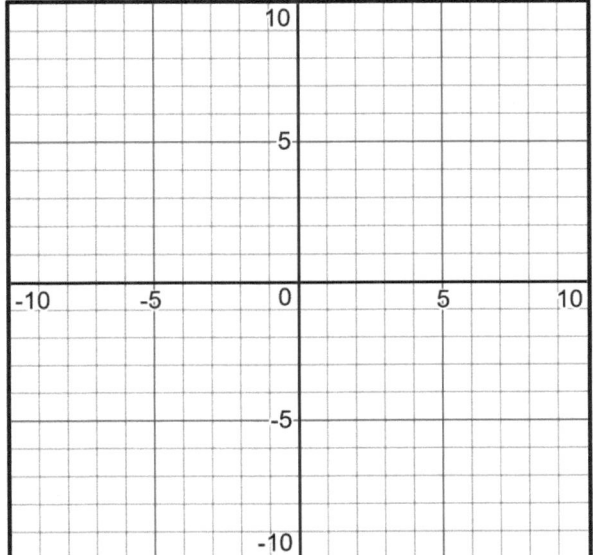

Regents Questions

MULTIPLE CHOICE

1. Two functions, $y = |x - 3|$ and $3x + 3y = 27$, are graphed on the same set of axes. Which statement is true about the solution to the system of equations?

    (1) (3,0) is the solution to the system because it satisfies the equation $y = |x - 3|$.

    (2) (9,0) is the solution to the system because it satisfies the equation $3x + 3y = 27$.

    (3) (6,3) is the solution to the system because it satisfies both equations.

    (4) (3,0), (9,0), and (6,3) are the solutions to the system of equations because they all satisfy at least one of the equations.

2. The line represented by the equation $4y + 2x = 33.6$ shares a solution point with the line represented by the table below.

x	y
−5	3.2
−2	3.8
2	4.6
4	5
11	6.4

    The solution for this system is

    (1) $(-14.0, -1.4)$ (3) $(1.9, 4.6)$

    (2) $(-6.8, 5.0)$ (4) $(6.0, 5.4)$

3. What is the solution to the system of equations below?
    $$y = 2x + 8$$
    $$3(-2x + y) = 12$$

    (1) no solution   (3) $(-1,6)$

    (2) infinite solutions   (4) $\left(\frac{1}{2}, 9\right)$

CONSTRUCTED RESPONSE

4. In attempting to solve the system of equations $y = 3x - 2$ and $6x - 2y = 4$, John graphed the two equations on his graphing calculator. Because he saw only one line, John wrote that the answer to the system is the empty set. Is he correct? Explain your answer.

5. Lydia wants to take art classes. She compares the cost at two art centers. Center *A* charges $25 per hour and a registration fee of $25. Center *B* charges $15 per hour and a registration fee of $75. Lydia plans to take *x* hours of classes.

Write an equation that models this situation, where *A* represents the total cost of Center *A*. Write an equation that models this situation, where *B* represents the total cost of Center *B*.

If Lydia wants to take 10 hours of classes, use your equations to determine which center will cost *less*.

Graph your equations for Center *A* and Center *B* on the set of axes below.

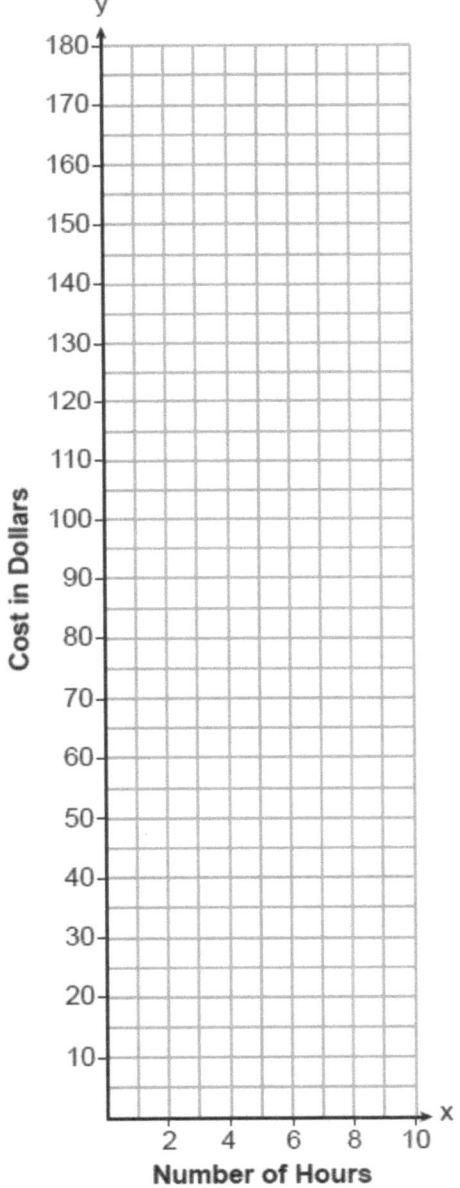

State the number of hours of classes when the centers will cost the same.

4.3   Solutions to Systems of Inequalities

If given a system of two inequalities, you may determine whether an ordered pair is in the solution set by substitution. Simply substitute the x- and y-coordinates of the ordered pair for the variables, x and y, in both inequalities. Then check if both inequalities are true.

MODEL PROBLEM

Which ordered pair is in the solution set of the following system of linear inequalities?

$$y + 2 > 3x$$
$$-2y \geq x - 2$$

(1) $(5,2)$ (2) $(-5,-2)$ (3) $(-2,5)$ (4) $(2,-5)$

Solution:

(A)

$(5,2)$	$(-5,-2)$	$(-2,5)$	$(2,-5)$
$y + 2 > 3x$ $(2) + 2 > 3(5)$? $4 > 15$? *false*	$y + 2 > 3x$ $(-2) + 2 > 3(-5)$? $0 > -15$? *true*	$y + 2 > 3x$ $(5) + 2 > 3(-2)$? $7 > -6$? *true*	$y + 2 > 3x$ $(-5) + 2 > 3(2)$? $-3 > 6$? *false*
	$-2y \geq x - 2$ $-2(-2) \geq (-5) - 2$? $4 \geq -7$? *true*	$-2y \geq x - 2$ $-2(5) \geq (-2) - 2$? $-10 \geq -4$? *false*	
NO	YES	NO	NO

(B)   The ordered pair $(-5, -2)$ is a solution.

Explanation of steps:

(A)   For each ordered pair, substitute the coordinates for the variables x and y in the inequalities. Check whether each inequality is true for these values. If either is *false*, the ordered pair is not a solution and you may go on to check the next ordered pair.

(B)   The ordered pair is in the solution set if it makes both inequalities *true*.

▦▢ CALCULATOR TIP

Checking can be done using a calculator instead. To check if $(5,2)$ is a solution:

5 STO▸ X,T,Θ,*n* ENTER   2 STO▸ ALPHA [Y] ENTER

ALPHA [Y] + 2 2nd [TEST] 3 3 X,T,Θ,*n* ENTER

Result is 0 for false, so $(5,2)$ is *not* a solution.

PRACTICE PROBLEMS

1. Which ordered pair is in the solution set of the following system of inequalities?

$$y < 2x + 1$$
$$y \leq -3x + 4$$

   (1) (1,1)   (3) (1,3)

   (2) (1,2)   (4) none of these

2. Which ordered pair is in the solution set of the following system of inequalities?

$$y \geq x + 7$$
$$2x + y \leq -5$$

   (1) (0,9)   (3) (9,0)

   (2) (0,−9)   (4) (−9,0)

3. Which ordered pair is in the solution set of the following system of inequalities?

$$y < \frac{1}{2}x + 4$$
$$y \geq -x + 1$$

(1) $(-5, 3)$ (3) $(3, -5)$

(2) $(0, 4)$ (4) $(4, 0)$

4. Which ordered pair is in the solution set of the following system of inequalities?

$$y < 2x + 2$$
$$y \geq -x - 1$$

(1) $(0, 3)$ (3) $(-1, 0)$

(2) $(2, 0)$ (4) $(-1, -4)$

5. Which ordered pair is in the solution set of the following system of inequalities?

$$y \leq \frac{1}{2}x + 13$$
$$4x + 2y > 3$$

(1) $(-4,1)$ (3) $(1,-4)$

(2) $(-2,2)$ (4) $(2,-2)$

6. Which ordered pair is in the solution set of the following system of inequalities?

$$y \leq 3x + 1$$
$$x - y > 1?$$

(1) $(-1,-2)$ (3) $(1,2)$

(2) $(2,-1)$ (4) $(-1,2)$

Regents Questions

MULTIPLE CHOICE

1. Which ordered pair is *not* in the solution set of $y > -\frac{1}{2}x + 5$ and $y \leq 3x - 2$?

   (1) (5,3)   (3) (3,4)

   (2) (4,3)   (4) (4,4)

2. Which point is a solution to the system below?
$$2y < -12x + 4$$
$$y < -6x + 4$$

   (1) $\left(1, \frac{1}{2}\right)$ (3) $\left(-\frac{1}{2}, 5\right)$

   (2) (0,6)   (4) (−3,2)

4.4   Solve Systems of Inequalities Graphically

To solve a system of inequalities, graph both inequalities on the same coordinate plane. Be sure to label both inequalities. The graph of each inequality will have a shaded region representing the solution set for that inequality. The region where the two shaded regions **overlap** (the region that is shaded twice or darker), including any points on a **solid line** (but *not* a dashed line) bordering the region, represents the solution set of the system. Any point in the solution set would solve *both* inequalities. The point of intersection of the two lines is a solution only if *both* lines are solid.

MODEL PROBLEM 1:   *GRAPHING*

Graph the following system of inequalities and label the solution set S:

$$y + 2 > 3x$$
$$-2y \geq x - 2$$

Solution:

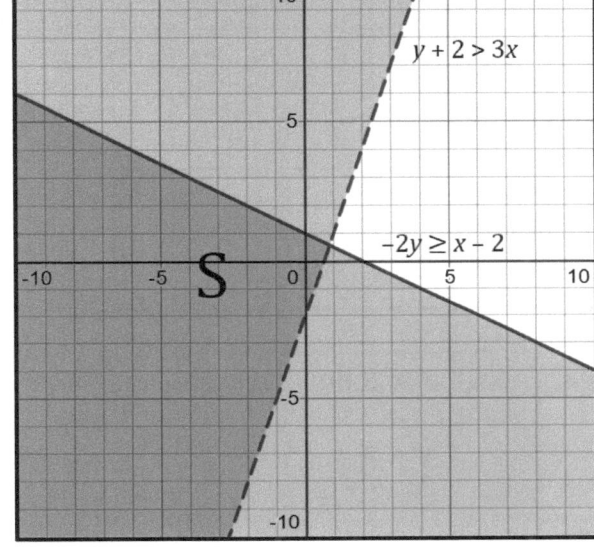

(A)
$$\begin{array}{ll} y + 2 > 3x \\ \underline{-2 \quad -2} \\ y > 3x - 2 \end{array}$$

$$\begin{array}{ll} \dfrac{-2y}{-2} \geq \dfrac{x - 2}{-2} \\ y \leq -\dfrac{1}{2}x + 1 \end{array}$$

(B)

Explanation of steps:

(A) If either inequality is not already in slope-intercept form, solve the inequality for y. Remember to reverse the inequality symbol if both sides of an inequality are multiplied or divided by a negative value. *[In the second inequality, $\geq$ becomes $\leq$]*

(B) Graph both inequalities on the same set of axes. Remember to use a solid line for inequalities with $\leq$ or $\geq$ symbols but a dashed line for those with $<$ or $>$ symbols. Label the double-shaded region "S" to represent the solution set.

PRACTICE PROBLEMS

1. Graph the following system and label the solution set S:
$$y \leq -x + 2$$
$$y < -1$$

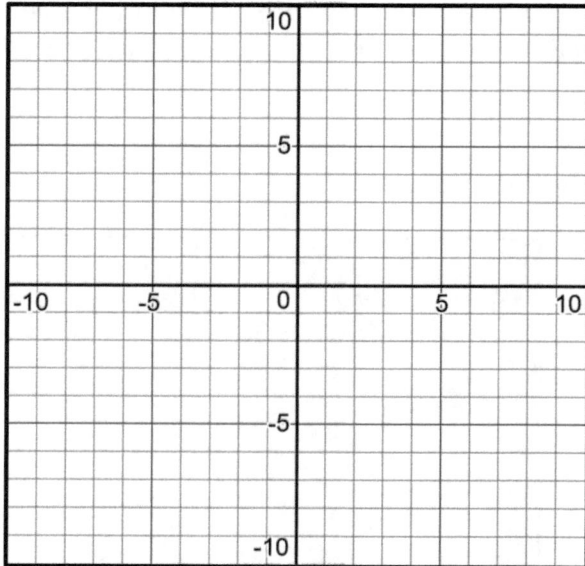

2. Graph the following system and label the solution set S:
$$y \geq 2x + 1$$
$$y \leq -x + 4$$

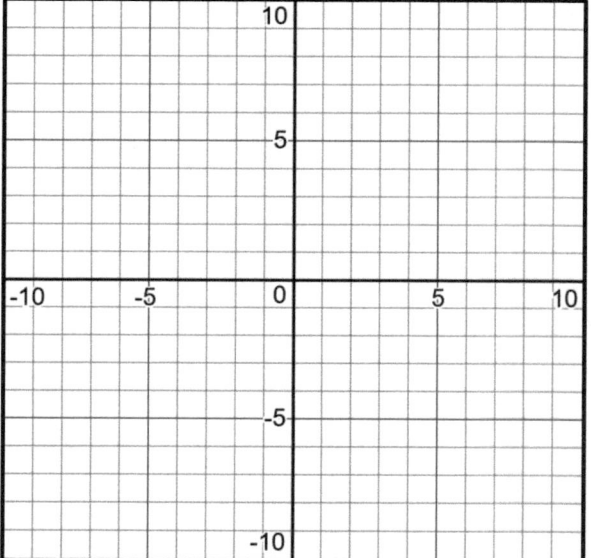

3. Graph the following system and label the solution set S:
$$y < x - 2$$
$$2x + y \geq 1$$

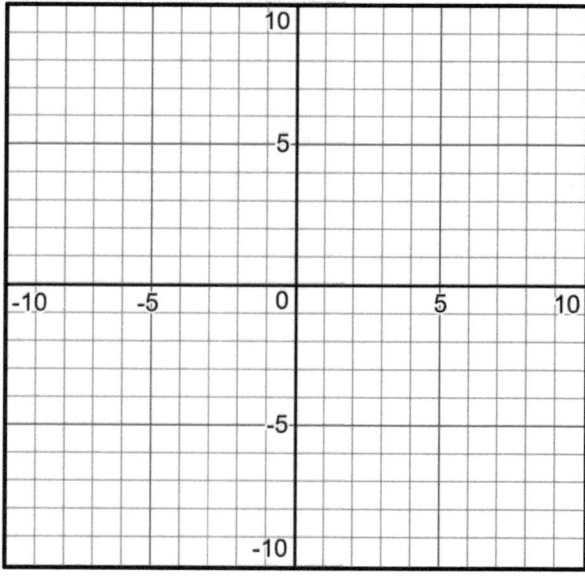

4. Graph the following system and label the solution set S:
$$2x + y \geq 3$$
$$x - 3y < -6$$

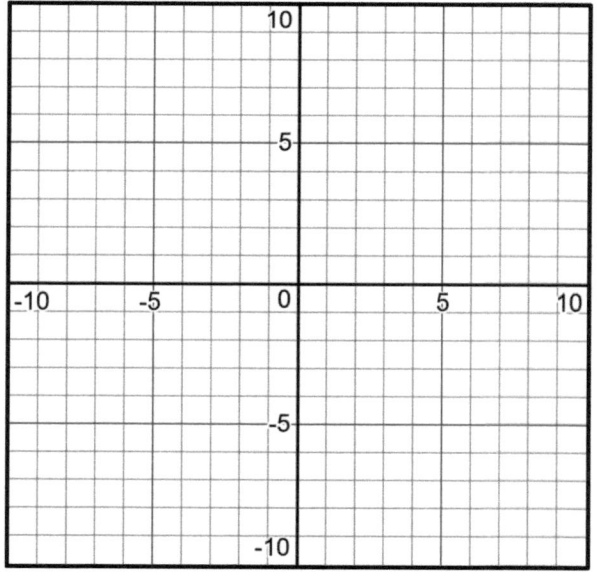

5. Graph the following systems of inequalities on the set of axes shown below and label the solution set S:

$$y > -x + 2$$
$$y \leq \frac{2}{3}x + 5$$

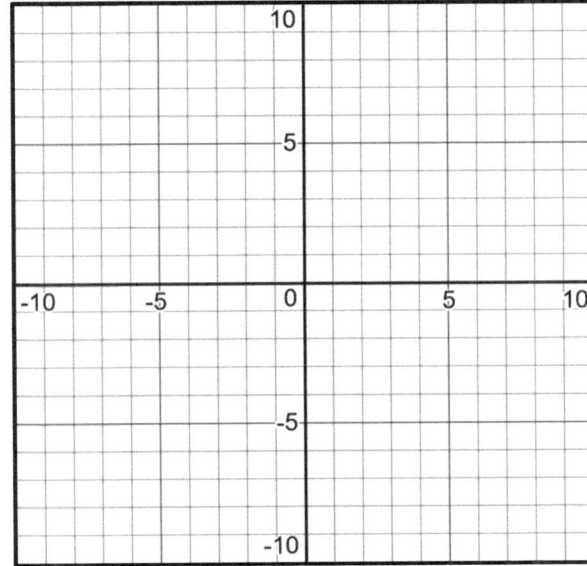

6. Graph the following systems of inequalities on the set of axes shown below and label the solution set S:

$$2x + 3y < -3$$
$$y - 4x \geq 2$$

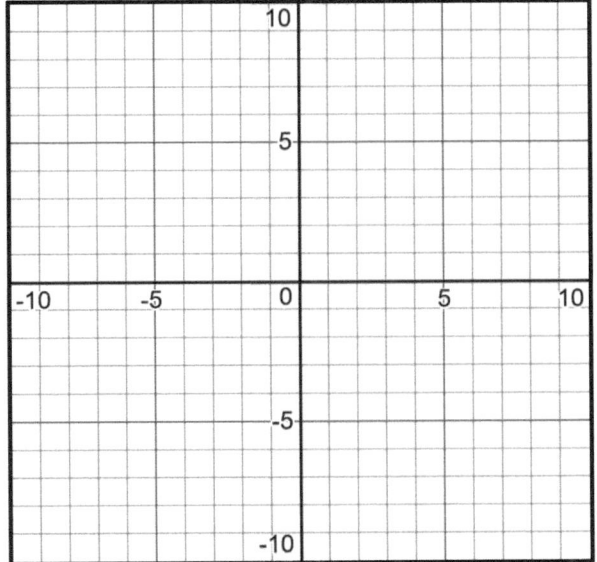

7. Graph the system of inequalities and state the coordinates of a point in the solution set.

$$2x - y \geq 6$$
$$x > 2$$

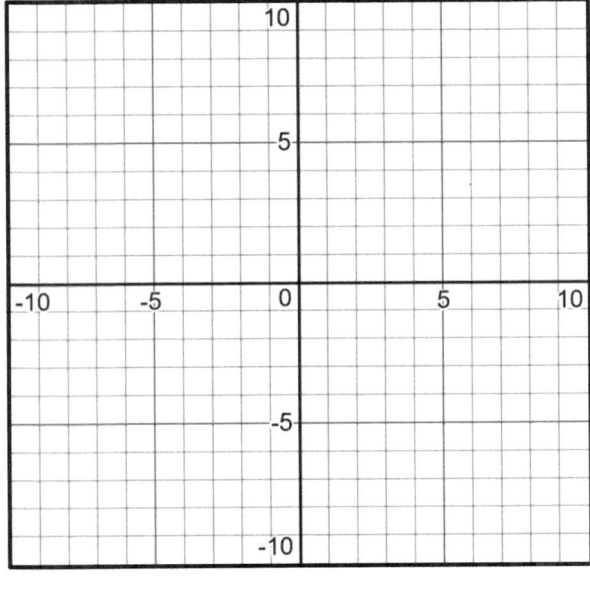

8. Graph the system of inequalities and state the coordinates of a point in the solution set.

$$y < 2x + 1$$
$$y \geq -\frac{1}{3}x + 4$$

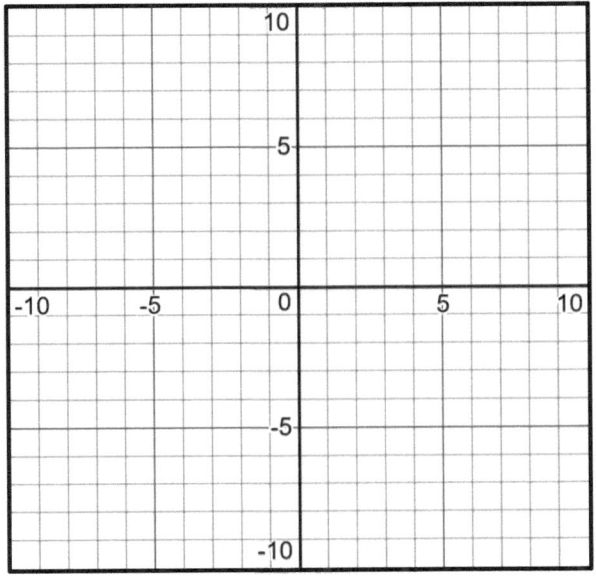

9. Graph $y < x$ and $x > 5$ on the axes below and state the coordinates of a point in the solution set.

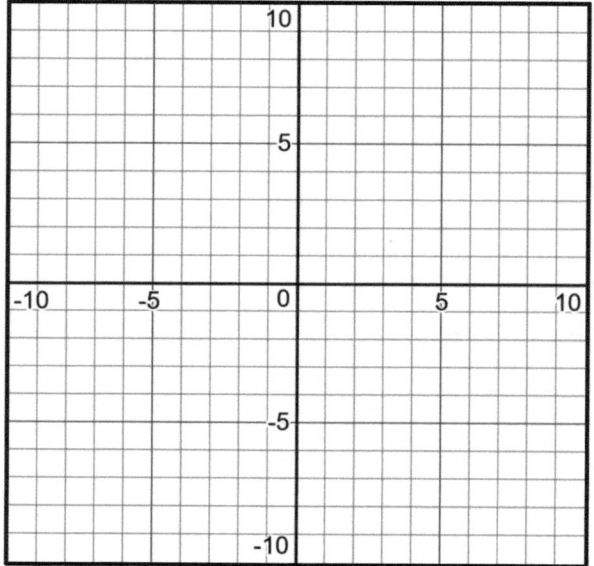

10. Graph the system of inequalities and state the coordinates of a point in the solution set.
$$y + 3 < 2x$$
$$-2y \leq 6x - 10$$

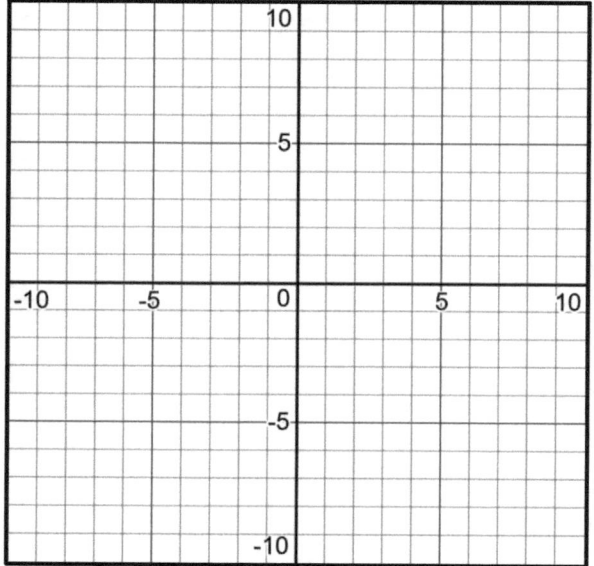

11. Graph the system of inequalities and state the coordinates of a point in the solution set.
$$3x + y < 7$$
$$y \geq \frac{2}{3}x - 4$$

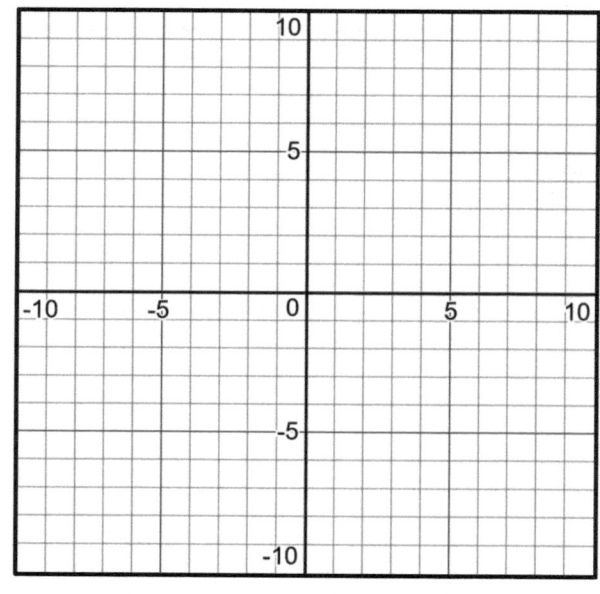

12. Graph the system of inequalities.
$$y + x \geq 3$$
$$5x - 2y > 10$$
State a point that satisfies $y + x \geq 3$, but does *not* satisfy $5x - 2y > 10$.

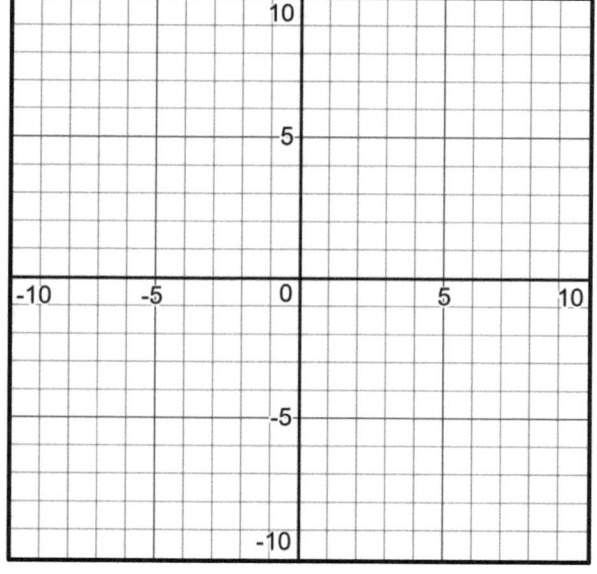

MODEL PROBLEM 2:   *SOLUTIONS*

Which ordered pair is in the solution set of the system of linear inequalities graphed at right?

     (1) $(5,2)$ (3) $(-2,5)$
     (2) $(-5,-2)$ (4) $(2,-5)$

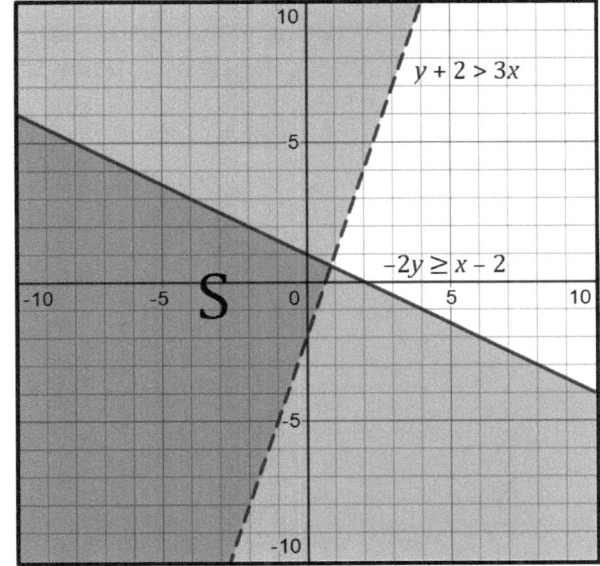

Solution:
The answer is (2). The ordered pair $(-5,-2)$ is in the solution set.

Explanation of steps:
Plot each point to determine whether it lies within the double-shaded or darker region. *[(5,2) is in the unshaded region. Both $(-2,5)$ and $(2,-5)$ are in single-shaded regions.]*

PRACTICE PROBLEMS

13. Is $(5,1)$ in the solution set of the system of inequalities graphed below? 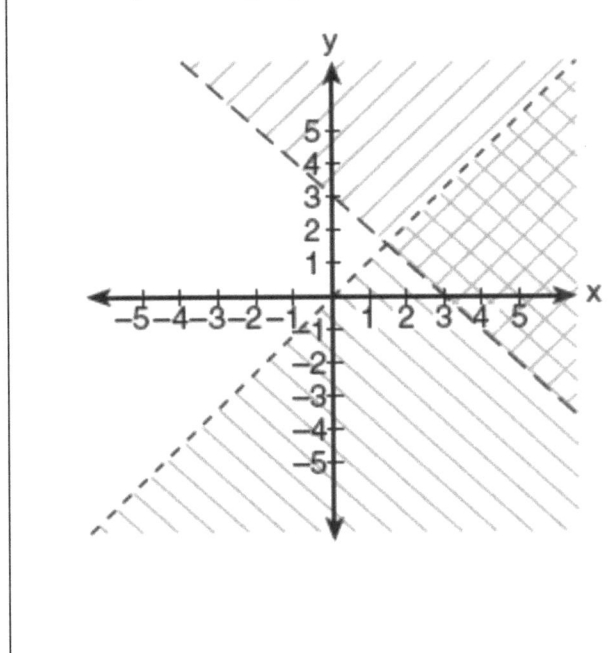

14. Is $(-1,-8)$ in the solution set of the system of inequalities graphed below?

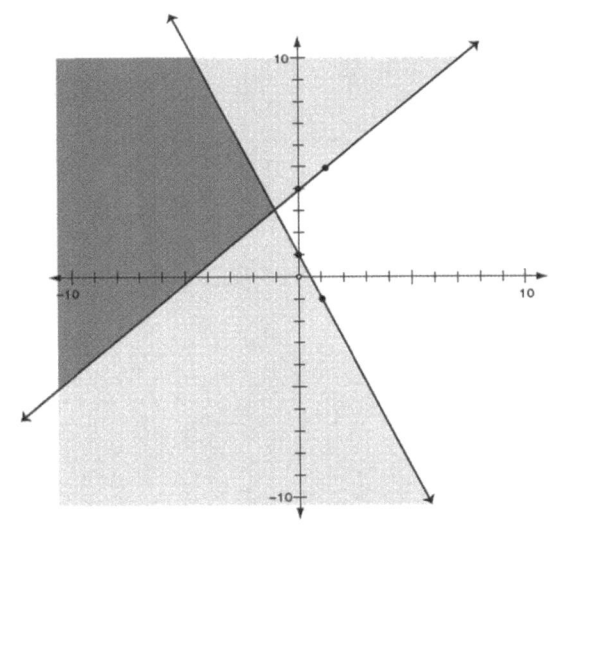

15. Which ordered pair is in the solution set of the system of inequalities shown in the graph below?

    (1) $(1, -4)$ (3) $(5, 3)$

    (2) $(-5, 7)$ (4) $(-7, -2)$

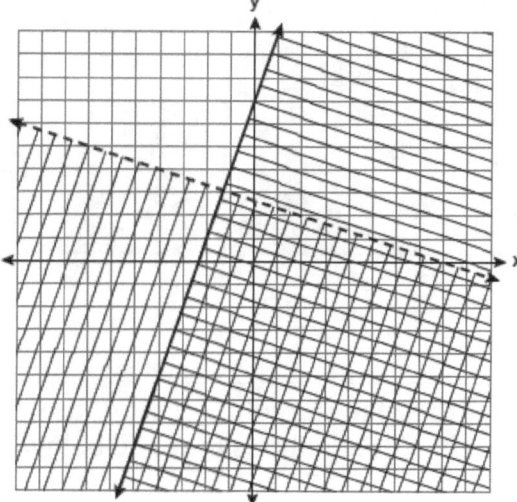

16. Which ordered pair is in the solution set of the system of inequalities shown in the graph below?

    (1) $(-2, -1)$ (3) $(-2, -4)$

    (2) $(-2, 2)$ (4) $(2, -2)$

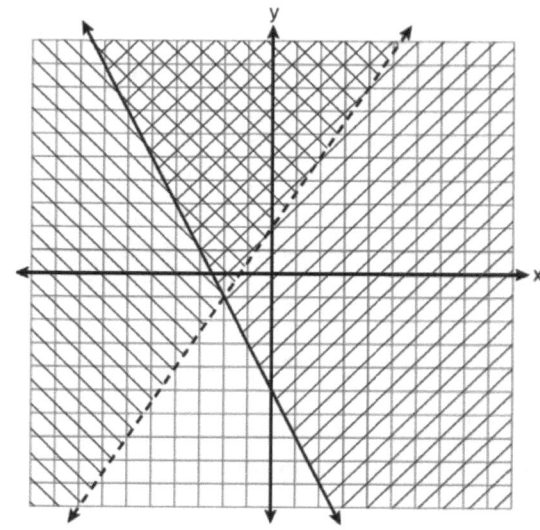

Regents Questions

MULTIPLE CHOICE

1. Given: $y + x > 2$
              $y \leq 3x - 2$
    Which graph shows the solution of the given set of inequalities?

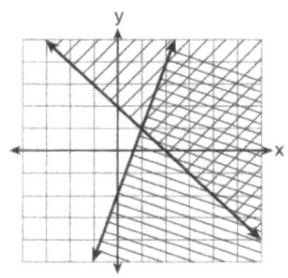

(1)

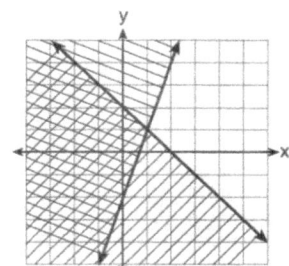

(3)

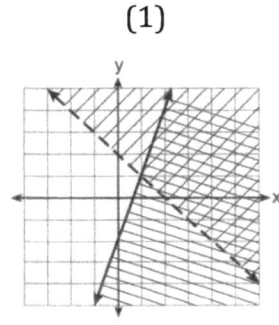

(2)

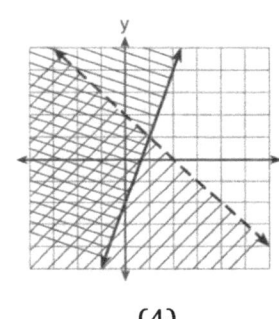

(4)

2. What is one point that lies in the solution set of the system of inequalities graphed below?

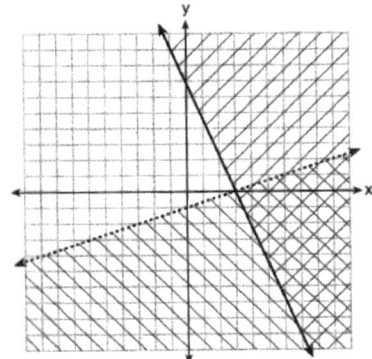

(1) (7,0)   (3) (0,7)
(2) (3,0)   (4) (−3,5)

3.   Which graph represents the solution of $y \leq x + 3$ and $y \geq -2x - 2$?

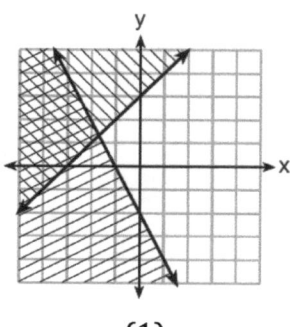

(1)

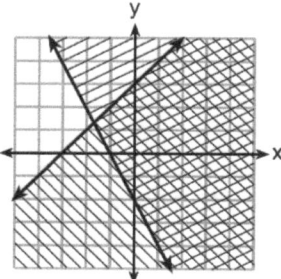

(3)

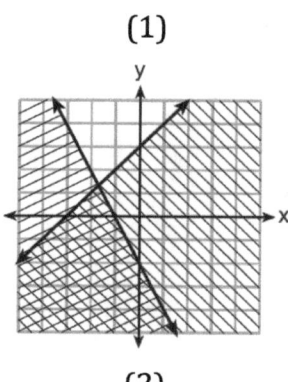

(2)

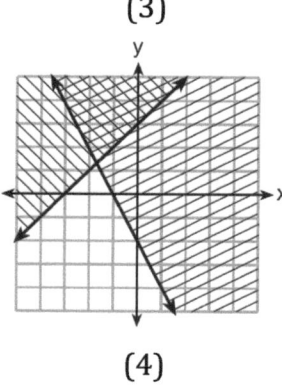

(4)

4.   First consider the system of equations $y = -\frac{1}{2}x + 1$ and $y = x - 5$. Then consider the system of inequalities $y > -\frac{1}{2}x + 1$ and $y < x - 5$. When comparing the number of solutions in each of these systems, which statement is true?

(1)   Both systems have an infinite number of solutions.

(2)   The system of equations has more solutions.

(3)   The system of inequalities has more solutions.

(4)   Both systems have only one solution.

CONSTRUCTED RESPONSE

5.   The graph of an inequality is shown below.

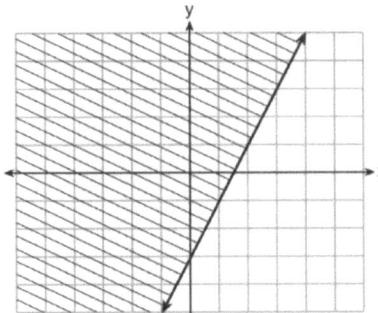

Write the inequality represented by the graph.

On the same set of axes, graph the inequality $x + 2y < 4$.

The two inequalities graphed on the set of axes form a system. Oscar thinks that the point (2,1) is in the solution set for this system of inequalities. Determine and state whether you agree with Oscar. Explain your reasoning.

6.   Solve the following system of inequalities graphically on the grid below and label the solution S.

$$3x + 4y > 20$$
$$x < 3y - 18$$

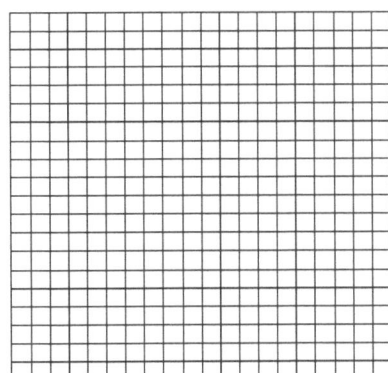

Is the point (3,7) in the solution set? Explain your answer.

7.   Determine if the point (0,4) is a solution to the system of inequalities graphed below. Justify your answer.

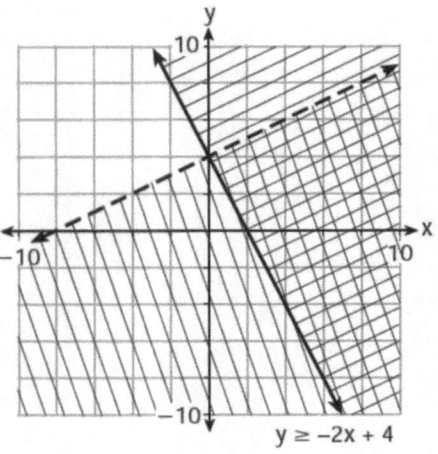

$y \geq -2x + 4$

8.   On the set of axes below, graph the following system of inequalities:

$$2y + 3x \leq 14$$
$$4x - y < 2$$

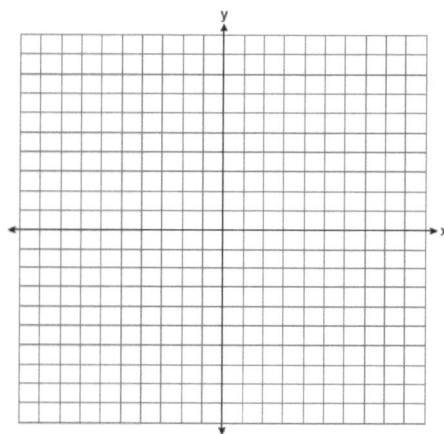

Determine if the point (1,2) is in the solution set. Explain your answer.

9.   Graph the following systems of inequalities on the set of axes below:

$$2y \geq 3x - 16$$
$$y + 2x > -5$$

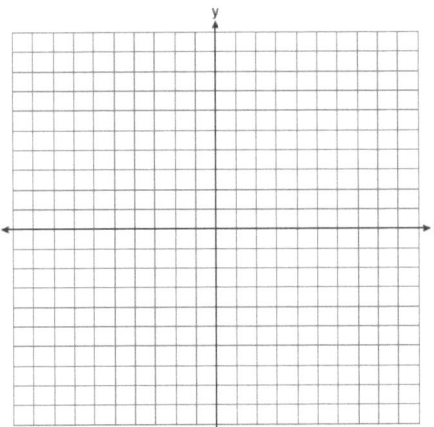

Based upon your graph, explain why $(6,1)$ is a solution to this system and why $(-6,7)$ is *not* a solution to this system.

10.   A system of inequalities is graphed on the set of axes below.

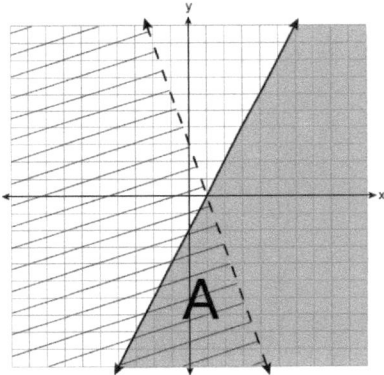

State the system of inequalities represented by the graph. State what region A represents. State what the entire gray region represents.

11.   On the set of axes below, graph the following system of inequalities:

$$2x + y \geq 8$$
$$y - 5 < 3x$$

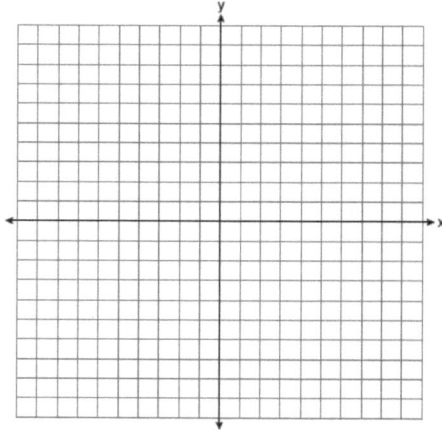

Determine if the point (1,8) is in the solution set. Explain your answer.

12.   Graph the system of inequalities:

$$-x + 2y - 4 < 0$$
$$3x + 4y + 4 \geq 0$$

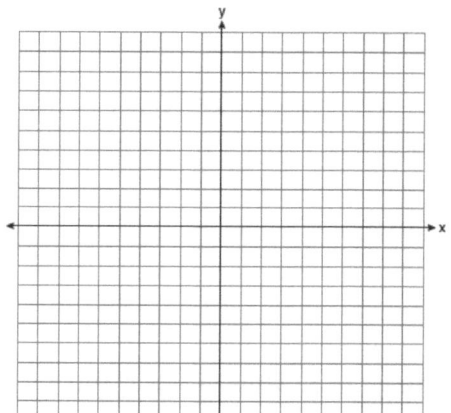

Stephen says the point (0,0) is a solution to this system. Determine if he is correct, and explain your reasoning.

13.   Graph the system of inequalities on the set of axes below:

$$y \le -\frac{3}{4}x + 5$$
$$3x - 2y > 4$$

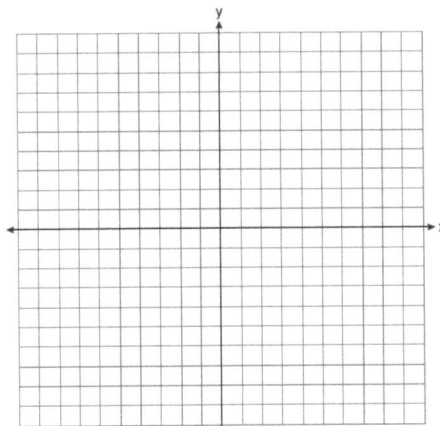

Is (6,3) a solution to the system of inequalities? Explain your answer.

14.   Solve the system of inequalities graphically on the set of axes below. Label the solution set S.

$$2x + 3y < 9$$
$$2y \ge 4x + 6$$

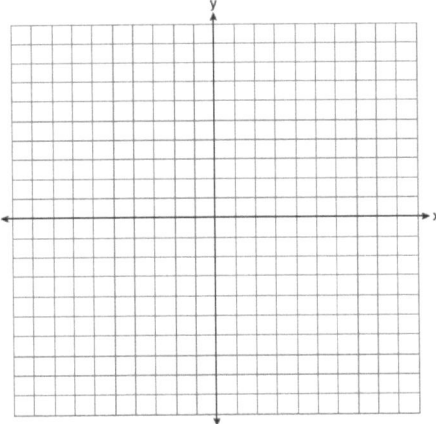

Determine if the point (0,3) is a solution to this system of inequalities. Justify your answer.

15.   Solve the system of inequalities graphically on the set of axes below. Label the solution set S.

$$y + 3x < 5$$
$$1 \geq 2x - y$$

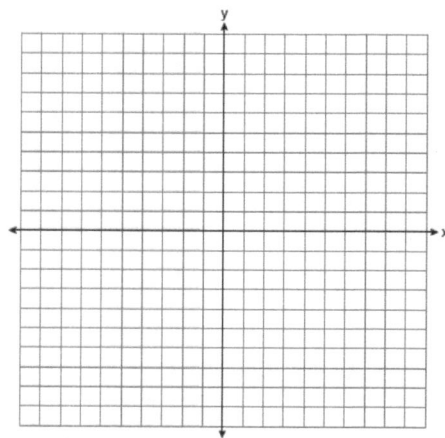

   Is the point $(-5,0)$ in the solution set? Explain your answer.

16.   Given: $3y - 9 \leq 12$
               $y < -2x - 4$

   Graph the system of inequalities on the set of axes below.

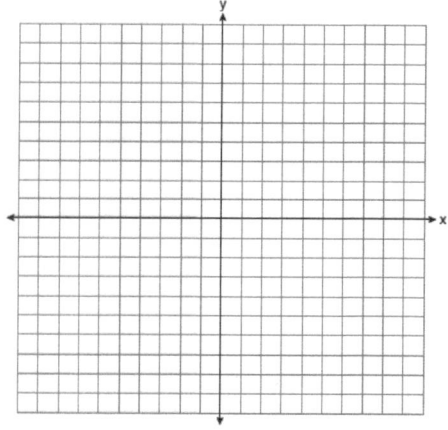

   State the coordinates of a point that satisfies both inequalities. Justify your answer.

17. Solve the following system of inequalities graphically on the set of axes below.

$$2x + 3y \geq -6$$
$$x < 3y + 6$$

Label the solution set S.

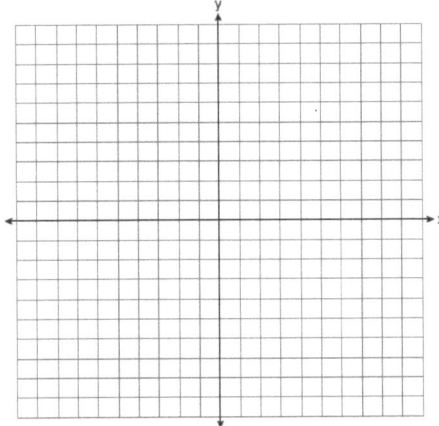

Is the point $(4, -2)$ in the solution set? Explain your answer.

18. Graph the following system of inequalities on the set of axes below:

$$-2y < 3x + 12$$
$$x \geq -3$$

Label the solution set S.

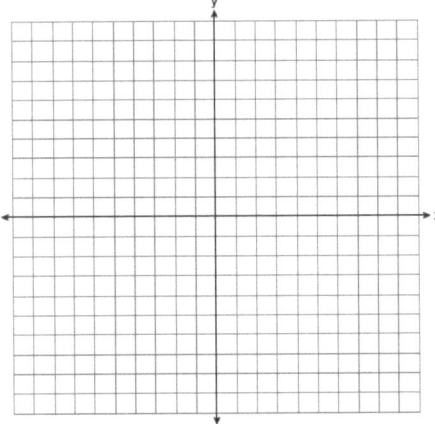

Allison thinks that $(2, -9)$ is a solution to this system. Determine if Allison is correct. Justify your answer.

19. On the set of axes below, graph the following system of inequalities:

$$2x - y > 4$$
$$x + 3y > 6$$

Label the solution set S.

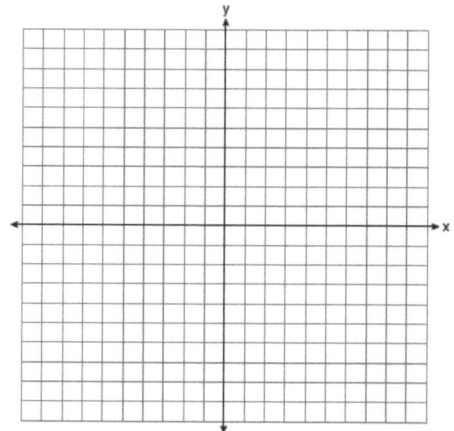

Is (4,2) a solution to this system? Justify your answer.

4.5   Word Problems – Linear Systems

Sometimes, the circumstances described by a word problem may lead more naturally to the use of **two variables** instead of one. If this is the case, we can find a solution by writing **two equations** in terms of the variables. We can solve for the two variables by using one of the methods learned – addition or substitution – for solving a system of equations.

MODEL PROBLEM

Brenda's school is selling tickets to a spring musical. On the first day of ticket sales the school sold 3 senior citizen tickets and 9 child tickets for a total of $75. On the second day the school sold 8 senior citizen tickets and 5 child tickets for a total of $67. What is the price of each senior citizen ticket and each child ticket?

Solution:

(A) Let s represent the cost of one *senior citizen* ticket.
      Let c represent the cost of one *child* ticket.

(B)
$$3s + 9c = 75$$
$$8s + 5c = 67$$

(C)
$$\times\, 8$$
$$\times\, (-3)$$

$$24s + 72c = 600$$
$$\underline{-24s - 15c = -201}$$
$$\frac{57c}{57} = \frac{399}{57}$$
$$c = 7$$

(D)
$$3s + 9(7) = 75$$
$$3s + 63 = 75$$
$$3s = 12$$
$$s = 4$$

(E) Senior citizen tickets cost $4 each and child tickets cost $7 each.

Explanation of steps:

(A) Write what each variable represents.

(B) Write equations using the given information.

(C) Solve the system of equations for one of the variables using either the addition method or the substitution method. *[Using the addition method, we need to create additive inverses. To make additive inverses of the s terms, we can multiply the first equation by 8 and the second equation by –3. Then, adding the equations eliminates the s terms, allowing us to solve the resulting equation for c.]*

(D) Substitute the value of one variable into either original equation in order to solve for the other variable. *[Substituting 7 for c in the first equation allows us to solve for s.]*

(E) Write your answer to the specific question posed by the word problem.

PRACTICE PROBLEMS

1. The difference of two numbers is 5. Their sum is 59. Find the numbers.	2. The sum of two numbers is 47, and their difference is 15. What is the larger number?
3. Justin went to the movie theater and bought one bag of popcorn and two cookies for $5.00. Martin went to the same theater and bought one bag of popcorn and four cookies for $6.00. How much does one cookie cost?	4. Alexandra purchases two doughnuts and three cookies at a doughnut shop and is charged $3.30. Briana purchases five doughnuts and two cookies at the same shop for $4.95. Find the cost of one doughnut and find the cost of one cookie.
5. Tanisha and Rachel had lunch at the mall. Tanisha ordered three slices of pizza and two colas. Rachel ordered two slices of pizza and three colas. Tanisha's bill was $6.00, and Rachel's bill was $5.25. What was the price of one slice of pizza? What was the price of one cola?	6. Ramón rented a sprayer and a generator. On his first job, he used each piece of equipment for 6 hours at a total cost of $90. On his second job, he used the sprayer for 4 hours and the generator for 8 hours at a total cost of $100. What was the hourly cost of *each* piece of equipment?

7. Kristin spent $131 on shirts. Fancy shirts cost $28 and plain shirts cost $15. If she bought a total of 7 shirts then how many of each kind did she buy?

8. The cost of three notebooks and four pencils is $8.50. The cost of five notebooks and eight pencils is $14.50. Determine the cost of one notebook and the cost of one pencil.

9. Last week, a fruit market sold a total of 108 apples and oranges. This week, five times the number of apples and three times the number of oranges were sold. A total of 452 apples and oranges were sold this week. Determine how many apples and how many oranges were sold last week.

10. The sum of the digits of a certain two-digit number is 7. Reversing its digits increases the number by 9. What is the number?

Regents Questions

MULTIPLE CHOICE

1. During the 2010 season, football player McGee's earnings, m, were 0.005 million dollars more than those of his teammate Fitzpatrick's earnings, f. The two players earned a total of 3.95 million dollars. Which system of equations could be used to determine the amount each player earned, in millions of dollars?

    (1) $m + f = 3.95$
        $m + 0.005 = f$

    (2) $m - 3.95 = f$
        $f + 0.005 = m$

    (3) $f - 3.95 = m$
        $m + 0.005 = f$

    (4) $m + f = 3.95$
        $f + 0.005 = m$

2. Mo's farm stand sold a total of 165 pounds of apples and peaches. She sold apples for $1.75 per pound and peaches for $2.50 per pound. If she made $337.50, how many pounds of peaches did she sell?

    (1) 11

    (2) 18

    (3) 65

    (4) 100

3. The Celluloid Cinema sold 150 tickets to a movie. Some of these were child tickets and the rest were adult tickets. A child ticket cost $7.75 and an adult ticket cost $10.25. If the cinema sold $1470 worth of tickets, which system of equations could be used to determine how many adult tickets, a, and how many child tickets, c, were sold?

    (1) $a + c = 150$
        $10.25a + 7.75c = 1470$

    (2) $a + c = 1470$
        $10.25a + 7.75c = 150$

    (3) $a + c = 150$
        $7.75a + 10.25c = 1470$

    (4) $a + c = 1470$
        $7.75a + 10.25c = 150$

4. Alicia purchased H half-gallons of ice cream for $3.50 each and P packages of ice cream cones for $2.50 each. She purchased 14 items and spent $43. Which system of equations could be used to determine how many of each item Alicia purchased?

    (1) $3.50H + 2.50P = 43$
        $H + P = 14$

    (2) $3.50P + 2.50H = 43$
        $P + H = 14$

    (3) $3.50H + 2.50P = 14$
        $H + P = 43$

    (4) $3.50P + 2.50H = 14$
        $P + H = 43$

5. Lizzy has 30 coins that total $4.80. All of her coins are dimes, D, and quarters, Q. Which system of equations models this situation?

    (1) $D + Q = 4.80$
        $.10D + .25Q = 30$

    (2) $D + Q = 30$
        $.10D + .25Q = 4.80$

    (3) $D + Q = 30$
        $.25D + .10Q = 4.80$

    (4) $D + Q = 4.80$
        $.25D + .10Q = 30$

CONSTRUCTED RESPONSE

6. An animal shelter spends $2.35 per day to care for each cat and $5.50 per day to care for each dog. Pat noticed that the shelter spent $89.50 caring for cats and dogs on Wednesday.

    Write an equation to represent the possible numbers of cats and dogs that could have been at the shelter on Wednesday.

    Pat said that there might have been 8 cats and 14 dogs at the shelter on Wednesday. Are Pat's numbers possible? Use your equation to justify your answer.

    Later, Pat found a record showing that there were a total of 22 cats and dogs at the shelter on Wednesday. How many cats were at the shelter on Wednesday?

7. Jacob and Zachary go to the movie theater and purchase refreshments for their friends. Jacob spends a total of $18.25 on two bags of popcorn and three drinks. Zachary spends a total of $27.50 for four bags of popcorn and two drinks.

    Write a system of equations that can be used to find the price of one bag of popcorn and the price of one drink.

    Using these equations, determine and state the price of a bag of popcorn and the price of a drink, to the nearest cent.

8. Franco and Caryl went to a bakery to buy desserts. Franco bought 3 packages of cupcakes and 2 packages of brownies for $19. Caryl bought 2 packages of cupcakes and 4 packages of brownies for $24. Let x equal the price of one package of cupcakes and y equal the price of one package of brownies. Write a system of equations that describes the given situation. On the set of axes below, graph the system of equations.

    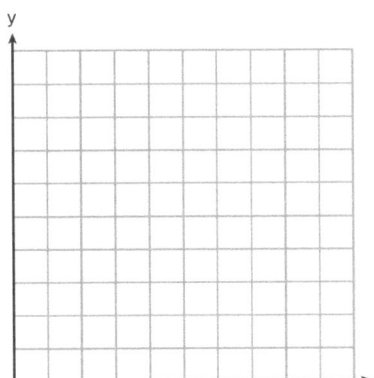

    Determine the exact cost of one package of cupcakes and the exact cost of one package of brownies in dollars and cents. Justify your solution.

9.   For a class picnic, two teachers went to the same store to purchase drinks. One teacher purchased 18 juice boxes and 32 bottles of water, and spent $19.92. The other teacher purchased 14 juice boxes and 26 bottles of water, and spent $15.76.

Write a system of equations to represent the costs of a juice box, j, and a bottle of water, w.

Kara said that the juice boxes might have cost 52 cents each and that the bottles of water might have cost 33 cents each. Use your system of equations to justify that Kara's prices are *not* possible.

Solve your system of equations to determine the actual cost, in dollars, of each juice box and each bottle of water.

10. Two friends went to a restaurant and ordered one plain pizza and two sodas. Their bill totaled $15.95. Later that day, five friends went to the same restaurant. They ordered three plain pizzas and each person had one soda. Their bill totaled $45.90. Write and solve a system of equations to determine the price of one plain pizza. [Only an algebraic solution can receive full credit.]

11. Ian is borrowing $1000 from his parents to buy a notebook computer. He plans to pay them back at the rate of $60 per month. Ken is borrowing $600 from his parents to purchase a snowboard. He plans to pay his parents back at the rate of $20 per month.

Write an equation that can be used to determine after how many months the boys will owe the same amount. Determine algebraically and state in how many months the two boys will owe the same amount. State the amount they will owe at this time.

Ian claims that he will have his loan paid off 6 months after he and Ken owe the same amount. Determine and state if Ian is correct. Explain your reasoning.

12. Central High School had five members on their swim team in 2010. Over the next several years, the team increased by an average of 10 members per year. The same school had 35 members in their chorus in 2010. The chorus saw an increase of 5 members per year. Write a system of equations to model this situation, where x represents the number of years since 2010. Graph this system of equations on the set of axes below.

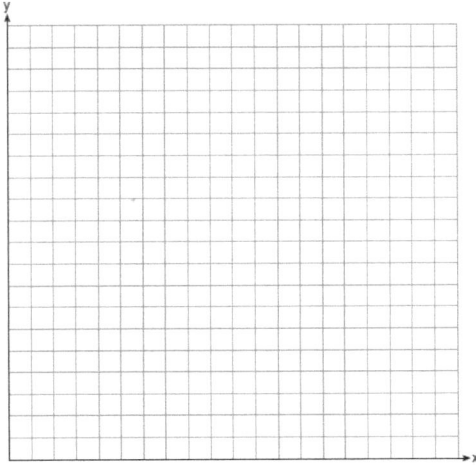

Explain in detail what each coordinate of the point of intersection of these equations means in the context of this problem.

13. The graph below models the cost of renting video games with a membership in Plan A and Plan B.

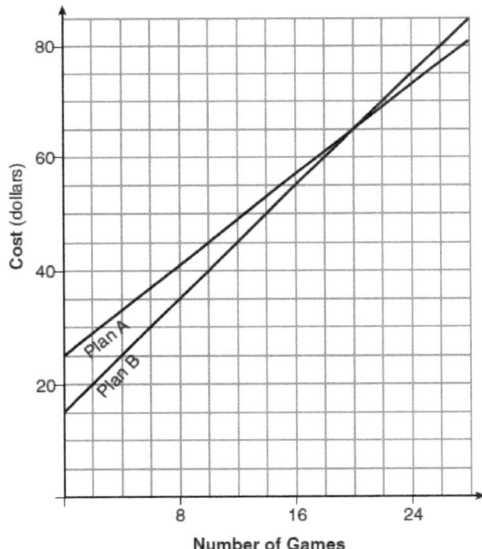

Explain why Plan B is the better choice for Dylan if he only has $50 to spend on video games, including a membership fee. Bobby wants to spend $65 on video games, including a membership fee. Which plan should he choose? Explain your answer.

14.   At Bea's Pet Shop, the number of dogs, d, is initially five less than twice the number of cats, c. If she decides to add three more of each, the ratio of cats to dogs will be $\frac{3}{4}$.

   Write an equation or system of equations that can be used to find the number of cats and dogs Bea has in her pet shop.

   Could Bea's Pet Shop initially have 15 cats and 20 dogs? Explain your reasoning.

   Determine algebraically the number of cats and the number of dogs Bea initially had in her pet shop.

15.   There are two parking garages in Beacon Falls. Garage A charges \$7.00 to park for the first 2 hours, and each additional hour costs \$3.00. Garage B charges \$3.25 per hour to park.

   When a person parks for at least 2 hours, write equations to model the cost of parking for a total of x hours in Garage A and Garage B.

   Determine algebraically the number of hours when the cost of parking at both garages will be the same.

16.   Dylan has a bank that sorts coins as they are dropped into it. A panel on the front displays the total number of coins inside as well as the total value of these coins. The panel shows 90 coins with a value of \$17.55 inside of the bank.

   If Dylan only collects dimes and quarters, write a system of equations in two variables or an equation in one variable that could be used to model this situation.

   Using your equation or system of equations, algebraically determine the number of quarters Dylan has in his bank.

   Dylan's mom told him that she would replace each one of his dimes with a quarter. If he uses all of his coins, determine if Dylan would then have enough money to buy a game priced at \$20.98 if he must also pay an 8% sales tax. Justify your answer.

17.   At the present time, Mrs. Bee's age is six years more than four times her son's age. Three years ago, she was seven times as old as her son was then. If b represents Mrs. Bee's age now and s represents her son's age now, write a system of equations that could be used to model this scenario.

   Use this system of equations to determine, algebraically, the ages of both Mrs. Bee and her son now.

   Determine how many years from now Mrs. Bee will be three times as old as her son will be then.

18.   A recreation center ordered a total of 15 tricycles and bicycles from a sporting goods store. The number of wheels for all the tricycles and bicycles totaled 38.

Write a linear system of equations that models this scenario, where t represents the number of tricycles and b represents the number of bicycles ordered.

On the set of axes below, graph this system of equations.

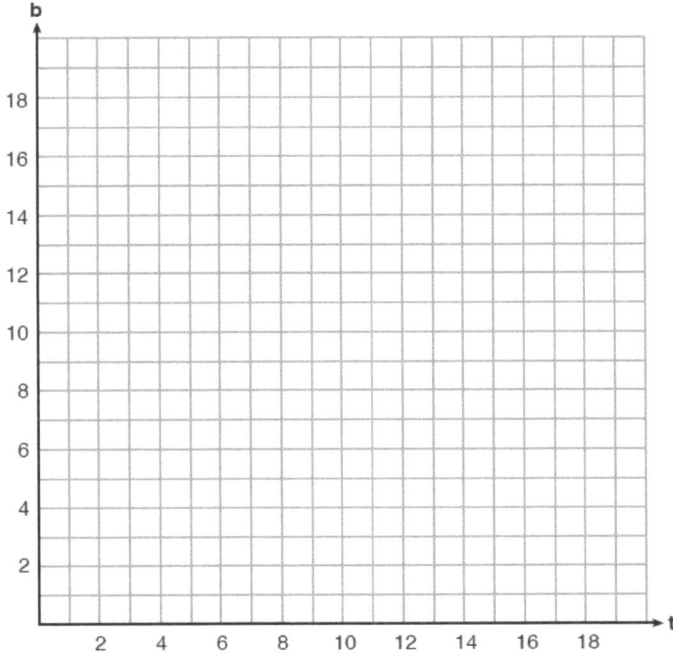

Based on your graph of this scenario, could the recreation center have ordered 10 tricycles? Explain your reasoning.

19.   When visiting friends in a state that has no sales tax, two families went to a fast-food restaurant for lunch. The Browns bought 4 cheeseburgers and 3 medium fries for $16.53. The Greens bought 5 cheeseburgers and 4 medium fries for $21.11. Using c for the cost of a cheeseburger and f for the cost of medium fries, write a system of equations that models this situation.

The Greens said that since their bill was $21.11, each cheeseburger must cost $2.49 and each order of medium fries must cost $2.87 each. Are they correct? Justify your answer.

Using your equations, algebraically determine both the cost of one cheeseburger and the cost of one order of medium fries.

20. Allysa spent $35 to purchase 12 chickens. She bought two different types of chickens. Americana chickens cost $3.75 each and Delaware chickens cost $2.50 each.

Write a system of equations that can be used to determine the number of Americana chickens, A, and the number of Delaware chickens, D, she purchased.

Determine algebraically how many of each type of chicken Allysa purchased.

Each Americana chicken lays 2 eggs per day and each Delaware chicken lays 1 egg per day. Allysa only sells eggs by the full dozen for $2.50. Determine how much money she expects to take in at the end of the first week with her 12 chickens.

21. Two families went to Rollercoaster World. The Brown family paid $170 for 3 children and 2 adults. The Peckham family paid $360 for 4 children and 6 adults.

If x is the price of a child's ticket in dollars and y is the price of an adult's ticket in dollars, write a system of equations that models this situation.

Graph your system of equations on the set of axes below.

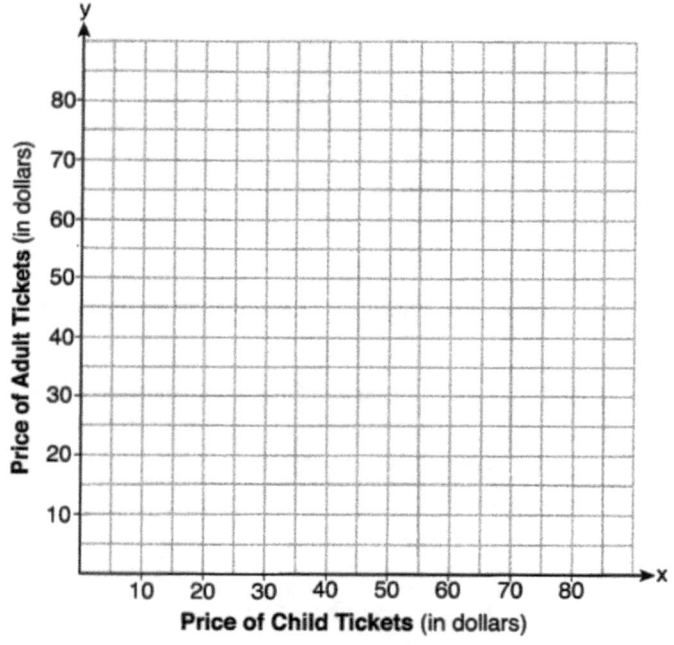

State the coordinates of the point of intersection. Explain what each coordinate of the point of intersection means in the context of the problem.

22. At a local garden shop, the price of plants includes sales tax. The cost of 4 large plants and 8 medium plants is $40. The cost of 5 large plants and 2 medium plants is $28. If l is the cost of a large plant and m is the cost of a medium plant, write a system of equations that models this situation.

Could the cost of one large plant be $5.50 and the cost of one medium plant be $2.25? Justify your answer.

Determine algebraically both the cost of a large plant and the cost of a medium plant.

23. At an amusement park, the cost for an adult admission is a, and for a child the cost is c. For a group of six that included two children, the cost was $325.94. For a group of five that included three children, the cost was $256.95. All ticket prices include tax.

Write a system of equations, in terms of a and c, that models this situation.

Use your system of equations to determine the exact cost of each type of ticket algebraically.

Determine the cost for a group of four that includes three children.

24. An ice cream shop sells small and large sundaes. One day, 30 small sundaes and 50 large sundaes were sold for $420. Another day, 15 small sundaes and 35 large sundaes were sold for $270. Sales tax is included in all prices.

If x is the cost of a small sundae and y is the cost of a large sundae, write a system of equations to represent this situation.

Peyton thinks that small sundaes cost $2.75 and large sundaes cost $6.75. Is Peyton correct? Justify your answer.

Using your equations, determine algebraically the cost of one small sundae and the cost of one large sundae.

25. A fence was installed around the edge of a rectangular garden. The length, l, of the fence was 5 feet less than 3 times its width, w. The amount of fencing used was 90 feet. Write a system of equations or write an equation using one variable that models this situation.

Determine algebraically the dimensions, in feet, of the garden.

26. Aidan and his sister Ella are having a race. Aidan runs at a rate of 10 feet per second. Ella runs at a rate of 6 feet per second. Since Ella is younger, Aidan is letting her begin 30 feet ahead of the starting line.

Let y represent the distance from the starting line and x represent the time elapsed, in seconds. Write an equation to model the distance Aidan traveled. Write an equation to model the distance Ella traveled.

On the set of axes below, graph your equations.

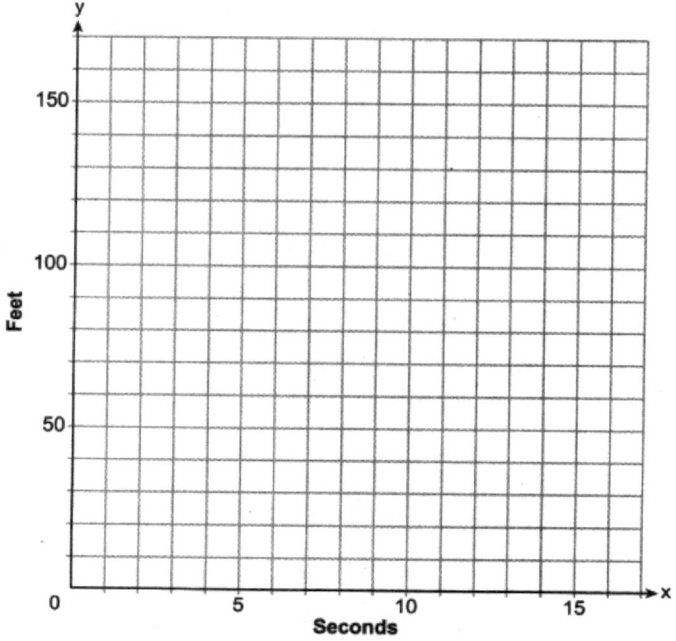

Exactly how many seconds does it take Aidan to catch up to Ella? Justify your answer.

27. Dana went shopping for plants to put in her garden. She bought three roses and two daisies for $31.88. Later that day, she went back and bought two roses and one daisy for $18.92.

If r represents the cost of one rose and d represents the cost of one daisy, write a system of equations that models this situation.

Use your system of equations to algebraically determine both the cost of one rose and the cost of one daisy.

If Dana had waited until the plants were on sale, she would have paid $4.50 for each rose and $6.50 for each daisy. Determine the total amount of money she would have saved by buying all of her flowers during the sale.

28.   Jim had a bag of coins. The number of nickels, n, and the number of quarters, q, totaled 28 coins. The combined value of the coins was \$4. Write a system of equations that models this situation.

Use your system of equations to algebraically determine both the number of quarters, q, and the number of nickels, n, that Jim had in the bag.

Jim was given an additional \$3.00 that was made up of equal numbers of nickels and quarters. How many of each coin was he given? Justify your answer.

4.6   Word Problems – Systems of Inequalities

There are times when a verbal problem requires more than one variable in one inequality. If the situation calls for it, use two variables and set up a system of two inequalitites. Be sure to label exactly what each variable represents.

Example: Don't write p = pencil. Instead, write p = number of pencils, or p = cost of a pencil in cents, or p = weight of a pencil in ounces, depending on the problem.

MODEL PROBLEM

A home-based company produces both hand-knitted scarves and sweaters. The scarves take 2 hours of labor to produce, and the sweaters take 14 hours. The labor available is limited to 40 hours per week, and the total production capacity is 5 items per week. Write a system of inequalities representing this situation, where x is the number of scarves and y is the number of sweaters.

Solution:	**Explanation of steps:**
(A) x = number of scarves y = number of sweaters	(A) Clearly label the variables used.
(B) $2x + 14y \leq 40$	(B) Write the first inequality based on given information *[constraint on total hours]*.
(C) $x + y \leq 5$	(C) Write the second inequality based on given information *[constraint on number of items]*.

PRACTICE PROBLEMS

1. You can work at most 20 hours next week. You need to earn at least $92 to cover you weekly expenses. Your dog- walking job pays $7.50 per hour and your job as a car wash attendant pays $6 per hour. Write a system of linear inequalities to model the situation.	2. Marsha is buying plants and soil for her garden. The soil cost $4 per bag, and the plants cost $10 each. She wants to buy at least 5 plants and can spend no more than $100. Write a system of linear inequalities to model the situation.

3. John is packing books into boxes. Each box can hold either 15 small books or 8 large books. He needs to pack at least 35 boxes and at least 350 books. Write a system of linear inequalities to model the situation.

4. During a family trip, you share the driving with your dad. At most, you are allowed to drive for three hours. While driving, your maximum speed is 55 miles per hour.

   Write a system of inequalities describing the possible numbers of hours t and distance d you may have to drive.

   Is it possible for you to have driven 160 miles?

5.   FlyGlyde is a company that manufactures drones and hoverboards. FlyGlyde's daily production of drones cannot exceed 10, and its daily production of hoverboards must be less than or equal to 12. The combined number of drones and hoverboards cannot be more than 16.

If *x* is the number of drones and *y* is the number of hoverboards, graph on the accompanying set of axes the region that contains the number of drones and hoverboards FlyGlyde can manufacture daily.

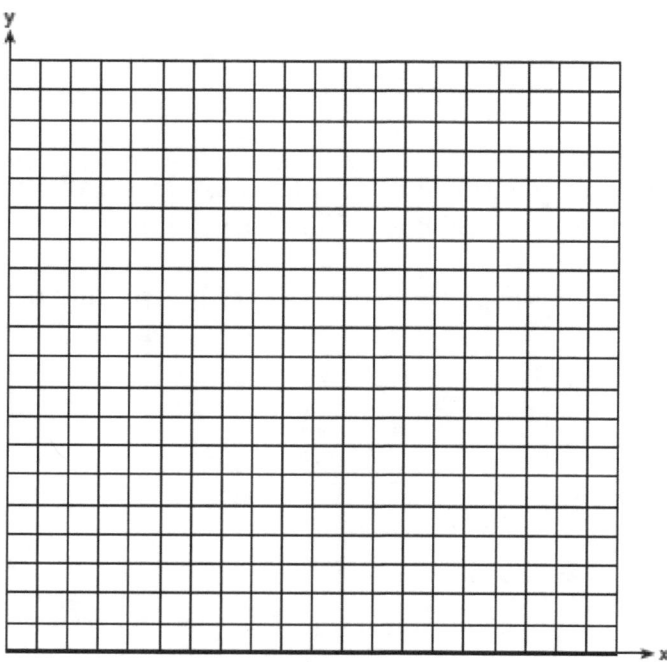

Regents Questions

MULTIPLE CHOICE

1. Jordan works for a landscape company during his summer vacation. He is paid $12 per hour for mowing lawns and $14 per hour for planting gardens. He can work a maximum of 40 hours per week, and would like to earn at least $250 this week. If m represents the number of hours mowing lawns and g represents the number of hours planting gardens, which system of inequalities could be used to represent the given conditions?

    (1) $m + g \leq 40$
         $12m + 14g \geq 250$

    (2) $m + g \geq 40$
         $12m + 14g \leq 250$

    (3) $m + g \leq 40$
         $12m + 14g \leq 250$

    (4) $m + g \geq 40$
         $12m + 14g \geq 250$

2. Gretchen has $50 that she can spend at the fair. Ride tickets cost $1.25 each and game tickets cost $2 each. She wants to go on a minimum of 10 rides and play at least 12 games. Which system of inequalities represents this situation when r is the number of ride tickets purchased and g is the number of game tickets purchased?

    (1) $1.25r + 2g < 50$
         $r \leq 10$
         $g > 12$

    (2) $1.25r + 2g \leq 50$
         $r \geq 10$
         $g \geq 12$

    (3) $1.25r + 2g \leq 50$
         $r \geq 10$
         $g > 12$

    (4) $1.25r + 2g < 50$
         $r \leq 10$
         $g \geq 12$

3. During summer vacation, Ben decides to sell hot dogs and pretzels on a food cart in Manhattan. It costs Ben $0.50 for each hot dog and $0.40 for each pretzel. He has only $100 to spend each day on hot dogs and pretzels. He wants to sell at least 200 items each day. If h is the number of hot dogs and p is the number of pretzels, which inequality would be part of a system of inequalities used to determine the total number of hot dogs and pretzels Ben can sell?

    (1) $h + p \leq 200$

    (2) $h + p \geq 200$

    (3) $0.50h + 0.40p \geq 200$

    (4) $0.50h + 0.40p \leq 200$

CONSTRUCTED RESPONSE

4. A high school drama club is putting on their annual theater production. There is a maximum of 800 tickets for the show. The costs of the tickets are $6 before the day of the show and $9 on the day of the show. To meet the expenses of the show, the club must sell at least $5,000 worth of tickets.

    Write a system of inequalities that represent this situation.

    The club sells 440 tickets before the day of the show. Is it possible to sell enough additional tickets on the day of the show to at least meet the expenses of the show? Justify your answer.

5. An on-line electronics store must sell at least $2500 worth of printers and computers per day. Each printer costs $50 and each computer costs $500. The store can ship a maximum of 15 items per day. On the set of axes below, graph a system of inequalities that models these constraints.

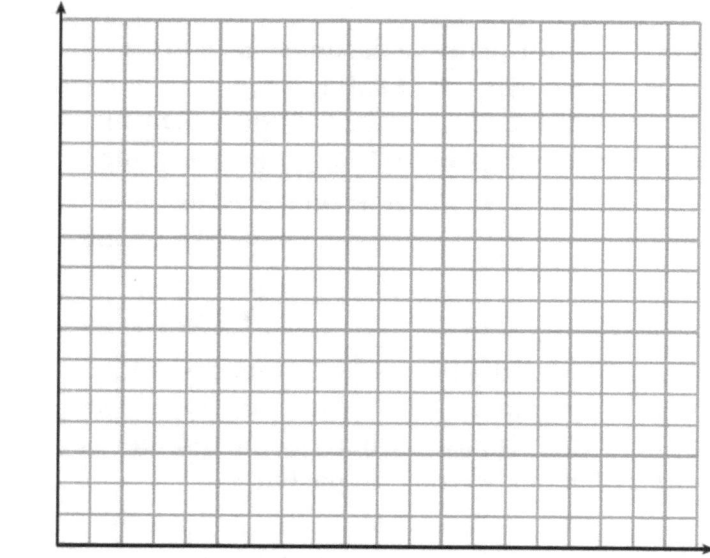

Number of Printers

Determine a combination of printers and computers that would allow the electronics store to meet all of the constraints. Explain how you obtained your answer.

6.   Edith babysits for *x* hours a week after school at a job that pays $4 an hour. She has accepted a job that pays $8 an hour as a library assistant working *y* hours a week. She will work both jobs. She is able to work *no more than* 15 hours a week, due to school commitments. Edith wants to earn *at least* $80 a week, working a combination of both jobs.

Write a system of inequalities that can be used to represent the situation.

Graph these inequalities on the set of axes below.

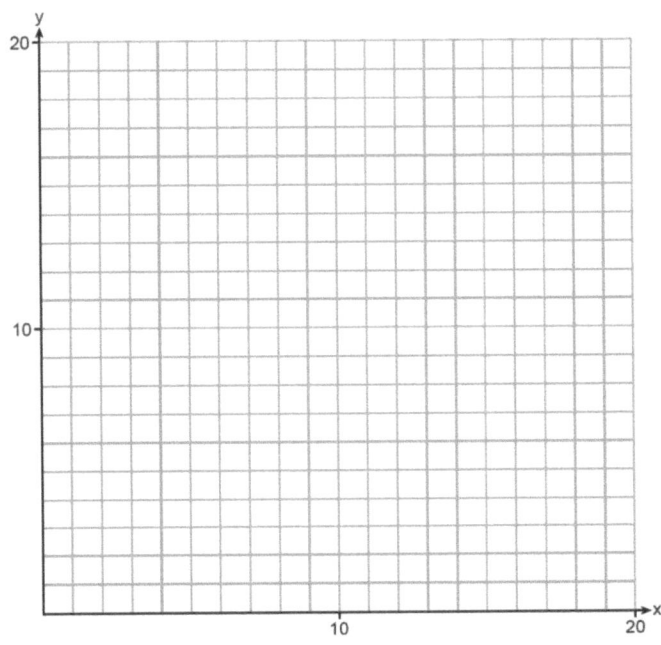

Determine and state one combination of hours that will allow Edith to earn *at least* $80 per week while working *no more than* 15 hours.

7.   The Reel Good Cinema is conducting a mathematical study. In its theater, there are 200 seats. Adult tickets cost $12.50 and child tickets cost $6.25. The cinema's goal is to sell at least $1500 worth of tickets for the theater.

Write a system of linear inequalities that can be used to find the possible combinations of adult tickets, x, and child tickets, y, that would satisfy the cinema's goal.

Graph the solution to this system of inequalities on the set of axes below. Label the solution with an *S*.

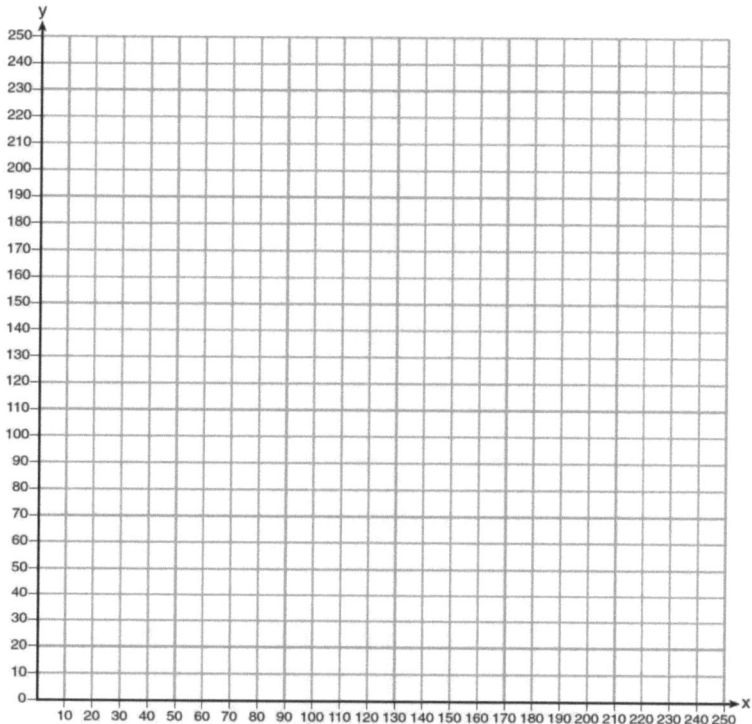

Marta claims that selling 30 adult tickets and 80 child tickets will result in meeting the cinema's goal. Explain whether she is correct or incorrect, based on the graph drawn.

8. The sum of two numbers, x and y, is more than 8. When you double x and add it to y, the sum is less than 14. Graph the inequalities that represent this scenario on the set of axes below.

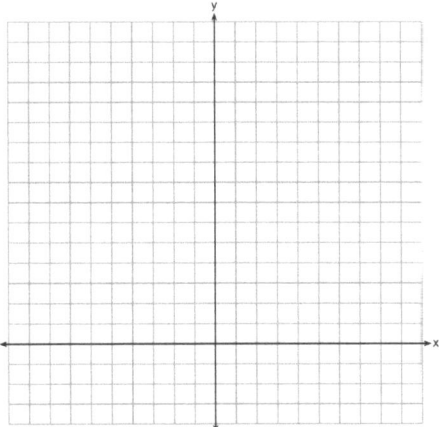

Kai says that the point (6,2) is a solution to this system. Determine if he is correct and explain your reasoning.

9. A drama club is selling tickets to the spring musical. The auditorium holds 200 people. Tickets cost $12 at the door and $8.50 if purchased in advance. The drama club has a goal of selling at least $1000 worth of tickets to Saturday's show. Write a system of inequalities that can be used to model this scenario.

If 50 tickets are sold in advance, what is the minimum number of tickets that must be sold at the door so that the club meets its goal? Justify your answer.

10. The drama club is running a lemonade stand to raise money for its new production. A local grocery store donated cans of lemonade and bottles of water. Cans of lemonade sell for $2 each and bottles of water sell for $1.50 each. The club needs to raise at least $500 to cover the cost of renting costumes. The students can accept a maximum of 360 cans and bottles. Write a system of inequalities that can be used to represent this situation.

The club sells 144 cans of lemonade. What is the *least* number of bottles of water that must be sold to cover the cost of renting costumes? Justify your answer.

CHAPTER 5.   POLYNOMIALS

5.1   Polynomial Expressions

A **term** is a number, a variable, or any product or quotient of numbers and variables.
A **monomial** is a single term without any variables in the denominator, such as $-2x^3y$.
A **polynomial** is a sum of any number of terms.
Example: $x^2 + y - 2xy$ is a polynomial with 3 terms: x^2, y, and $-2xy$.
A **binomial** is a polynomial of 2 unlike terms, and a **trinomial** is a polynomial of 3 unlike terms.
Examples: $2x^2 + 5$ is a binomial and $2x^2 - 3x + 5$ is a trinomial.

The prefix "mono" means one, "poly" means many, "bi" means two, and "tri" means three.

A **constant term** is a term with no variable part.
Example: In the expression $2x^2 - 3 + x + 5$, the constant terms are -3 and 5.

The **degree of a term** is the sum of its variables' exponents.
Examples: $8x^2$ has a degree of 2, $4x^3$ has a degree of 3, and $2x$ has a degree of 1.
              The term $12x^2y^3z$ has a degree of 6 (add the exponents $2 + 3 + 1$).

A constant term has a degree of 0. Think of its missing variable part as a variable to a 0 power.
Example: The constant 5 is equivalent to $5x^0$, which is $5(1)$, so its degree is 0.

The **degree of a polynomial** is the largest degree of its terms.
Example: The degree of $9x^2 + x^4 - x$ is 4, since x^4 has the highest degree of 4.

To write a polynomial in standard form:
    1.   Combine all like terms (*simplify*)
    2.   Write terms in descending order (*exponents of a variable decrease*)
Example: $2a - 3a^2 + 4 + 9a$ → $11a - 3a^2 + 4$ → $-3a^2 + 11a + 4$

The **leading coefficient** of a polynomial is the coefficient of the term with the highest degree.
When written in standard form, this is the coefficient of the first term of the polynomial.
Example: The leading coefficient of $-3a^2 + 11a + 4$ is -3.

MODEL PROBLEM

Write $2x + 3 - 4x^2 - x + 1$ in standard form.

Solution:
$-4x^2 + x + 4$

Explanation of steps:
Combine like terms *[$2x - x = x$ and $3 + 1 = 4$]*, then write terms in descending order
[the x^2 term, then the x term, then the constant].

PRACTICE PROBLEMS

1.
a) How many terms does the polynomial $3x^4 - 2x^3 - 1$ have?

b) What is the degree of this polynomial?

c) What is its leading coefficient?

d) What is its constant term?

2.
a) Write $x - 2 + x^2 - 3x + 5$ in standard form.

b) What is the degree of this polynomial?

c) What is its leading coefficient?

d) What is its constant term?

3. Write the expression $2x - 5x^3 - 2x^2 - 10x + 15 - x^2$ as a polynomial in standard form.

Regents Questions

MULTIPLE CHOICE

1. An expression of the fifth degree is written with a leading coefficient of seven and a constant of six. Which expression is correctly written for these conditions?

    (1) $6x^5 + x^4 + 7$ (3) $6x^7 - x^5 + 5$

    (2) $7x^6 - 6x^4 + 5$ (4) $7x^5 + 2x^2 + 6$

2. Mrs. Allard asked her students to identify which of the polynomials below are in standard form and explain why.

    I. $15x^4 - 6x + 3x^2 - 1$

    II. $12x^3 + 8x + 4$

    III. $2x^5 + 8x^2 + 10x$

    Which student's response is correct?

    (1) Tyler said I and II because the coefficients are decreasing.

    (2) Susan said only II because all the numbers are decreasing.

    (3) Fred said II and III because the exponents are decreasing.

    (4) Alyssa said II and III because they each have three terms.

3. Students were asked to write $6x^5 + 8x - 3x^3 + 7x^7$ in standard form. Shown below are four student responses.

    Anne: $7x^7 + 6x^5 - 3x^3 + 8x$

    Bob: $-3x^3 + 6x^5 + 7x^7 + 8x$

    Carrie: $8x + 7x^7 + 6x^5 - 3x^3$

    Dylan: $8x - 3x^3 + 6x^5 + 7x^7$

    Which student is correct?

    (1) Anne   (3) Carrie

    (2) Bob   (4) Dylan

4. Which polynomial has a leading coefficient of 4 and a degree of 3?

    (1) $3x^4 - 2x^2 + 4x - 7$ (3) $4x^4 - 3x^3 + 2x^2$

    (2) $4 + x - 4x^2 + 5x^3$ (4) $2x + x^2 + 4x^3$

5. Students were asked to write an expression which had a leading coefficient of 3 and a constant term of –4. Which response is correct?

    (1) $3 - 2x^3 - 4x$ (3) $4 - 7x + 3x^3$

    (2) $7x^3 - 3x^5 - 4$ (4) $-4x^2 + 3x^4 - 4$

6. An example of a sixth-degree polynomial with a leading coefficient of seven and a constant term of four is

    (1) $6x^7 - x^5 + 2x + 4$ (3) $7x^4 + 6 + x^2$

    (2) $4 + x + 7x^6 - 3x^2$ (4) $5x + 4x^6 + 7$

7. What is the constant term of the polynomial $4d + 6 + 3d^2$?

    (1)   6   (3)   3

    (2)   2   (4)   4

8. When $3x^2 + 7x - 6 + 2x^3$ is written in standard form, the leading coefficient is

    (1)   7   (3)   3

    (2)   2   (4)   –6

9. Students were asked to write $2x^3 + 3x + 4x^2 + 1$ in standard form. Four student responses are shown below.

            Alexa: $4x^2 + 3x + 2x^3 + 1$

            Carol: $2x^3 + 3x + 4x^2 + 1$

            Ryan: $2x^3 + 4x^2 + 3x + 1$

            Eric: $1 + 2x^3 + 3x + 4x^2$

    Which student's response is correct?

    (1)   Alexa   (3)   Ryan

    (2)   Carol   (4)   Eric

10. Which statement is correct about the polynomial $3x^2 + 5x - 2$?

    (1)   It is a third-degree polynomial with a constant term of –2.

    (2)   It is a third-degree polynomial with a leading coefficient of 3.

    (3)   It is a second-degree polynomial with a constant term of 2.

    (4)   It is a second-degree polynomial with a leading coefficient of 3.

11. What is the degree of the polynomial $2x + x^3 + 5x^2$?

    (1)   1   (3)   3

    (2)   2   (4)   4

12. What is the degree of the polynomial $5x - 3x^2 - 1 + 7x^3$?

    (1)   1   (3)   3

    (2)   2   (4)   5

CONSTRUCTED RESPONSE

13. When multiplying polynomials for a math assignment, Pat found the product to be $-4x + 8x^2 - 2x^3 + 5$. He then had to state the leading coefficient of this polynomial. Pat wrote down –4. Do you agree with Pat's answer? Explain your reasoning.

5.2   Add and Subtract Polynomials

In this section, remember the rules for adding or subtracting terms with exponents:
- Must be like terms (same variable parts, including the same exponents)
- Add or subtract the coefficients and keep the same variable part

Examples: (a) $x^5 + x^3$ cannot be combined; not like terms
(b) $x^3 + x^3 = 2x^3$
(c) $4x^3 - x^3 = 3x^3$
(d) $2x^2y + 3x^2y = 5x^2y$

To add polynomials: Join the polynomials and simplify into standard form.
Example: $(x^2 + 5x - 24) + (2x^2 + 10) =$
$x^2 + 5x - 24 + 2x^2 + 10 =$
$3x^2 + 5x - 14$

To subtract polynomials: Negate all the signs of the second polynomial and then add.
Example: $(2x^2 - 5x + 4) - (x^2 + 6x - 5) =$
$2x^2 - 5x + 4 - x^2 - 6x + 5 =$
$x^2 - 11x + 9$

You may find it easier to arrange the polynomials vertically before adding or subtracting. Just be sure to line up the terms that have the same degree.
Examples: The above problems could be written as shown below.

$$x^2 + 5x - 24 \qquad\qquad 2x^2 - 5x + 4$$
$$\underline{2x^2 \qquad\ + 10} \qquad\qquad \underline{-(x^2 + 6x - 5)}$$
$$3x^2 + 5x - 14 \qquad\qquad x^2 - 11x + 9$$

When we add or subtract polynomials, the result is always a polynomial. Therefore, we can say that the set of polynomials is **closed** under addition and subtraction.

MODEL PROBLEM

Subtract $-4x^2 - 12x + 5$ from $x^2 + 9x - 5$ and write the result in standard form.

Solution:
(A) $(x^2 + 9x - 5) - (-4x^2 - 12x + 5) =$
(B) $x^2 + 9x - 5 + 4x^2 + 12x - 5 =$
(C) $5x^2 + 21x - 10$

Explanation of steps:
(A)   Set up the problem. *[Remember the rule: "Subtract x from y" means y – x.]*
(B)   In subtraction, negate all the terms in the second polynomial and then add.
(C)   Combine like terms and express in standard form.

PRACTICE PROBLEMS

1. Write the sum of $8x^2 - x + 4$ and $x - 5$ in standard form.	2. Write the sum of $3x^2 + x + 8$ and $x^2 - 9$ in standard form.
3. Write the sum of $4x^3 + 6x^2 + 2x - 3$ and $3x^3 + 3x^2 - 5x - 5$ in standard form.	4. What is the sum of $-3x^2 - 7x + 9$ and $-5x^2 + 6x - 4$ in standard form?
5. Add $3x^2 + 5x - 6$ and $-x^2 + 3x + 9$.	6. Add $8n^2 - 3n + 10$ and $-3n^2 - 6n - 7$.
7. Write in standard form: $(3x^2 + 2xy + 7) - (6x^2 - 4xy + 3)$	8. What is the result when $3a^2 - 2a + 5$ is subtracted from $a^2 + a - 1$?

9. What is the difference when $2x^2 - x + 6$ is subtracted from $x^2 - 3x - 2$?	10. Subtract $3x^2 + 4x - 1$ from $x^2 + 1$ and write the result in standard form.
11. Subtract $4x^2 + 7x - 5$ from $9x^2 - 2x + 3$. Write the result in standard form.	12. Subtract $5x^2 - 7x - 6$ from $9x^2 + 3x - 4$. Express the answer in standard form.
13. Subtract $2x^2 - 5x + 8$ from $6x^2 + 3x - 2$ and express the answer as a trinomial in standard form.	14. What is the result when $6x^2 - 13x + 12$ is subtracted from $-3x^2 + 6x + 7$?
15. What is the difference when $x^2 + 3x - 4$ is subtracted from $x^3 + 3x^2 - 2x$?	16. What is the difference when $5x + 4$ is subtracted from $5x - 4$?

Regents Questions

MULTIPLE CHOICE

1. If $A = 3x^2 + 5x - 6$ and $B = -2x^2 - 6x + 7$, then $A - B$ equals
   - (1) $-5x^2 - 11x + 13$
   - (2) $5x^2 + 11x - 13$
   - (3) $-5x^2 - x + 1$
   - (4) $5x^2 - x + 1$

2. The expression $3(x^2 - 1) - (x^2 - 7x + 10)$ is equivalent to
   - (1) $2x^2 - 7x + 7$
   - (2) $2x^2 + 7x - 13$
   - (3) $2x^2 - 7x + 9$
   - (4) $2x^2 + 7x - 11$

3. Which expression is equivalent to $2(3g - 4) - (8g + 3)$?
   - (1) $-2g - 1$
   - (2) $-2g - 5$
   - (3) $-2g - 7$
   - (4) $-2g - 11$

4. If $C = 2a^2 - 5$ and $D = 3 - a$, then $C - 2D$ equals
   - (1) $2a^2 + a - 8$
   - (2) $2a^2 - a - 8$
   - (3) $2a^2 + 2a - 11$
   - (4) $2a^2 - a - 11$

5. The expression $(3x^2 + 4x - 8) + 2(11 - 5x)$ is equivalent to
   - (1) $3x^2 - x + 5$
   - (2) $3x^2 - x + 14$
   - (3) $3x^2 - 6x + 14$
   - (4) $3x^2 + 14x + 14$

6. The expression $(-x^2 + 3x - 7) - (4x^2 + 5x - 2)$ is equivalent to
   - (1) $-5x^2 - 2x - 9$
   - (2) $-5x^2 - 2x - 5$
   - (3) $-5x^2 + 8x - 9$
   - (4) $-5x^2 + 8x - 5$

CONSTRUCTED RESPONSE

7. Subtract $5x^2 + 2x - 11$ from $3x^2 + 8x - 7$. Express the result as a trinomial.

8. Express in simplest form: $(3x^2 + 4x - 8) - (-2x^2 + 4x + 2)$

5.3   Multiply Polynomials

In this section, remember the multiplication rule for exponents:
Multiply the coefficients and, for common bases, add the exponents.

Examples: (a) $x^5 \cdot x^3 = x^{5+3} = x^8$

               (b) $(3a^2)(4ac) = 12a^3c$

Factors are any parts of an expression that are multiplied to produce a product.

Examples: (a) In $2 \cdot 3$, both 2 and 3 are factors.

               (b) In $2x^2y$, the factors are 2, x^2 and y.

               (c) In $3(x + 1)$, both 3 and $(x + 1)$ are factors.

               (d) The expression $(a - 2)(a + 3)$ has two binomial factors.

Multiplying a Monomial by a Polynomial: Apply the Distributive Property.

Examples: (a) $3(x^2 + x) \;\rightarrow\; 3x^2 + 3x$

               (b) $-(2m^4 - n^2) \;\rightarrow\; -2m^4 + n^2$

               (c) $5a(a^3 - 3a + 1) \;\rightarrow\; 5a^4 - 15a^2 + 5a$

Multiplying Binomials

Method 1: Distribution – Multiply term 1 by the second binomial, then term 2 by the second binomial

$$(a - 2)(a + 3) = a(a + 3) - 2(a + 3) =$$
$$a^2 + 3a - 2a - 6 =$$
$$a^2 + a - 6$$

Method 2: Vertically – Multiply terms like they were digits in whole number multiplication

$$
\begin{array}{r}
a - 2 \\
\times \quad a + 3 \\
\hline
3a - 6 \\
a^2 - 2a \quad\;\; \\
\hline
a^2 + a - 6
\end{array}
$$

$\leftarrow$ multiply $a - 2$ times $+3$

$\leftarrow$ shift left, then multiply $a - 2$ times a

$\leftarrow$ add like terms

Method 3: Rectangle diagram (Box Method) – Write the terms of each factor outside the side of a rectangle, find the algebraic areas (products) of each inner rectangle, and add the inner areas

$(a - 2)(a + 3)$

	a	-2
a	a^2	$-2a$
3	$3a$	-6

$$a^2 + 3a - 2a - 6 =$$
$$a^2 + a - 6$$

Method 4: "FOIL" – Multiply these pairs of terms: Firsts, Outers, Inners, Lasts

$$(a - 2)(a + 3) =$$

F   O   I   L
$$a^2 + 3a - 2a - 6 =$$
$$a^2 + a - 6$$

Firsts   Lasts

$(a + b)(c + d)$

Outers   Inners

Squaring a Binomial: multiply the binomial by itself.
Example: $(x - 3)^2 = (x - 3)(x - 3) = x^2 - 3x - 3x + 9 = x^2 - 6x + 9$

Multiplying a Binomial by a Trinomial
Expand one of the methods above except Method 4 (FOIL). Note that the "FOIL" method only works for multiplying two binomials.

Method 1: *Distribution* – Multiply term 1 by the trinomial, then term 2 by the trinomial.

$$(a + 2)(2a^2 - 5a + 3) = a(2a^2 - 5a + 3) + 2(2a^2 - 5a + 3) =$$
$$2a^3 - 5a^2 + 3a + 4a^2 - 10a + 6 =$$
$$2a^3 - a^2 - 7a + 6$$

Method 2: *Vertically* – Multiply terms like they were digits in whole number multiplication

$$
\begin{array}{r}
a + 2 \\
\times \quad 2a^2 - 5a + 3 \\
\hline
3a + 6 \\
-5a^2 - 10a \\
2a^3 + 4a^2 \\
\hline
2a^3 - a^2 - 7a + 6
\end{array}
$$

← multiply $a + 2$ times $+3$
← shift left, then multiply $a + 2$ times $-5a$
← shift left, then multiply $a + 2$ times $2a^2$
← add like terms

Method 3: *Rectangle diagram (Box Method)* – Write the terms of each factor outside a side of a rectangle, find the algebraic areas of each inner rectangle, and add the inner areas

$$(a + 2)(2a^2 - 5a + 3)$$

	a	2
$2a^2$	$2a^3$	$4a^2$
$-5a$	$-5a^2$	$-10a$
3	$3a$	6

$$2a^3 - 5a^2 + 3a + 4a^2 - 10a + 6 =$$
$$2a^3 - a^2 - 7a + 6$$

When we multiply polynomials, the result is always a polynomial. Therefore, we can say that the set of polynomials is **closed** under multiplication. This is not true when we divide polynomials, since we may end up with an algebraic fraction (rational expression) that is not a polynomial.

MODEL PROBLEM 1: *MULTIPLY A MONOMIAL BY A POLYNOMIAL*

Write the product of $4x^2$ and $(2x^2 - 3x + 10)$ as a polynomial in standard form.

Solution: $(4x^2)(2x^2 - 3x + 10) = 8x^4 - 12x^3 + 40x^2$

Explanation of steps:
Multiply the monomial by each term of the polynomial.
[The solution is already in standard form. If it weren't, you may need to rearrange the terms.]

PRACTICE PROBLEMS

1. Multiply: $7x(1 - x^3)$	2. What is the product of $2r^2 - 5$ and $3r$?
3. What is the product of $-3x^2y$ and $(5xy^2 + xy)$?	4. The length of a rectangle is represented by $x^2 + 3x + 2$, and the width is represented by $4x$. Express the area of the rectangle as a trinomial.

5. The length of a rectangular room is 7 less than three times the width, w, of the room. Write an expression, in simplest form, for the area of the room.

MODEL PROBLEM 2: *MULTIPLY BINOMIALS*

Write the product of $x + 8$ and $x - 2$ as a polynomial in standard form.

Solution:

(A) $(x + 8)(x - 2) =$
(B) $x(x - 2) + 8(x - 2) =$
(C) $x^2 - 2x + 8x - 16 =$
(D) $x^2 + 6x - 16$

Explanation of steps:

(A) Set up the problem.
(B) Using distribution, multiply term 1 *[x]* by the second binomial, then term 2 *[8]* by the second binomial.
(C) Distribute.
(D) Combine like terms and express in standard form.

PRACTICE PROBLEMS

6. What is the product of $(c + 8)$ and $(c - 5)$?	7. What is the product of $(3x + 2)$ and $(x - 7)$?
8. Express $(x - 7)(2x + 3)$ as a trinomial.	9. Simplify the expression $(x - 6)^2$.
10. Use the following diagram to expand $(a + b)^2$ into an equivalent polynomial. 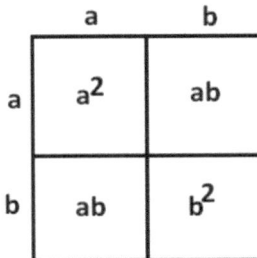	11. Which of the following expressions is *not* equivalent to $(a + b)(x + y)$? (1) $(a + b)x + (a + b)y$ (2) $a(x + y) + b(x + y)$ (3) $ax + by$ (4) $ax + bx + ay + by$

MODEL PROBLEM 3: *MULTIPLY A BINOMIAL BY A TRINOMIAL*

Write the product of $x + 4$ and $3x^2 + x - 2$ as a polynomial in standard form.

Solution: **Explanation of steps:**

(A)

	x	4
$3x^2$	$3x^3$	$12x^2$
x	x^2	$4x$
-2	$-2x$	-8

(A) Using the rectangle diagram method, place the terms of each polynomial on the outside of a rectangle, then find the products of each pair of terms and place them in the inner boxes.

(B) Add all the terms within the inner boxes.

(C) Combine like terms and write the polynomial in standard form.

(B) $3x^3 + 12x^2 + x^2 + 4x - 2x - 8 =$
(C) $3x^3 + 13x^2 + 2x - 8$

PRACTICE PROBLEMS

12. Create a rectangle diagram for $(x + 3)(x - y - 1)$ and fill in the inner boxes.

13. Multiply $(x - 1)(2x^2 + x - 2)$ and write the product in standard form.

14. Multiply $(x^2 + 2)(x^2 - 2x + 1)$ and write the product in standard form.

Regents Questions

MULTIPLE CHOICE

1. Fred is given a rectangular piece of paper. If the length of Fred's piece of paper is represented by $2x - 6$ and the width is represented by $3x - 5$, then the paper has a total area represented by
   (1) $5x - 11$ (3) $10x - 22$
   (2) $6x^2 - 28x + 30$ (4) $6x^2 - 6x - 11$

2. Four expressions are shown below.
   I $2(2x^2 - 2x - 60)$
   II $4(x^2 - x - 30)$
   III $4(x + 6)(x - 5)$
   IV $4x(x - 1) - 120$
   The expression $4x^2 - 4x - 120$ is equivalent to
   (1) I and II, only   (3) I, II, and IV
   (2) II and IV, only   (4) II, III, and IV

3. Which trinomial is equivalent to $3(x - 2)^2 - 2(x - 1)$?
   (1) $3x^2 - 2x - 10$ (3) $3x^2 - 14x + 10$
   (2) $3x^2 - 2x - 14$ (4) $3x^2 - 14x + 14$

4. When $(2x - 3)^2$ is subtracted from $5x^2$, the result is
   (1) $x^2 - 12x - 9$ (3) $x^2 + 12x - 9$
   (2) $x^2 - 12x + 9$ (4) $x^2 + 12x + 9$

5. What is the product of $2x + 3$ and $4x^2 - 5x + 6$?
   (1) $8x^3 - 2x^2 + 3x + 18$ (3) $8x^3 + 2x^2 - 3x + 18$
   (2) $8x^3 - 2x^2 - 3x + 18$ (4) $8x^3 + 2x^2 + 3x + 18$

6. Which polynomial is twice the sum of $4x^2 - x + 1$ and $-6x^2 + x - 4$?
   (1) $-2x^2 - 3$ (3) $-4x^2 - 6$
   (2) $-4x^2 - 3$ (4) $-2x^2 + x - 5$

7. The expression $3(x^2 + 2x - 3) - 4(4x^2 - 7x + 5)$ is equivalent to
   (1) $-13x - 22x + 11$ (3) $19x^2 - 22x + 11$
   (2) $-13x^2 + 34x - 29$ (4) $19x^2 + 34x - 29$

8. If $y = 3x^3 + x^2 - 5$ and $z = x^2 - 12$, which polynomial is equivalent to $2(y + z)$?
   (1) $6x^3 + 4x^2 - 34$ (3) $6x^3 + 3x^2 - 22$
   (2) $6x^3 + 3x^2 - 17$ (4) $6x^3 + 2x^2 - 17$

9. The length, width, and height of a rectangular box are represented by $2x$, $3x + 1$, and $5x - 6$, respectively. When the volume is expressed as a polynomial in standard form, what is the coefficient of the 2nd term?

    (1) -13 (3) -26
    (2) 13 (4) 26

10. Which expression is equivalent to $2(x^2 - 1) + 3x(x - 4)$?

    (1) $5x^2 - 5$ (3) $5x^2 - 12x - 1$
    (2) $5x^2 - 6$ (4) $5x^2 - 12x - 2$

11. When $(x)(x - 5)(2x + 3)$ is expressed as a polynomial in standard form, which statement about the resulting polynomial is true?

    (1) The constant term is 2.   (3) The degree is 2.
    (2) The leading coefficient is 2.   (4) The number of terms is 2.

12. When written in standard form, the product of $(3 + x)$ and $(2x - 5)$ is

    (1) $3x - 2$ (3) $2x^2 - 11x - 15$
    (2) $2x^2 + x - 15$ (4) $6x - 15 + 2x^2 - 5x$

13. The expression $(m - 3)^2$ is equivalent to

    (1) $m^2 + 9$ (3) $m^2 - 6m + 9$
    (2) $m^2 - 9$ (4) $m^2 - 6m - 9$

14. The expression $\frac{1}{3}x(6x^2 - 3x + 9)$ is equivalent to

    (1) $2x^2 - x + 3$ (3) $2x^3 - x^2 + 3x$
    (2) $2x^2 + 3x + 3$ (4) $2x^3 + 3x^2 + 3x$

15. When the expression $2x(x - 4) - 3(x + 5)$ is written in simplest form, the result is

    (1) $2x^2 - 11x - 15$ (3) $2x^2 - 3x - 19$
    (2) $2x^2 - 11x + 5$ (4) $2x^2 - 3x + 1$

16. The expression $(5x^2 - x + 4) - 3(x^2 - x - 2)$ is equivalent to

    (1) $2x^2 - 2x + 2$ (3) $2x^4 - 2x^2 + 2$
    (2) $2x^2 + 2x + 10$ (4) $2x^4 - 2x^2 + 10$

17. What is the product of $(2x + 7)$ and $(x - 3)$?

    (1) $2x^2 - 21$ (3) $2x^2 + 4x - 21$
    (2) $2x^2 + x - 21$ (4) $2x^2 + 13x - 21$

18. The product of $(x^2 + 3x + 9)$ and $(x - 3)$ is

    (1) $x^3 - 27$ (3) $x^3 - 6x^2 - 18x - 27$
    (2) $x^2 + 4x + 6$ (4) $-6x^4 + x^3 - 18x^2 - 27$

CONSTRUCTED RESPONSE

19. Express the product of $2x^2 + 7x - 10$ and $x + 5$ in standard form.

20. If the difference $(3x^2 - 2x + 5) - (x^2 + 3x - 2)$ is multiplied by $\frac{1}{2}x^2$, what is the result, written in standard form?

21. Write the expression $5x + 4x^2(2x + 7) - 6x^2 - 9x$ as a polynomial in standard form.

22. If $C = G - 3F$, find the trinomial that represents C when $F = 2x^2 + 6x - 5$ and $G = 3x^2 + 4$.

23. Express $(3x - 4)(x + 7) - \frac{1}{4}x^2$ as a trinomial in standard form.

24. Subtract $3x(x - 2y)$ from $6(x^2 - xy)$ and express your answer as a monomial.

25. Given:
$$A = x + 5$$
$$B = x^2 - 18$$
Express $A^2 + B$ in standard form.

5.4   Divide a Polynomial by a Monomial

In this section, remember the division rule for exponents:
Divide the coefficients and, for common bases, subtract the exponents.

Example: $\dfrac{30x^6y^3z^2}{2x^4y^3z} = \dfrac{30}{2} \cdot x^{6-4} \cdot y^{3-3} \cdot z^{2-1} = 15x^2z$

When **dividing a polynomial by a monomial**, we may be able to simplify by dividing each term of the polynomial in the numerator by the monomial in the denominator. For each new fraction, divide the coefficients and subtract exponents of the same base. The result should have as many terms as the original polynomial.

Example: $\dfrac{15x^2 + 20x}{5x} = \dfrac{15x^2}{5x} + \dfrac{20x}{5x} = 3x + 4$

Note: In Algebra II, we will learn how to divide by a polynomial of more than one term.

MODEL PROBLEM

Divide $21a^2b - 3ab$ by $3ab$.

Solution:

$\quad$ (A) $\qquad\qquad$ (B) $\qquad\qquad$ (C)

$$\frac{21a^2b - 3ab}{3ab} = \frac{21a^2b}{3ab} - \frac{3ab}{3ab} = 7a - 1$$

Explanation of steps:
  (A) Write the division as an algebraic fraction.
  (B) Split the expression into separate fractions, each with one term of the numerator written over the same denominator. Be sure to keep the correct operation between the terms (addition or subtraction).
  (C) If each term of the numerator is divisible by the monomial in the denominator, simplify by dividing. For the variable parts, subtract exponents of the same base. *[$21a^2b \div 3ab = 7a$ and $3ab \div 3ab = 1$.]*

PRACTICE PROBLEMS

1. Divide: $\dfrac{2x + 4}{2}$	2. Write $\dfrac{x^2 + 2x}{x}$ in simplest form.
3. Divide: $\dfrac{14ab + 28b}{14b}$	4. Simplify: $\dfrac{6x^3 + 9x^2 + 3x}{3x}$
5. Simplify: $\dfrac{12x^3 - 6x^2 + 2x}{2x}$	6. What is the quotient when $16x^3 - 12x^2 + 4x$ is divided by $4x$?

7. Divide: $\dfrac{2x^6 - 18x^4 + 2x^2}{2x^2}$	8. Divide: $\dfrac{8x^5 - 2x^4 + 4x^3 - 6x^2}{2x^2}$
9. Simplify: $\dfrac{45a^4b^3 - 90a^3b}{15a^2b}$	10. Divide $24x^2y^6 - 16x^6y^2 + 4xy^2$ by $4xy^2$.

Regents Questions

There are no Regents exam questions on this topic.

CHAPTER 6.   INTRODUCTION TO FUNCTIONS

6.1   Recognize Functions

An **ordered pair** is a pair of two values (*entries*) written in a certain order. The coordinates of a point on a graph are usually represented by an ordered pair in parentheses, with the *x*-coordinate (*abscissa*) written first and the *y*-coordinate (*ordinate*) written second, as in $(3, -2)$.

A **relation** is a set of ordered pairs. It may be graphed as a set of points.

A **function** is a type of relation in which every first entry is mapped to exactly one second entry. In a function, no two ordered pairs can have the same first entries but different second entries. Represented as a graph, a relation is a function only if no points have the same *x*-coordinates but different *y*-coordinates.

Given its graph, we can test whether a relation is a function by using the **vertical line test**. If we can draw a vertical line that intersects the graph at two or more points, then these points have the same *x*-values but different *y*-values, and therefore the relation is not a function. It is a function only if it is *impossible* to draw a vertical line that would intersect the graph at multiple points.

            function   not a function

MODEL PROBLEM 1:   *DETERMINING IF A RELATION IS A FUNCTION*

Determine whether the relation, $\{(1,-1),(0,0),(1,1),(4,2)\}$, is a function or is *not* a function.

Solution:
It is *not* a function.

Explanation of steps:
Look for any two ordered pairs that have the same x-coordinate but different y-coordinates *[(1, −1) and (1,1)]*. A relation is a function only if no such cases can be found.

PRACTICE PROBLEMS

1. Which relation is *not* a function? (1) $\{(1,2),(3,4),(4,5),(5,6)\}$ (2) $\{(3,1),(2,1),(1,2),(3,2)\}$ (3) $\{(4,1),(5,1),(6,1),(7,1)\}$ (4) $\{(0,0),(1,1),(2,2),(3,3)\}$	2. Which relation is *not* a function? (1) $\{(3,-2),(4,-3),(5,-4),(6,-5)\}$ (2) $\{(3,-2),(3,-4),(4,-1),(4,-3)\}$ (3) $\{(3,-2),(5,-2),(4,-2),(-1,-2)\}$ (4) $\{(3,-2),(-2,3),(4,-1),(-1,4)\}$
3. Which relation is a function? (1) $\{(2,1),(3,1),(4,1),(5,1)\}$ (2) $\{(1,2),(1,3),(1,4),(1,5)\}$ (3) $\{(2,3),(3,2),(4,2),(2,4)\}$ (4) $\{(1,6),(2,8),(3,9),(3,12)\}$	4. Which relation is a function? (1) $\{(3,4),(3,5),(3,6),(3,7)\}$ (2) $\{(1,2),(3,4),(4,3),(2,1)\}$ (3) $\{(6,7),(7,8),(8,9),(6,5)\}$ (4) $\{(0,2),(3,4),(0,8),(5,6)\}$

5. The table to the right shows all the ordered pairs (x, y) that define a relation between the variables x and y.

Is y a function of x? Justify your answer.

x	y
−2	3
−1	0
0	−1
1	0
2	3
3	8

MODEL PROBLEM 2: *DETERMINING IF A GRAPH REPRESENTS A FUNCTION*

Determine whether the graph below represents a function.

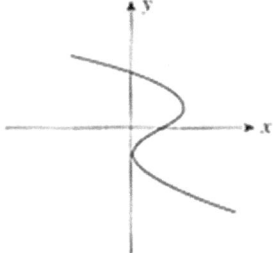

Solution:
The graph is *not* a function.

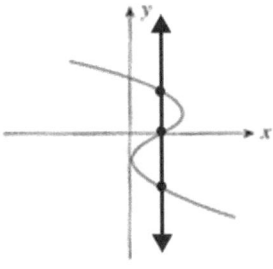

Explanation of steps:
A graph is not a function if we are able to draw any vertical line that intersects the graph at two or more points. *[We can draw the vertical line shown to the right, for just one example. Since it crosses the graph in three points, this cannot be a function.]*

PRACTICE PROBLEMS

6. Which graph represents a function?	7. Which graph represents a function?
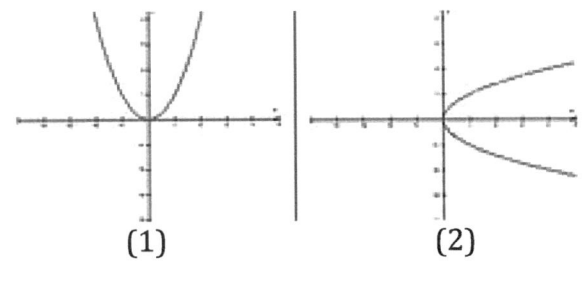 (1) (2)	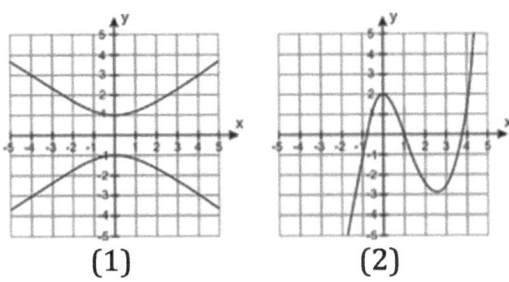 (1) (2)

8. Which graph does *not* represent a function?

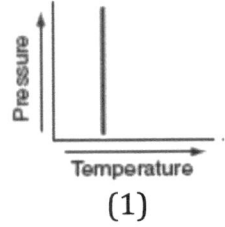   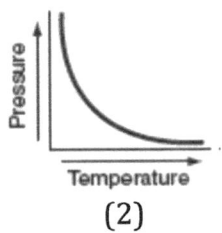   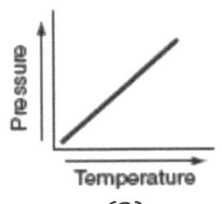   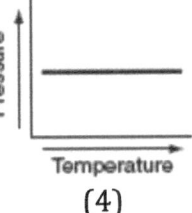

(1)   (2)   (3)   (4)

9.   Which graph represents a function?

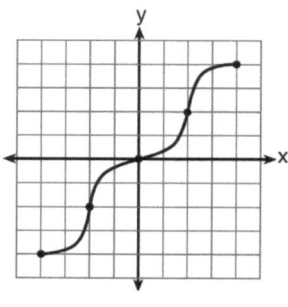

(1)

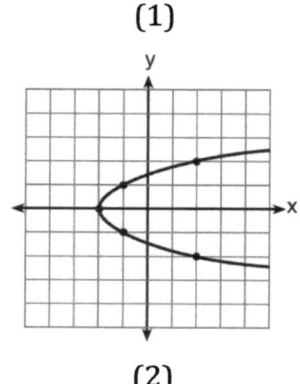

(2)

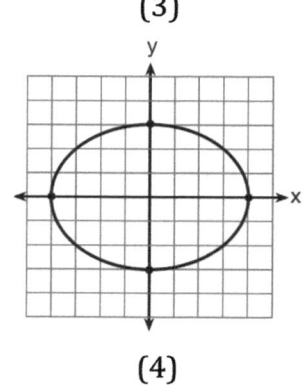

(3)

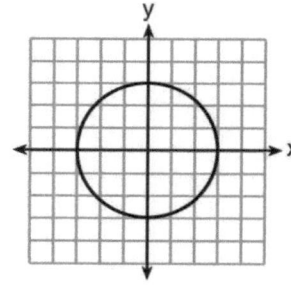

(4)

10. Which graph represents a function?

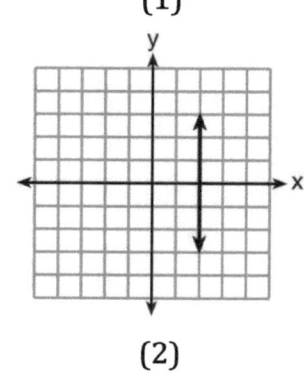

(1)

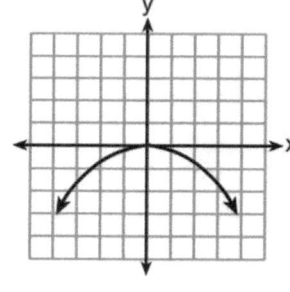

(2)

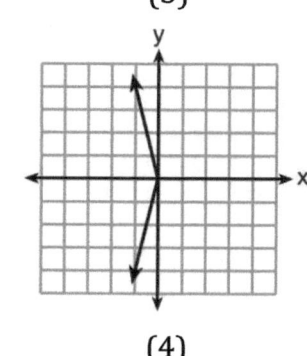

(3)

(4)

Regents Questions

MULTIPLE CHOICE

1. Which table represents a function?

x	2	4	2	4
f(x)	3	5	7	9

(1)

x	3	5	7	9
f(x)	2	4	2	4

(3)

x	0	−1	0	1
f(x)	0	1	−1	0

(2)

x	0	1	−1	0
f(x)	0	−1	0	1

(4)

2. Which representations are functions?

I
x	y
2	6
3	−12
4	7
5	5
2	−6

III

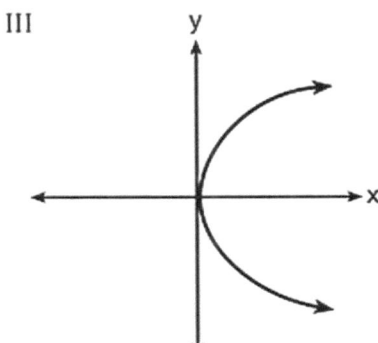

II $\{(1,1), (2,1), (3,2), (4,3), (5,5), (6,8), (7,13)\}$

IV $y = 2x + 1$

(1) I and II

(2) II and IV

(3) III, only

(4) IV, only

3.　A mapping is shown in the diagram below.

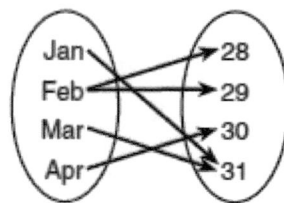

This mapping is
- (1) a function, because Feb has two outputs, 28 and 29
- (2) a function, because two inputs, Jan and Mar, result in the output 31
- (3) not a function, because Feb has two outputs, 28 and 29
- (4) not a function, because two inputs, Jan and Mar, result in the output 31

4.　A relation is graphed on the set of axes below.

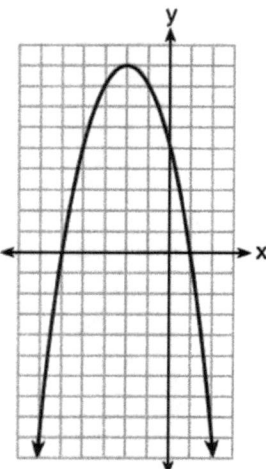

Based on this graph, the relation is
- (1) a function because it passes the horizontal line test
- (2) a function because it passes the vertical line test
- (3) not a function because it fails the horizontal line test
- (4) not a function because it fails the vertical line test

5.　A function is defined as $\{(0,1), (2,3), (5,8), (7,2)\}$. Isaac is asked to create one more ordered pair for the function. Which ordered pair can he add to the set to keep it a function?
- (1) (0,2)
- (3) (7,0)
- (2) (5,3)
- (4) (1,3)

6.   Which relation does *not* represent a function?

x	1	2	3	4	5	6
y	3.2	4	5.1	6	7.4	8.8

(1)

$$y = 3\sqrt{x+1} - 2$$

(3)

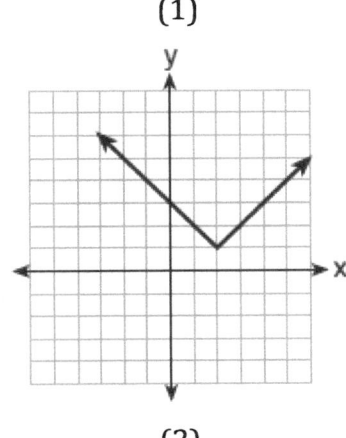

(2)

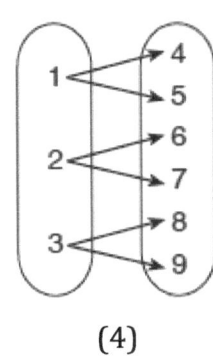

(4)

7.   Which relation is *not* a function?

x	y
−10	−2
−6	2
−2	6
1	9
5	13

(1)

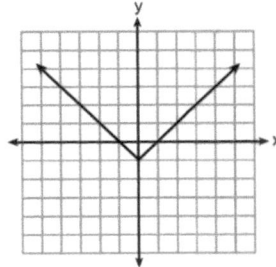

(3)

$$3x + 2y = 4$$

(2)

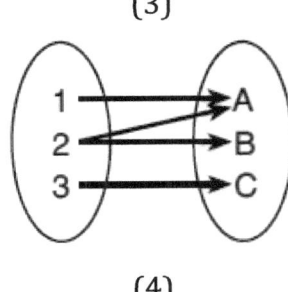

(4)

8.   Which table represents a function?

x	y
2	−3
3	0
4	−3
2	1

(1)

x	y
−3	0
−2	1
−3	2
2	3

(3)

x	y
1	2
1	3
1	4
1	5

(2)

x	y
−2	−4
0	2
2	4
4	6

(4)

9.   Which table could represent a function?

x	f(x)
1	4
2	2
3	4
2	6

(1)

x	h(x)
2	6
0	4
1	6
2	2

(3)

x	g(x)
1	2
2	4
3	6
4	2

(2)

x	k(x)
2	2
3	2
4	6
3	6

(4)

10. Which relation is *not* a function?

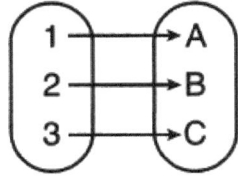

(1)

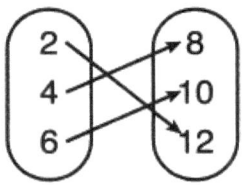

(3)

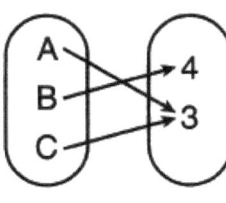

(2)

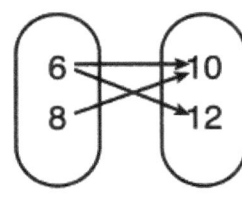

(4)

11. Given the relation $R = \{(-4,2), (3,6), (x, 8), (-1,4)\}$
     Which value of x would make this relation a function?

    (1) −4   (3) 3

    (2) −1   (4) 0

12. Which relation is a function?

$\{(1,3), (2,1), (3,1), (4,7)\}$

(1)

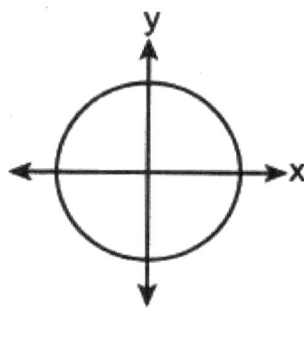

(3)

Input	Output
−6	−2
−4	2
7	3
7	5

(2)

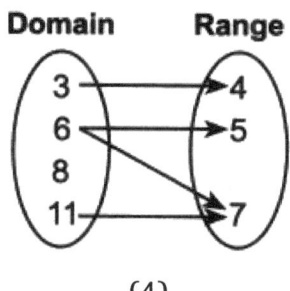

(4)

13. Given the relation: $\{(0,4), (2,6), (4,8), (x, 7)\}$
     Which value of x will make this relation a function?

    (1) 0   (3) 6

    (2) 2   (4) 4

CONSTRUCTED RESPONSE

14.   A function is shown in the table below.

x	$f(x)$
-4	2
-1	-4
0	-2
3	16

If included in the table, which ordered pair, $(-4,1)$ or $(1,-4)$, would result in a relation that is no longer a function? Explain your answer.

15.   Marcel claims that the graph below represents a function.

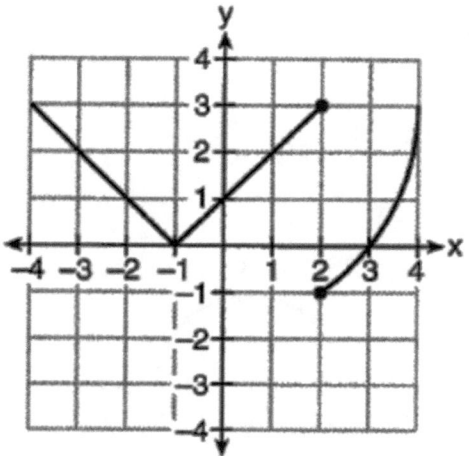

State whether Marcel is correct. Justify your answer.

16.   Nora says that the graph of a circle is a function because she can trace the whole graph without picking up her pencil. Mia says that a circle graph is *not* a function because multiple values of x map to the same y-value. Determine if either one is correct, and justify your answer completely.

17.   Four relations are shown below.

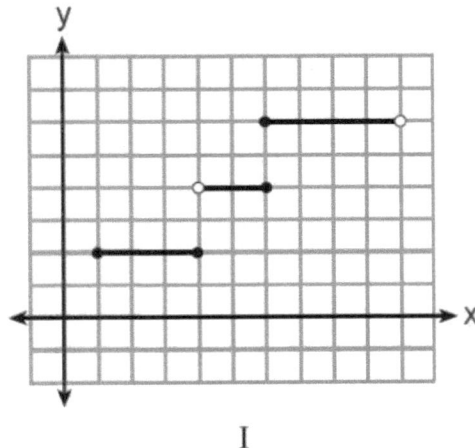

x	y
−4	1
0	3
4	5
6	6

I   III

$\{(1,2),(2,5),(3,8),(2,-5),(1,-2)\}$ $y = x^2$

II   IV

State which relation(s) are functions. Explain why the other relation(s) are *not* functions.

6.2 Function Graphs

As we have seen, a function may be represented as a set of ordered pairs (x, y) such that each x is paired with a unique y. Often, the relationship between each x and its corresponding y can be represented by an algebraic expression.

Example: Considering the function represented by the expression $3x + 5$, for each x-value, the corresponding y-value is 5 more than 3 times the x-value.
Note that for each x-value, there is only one possible y-value.

A function is named using a letter (often f) followed by the expression's independent variable (often x) in parentheses, as in $f(x)$. This is read as "f of x". We could **define** the function by writing that it is equal to an expression in terms of the independent variable.

Examples: $f(x) = 3x + 5$ $A(n) = n^2 - 2n + 1$

When a function $f(x)$ is graphed on a coordinate graph, the y-values are the values produced by the function. For this reason, the y-axis is often labelled $f(x)$.

Example: $f(x) = 3x + 5$ is graphed as the line $y = 3x + 5$. So, $f(x) = y$.

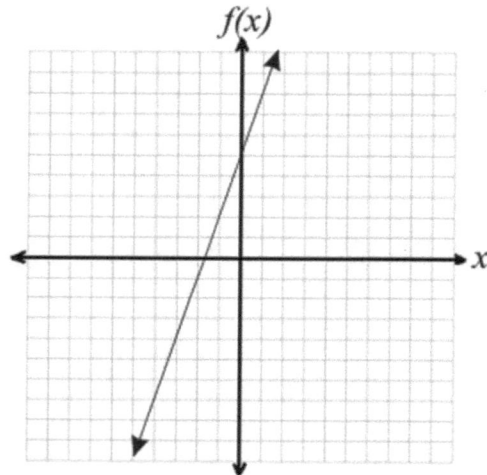

Linear Function
$$y = mx + b$$
$$f(x) = mx + b$$

A function's value, $f(x)$, for a given value of x can be determined by finding the y-value of a point (x, y) for the given x on a graph of the function.

Example: For the function $f(x) = 3x + 5$ shown above, $f(-1) = 2$, because $(-1, 2)$ is a point on the line.

As we have seen, you can **determine whether a point is in the solution set of an equation** by substituting the x and y coordinates for the variables x and y in the equation and then checking if these values make the equation true. When the equation is written as a function definition, we can simply replace $f(x)$ with y to give us an equation in x and y.

Example: $(-3, 1)$ is a point that lies on the graph of the function $f(x) = x^2 + 3x + 1$. We can check this by substituting -3 for x and 1 for y in the equation $y = x^2 + 3x + 1$, giving us $1 = (-3)^2 + 3(-3) + 1$, which is true.

MODEL PROBLEM

Below is a graph of $f(x)$. *[Assume the points graphed with closed circles have integer coordinates and the lines between them are straight.]*

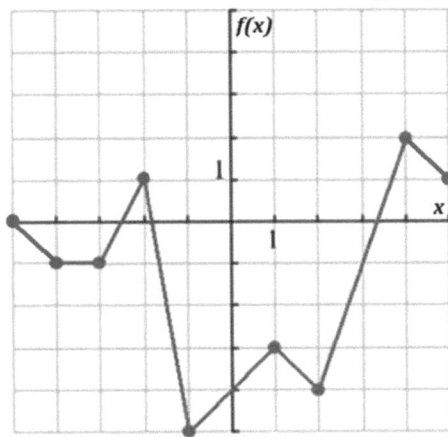

(A) What is $f(2)$?
(B) What is $f(0)$?
(C) For what value of x does $f(x) = -5$?
(D) Which of the following does *not* equal -1?
  (1) $f(-4)$ (2) $f(-3)$ (3) $f(3)$ (4) $f(4)$

Solution:
(A) -4
(B) -4
(C) $x = -1$
(D) (4)

Explanation of steps:
(A) $f(2) = -4$ because $(2, -4)$ is a point on the graph of $f(x)$.
(B) $f(0) = -4$ because the graph of $f(x)$ crosses the y-axis at the point $(0, -4)$.
(C) $f(-1) = -5$ because $(-1, -5)$ is a point on the graph of $f(x)$.
(D) $f(4) = 2$ because $(4, 2)$ is a point on the graph of $f(x)$.

PRACTICE PROBLEMS

1. Complete the table based on the graph of the function $f(x)$ below.

x	$f(x)$
0	
1	
2	
3	
4	
5	

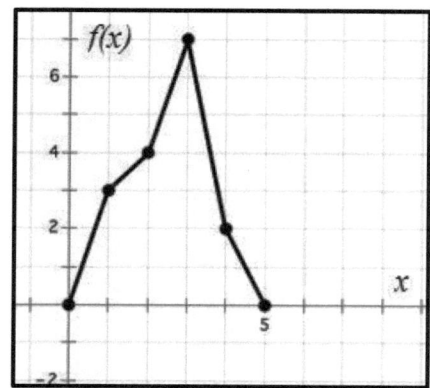

2. The function $f(x)$ is graphed below.

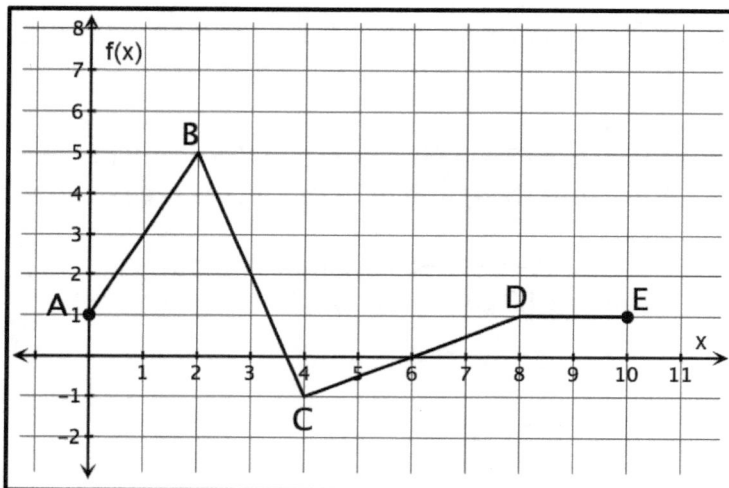

a) What is $f(9)$?

b) For what values of x does $f(x) = 2$?

Regents Questions

MULTIPLE CHOICE

1. The graph of $y = f(x)$ is shown below.

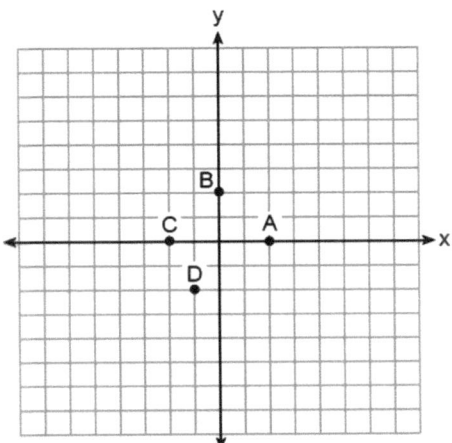

Which point could be used to find $f(2)$?

(1) A

(3) C

(2) B

(4) D

2. The graph of $f(x)$ is shown below.

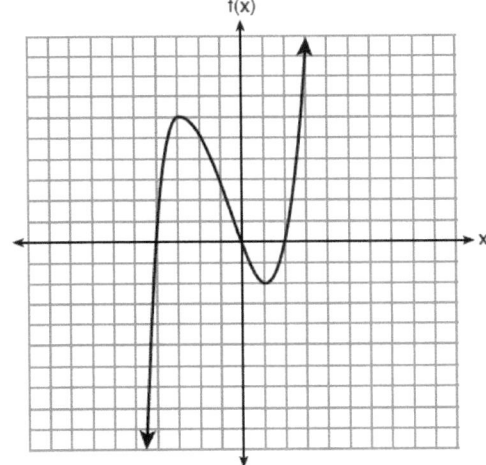

What is the value of $f(-3)$?

(1) 6

(3) -2

(2) 2

(4) -4

3. Which point is *not* in the solution set of the equation $3y + 2 = x^2 - 5x + 17$?

(1) $(-2,10)$

(3) $(2,3)$

(2) $(-1,7)$

(4) $(5,5)$

4. Given: $f(x) = \frac{2}{3}x - 4$ and $g(x) = \frac{1}{4}x + 1$

Four statements about this system are written below.

I. $f(4) = g(4)$
II. When $x = 12$, $f(x) = g(x)$.
III. The graphs of $f(x)$ and $g(x)$ intersect at $(12, 4)$.
IV. The graphs of $f(x)$ and $g(x)$ intersect at $(4, 12)$.

Which statement(s) are true?

(1) II, only   (3) I and IV

(2) IV, only   (4) II and III

6.3   Evaluate Functions

We saw in the previous section that we can find the value of $f(x)$ for a given x by looking at a graph of the function. More frequently, however, we will use the function's definition. We can find $f(x)$ for a given value of x by **substituting and evaluating** the expression.

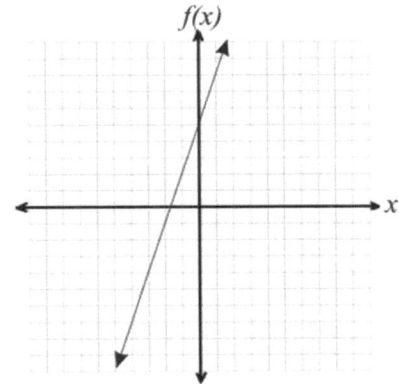

Example: For the function $f(x) = 3x + 5$ shown to the right, $f(-1) = 2$,
(a) because $(-1,2)$ is a point on the line, and also
(b) because $f(x) = 3x + 5 = 3(-1) + 5 = 2$.

 CALCULATOR TIP

We can also evaluate a function on the calculator:

1.   Press Y= and type the equation of the function as Y1. Then, press 2nd QUIT.
2.   On the TI-84 models, press ALPHA F4 1 to select Y1. On the TI-83 models, press VARS <Y-VARS> 1 1 instead.
3.   Enter the given value and press) ENTER.

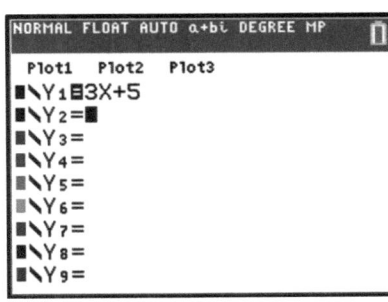 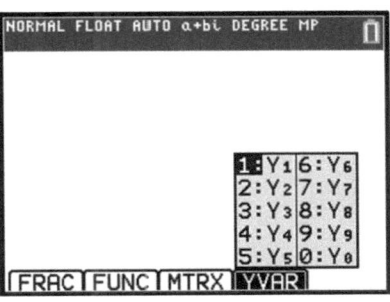 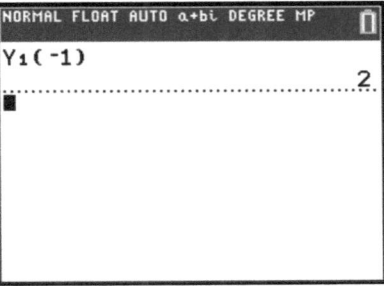

The **y-intercept** of a function $f(x)$ is the value of $f(0)$.
Example: For the function $f(x) = 3x + 5$ shown above, $f(0) = 3(0) + 5 = 5$, so the y-intercept is 5.

We can also **evaluate a function at a given expression** by replacing the variable in the original function with the expression, using parentheses to avoid errors.
Example: If $f(x) = 3x + 5$, find $f(n + 1)$.
$$f(n + 1) = 3(n + 1) + 5$$
$$= 3n + 3 + 5$$
$$= 3n + 8$$

MODEL PROBLEM

Given $f(x) = 3x^2 - 2x + 5$, find $f(1) + f(2)$.

Solution:
$f(1) = 3(1)^2 - 2(1) + 5 = 6$ and $f(2) = 3(2)^2 - 2(2) + 5 = 13$, so $f(1) + f(2) = 6 + 13 = 19$

Explanation of steps:
To evaluate a function of x for a given value, substitute the value for x in the expression. *[Find $f(1)$ by substituting 1 for x in $3x^2 - 2x + 5$ and find $f(2)$ by substituting 2 for x in $3x^2 - 2x + 5$.]*

PRACTICE PROBLEMS

1. Find $f(3)$ given $f(x) = -2x^2 - 3x - 6$.	2. If $f(a) = a^2 - 2a + 1$, find $f(-3)$.
3. If $f(x) = (x - 3)^2$, find $f(0)$.	4. $f(m) = 0.5^m$. Evaluate the function for $m = 2$.
5. If $f(x) = 3x - 4$ and $g(x) = x^2$, find the value of $f(3) - g(2)$.	6. If $h(x) = 2x - 1$, find the product $h(0) \cdot h(-2)$.

7. For what integer value of x is $f(x) = -10$ if $f(x) = -4x + 2$?

8. If $f(x) = kx^2$, and $f(2) = 12$, then what is the value of k?

9. If $g(x) = 2x^2 + 6x - 3$, find $g(4a)$ in terms of a.

10. Find $f(a + 2)$ in terms of a, given $f(x) = x^2 + 2x - 1$.

11. $P(t) = 0.0089t^2 + 1.1149t + 78.4491$ models the U. S. population, P, in millions since 1900. If t represents the number of years after 1900, then what is the estimated population in 2025 to the *nearest tenth of a million*?

Regents Questions

MULTIPLE CHOICE

1.   If $f(x) = 3^x$ and $g(x) = 2x + 5$, at which value of x is $f(x) < g(x)$?
   (1) −1   (3) −3
   (2) 2   (4) 4

2.   If $f(x) = \dfrac{\sqrt{2x+3}}{6x-5}$, then $f\left(\frac{1}{2}\right) =$
   (1) 1   (3) −1
   (2) −2   (4) $-\frac{13}{3}$

3.   If $f(n) = (n − 1)^2 + 3n$, which statement is true?
   (1) $f(3) = −2$ (3) $f(−2) = −15$
   (2) $f(−2) = 3$ (4) $f(−15) = −2$

4.   Which function has the greatest y-intercept?
   (1) $f(x) = 3x$
   (2) $2x + 3y = 12$
   (3) the line that has a slope of 2 and passes through $(1, −4)$
   (4)

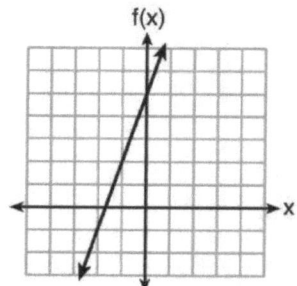

5.   Faith wants to use the formula $C(f) = \frac{5}{9}(f − 32)$ to convert degrees Fahrenheit, f, to degrees Celsius, $C(f)$. If Faith calculated $C(68)$, what would her result be?
   (1) 20° Celsius   (3) 154° Celsius
   (2) 20° Fahrenheit   (4) 154° Fahrenheit

6.   Lynn, Jude, and Anne were given the function $f(x) = −2x^2 + 32$, and they were asked to find $f(3)$. Lynn's answer was 14, Jude's answer was 4, and Anne's answer was ±4. Who is correct?
   (1) Lynn, only   (3) Anne, only
   (2) Jude, only   (4) Both Lynn and Jude

7.   If $f(x) = \frac{1}{2}x^2 - \left(\frac{1}{4}x + 3\right)$, what is the value of $f(8)$?

    (1)   11   (3)   27

    (2)   17   (4)   33

8.   If $k(x) = 2x^2 - 3\sqrt{x}$, then $k(9)$ is

    (1)   315   (3)   159

    (2)   307   (4)   153

9.   Which ordered pair below is *not* a solution to $f(x) = x^2 - 3x + 4$?

    (1)   (0,4)   (3)   (5,14)

    (2)   (1.5,1.75)   (4)   (−1,6)

10.   Three functions are shown below.

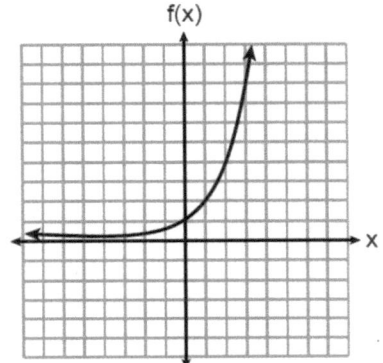

f(x)

x	h(x)
−5	30
−4	14
−3	6
−2	2
−1	0
0	−1
1	−1.5
2	−1.75

$$g(x) = 3^x + 2$$

Which statement is true?

    (1)   The y-intercept for $h(x)$ is greater than the y-intercept for $f(x)$.

    (2)   The y-intercept for $f(x)$ is greater than the y-intercept for $g(x)$.

    (3)   The y-intercept for $h(x)$ is greater than the y-intercept for both $g(x)$ and $f(x)$.

    (4)   The y-intercept for $g(x)$ is greater than the y-intercept for both $f(x)$ and $h(x)$.

11.   The function $g(x)$ is defined as $g(x) = -2x^2 + 3x$. The value of $g(-3)$ is

    (1)   −27   (3)   27

    (2)   −9   (4)   45

12.   The functions $f(x)$, $q(x)$, and $p(x)$ are shown below.

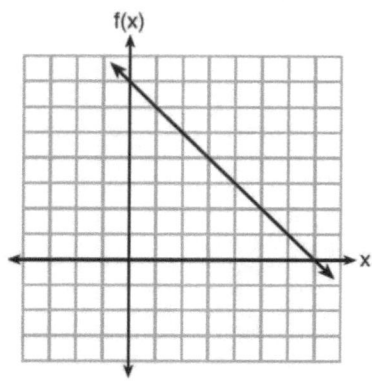

x	p(x)
2	5
3	4
4	3
5	4
6	5

$$q(x) = (x - 1)^2 - 6$$

When the input is 4, which functions have the same output value?
(1) $f(x)$ and $q(x)$ only   (3) $q(x)$ and $p(x)$, only
(2) $f(x)$ and $p(x)$, only   (4) $f(x)$, $q(x)$, and $p(x)$

13.   If $f(x) = 4x + 5$, what is the value of $f(-3)$?
(1) -2 (3) 17
(2) -7 (4) 4

14.   If $f(x) = 2(3^x) + 1$, what is the value of $f(2)$?
(1) 13   (3) 37
(2) 19   (4) 54

15.   A function is defined as $K(x) = 2x^2 - 5x + 3$. The value of $K(-3)$ is
(1) 54   (3) 0
(2) 36   (4) -18

16.   Which function has the *smallest y*-intercept?

$$g(x) = 2x - 6$$ $$f(x) = \sqrt{x} - 2$$
(1)   (3)

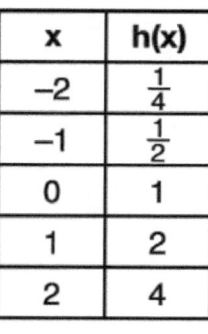

x	h(x)
-2	$\frac{1}{4}$
-1	$\frac{1}{2}$
0	1
1	2
2	4

(2)

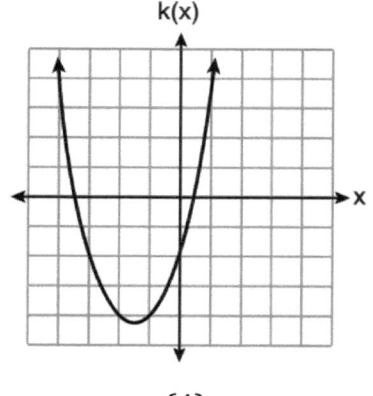

(4)

17. Given $f(x) = 3x - 5$, which statement is true?
    - (1) $f(0) = 0$
    - (2) $f(3) = 4$
    - (3) $f(4) = 3$
    - (4) $f(5) = 0$

18. If $f(x) = \dfrac{3x + 4}{2}$, then $f(8)$ is
    - (1) 21
    - (2) 16
    - (3) 14
    - (4) 4

19. Given $f(x) = -3x^2 + 10$, what is the value of $f(-2)$?
    - (1) −26
    - (2) −2
    - (3) 22
    - (4) 46

20. If $g(x) = -x^2 - x + 5$, then $g(-4)$ is equal to
    - (1) −15
    - (2) −7
    - (3) 17
    - (4) 25

21. For which function is the value of the *y*-intercept the *smallest*?

x	f(x)
−4	5
−2	4
0	3
2	2
4	1

(1)

x	h(x)
−1	3
0	2
1	3
2	6
3	11

(3)

$g(x) = |x| + 4$

(2)

$k(x) = 5^x$

(4)

22. If $f(x) = x^2 + 3x$, then which statement is true?
    - (1) $f(1) = f(-1)$
    - (2) $f(2) = f(-2)$
    - (3) $f(1) = f(2)$
    - (4) $f(-1) = f(-2)$

23. If $f(x) = x^2 + 2x + 1$ and $g(x) = 3x + 5$, then what is the value of $f(1) - g(3)$?
    - (1) 10
    - (2) 8
    - (3) −10
    - (4) −8

CONSTRUCTED RESPONSE

24. Jacob and Jessica are studying the spread of dandelions. Jacob discovers that the growth over t weeks can be defined by the function $f(t) = (8) \cdot 2^t$. Jessica finds that the growth function over t weeks is $g(t) = 2^{t+3}$.

    Calculate the number of dandelions that Jacob and Jessica will each have after 5 weeks.

    Based on the growth from both functions, explain the relationship between $f(t)$ and $g(t)$.

25. If $g(x) = -4x^2 - 3x + 2$, determine $g(-2)$.

6.4   Features of Function Graphs

When looking at the behavior of a function on a graph, it is helpful to identify key features of the graph. We should recognize where the function is positive or negative, increasing or decreasing, and its end behavior. We should also recognize its extrema.

Example: For the discussion to follow, we will refer to the graph below.

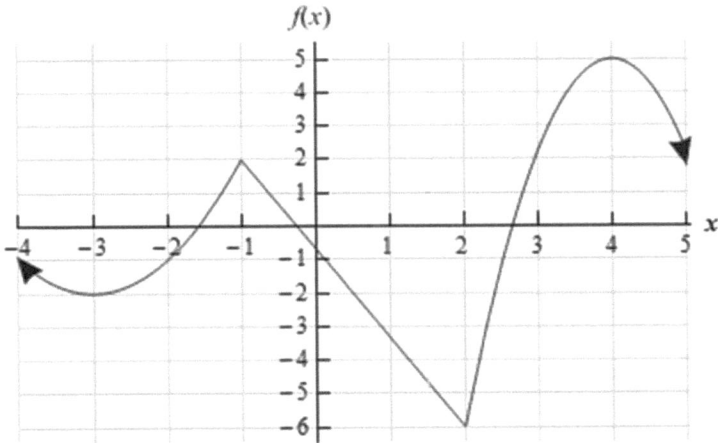

A function is **positive** where its graph lies above the *x*-axis, and **negative** where its graph lies below the *x*-axis. It is **increasing** where the graph goes up, when moving from left to right, and **decreasing** where it goes down. Its **end behavior** describes the function at the arrowheads; that is, at the leftmost or rightmost extremes of the graph.

Example:

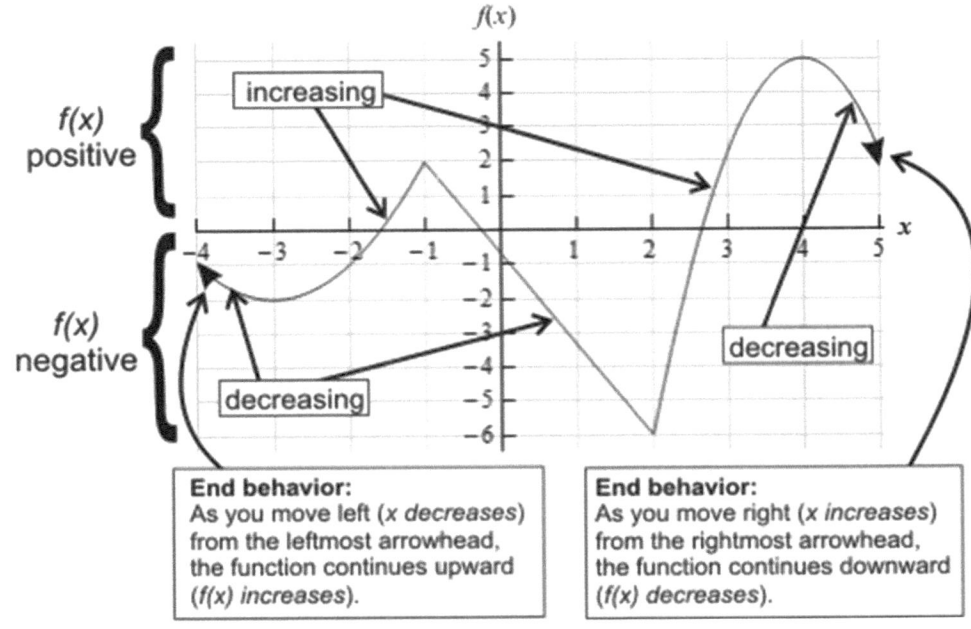

Some points on the graph of a function can be described as a relative maximum (plural, *maxima*) or a relative minimum (plural, *minima*). A **relative maximum** is a point where no other nearby points have a greater function value (*y*-coordinate), and a **relative minimum** is a point where no other nearby points have a lesser function value. These points are also called **extrema**. The function is *neither increasing nor decreasing* at the extrema.

Example:

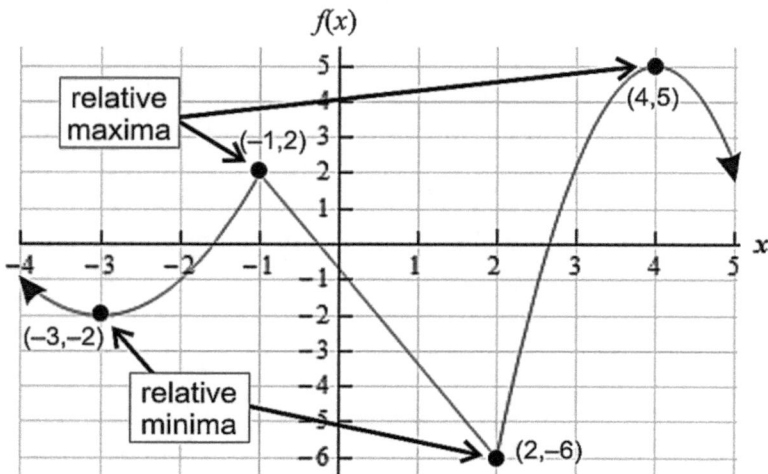

MODEL PROBLEM

Describe the features of the function graph below.

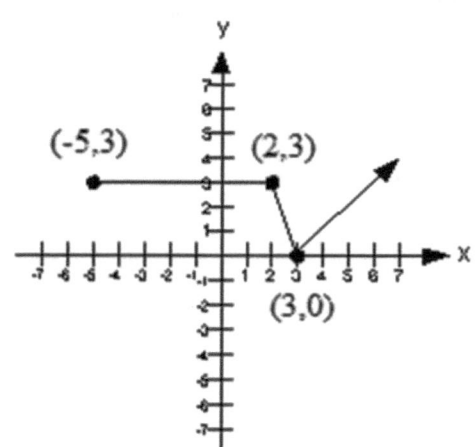

Solution:

The function is constant in the interval $-5 < x < 2$, decreasing where $2 < x < 3$, and increasing at $x > 3$. There is a relative minimum at $(3,0)$. There are no relative maxima.

PRACTICE PROBLEMS

1. The function $f(x)$ is graphed below.

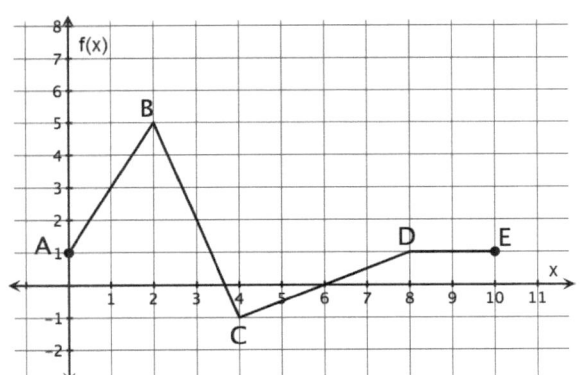

a) Over which interval(s) is $f(x)$ increasing?

b) Over which interval(s) is $f(x)$ decreasing?

c) Over which interval(s) is $f(x)$ constant?

2. For the graph of the function below,

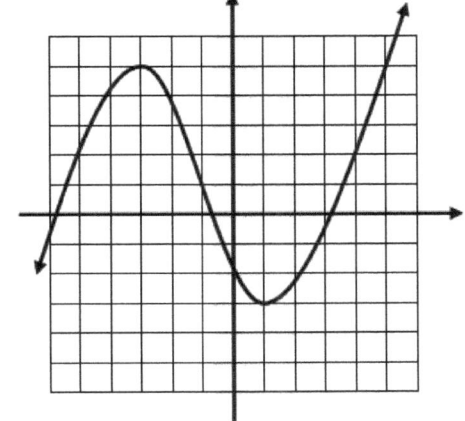

a) which point is a relative maximum?

b) which point is a relative minimum?

Note: you may assume integer coordinates.

3. For the graph of the function below,

   a) state all the intervals where the function is positive or negative
   b) state all the intervals where the function is increasing, decreasing, or constant
   c) state the coordinates of any relative maxima or minima

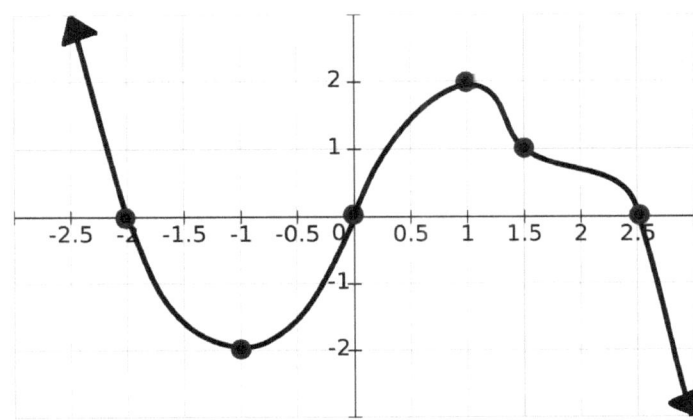

Regents Questions

MULTIPLE CHOICE

1. A ball is thrown into the air from the edge of a 48-foot-high cliff so that it eventually lands on the ground. The graph below shows the height, y, of the ball from the ground after x seconds.

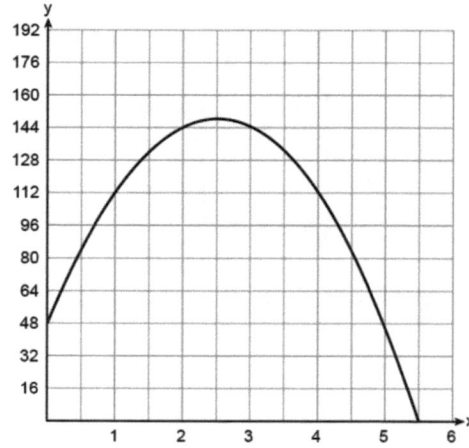

For which interval is the ball's height always *decreasing*?

    (1) $0 \leq x \leq 2.5$ (3) $2.5 < x < 5.5$

    (2) $0 < x < 5.5$ (4) $x \geq 2$

2. A graph of average resting heart rates is shown below. The average resting heart rate for adults is 72 beats per minute, but doctors consider resting rates from 60–100 beats per minute within normal range.

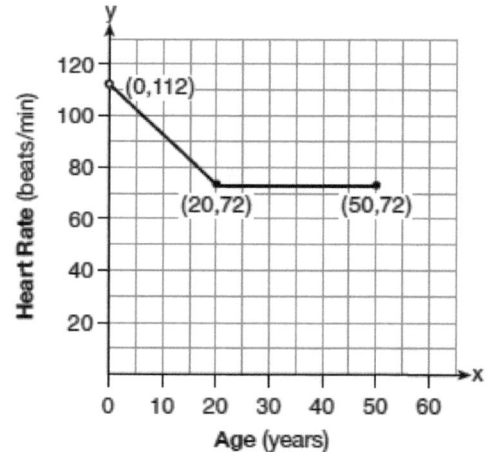

Which statement about average resting heart rates is *not* supported by the graph?

    (1)   A 10-year-old has the same average resting heart rate as a 20-year-old.

    (2)   A 20-year-old has the same average resting heart rate as a 30-year-old.

    (3)   A 40-year-old may have the same average resting heart rate for ten years.

    (4)   The average resting heart rate for teenagers steadily decreases.

3.   To keep track of his profits, the owner of a carnival booth decided to model his ticket sales on a graph. He found that his profits only declined when he sold between 10 and 40 tickets. Which graph could represent his profits?

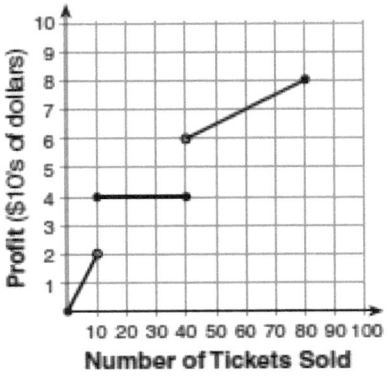

(1)

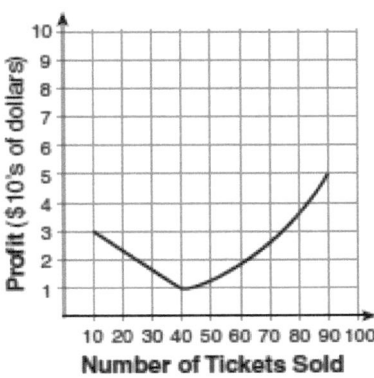

(3)

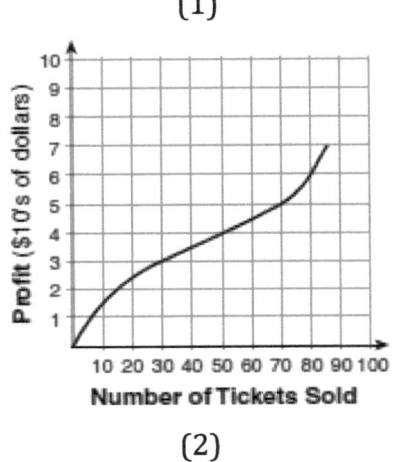

(2)

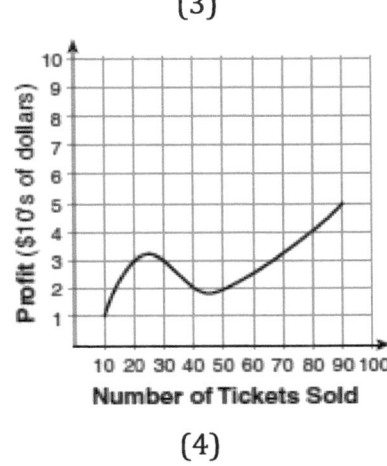

(4)

4. Which graph does *not* represent a function that is always increasing over the entire interval $-2 < x < 2$?

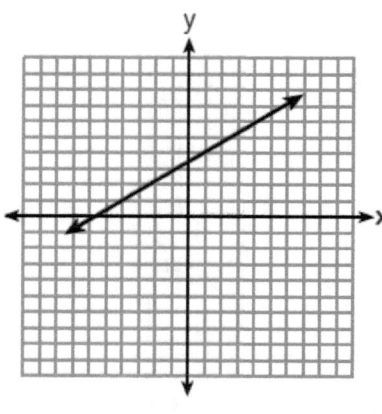

(1)

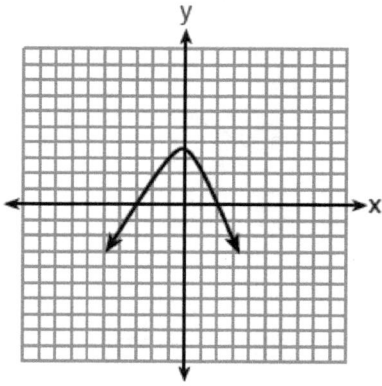

(3)

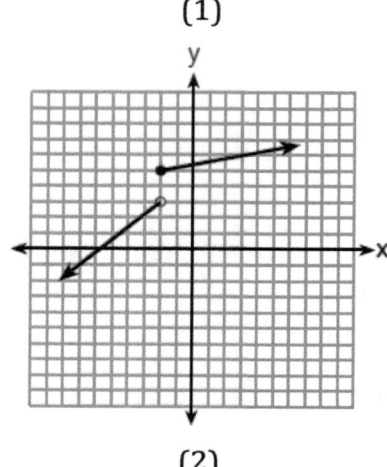

(2)

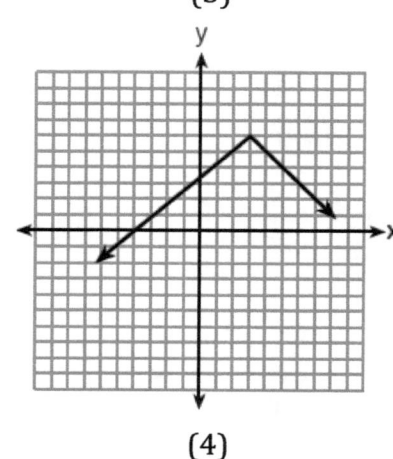

(4)

6.5   Domain and Range

For a function, the set of possible *x*-values is called the **domain**, and the set of *y*-values that are produced by the function is called the **range**.

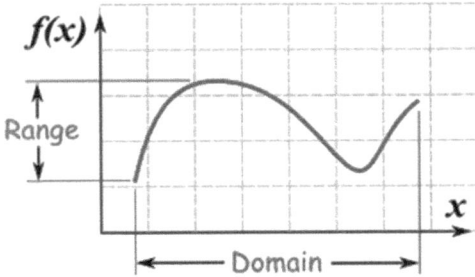

The function is like a machine in that for any given **input** value from the *domain*, the function produces a unique **output** value in the *range*.

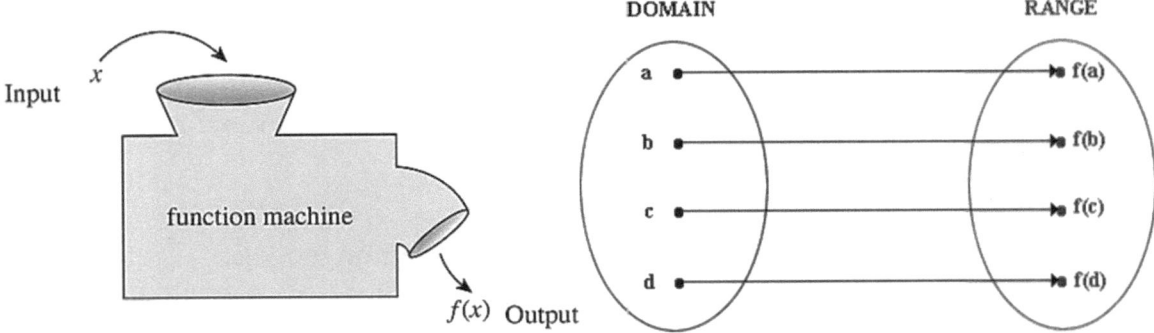

A **discrete domain** is a set of input values that consist of only certain numbers in an interval, whereas a **continuous domain** consists of all numbers in an interval. Likewise, the range of a function may be either discrete or continuous.

Example: A domain of *integers* from 1 to 5 would be *discrete*.
                   A domain of *real numbers* from 1 to 5 would be *continuous*.

Interval notation uses parentheses ( ) and/or brackets [ ] to name the endpoints (lower and upper bounds) of a set of all real numbers between those endpoints. These correspond to the open and closed circles in the graph of an inequality on a number line.

- A parenthesis represents an "open" endpoint (> or <; not included in the set)
- A bracket represents a "closed" endpoint ($\geq$ or $\leq$; included in the set)

Example: $(-1,5]$ represents all real numbers *x* such that $-1 < x \leq 5$.

Set-builder notation uses braces { } and a vertical bar | to define a set by the properties that its members must satisfy. It will often start with "{x|" which is read as "the set of all x such that."
Example: {3, 4, 5} can be written as {x| 3 ≤ x ≤ 5, where x is a whole number}
We can express a continuous domain or range as a compound inequality, or by using interval notation or set-builder notation.
Example: A domain of real numbers from 3 to 7, including 3 but excluding 7, could be written as $3 \le x < 7$, or as $[3, 7)$, or as $\{x \mid x \ge 3 \text{ and } x < 7\}$.

If the domain or range extends *infinitely* in the negative or positive direction, we may use the symbols $-\infty$ or ∞, respectively.
Example: If $f(x) = x$ is defined for the domain of all real numbers $-\infty < x < \infty$, then the range would also include all real numbers, $-\infty < y < \infty$.

When $-\infty$ or ∞ is used in interval notation, always use parentheses (not brackets) on that end.
Examples: $[3, \infty)$ for $x \ge 3$ $(-\infty, 2)$ for $x < 2$ $(-\infty, \infty)$ for all real numbers

Restrictions on the domain:
If the domain is not specified, it is assumed to be the set of real numbers. However, the domain may be restricted to only certain intervals of the real numbers such as $0 \le x < \infty$ or $-2 \le x \le 2$, or to *discrete* sets such as the set of integers, or even to *finite* sets such as {-2, -1, 0, 1, 2}. Or the domain may simply exclude certain values of x; for example, the domain for the function $f(x) = \frac{1}{x}$ for $x \ne 0$ excludes 0 from the domain because $\frac{1}{0}$ is undefined.

Restrictions on the domain may derive from the situation that the function models.
Examples: (a) If x represents the length of a side of a triangle, then $0 < x < \infty$ would be an appropriate domain for $f(x)$.
                (b) If x represents a number of people, then *the set of whole numbers* would be an appropriate domain for $f(x)$.
                (c) If x is the result of rolling a six-sided die, then the domain {1, 2, 3, 4, 5, 6} would be appropriate for $f(x)$.

The properties of the set of real numbers may also dictate restrictions on the domain.
Example: If $f(x) = \frac{\sqrt{x-2}}{x-10}$, then we need to restrict the domain in two ways. To avoid a square root of a negative number in the numerator, we need $x - 2 \ge 0$, or $x \ge 2$. Also, to avoid division by zero, we need $x - 10 \ne 0$, or $x \ne 10$. So, our domain for f can be written as $\{x \mid x \ge 2, x \ne 10\}$.

On a graph, and **open circle** can be used to show an endpoint that is not included. Just as we saw when graphing inequalities, an **open circle** means the point is *excluded*, but a **closed circle** means the point is *included*.

Example: The linear graph below shows the function $f(x) = 2x - 1$ restricted to the domain $-2 \le x < 4$. Note the open circle shows that the graph does not include $f(4)$.

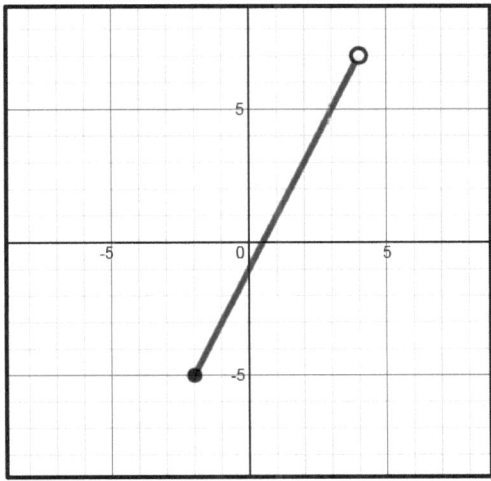

Determining the range:

Not all functions map to all real values of y.

Example: Suppose $f(x) = |x|$ (the absolute value of x) for the domain of all real numbers. The range cannot include negative numbers, so the range is $y \ge 0$.

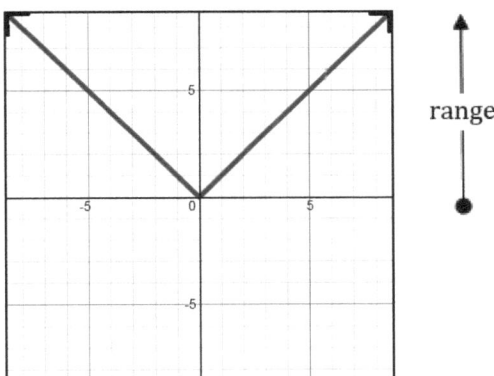

For a simple linear function with a restricted domain, we can find the range by finding the y-values (ie, evaluating the function) at the *endpoints* of the line segment.

Example: If $f(x) = 2x - 1$ is defined on the domain $-2 \leq x < 4$, we can find the range by finding $f(-2) = 2(-2) - 1 = -5$ and $f(4) = 2(4) - 1 = 7$. The range will include all y-values between –5 and 7, but *excluding* 7, which is written as $-5 \leq y < 7$.

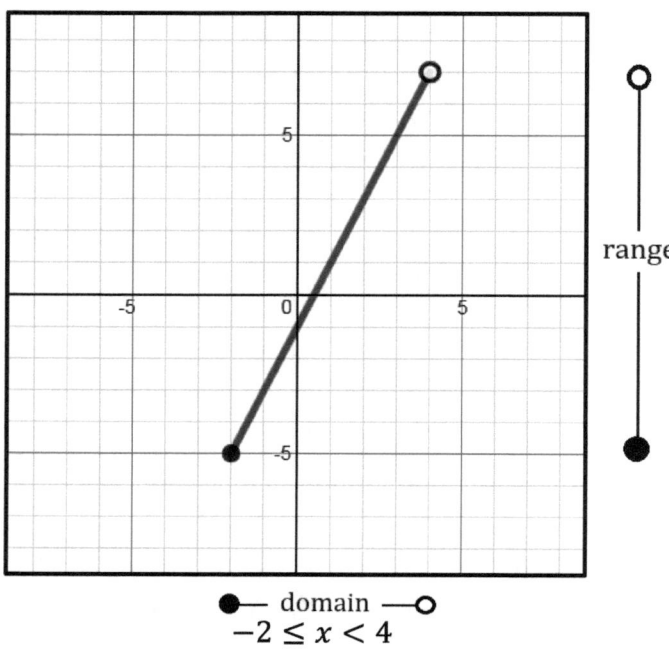

●— domain —○
$-2 \leq x < 4$

This method for determining the range works for linear functions, but *other types of functions* may require more work.

Example: Here's the graph of the quadratic function $f(x) = x^2 - 5$ defined on the domain $-2 \leq x \leq 3$. We cannot assume the range includes only those values between $f(-2) = -1$ and $f(3) = 4$, or $-1 \leq y \leq 4$. As we can see by the graph, the range includes y-values as low as -5, so the actual range is $-5 \leq y \leq 4$.

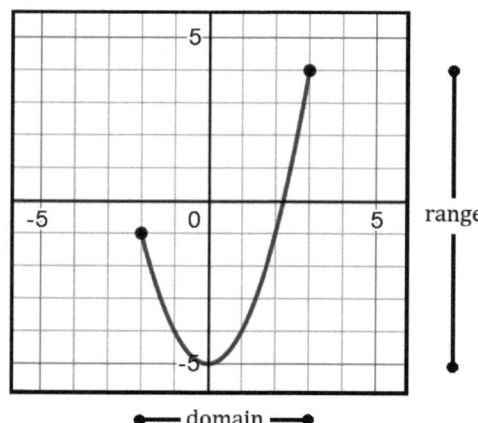

●— domain —●

Note: In an upcoming unit on quadratic functions, you'll learn how to find the minimum or maximum value in the range, without having to graph, by finding the y-value of the vertex.

MODEL PROBLEM

The graph below represents the function $f(x)$ on the domain, $-5 \leq x < 5$. Based on the graph, describe the range.

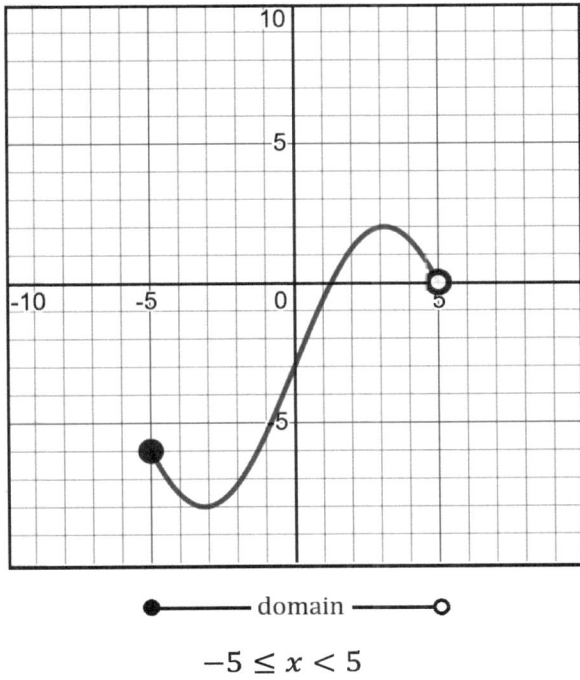

domain

$$-5 \leq x < 5$$

Solution:
The range is approximately $-8 \leq y \leq 2$.

Explanation of steps:
The range of this continuous function extends from the lowest point (minimum, or least value of y) to the highest point (maximum, or greatest value of y). Without knowing the equation of the function, we can only approximate these values from the graph.
[The graph appears to go as low as –8 at $(-3, -8)$ and as high as 2 at $(3,2)$.]

PRACTICE PROBLEMS

1. State the range of the following function. $\{(-1,2),(1,3),(2,51),(8,22),(9,51)\}$	2. What is an appropriate domain of the function $f(x) = \frac{1}{x}$?
3. State the domain and range of the function shown in the graph. 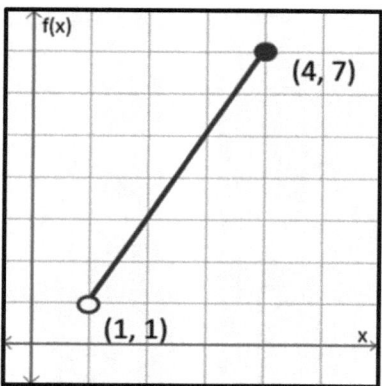	4. State the domain and range of the function shown in the graph. 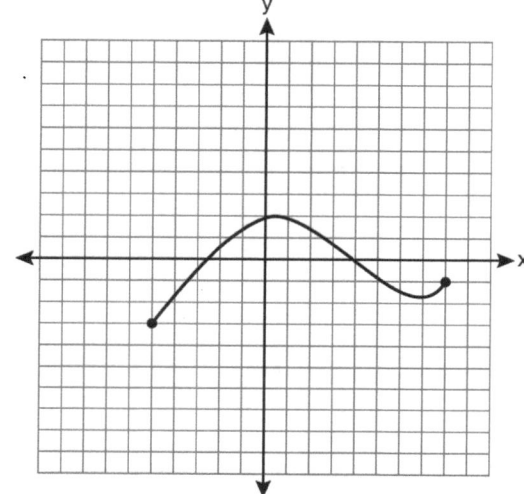

5. The graph below shows the effect of pH on the action of a certain enzyme.

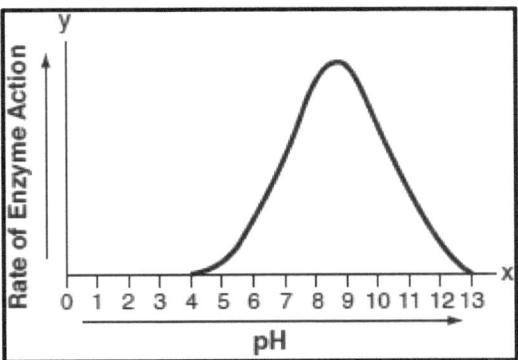

What is the approximate domain of this function?

6. Data collected from an experiment are shown in the graph below.

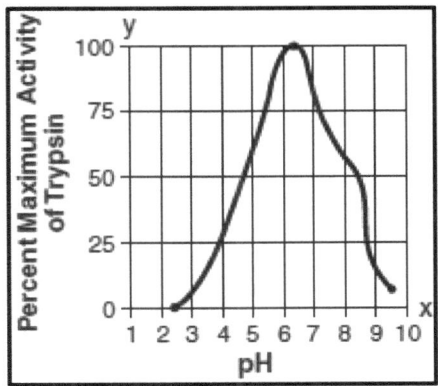

What is the approximate range of this function?

7. The graph shows the elevation of a region along a 12-mile hiker's trail.

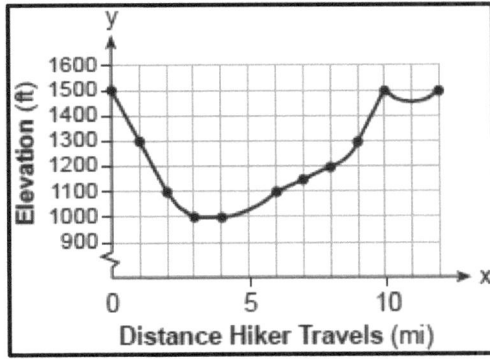

What is the approximate domain of this function?

8. The graph below shows the relative humidity during a 24-hour period.

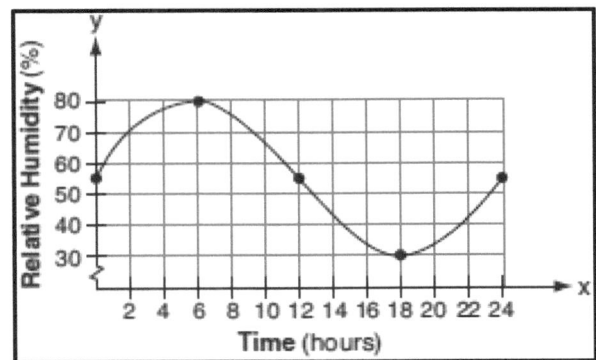

What is the approximate range of this function?

9. Suppose $g(n) = n + 1$ for the domain of whole numbers, n. Describe the range of this function.

10. Find the range of $f(x) = 3x + 10$ where $f(x)$ is defined on the domain $5 \leq x < 10$.

11. Given $f(x) = x^2$ for all real numbers, x.

    a) What is the range of this function?

    b) Suppose we restrict the domain of this function to $-3 \leq x \leq 3$. How does this affect the range?

12. Suppose n represents the number of multiple-choice questions answered correctly on a 20-question test. The function $f(n)$ represents the points earned on the test, where each question is worth 5 points with no partial credit.

    a) Define the function $f(n)$.

    b) What is an appropriate domain?

    c) What is the range?

Regents Questions

MULTIPLE CHOICE

1. Officials in a town use a function, C, to analyze traffic patterns. $C(n)$ represents the rate of traffic through an intersection where n is the number of observed vehicles in a specified time interval. What would be the most appropriate domain for the function?

    (1) $\{...-2,-1,0,1,2,3,...\}$

    (2) $\{-2,-1,0,1,2,3\}$

    (3) $\{0,\frac{1}{2},1,1\frac{1}{2},2,2\frac{1}{2}\}$

    (4) $\{0,1,2,3,...\}$

2. If $f(x) = \frac{1}{3}x + 9$, which statement is always true?

    (1) $f(x) < 0$

    (2) $f(x) > 0$

    (3) If $x < 0$, then $f(x) < 0$

    (4) If $x > 0$, then $f(x) > 0$

3. Let f be a function such that $f(x) = 2x - 4$ is defined on the domain $2 \le x \le 6$. The range of this function is

    (1) $0 \le y \le 8$

    (2) $0 \le y < \infty$

    (3) $2 \le y \le 6$

    (4) $-\infty < y < \infty$

4. The function $h(t) = -16t^2 + 144$ represents the height, $h(t)$, in feet, of an object from the ground at t seconds after it is dropped. A realistic domain for this function is

    (1) $-3 \le t \le 3$

    (2) $0 \le t \le 3$

    (3) $0 \le h(t) \le 144$

    (4) all real numbers

5. Which domain would be the most appropriate set to use for a function that predicts the number of household online-devices in terms of the number of people in the household?

    (1) integers

    (2) whole numbers

    (3) irrational numbers

    (4) rational numbers

6. The graph of the function $f(x) = \sqrt{x + 4}$ is shown below.

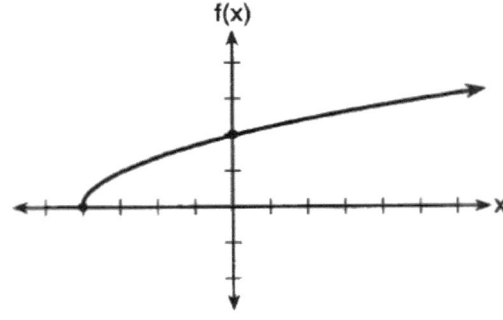

    The domain of the function is

    (1) $\{x|x > 0\}$

    (2) $\{x|x \ge 0\}$

    (3) $\{x|x > -4\}$

    (4) $\{x|x \ge -4\}$

7.   A construction company uses the function $f(p)$, where p is the number of people working on a project, to model the amount of money it spends to complete a project. A reasonable domain for this function would be

(1)   positive integers   (3)   both positive and negative integers
(2)   positive real numbers   (4)   both positive and negative real numbers

8.   The range of the function defined as $y = 5^x$ is

(1) $y < 0$ (3) $y \leq 0$
(2) $y > 0$ (4) $y \geq 0$

9.   A store sells self-serve frozen yogurt sundaes. The function $C(w)$ represents the cost, in dollars, of a sundae weighing w ounces. An appropriate domain for the function would be

(1)   integers   (3)   nonnegative integers
(2)   rational numbers   (4)   nonnegative rational numbers

10.   An online company lets you download songs for $0.99 each after you have paid a $5 membership fee. Which domain would be most appropriate to calculate the cost to download songs?

(1)   rational numbers greater than zero
(2)   whole numbers greater than or equal to one
(3)   integers less than or equal to zero
(4)   whole numbers less than or equal to one

11.   The daily cost of production in a factory is calculated using $c(x) = 200 + 16x$, where x is the number of complete products manufactured. Which set of numbers best defines the domain of $c(x)$?

(1)   integers   (3)   positive rational numbers
(2)   positive real numbers   (4)   whole numbers

12.   What is the domain of the relation shown below?

$$\{(4,2), (1,1), (0,0), (1,-1), (4,-2)\}$$

(1) $\{0, 1, 4\}$ (3) $\{-2, -1, 0, 1, 2, 4\}$
(2) $\{-2, -1, 0, 1, 2\}$ (4) $\{-2, -1, 0, 0, 1, 1, 1, 2, 4, 4\}$

13.   If the domain of the function $f(x) = 2x^2 - 8$ is $\{-2, 3, 5\}$, then the range is

(1) $\{-16, 4, 92\}$ (3) $\{0, 10, 42\}$
(2) $\{-16, 10, 42\}$ (4) $\{0, 4, 92\}$

14.   If $f(x) = x^2 + 2$, which interval describes the range of this function?

(1) $(-\infty, \infty)$ (3) $[2, \infty)$
(2) $[0, \infty)$ (4) $(-\infty, 2]$

15.   At an ice cream shop, the profit, $P(c)$, is modeled by the function $P(c) = 0.87c$, where c represents the number of ice cream cones sold. An appropriate domain for this function is

(1)   an integer ≤ 0 (3)   a rational number ≤ 0
(2)   an integer ≥ 0 (4)   a rational number ≥ 0

16. If the function $f(x) = x^2$ has the domain $\{0, 1, 4, 9\}$, what is its range?

     (1) $\{0, 1, 2, 3\}$ (3) $\{0, -1, 1, -2, 2, -3, 3\}$
     (2) $\{0, 1, 16, 81\}$ (4) $\{0, -1, 1, -16, 16, -81, 81\}$

17. The function $f(x) = 2x^2 + 6x - 12$ has a domain consisting of the integers from -2 to 1, inclusive. Which set represents the corresponding range values for $f(x)$?

     (1) $\{-32, -20, -12, -4\}$ (3) $\{-32, -4\}$
     (2) $\{-16, -12, -4\}$ (4) $\{-16, -4\}$

18. The function $f(x)$ is graphed below.

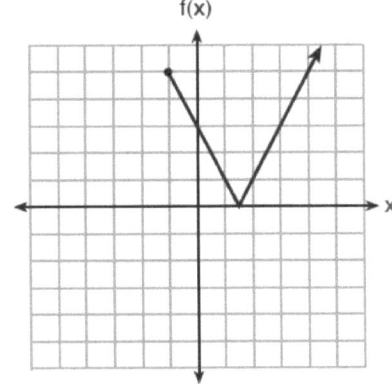

     The domain of this function is

     (1) all positive real numbers   (3) $x \geq 0$
     (2) all positive integers   (4) $x \geq -1$

19. A grocery store sells packages of beef. The function $C(w)$ represents the cost, in dollars, of a package of beef weighing w pounds. The most appropriate domain for this function would be

     (1) integers   (3) positive integers
     (2) rational numbers   (4) positive rational numbers

20. A dolphin jumps out of the water and then back into the water. His jump could be graphed on a set of axes where x represents time and y represents distance above or below sea level. The domain for this graph is best represented using a set of

     (1) integers   (3) real numbers
     (2) positive integers   (4) positive real numbers

21. The range of the function $f(x) = |x + 3| - 5$ is

     (1) $[-5, \infty)$ (3) $[3, \infty)$
     (2) $(-5, \infty)$ (4) $(3, \infty)$

22. A population of paramecia, P, can be modeled using the exponential function $P(t) = 3(2)^t$, where t is the number of days since the population was first observed. Which domain is most appropriate to use to determine the population over the course of the first two weeks?

     (1) $t \geq 0$ (3) $0 \leq t \leq 2$
     (2) $t \leq 2$ (4) $0 \leq t \leq 14$

23. Which domain would be the most appropriate to use for a function that compares the number of emails sent (x) to the amount of data used for a cell phone plan (y)?

(1) integers   (3) rational numbers

(2) whole numbers   (4) irrational numbers

24. Which domain is most appropriate for a function that represents the number of items, $f(x)$, placed into a laundry basket each day, x, for the month of January?

(1) integers   (3) rational numbers

(2) whole numbers   (4) irrational numbers

25. The diagram below shows the graph of $h(t)$, which models the height, in feet, of a rocket t seconds after it was shot into the air.

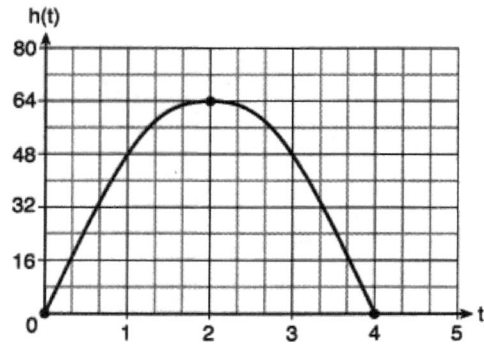

The domain of $h(t)$ is

(1) $(0, 4)$ (3) $(0, 64)$

(2) $[0, 4]$ (4) $[0, 64]$

26. Skyler mows lawns in the summer. The function $f(x)$ is used to model the amount of money earned, where x is the number of lawns completely mowed. A reasonable domain for this function would be

(1) real numbers   (3) irrational numbers

(2) rational numbers   (4) natural numbers

27. The domain of the function $f(x) = x^2 + x - 12$ is

(1) $(-\infty, -4]$ (3) $[-4, 3]$

(2) $(-\infty, \infty)$ (4) $[3, \infty)$

28. A store manager is trying to determine if they should continue to sell a particular brand of nails. To model their profit, they use the function $p(n)$, where n is the number of boxes of these nails sold in a day. A reasonable domain for this function would be

(1) nonnegative integers   (3) real numbers

(2) rational numbers   (4) integers

29. The function $G(m)$ represents the amount of gasoline consumed by a car traveling m miles. An appropriate domain for this function would be

(1) integers   (3) nonnegative integers

(2) rational numbers   (4) nonnegative rational numbers

30. What is the range of the function $f(x) = (x-4)^2 + 1$?

    (1) $x > 4$ (3) $f(x) > 1$

    (2) $x \geq 4$ (4) $f(x) \geq 1$

CONSTRUCTED RESPONSE

31. The function f has a domain of $\{1, 3, 5, 7\}$ and a range of $\{2, 4, 6\}$. Could f be represented by $\{(1,2), (3,4), (5,6), (7,2)\}$? Justify your answer.

32. Samantha purchases a package of sugar cookies. The nutrition label states that each serving size of 3 cookies contains 160 Calories. Samantha creates the graph below showing the number of cookies eaten and the number of Calories consumed.

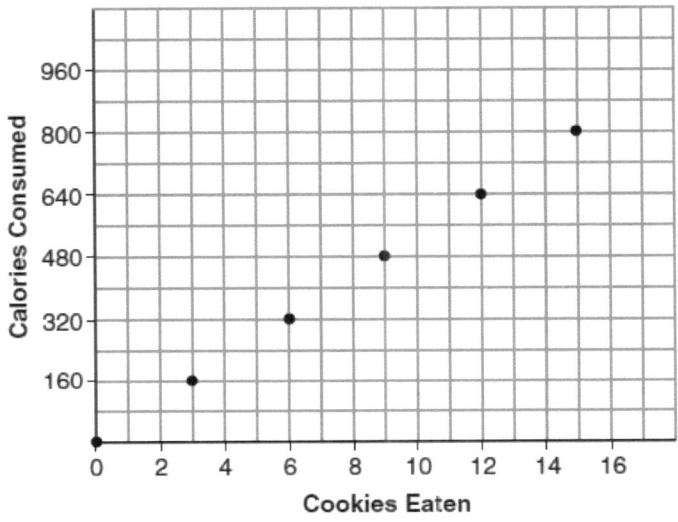

Explain why it is appropriate for Samantha to draw a line through the points on the graph.

33. A function is graphed on the set of axes below.

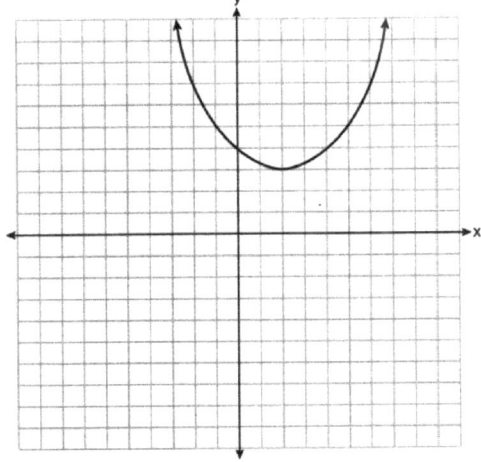

State the domain of this function. State the range of this function.

257

6.6 Absolute Value Functions

The **absolute value** of a number n is the distance between n and 0, written using the vertical symbols, $|n|$. The absolute value of a positive number (or 0) is the number itself; the absolute value of a negative number is its opposite.

Examples: $|5| = 5$ $|-8| = 8$ $|-0.25| = 0.25$ $|0| = 0$

An **absolute value function** can be graphed using a table or a calculator.
Example: We can graph $y = |x|$ as follows.

x	$\|x\|$	y	(x, y)
-2	$\|-2\|$	2	$(-2,2)$
-1	$\|-1\|$	1	$(-1,1)$
0	$\|0\|$	0	$(0,0)$
1	$\|1\|$	1	$(1,1)$
2	$\|2\|$	2	$(2,2)$

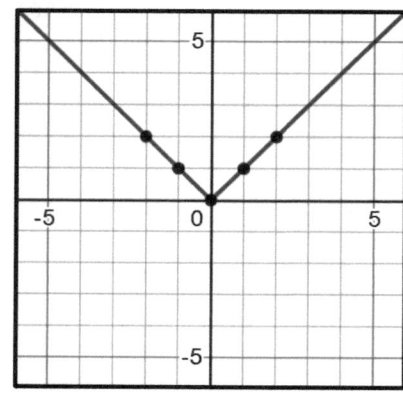

  CALCULATOR TIP

To graph $y = |x|$ on the calculator:

On the TI-84, enter: Y= ALPHA [F2] 1 X,T,Θ,n ▶ GRAPH.

On the TI-83, enter: Y= MATH <NUM> 1 X,T,Θ,n) GRAPH.

An absolute value function will have a **V shape** (or an upside down V shape).

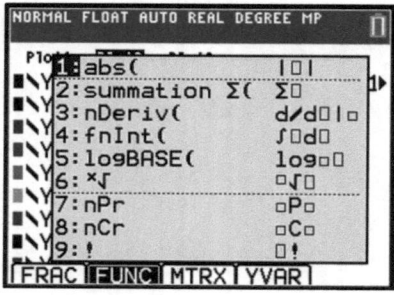 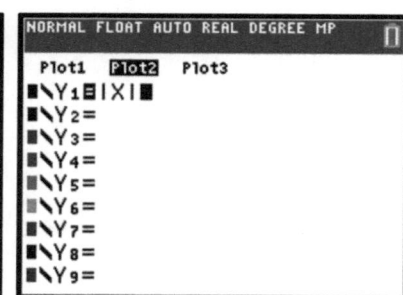 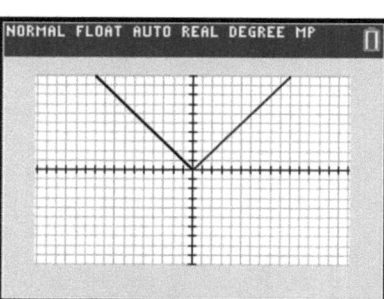

MODEL PROBLEM

Use a table to graph the function $y = 2|x + 1|$.

Solution:

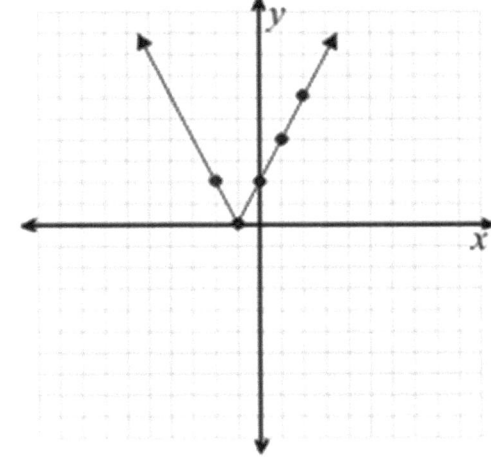

(A)	(B)	(C)	(D)
x	$2\|x + 1\|$	y	(x, y)
-2	$2\|-2 + 1\|$	2	$(-2,2)$
-1	$2\|-1 + 1\|$	0	$(-1,0)$
0	$2\|0 + 1\|$	2	$(0,2)$
1	$2\|1 + 1\|$	4	$(1,4)$
2	$2\|2 + 1\|$	6	$(2,6)$

Explanation of steps:

(A)   Pick values of x that will evaluate to both positive and negative expressions inside the absolute value sign, allowing you to see both sides of the V shape in the graph.

(B)   Substitute the values of x into the expression on the right side of the equation.

(C)   Evaluate for y.

(D)   Plot the resulting points on the graph and extend the rays infinitely with arrow heads.

PRACTICE PROBLEMS

1.   Which graph represents the equation $y = |x - 2|$?

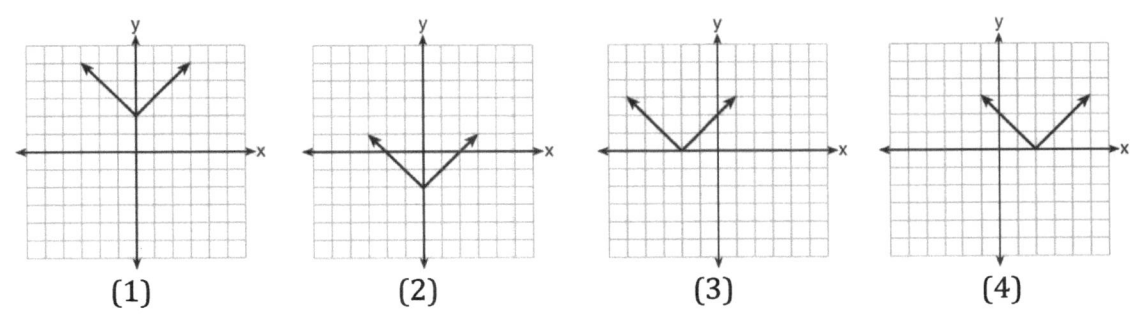

(1)   (2)   (3)   (4)

2. The graph below represents $f(x)$.

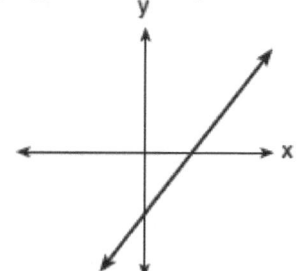

Which graph best represents $|f(x)|$?

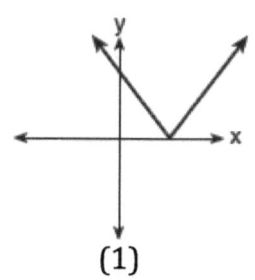   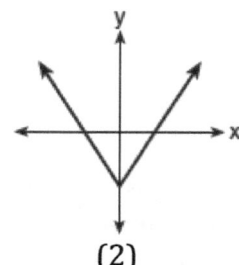   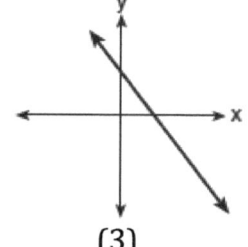   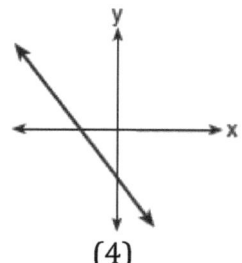

(1)   (2)   (3)   (4)

3. Graph $y = |x| - 3$.

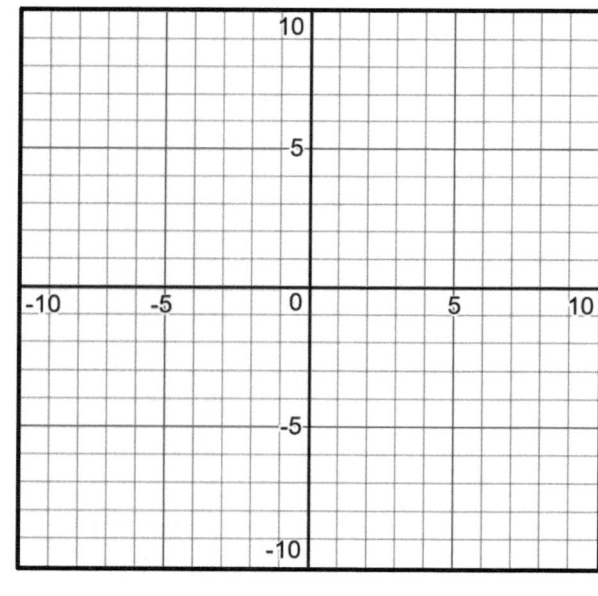

4. Graph $y = -|x|$.

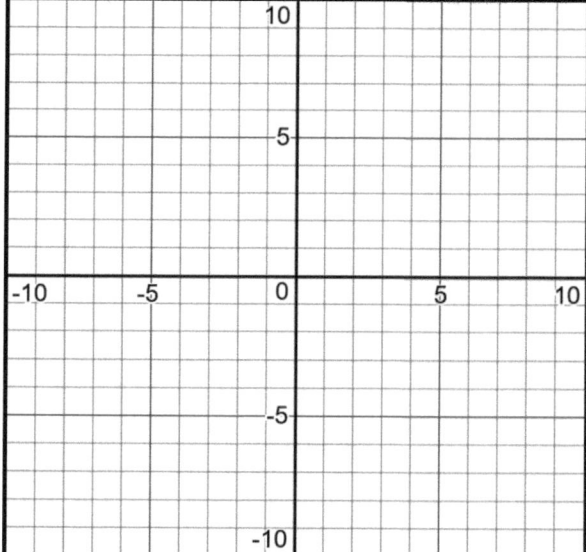

5. Graph $y = 3|x|$.

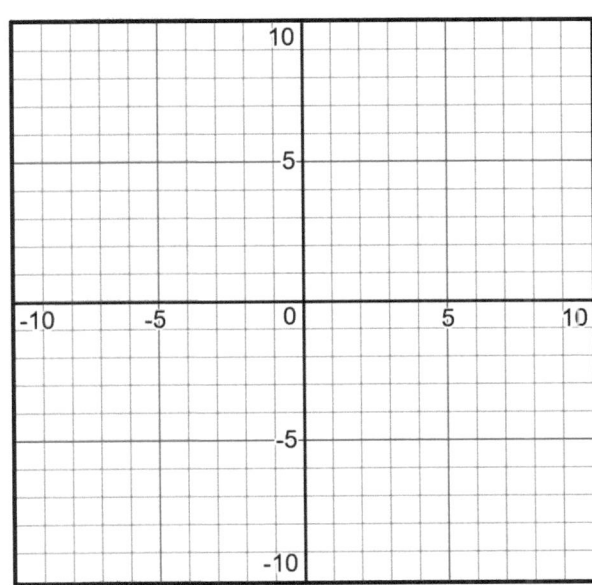

6. Graph $y = \frac{1}{2}|x - 1|$.

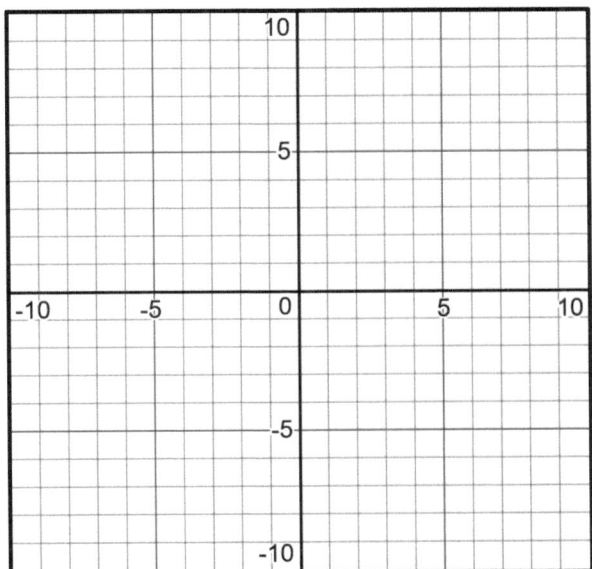

7. Graph $y = 2|x + 3|$ over the interval $-7 \leq x \leq 1$.

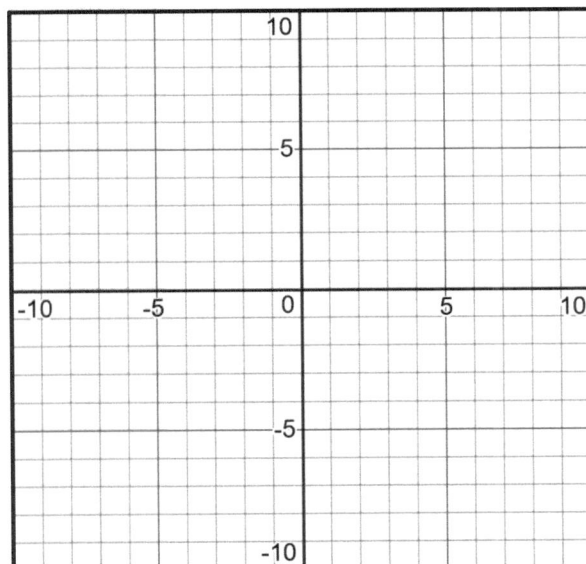

Regents Questions

MULTIPLE CHOICE

1. The graphs of the functions $f(x) = |x - 3| + 1$ and $g(x) = 2x + 1$ are drawn. Which statement about these functions is true?

    (1) The solution to $f(x) = g(x)$ is 3.   (3) The graphs intersect when $y = 1$.
    (2) The solution to $f(x) = g(x)$ is 1.   (4) The graphs intersect when $x = 3$.

2. What is the *minimum* value of the function $y = |x + 3| - 2$?

    (1) -2 (3) 3
    (2) 2   (4) -3

3. Which value of x results in equal outputs for $f(x) = 3x - 2$ and $b(x) = |x + 2|$?

    (1) -2 (3) $\dfrac{2}{3}$
    (2) 2   (4) 4

4. The function $h(x)$, which is graphed below, and the function $g(x) = 2|x + 4| - 3$ are given.

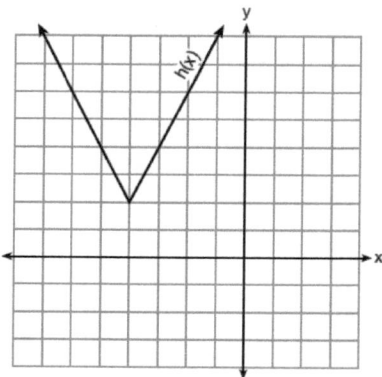

Which statements about these functions are true?
    I. $g(x)$ has a lower minimum value than $h(x)$.
    II. For all values of x, $h(x) < g(x)$.
    III. For any value of x, $g(x) \neq h(x)$.

    (1) I and II, only   (3) II and III, only
    (2) I and III, only   (4) I, II, and III

CONSTRUCTED RESPONSE

5.   On the set of axes below, graph the function $y = |x + 1|$.

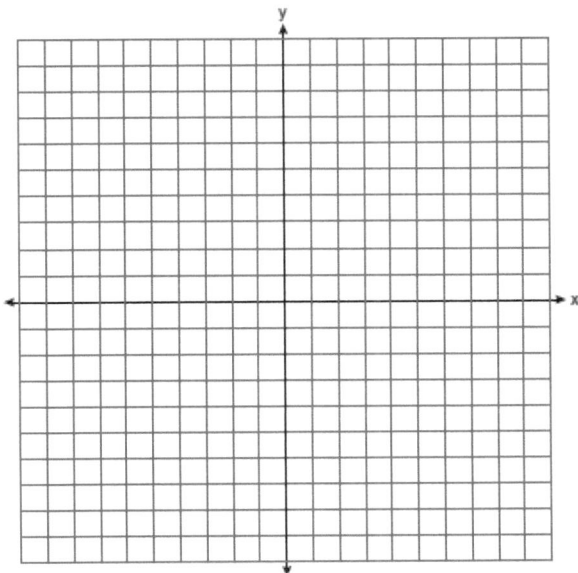

State the range of the function. State the domain over which the function is increasing.

6.   Graph $f(x) = |x|$ and $g(x) = -x^2 + 6$ on the grid below. Does $f(-2) = g(-2)$? Use your graph to explain why or why not.

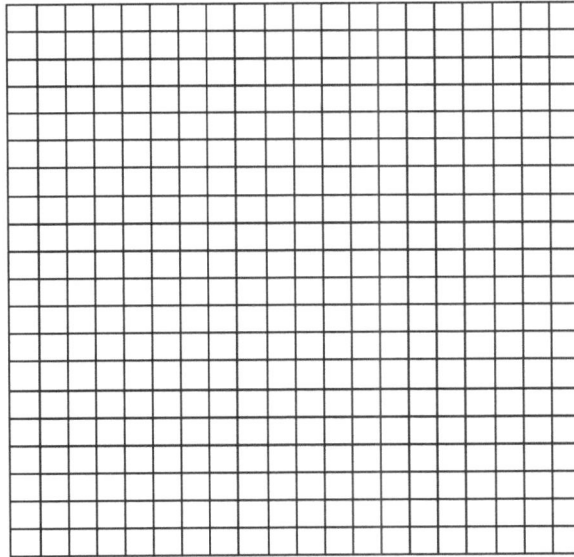

7.   On the set of axes below, graph $f(x) = |x - 3| + 2$.

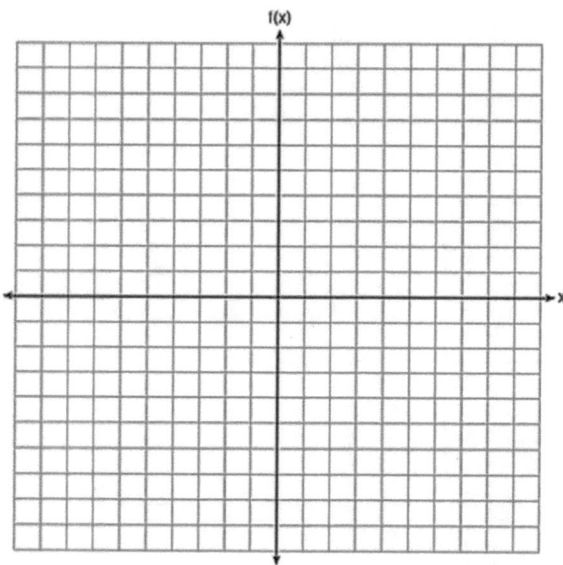

8.   Graph the function $f(x) = \left|\frac{1}{2}x + 3\right|$ over the interval $-8 \le x \le 0$.

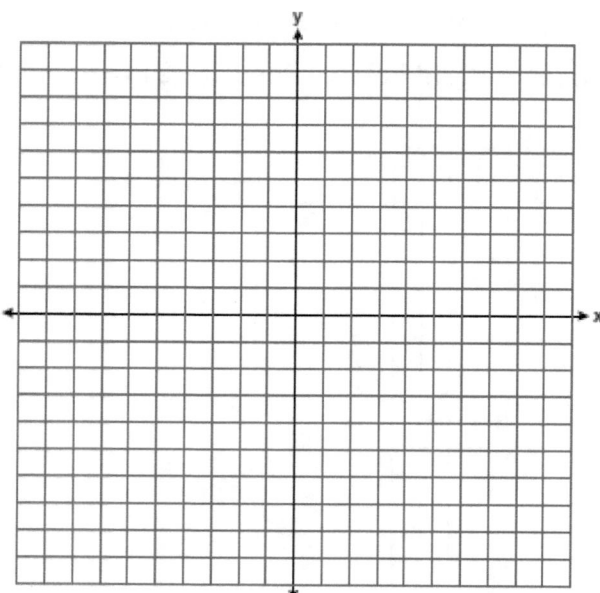

9.   Graph $f(x) = |x + 1|$ on the set of axes below.

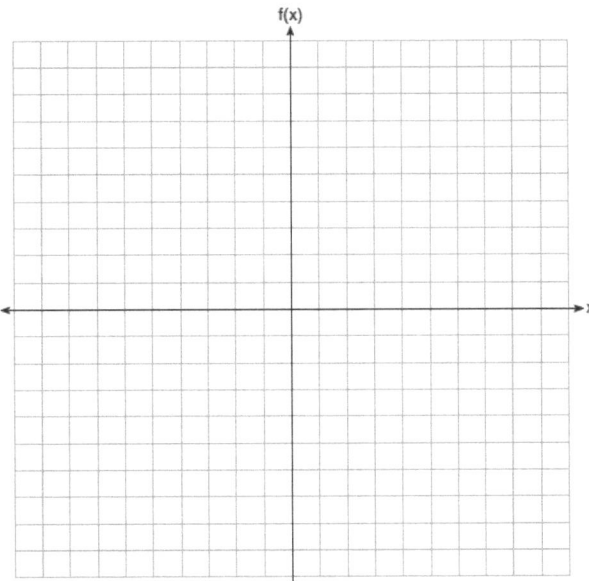

CHAPTER 7.   FUNCTIONS AS MODELS

7.1   Write a Function from a Table

We have seen that a linear equation may be written in slope-intercept form as $y = mx + b$ or in point-slope form as $y - y_1 = m(x - x_1)$.

We can use these forms to write linear functions for tables of coordinates with x-values given at equal intervals. If a table includes $x = 0$, we can use the simpler slope-intercept form, $f(x) = mx + b$. This is because the table already supplies us with the y-intercept, b, which is $f(0)$.

If a table doesn't include $x = 0$, it is easier to use the point-slope form and then isolate y, which is the method we'll use in this section. Alternatively, you could use the slope-intercept form, but then you'll need to solve for b by substituting one of the points' coordinates for x and y.

In either case, you'll first need to find the slope using the slope formula, $m = \dfrac{y_2 - y_1}{x_2 - x_1}$.

To write an equation for a linear function given a table including $x = 0$:
1. Find the slope m using the first two given points.
2. The value of b would be the y-intercept, $f(0)$.
3. Write the function in the form, $f(x) = mx + b$.

Example: In the table below, the slope $m = \dfrac{12 - 10}{5 - 0} = \dfrac{2}{5}$ and the y-intercept, $f(0) = 10$, so the linear equation is $f(x) = \dfrac{2}{5}x + 10$.

x	0	5	10	15
$f(x)$	10	12	14	16

To write an equation for a linear function given a table that *does not* include $x = 0$:
1. Find the slope m using the first two given points.
2. Write the equation in point-slope form. The first given point gives us x_1 and y_1.
3. Isolate y, then change y into $f(x)$ and simplify.
4. Write the function in the form, $f(x) = mx + b$.

Example: In the table below, the slope $m = \dfrac{14 - 8}{3 - 1} = \dfrac{6}{2} = 3$. Using the first point (1,8) as (x_1, y_1), we can write the equation in point-slope form as $y - 8 = 3(x - 1)$. Isolating y gives us $y = 3(x - 1) + 8$, so the function is $f(x) = 3(x - 1) + 8$, which simplifies to $f(x) = 3x + 5$.

x	1	3	5	7
$f(x)$	8	14	20	26

MODEL PROBLEM

Write an equation of the linear function represented by the table below.

x	2	4	6	8
$f(x)$	13	21	29	37

Solution:

(A) $m = \frac{21-13}{4-2} = \frac{8}{2} = 4$

(B) $y - 13 = 4(x - 2)$

(C) $y = 4(x - 2) + 13$

     $f(x) = 4(x - 2) + 13$

     $f(x) = 4x - 8 + 13$

(D) $f(x) = 4x + 5$

Explanation of steps:

(A) Find the slope. *[Use the points, (2,13) and (4,21).]*

(B) If $b = f(0)$ is not given, write the equation of the line in point-slope form, using the first point as (x_1, y_1). *[Using the first point, (2,13), substitute 2 for x_1 and 13 for y_1.]*

(C) Isolate y, then change y into $f(x)$ and simplify.

(D) Write the function in the form, $f(x) = mx + b$.

PRACTICE PROBLEMS

1. Write the linear function represented by the table below.

x	0	1	2	3	4
$f(x)$	9	13	17	21	25

2. Write the linear function represented by the table below.

x	0	3	6	9	12
$f(x)$	10	15	20	25	30

3. Write the linear function represented by the table below.

x	1	2	3	4	5
$f(x)$	7	10	13	16	19

4. Write the linear function represented by the table below.

x	1	5	9	13	17
$f(x)$	−5	−3	−1	1	3

5. Write the linear function represented by the table below.

x	2	4	6	8	10
$f(x)$	9	5	1	−3	−7

6. Write the linear function represented by the table below.

x	11	12	13	14	15
$f(x)$	0	5	10	15	20

Regents Questions

MULTIPLE CHOICE

1. Which chart could represent the function $f(x) = -2x + 6$?

x	f(x)
0	6
2	10
4	14
6	18

(1)

x	f(x)
0	8
2	10
4	12
6	14

(3)

x	f(x)
0	4
2	6
4	8
6	10

(2)

x	f(x)
0	6
2	2
4	-2
6	-6

(4)

CONSTRUCTED RESPONSE

2. Each day Toni records the height of a plant for her science lab. Her data are shown in the table below.

Day (n)	1	2	3	4	5
Height (cm)	3.0	4.5	6.0	7.5	9.0

The plant continues to grow at a constant daily rate. Write an equation to represent $h(n)$, the height of the plant on the nth day.

3. Jackson is starting an exercise program. The first day he will spend 30 minutes on a treadmill. He will increase his time on the treadmill by 2 minutes each day. Write an equation for $T(d)$, the time, in minutes, on the treadmill on day d.

Find $T(6)$, the minutes he will spend on the treadmill on day 6.

4. Tanya is making homemade greeting cards. The data table below represents the amount she spends in dollars, $f(x)$, in terms of the number of cards she makes, x.

x	$f(x)$
4	7.50
6	9
9	11.25
10	12

Write a linear function, $f(x)$, that represents the data. Explain what the slope and y-intercept of $f(x)$ mean in the given context.

7.2   Graph Linear Functions

Graphs of functions are often used to model real world situations. The **independent** variable is represented by **x-values** in a horizontal axis and the **dependent** variable is represented by **y-values** in a vertical axis. Very often in a real event, time is the independent variable.

An important first step in creating graphs of functions to model a situation is to determine what **units of measure** are used for the horizontal and vertical axes.

Example: To graph a function representing the gasoline in a car's tank during a trip, we could use distance (miles) for one axis and gasoline (gallons) for the other.

The real world situation may also require certain **contraints**, such as minimum or maximum values of the variables. A linear graph, therefore, may be a line segment rather than a line.

Example: The most commonly used units for the independent variables are units of time, which are generally constrained to non-negative real numbers.

Also, there may be **restrictions on the domain**.

Example: If an axis represents a number of people, or a number of items produced, we would restrict its values to counting numbers only.

Since the measurements or values in real world problems are not always small integers, we may need to **scale** our coordinate axes to fit the situation. It is very possible that a grid square in our graph may not represent a one unit by one unit square. No matter what scale we choose to use for an axis, we must use consistent intervals on that axis.

Example: The graph below uses a grid square of 1 hour by 25 pages.

Number of Pages Read by Time

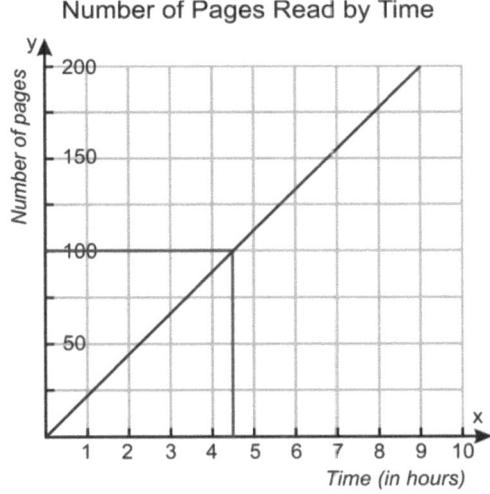

271

MODEL PROBLEM

A swimming pool with a maximum capacity of 450 gallons contains 100 gallons of water before a hose begins to fill the pool by depositing 50 gallons of water each minute. Write and graph an equation that relates x, the number of minutes, to $g(x)$, the number of gallons, for the interval $100 \leq g(x) \leq 450$ only.

Solution:

(A) $g(x) = 50x + 100$

(B)   (C)

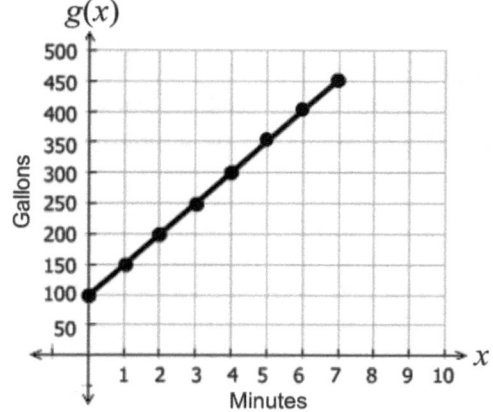

Explanation of steps:

(A) Write an equation.

(B) Create a grid with appropriately scaled axes.

(C) Graph the line. *[Due to the given constraints, the line segment should start at (0, 100) and stop when g(x) reaches 450 at (7, 450).]*

PRACTICE PROBLEMS

1. A cell phone company charges a monthly rate of $25 for a data plan plus $5 per gigabyte of data used. Write an equation for $c(g)$, the cost of the plan with g gigabytes of data usage. Graph $c(g)$.

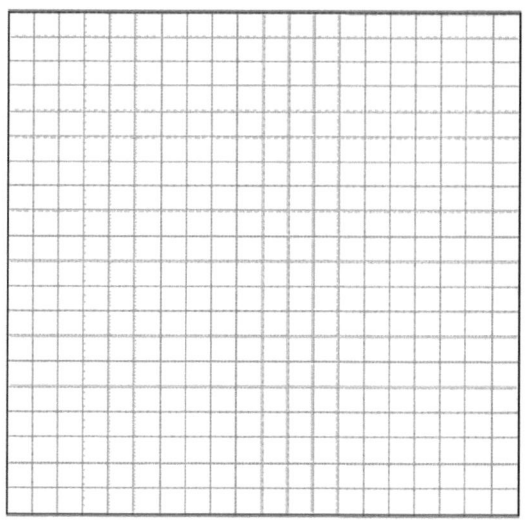

2. A handyman charges $1.00 per square foot plus an additional fee of $25.00 to paint a deck floor. Write an equation for $c(x)$, the cost, in dollars, for painting a deck floor that is x square feet in area. Graph $c(x)$.

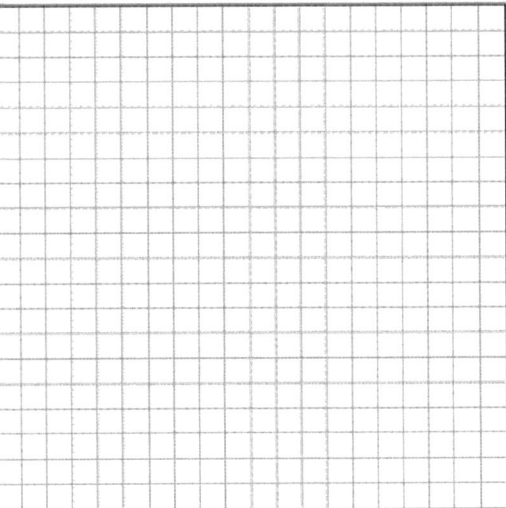

3. An elementary school is sponsoring a dance. The cost of a disk jockey is $40, and tickets sell for $2 each. Write a linear equation and, on the grid below, graph the equation to represent the relationship between the number of tickets sold and the profit from the dance.

4. The rate at which crickets chirp is a linear function of temperature. At 59° F they make 76 chirps per minute, and at 65° F they make 100 chirps per minute. Write an equation for $c(t)$, the chirping rate at temperature t. Then, graph $c(t)$ over the domain $50 \le t \le 75$ on the grid below.

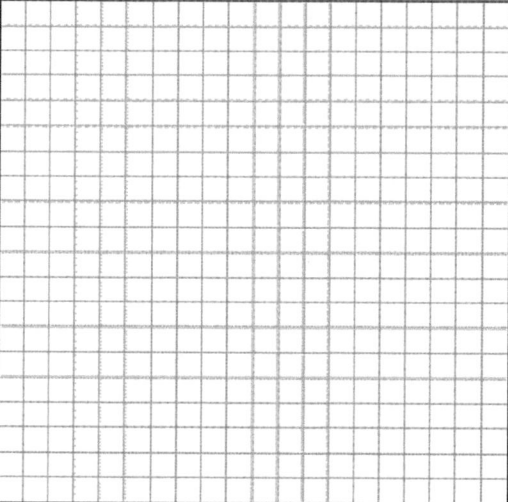

Regents Questions

CONSTRUCTED RESPONSE

1. Max purchased a box of green tea mints. The nutrition label on the box stated that a serving of three mints contains a total of 10 Calories. On the axes below, graph the function, C, where $C(x)$ represents the number of Calories in x mints. Write an equation that represents $C(x)$.

    A full box of mints contains 180 Calories. Use the equation to determine the total number of mints in the box.

    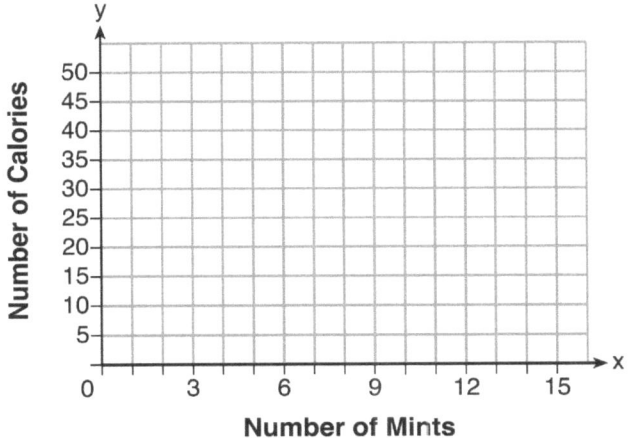

2. Zeke and six of his friends are going to a baseball game. Their combined money totals $28.50. At the game, hot dogs cost $1.25 each, hamburgers cost $2.50 each, and sodas cost $0.50 each. Each person buys one soda. They spend all $28.50 on food and soda.

    Write an equation that can determine the number of hot dogs, x, and hamburgers, y, Zeke and his friends can buy. Graph your equation on the grid below.

    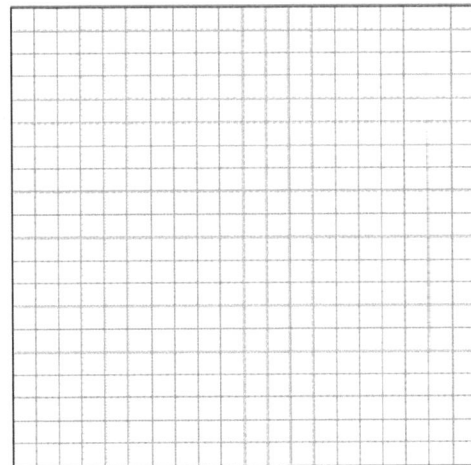

    Determine how many different combinations, including those combinations containing zero, of hot dogs and hamburgers Zeke and his friends can buy, spending all $28.50. Explain your answer.

7.3   Rate of Change for Linear Functions

A **rate of change** is a rate that describes how one quantity changes in relation to another quantity. If a graph of a function forms a straight line, it is called a **linear** function, and the rate of change can be calculated as the **slope of the line**. If the line has a **positive slope**, there is a **positive rate of change**. If the line has a **negative slope**, there is a **negative rate of change**. If the line is horizontal, the slope is zero and therefore the rate of change is zero.

Example: Every two hours, a driver records the total time and distance traveled. The table and graph below show the results. A positive slope indicates a *positive rate of change*. Since the slope is 40, the rate is 40 miles per hour.

Hours Driven x	Miles Traveled y
2	80
4	160
6	240

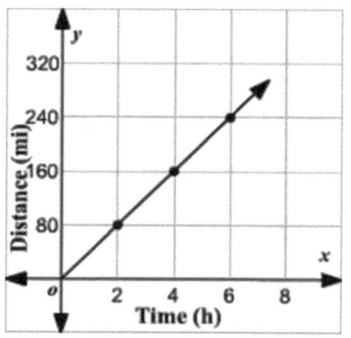

The **absolute value of the slope** will determine how **steep** it is. A graph with a slope of 5 will be steeper than a graph with a slope of 2, while a graph with a -5 slope will be steeper than a graph with a -2 slope.

If the line *passes through the origin*, then $y = mx$ and the variables are in **direct variation**. The slope m represents the **constant of variation** when comparing y to x. The above graph is an example of direct variation. The slope, 40, is the constant of variation of *miles* to *hours*.

If the line does not pass through the origin, then the constant term b (the y-intercept) in the equation $y = mx + b$ often represents the **starting value** in the model, especially where the x axis represents time passed.

MODEL PROBLEM

A candle has a starting length of 10 inches. Thirty minutes after lighting it, the length is 7 inches. The candle continues to get shorter at the same rate over time, as shown by the graph below. Is there a positive or negative rate of change in the length of the candle over time?

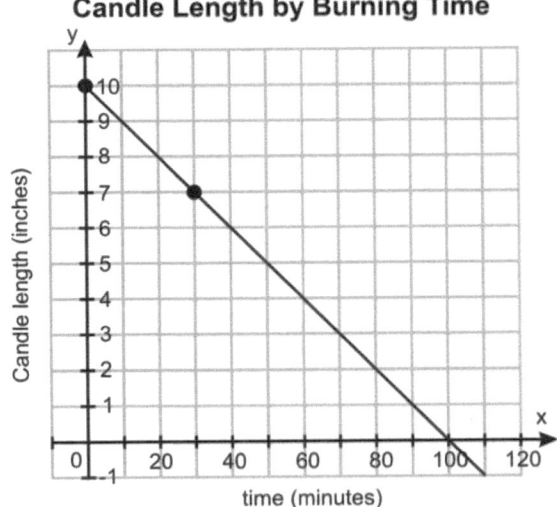

Candle Length by Burning Time

Solution:
    Negative

Explanation of steps:
    If the graph shows a line with a positive slope, the rate of change is positive. But if the slope of the line is negative, the rate of change is negative. *[Note that the y-intercept of 10 in the graph represents the starting length of the candle.]*

PRACTICE PROBLEMS

1. In a linear equation, the independent variable *increases* at a constant rate while the dependent variable *decreases* at a constant rate. The slope of this line is (1) zero (3) positive (2) negative (4) undefined	2. In a linear equation, the independent variable *increases* at a constant rate while the dependent variable *increases* at a constant rate. The slope of this line is (1) zero (3) positive (2) negative (4) undefined

3. Identify the rate of change in the following graph as positive or negative.

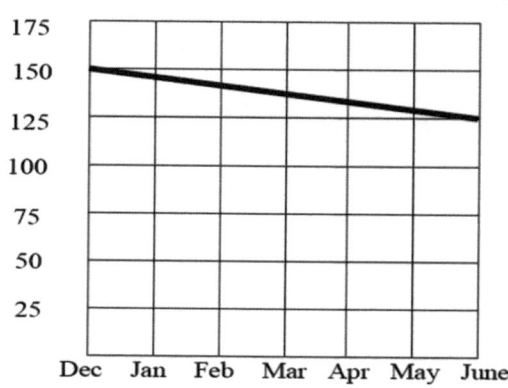

4. Identify the rate of change in the following table as positive or negative.

x	y
0.5	9.0
1	8.75
1.5	8.5
2	8.25
2.5	8.0

5. The following table shows a constant rate of change in distance over time. Is the rate of change positive or negative? Calculate the rate.

Time (hours)	Distance (miles)
4	232
6	348
8	464
10	580

6. In a linear equation, x represents the distance that a car travels, in miles, and y represents the amount of gas in the car's gas tank, in gallons. As the car travels, is the constant rate of change positive or negative?

Regents Questions

MULTIPLE CHOICE

1. Rowan has $50 in a savings jar and is putting in $5 every week. Jonah has $10 in his own jar and is putting in $15 every week. Each of them plots his progress on a graph with time on the horizontal axis and amount in the jar on the vertical axis. Which statement about their graphs is true?

    (1) Rowan's graph has a steeper slope than Jonah's.

    (2) Rowan's graph always lies above Jonah's.

    (3) Jonah's graph has a steeper slope than Rowan's.

    (4) Jonah's graph always lies above Rowan's.

2. The graph below was created by an employee at a gas station.

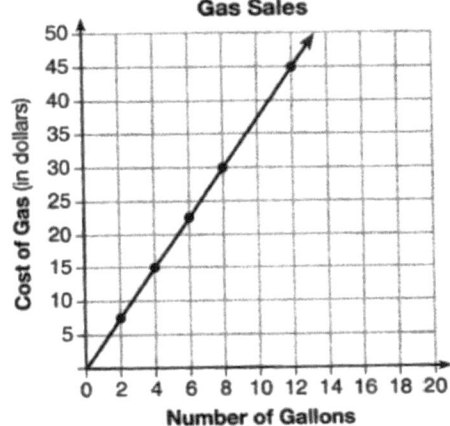

    Which statement can be justified by using the graph?

    (1) If 10 gallons of gas was purchased, $35 was paid.

    (2) For every gallon of gas purchased, $3.75 was paid.

    (3) For every 2 gallons of gas purchased, $5.00 was paid.

    (4) If zero gallons of gas were purchased, zero miles were driven.

3.   Which function has a constant rate of change equal to -3?

x	y
0	2
1	5
2	8
3	11

(1)

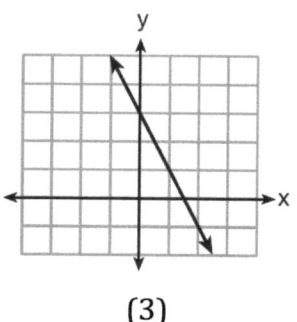

(3)

$\{(1,5), (2,2), (3,-5), (4,4)\}$

(2)

$2y = -6x + 10$

(4)

4.   A student plotted the data from a sleep study as shown in the graph below.

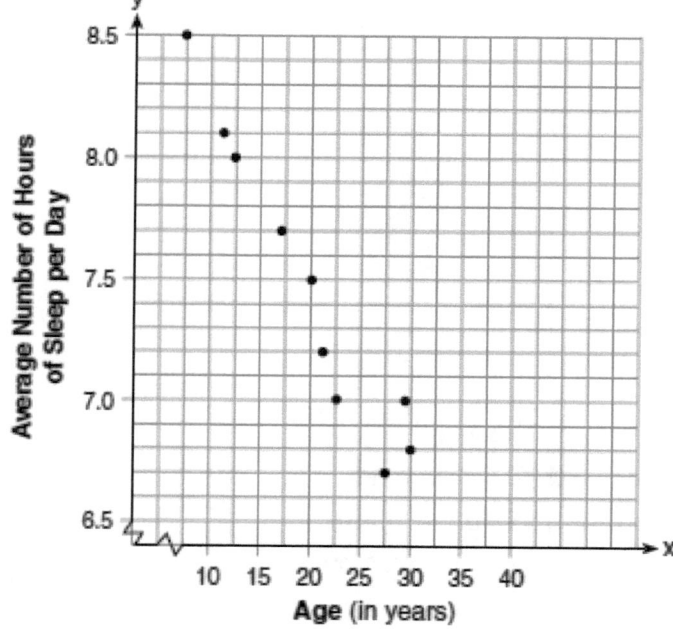

The student used the equation of the line $y = -0.09x + 9.24$ to model the data. What does the rate of change represent in terms of these data?

   (1)   The average number of hours of sleep per day increases 0.09 hour per year of age.
   (2)   The average number of hours of sleep per day decreases 0.09 hour per year of age.
   (3)   The average number of hours of sleep per day increases 9.24 hours per year of age.
   (4)   The average number of hours of sleep per day decreases 9.24 hours per year of age.

CONSTRUCTED RESPONSE

5. The cost of belonging to a gym can be modeled by $C(m) = 50m + 79.50$, where $C(m)$ is the total cost for m months of membership. State the meaning of the slope and y-intercept of this function with respect to the costs associated with the gym membership.

6. During a recent snowstorm in Red Hook, NY, Jaime noted that there were 4 inches of snow on the ground at 3:00 p.m., and there were 6 inches of snow on the ground at 7:00 p.m. If she were to graph these data, what does the slope of the line connecting these two points represent in the context of this problem?

7. Loretta and her family are going on vacation. Their destination is 610 miles from their home. Loretta is going to share some of the driving with her dad. Her average speed while driving is 55 mph and her dad's average speed while driving is 65 mph.

   The plan is for Loretta to drive for the first 4 hours of the trip and her dad to drive for the remainder of the trip. Determine the number of hours it will take her family to reach their destination.

   After Loretta has been driving for 2 hours, she gets tired and asks her dad to take over. Determine, to the _nearest tenth of an hour_, how much time the family will save by having Loretta's dad drive for the remainder of the trip.

7.4   Average Rate of Change

When we calculated the rate of change for linear functions, we simply calculated the slope of the line. For a linear function, the rate of change is constant because the slope is constant. But not all functions are linear. Nevertheless, we can still find an **average rate of change** between any two points on a curve by calculating the slope of the line through those two points, which is called the **secant line**.

As we know, the formula for the slope of a line through points (x_1, y_1) and (x_2, y_2) is $m = \dfrac{y_2 - y_1}{x_2 - x_1}$.
So, using function notation, we could say the average rate of change R over the interval $a \leq x \leq b$ is the slope of the secant line through points $(a, f(a))$ and $(b, f(b))$, which is $R = \dfrac{f(b) - f(a)}{b - a}$. We will use this formula to calculate average rate of change.

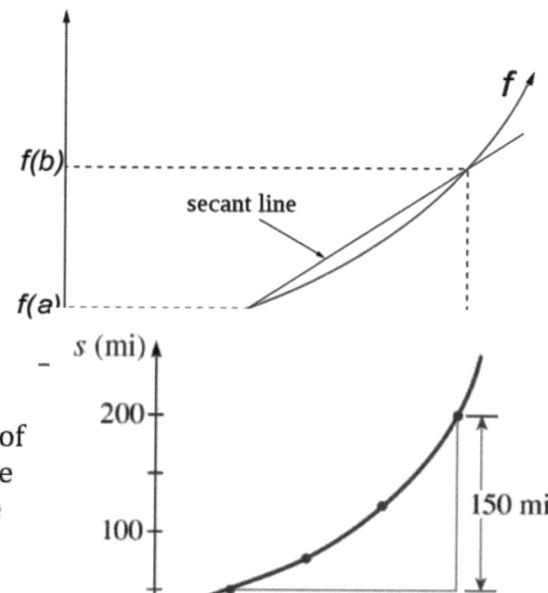

For example, suppose you take a car trip and record the distance that you travel every hour. The graph to the right shows the distance s (in miles) as a function of time t (in hours). If we want to calculate the average rate of change over the interval $1 \leq t \leq 4$, we simply find the slope between the points $(1, 50)$ and $(4, 200)$. So, the average rate of change (or average speed in this case) is
$$\frac{s(b) - s(a)}{b - a} = \frac{200 - 50}{4 - 1} = \frac{150}{3} = 50 \text{ mph.}$$

For non-linear functions, the **average rate of change may vary** for different intervals.

Example: In the example above, the average rate of change (ie, average speed) over the interval $2 \leq t \leq 4$ is $\dfrac{200 - 75}{4 - 2} = \dfrac{125}{2} = 62.5$ mph.

MODEL PROBLEM 1:   *FROM A GRAPH OR SET OF POINTS*

A ball is shot straight up in the air from ground level and its height is recorded every 0.5 seconds until it lands 4 seconds later. A graph of the height of the ball over time (in the shape of a parabola) is shown to the right. Find the average rate of change in the ball's height between seconds 2 and 3.

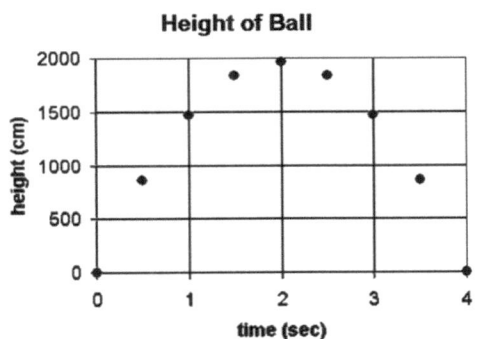

Solution:
  (A) (2,2000) and (3,1500)
  (B) $\dfrac{1500 - 2000}{3 - 2} = \dfrac{-500}{1} = -500$ cm/sec

Explanation of steps:
  (A) Find the points at the start and end of the given interval *[at 2 seconds, the height is at its maximum of 2000cm; at 3 seconds, the height is 1500cm]*.
  (B) Find the slope of the line between the two points *[a negative slope as it is falling]*.

PRACTICE PROBLEMS

1.   From the graph below,

   a) find the average rate of change in the interval
       $1 \leq x \leq 3$

   b) find the average rate of change in the interval
       $-1 \leq x \leq 2$

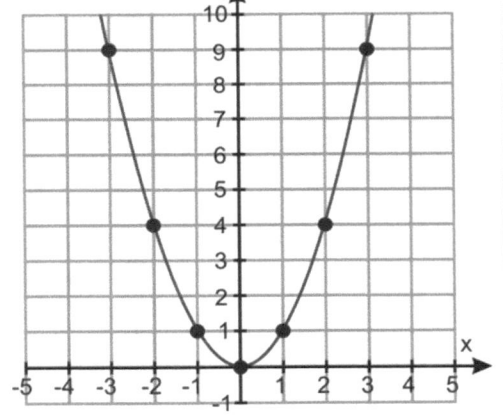

2.   The following table shows the average prices of movie tickets over a period of years.

Year	1987	1991	1995	1999	2003	2007	2009
Price ($)	3.91	4.21	4.35	5.06	6.03	6.88	7.50

   Determine which interval has the higher average rate of change:
       (1) 1987 – 1999   (2) 1999 – 2009

MODEL PROBLEM 2: *FROM A FUNCTION DEFINITION*

If an object is dropped from a tall building, then the distance it has fallen after t seconds is given by the function $d(t) = 16t^2$. Find its average speed (average rate of change) over the interval between 1 second and 5 seconds.

Solution:
  (A)   At 1 second, $d(1) = 16(1)^2 = 16$. At 5 seconds, $d(5) = 16(5)^2 = 16 \cdot 25 = 400$.

  (B)   Slope of the line through points (1,16) and (5,400) is $\dfrac{400 - 16}{5 - 1} = \dfrac{384}{4} = 96$ ft/sec.

Explanation of steps:
  (A)   Find the points at the start and end of the given interval.
        [Think of each point as the ordered pair of t and d(t). At t = 1, d(t) = 16,
        and at t = 5, d(t) = 400, so the points are (1,16) and (5,400)]
  (B)   Find the slope of the line between the two points.

PRACTICE PROBLEMS

3. Calculate the average rate of change of a function, $f(x) = x^2 + 2$ as x changes from 5 to 15?	4. Find the average rate of change for the function $f(x) = x^2 + 10x + 16$ over the interval $-3 \le x \le 3$.

Regents Questions

MULTIPLE CHOICE

1. Given the functions $g(x)$, $f(x)$, and $h(x)$ shown below:

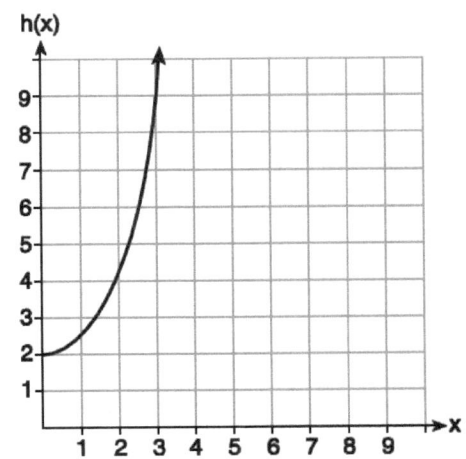

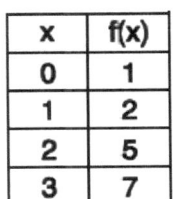

$g(x) = x^2 - 2x$

x	f(x)
0	1
1	2
2	5
3	7

The correct list of functions ordered from greatest to least by average rate of change over the interval $0 \le x \le 3$ is

(1) $f(x), g(x), h(x)$

(3) $g(x), f(x), h(x)$

(2) $h(x), g(x), f(x)$

(4) $h(x), f(x), g(x)$

2. The Jamison family kept a log of the distance they traveled during a trip, as represented by the graph below.

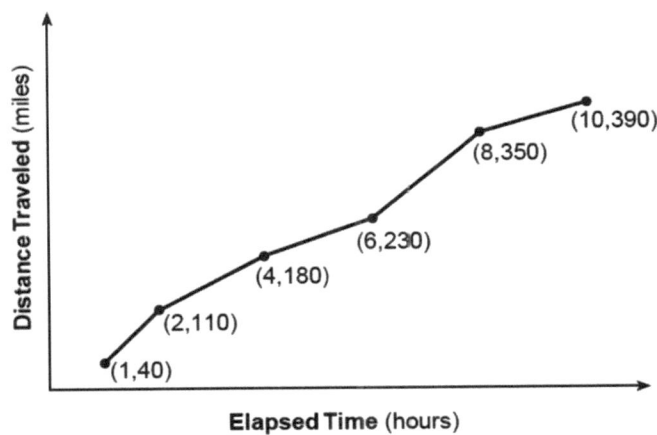

During which interval was their average speed the greatest?

(1) the first hour to the second hour

(3) the sixth hour to the eighth hour

(2) the second hour to the fourth hour

(4) the eighth hour to the tenth hour

3. The table below shows the average diameter of a pupil in a person's eye as he or she grows older.

Age (years)	Average Pupil Diameter (mm)
20	4.7
30	4.3
40	3.9
50	3.5
60	3.1
70	2.7
80	2.3

What is the average rate of change, in millimeters per year, of a person's pupil diameter from age 20 to age 80?

(1) 2.4
(2) 0.04
(3) −2.4
(4) −0.04

4. An astronaut drops a rock off the edge of a cliff on the Moon. The distance, $d(t)$, in meters, the rock travels after t seconds can be modeled by the function $d(t) = 0.8t^2$. What is the average speed, in meters per second, of the rock between 5 and 10 seconds after it was dropped?

(1) 12
(2) 20
(3) 60
(4) 80

5. Joey enlarged a 3-inch by 5-inch photograph on a copy machine. He enlarged it four times. The table below shows the area of the photograph after each enlargement.

Enlargement	0	1	2	3	4
Area (square inches)	15	18.8	23.4	29.3	36.6

What is the average rate of change of the area from the original photograph to the fourth enlargement, to the *nearest tenth*?

(1) 4.3
(2) 4.5
(3) 5.4
(4) 6.0

6.   Firing a piece of pottery in a kiln takes place at different temperatures for different amounts of time. The graph below shows the temperatures in a kiln while firing a piece of pottery after the kiln is preheated to 200ºF.

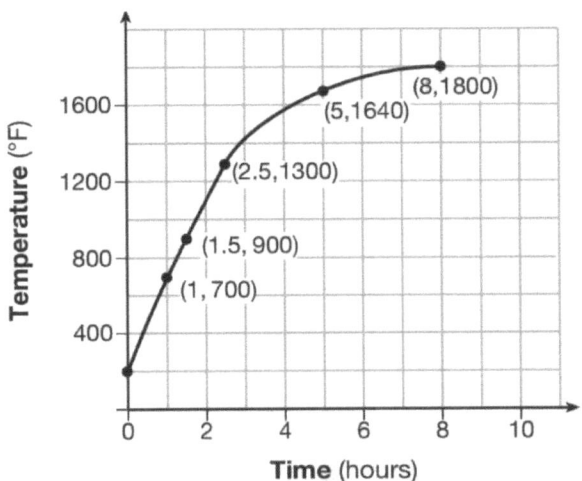

During which time interval did the temperature in the kiln show the greatest average rate of change?

(1)   0 to 1 hour   (3)   2.5 hours to 5 hours

(2)   1 hour to 1.5 hours   (4)   5 hours to 8 hours

7.   The table below shows the cost of mailing a postcard in different years. During which time interval did the cost increase at the greatest average rate?

Year	1898	1971	1985	2006	2012
Cost (¢)	1	6	14	24	35

(1)   1898–1971   (3)   1985–2006

(2)   1971–1985   (4)   2006–2012

8.   The table below shows the year and the number of households in a building that had high-speed broadband internet access.

Number of Households	11	16	23	33	42	47
Year	2002	2003	2004	2005	2006	2007

For which interval of time was the average rate of change the *smallest*?

(1)   2002 – 2004   (3)   2004 – 2006

(2)   2003 – 2005   (4)   2005 – 2007

9.   The graph below shows the distance in miles, m, hiked from a camp in h hours.

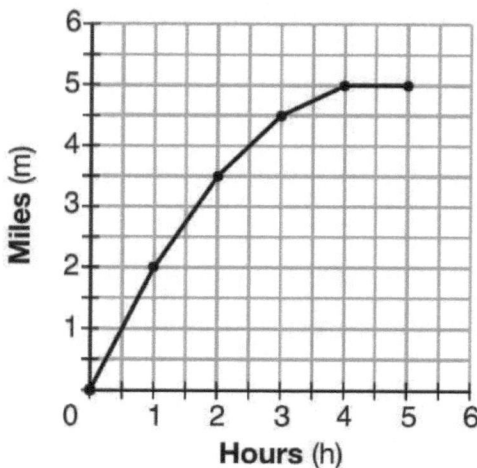

Which hourly interval had the greatest rate of change?
   (1)   hour 0 to hour 1   (3)   hour 2 to hour 3
   (2)   hour 1 to hour 2   (4)   hour 3 to hour 4

10.   The graph below models the height of a remote-control helicopter over 20 seconds during flight.

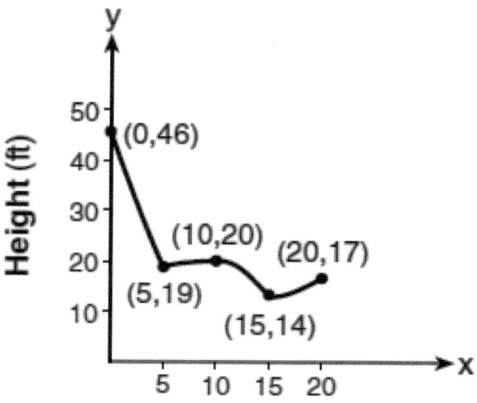

Over which interval does the helicopter have the *slowest* average rate of change?
   (1)   0 to 5 seconds   (3)   10 to 15 seconds
   (2)   5 to 10 seconds   (4)   15 to 20 seconds

11. Voting rates in presidential elections from 1996–2012 are modeled below.

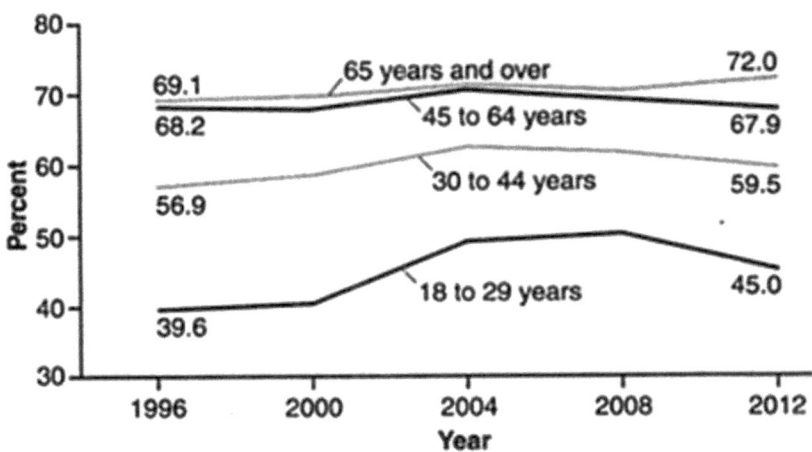

Voting Rates in Presidential Elections, by Age, for the Voting-Age Citizen Population: 1996-2012

Which statement does *not* correctly interpret voting rates by age based on the given graph?
  (1) For citizens 18–29 years of age, the rate of change in voting rate was greatest between years 2000–2004.
  (2) From 1996–2012, the average rate of change was positive for only two age groups.
  (3) About 70% of people 45 and older voted in the 2004 election.
  (4) The voting rates of eligible age groups lies between 35 and 75 percent during presidential elections every 4 years from 1996-2012.

12. The value of Tony's investment was $1140 on January 1st. On this date three years later, his investment was worth $1824. The average rate of change for this investment was $19 per
  (1) day   (3) quarter
  (2) month   (4) year

13. The table below shows the number of reported polio cases in Nigeria from 2006 to 2015.

Year	2006	2007	2008	2009	2010	2011	2012	2013	2014	2015
Number of Cases	1129	285	798	388	21	62	122	53	60	0

What is the average rate of change, to the *nearest hundredth*, of the number of reported polio cases per year in Nigeria from 2006 to 2013?
  (1) –0.01   (3) –134.50
  (2) –125.44   (4) –153.71

CONSTRUCTED RESPONSE

14. The graph below shows the variation in the average temperature of Earth's surface from 1950–2000, according to one source.

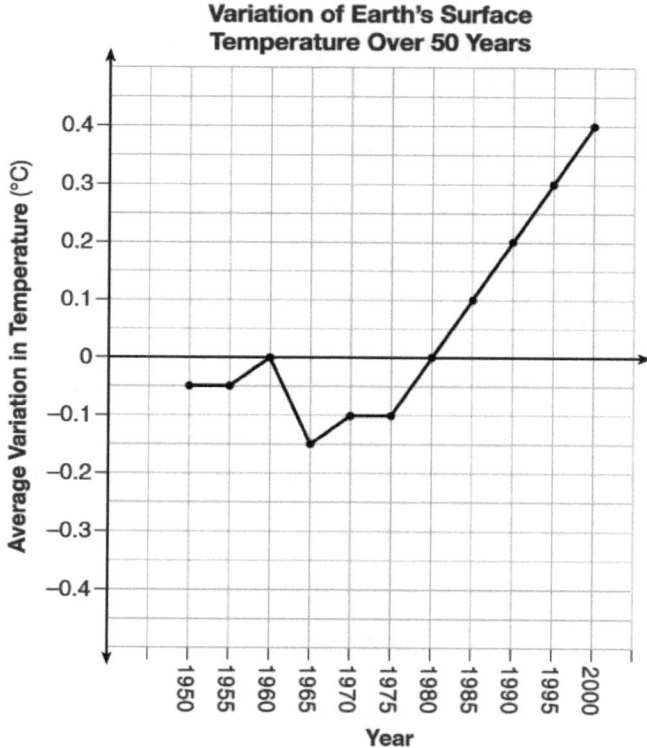

During which years did the temperature variation change the most per unit time? Explain how you determined your answer.

15. A family is traveling from their home to a vacation resort hotel. The table below shows their distance from home as a function of time.

Time (hrs)	0	2	5	7
Distance (mi)	0	140	375	480

Determine the average rate of change between hour 2 and hour 7, including units.

16. A manager wanted to analyze the online shoe sales for his business. He collected data for the number of pairs of shoes sold each hour over a 14-hour time period. He created a graph to model the data, as shown below.

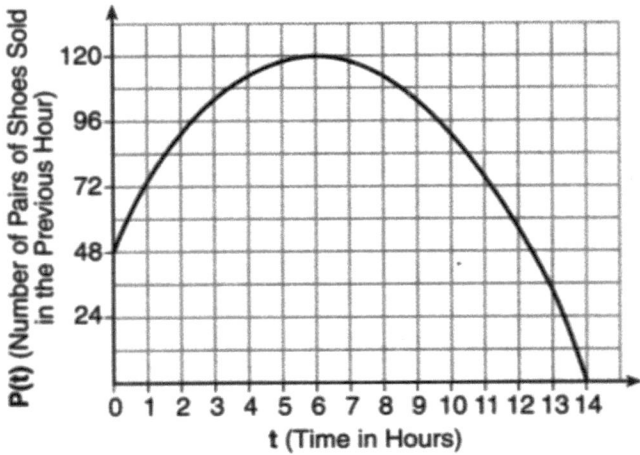

The manager believes the set of integers would be the most appropriate domain for this model. Explain why he is *incorrect*.

State the entire interval for which the function shown in the graph, $P(t)$, is increasing.

Determine the average rate of change between the sixth and fourteenth hours, and explain what it means in the context of the problem.

17. The table below represents the height of a bird above the ground during flight, with $P(t)$ representing height in feet and t representing time in seconds.

t	$P(t)$
0	6.71
3	6.26
4	6
9	3.41

Calculate the average rate of change from 3 to 9 seconds, in feet per second.

18. A blizzard occurred on the East Coast during January, 2016. Snowfall totals from the storm were recorded for Washington, D.C. and are shown in the table below.

Washington, D.C.	
Time	Snow (inches)
1 a.m.	1
3 a.m.	5
6 a.m.	11
12 noon	33
3 p.m.	36

Which interval, 1 a.m. to 12 noon or 6 a.m. to 3 p.m., has the greater rate of snowfall, in inches per hour? Justify your answer.

19. The total profit earned at a garage sale during the first five hours is modeled by the graph shown below.

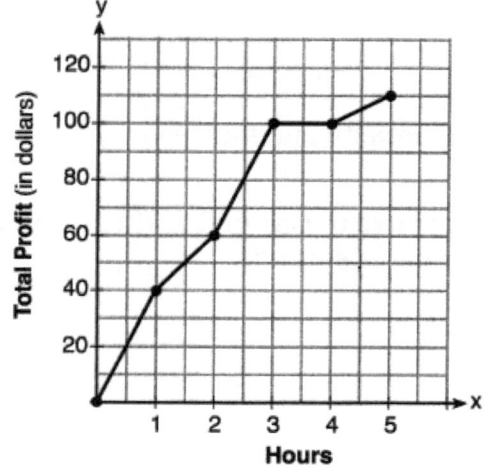

Determine the average rate of change, in dollars per hour, over the interval $1 \leq x \leq 4$.

20. The table below shows data from a recent car trip for the Burke family.

Hours After Leaving (x)	1	2	3	4	5
Miles from Home (y)	45	112	178	238	305

State the average rate of change for the distance traveled between hours 2 and 4. Include appropriate units.

7.5   Functions of Time

Distance-Time functions: One of the most common types of graphs shows distance over time. Before being able to create such graphs, one needs to know how to interpret them.

If an object moves at a constant speed (*constant rate of change*) away from a starting base, the graph will show a straight line with a positive slope (*the distance steadily increases as time increases*). If it returns to base at a constant speed, the graph will show a straight line with a negative slope (*the distance is steadily decreasing over time*). If the object is stationary, we would see a horizontal line with a zero slope (*the distance remains the same as time passes*).

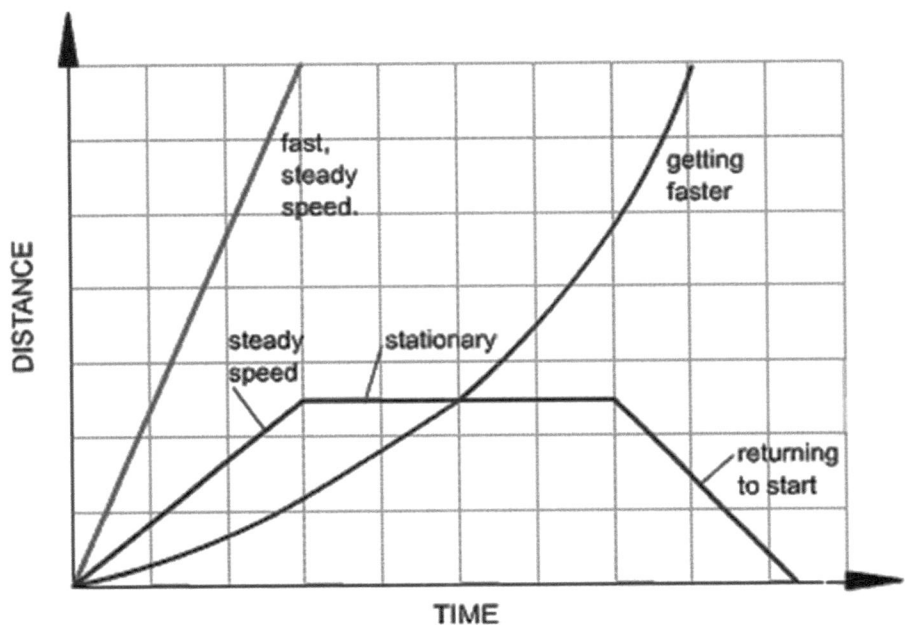

Generally, the horizontal (*x*, or often *t*) axis represents the amount of time passed. A negative value of time would be impossible, so the *domain* of the function is restricted to non-negative values. The vertical (*y*, or often *d*) axis represents the distance from a starting location. A measure of distance, and thus the *range* of the function, is also limited to non-negative values.

Speed-Time functions: Another type of graph shows an object's speed (or *velocity*) over time. In these graphs, the *y* axis represents the object's velocity, not its distance.

Example: The graph below shows the speed of a sprinter in a race. From a standstill (*zero velocity*) at the start of the race, the runner increases his speed (*accelerates*) until he reaches his maximum speed of 12.5 m/s about 4 seconds into the race. He continues to run at this same speed for more than four seconds before he slows down (*decelerates*) near the finish line.

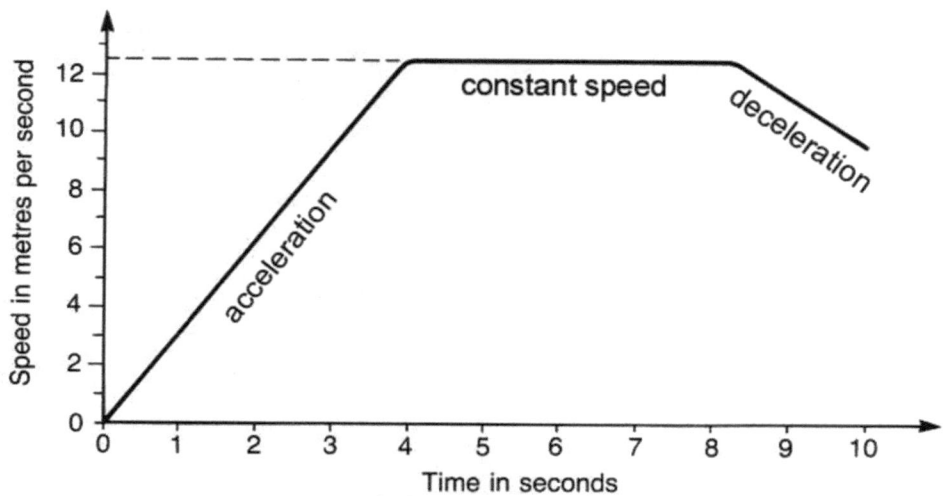

The slopes in this type of graph need to be interpreted differently than in a distance-time graph:

Feature of Graph	Distance-Time Graph	Speed-Time Graph
line with positive slope	constant speed (away from base)	steadily increasing speed
horizontal line (0 slope)	standing still	constant speed
line with negative slope	constant speed (back to base)	steadily decreasing speed

A **piecewise linear function** is one whose graph is made up of multiple line segments, like the one shown above. Some of the examples in this section will involve these types of graphs.

MODEL PROBLEM

A car travels away from its starting location at a constant rate of 2 km every 5 minutes for 15 minutes. The car stops and remains idle for the next 20 minutes. It then continues in the same direction at a constant rate of 1 km every 5 minutes for the next 10 minutes. Graph this event on a distance-time graph.

Solution:

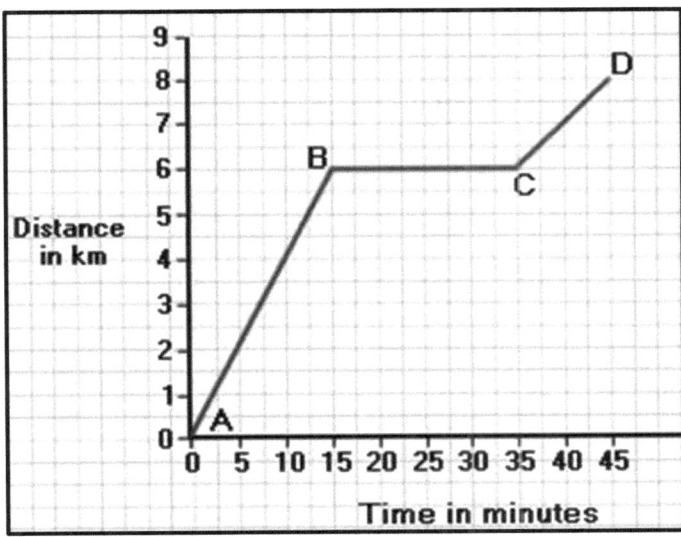

Explanation of steps:

(A)   The *y*-intercept of the graph is the distance at the start time *[we can consider the starting location as a distance of zero].*

(B)   An object moving at a constant rate will show as a straight line. The slope can be determined by moving up (the rise) by a certain distance *[2 km]* and to the right (the run) by a certain amount of time *[5 minutes].*

(C)   A stationary object will not change its distance (*y* stays the same), but time will pass (*x* increases), resulting in a horizontal line *[15 + 20 mins = 35 mins].*

(D)   If the distance increases at a slower rate, the line will have a smaller slope *[rise of 1 km and run of 5 minutes].*

PRACTICE PROBLEMS

1. A caterpillar travels up a tree, from the ground, over a 3-minute interval. It travels fast at first and then slows down. It stops for a minute, then proceeds slowly, speeding up as it goes. Which sketch best illustrates the caterpillar's distance (d) from the ground over the 3-minute interval (t)?

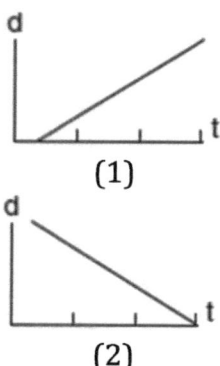

(1)

(3)

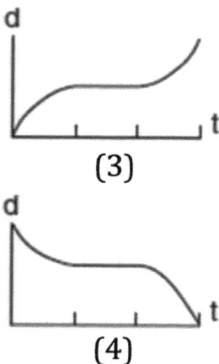

(2)

(4)

2. John walked 3 blocks from home to his school, as shown in the graph below.

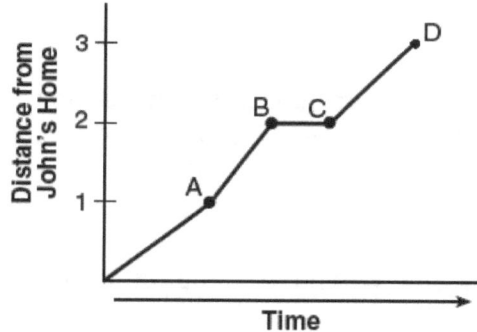

Which of the following may have happened between points B and C of the graph?

(1) John arrived at school and stayed throughout the day.
(2) John waited before crossing a busy street.
(3) John returned home to get his mathematics homework.
(4) John reached the top of a hill and began walking on level ground.

3. Rover's electronic water dish measures and records the amount of water in his dish over a period of time, as shown by the graph below.

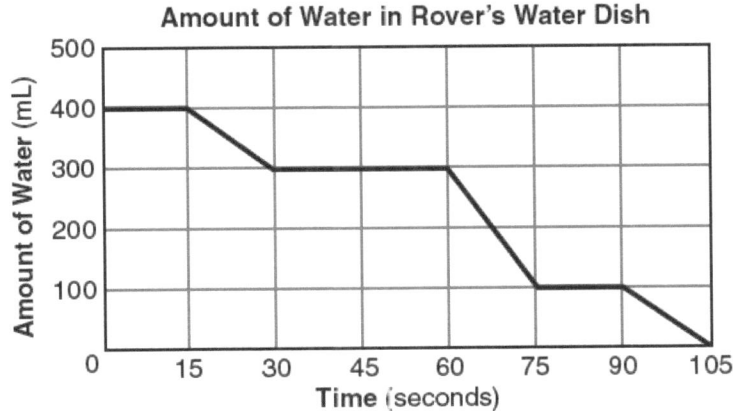

According to the graph, how long did Rover wait from the end of his first drink to the start of his second drink of water?

4. Tom went to the grocery store. The graph below shows Tom's distance from home during his trip. Tom stopped twice to rest on his trip to the store. What is the total amount of time, in minutes, that he spent resting?

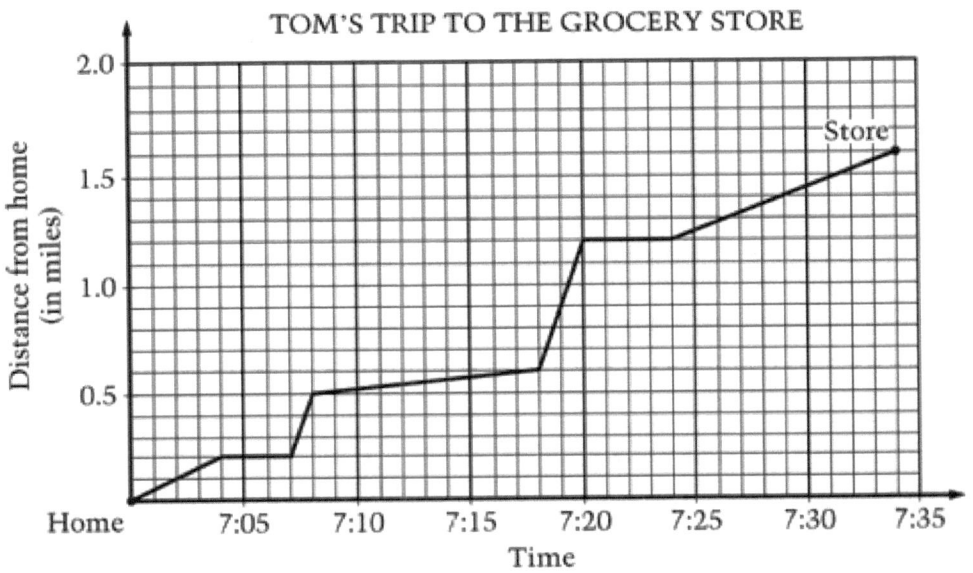

5. Marie works as a doctor at a suburban hospital that is 20 miles from her home. The graph below depicts one of her morning commutes from home to work.

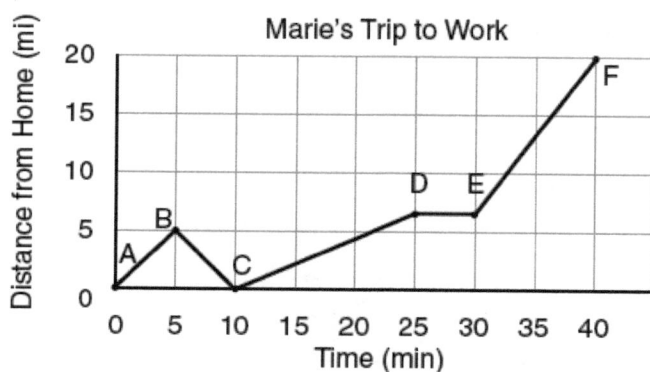

Marie left some patients' files at home and had to return home to get them. Which point represents when she turned back around to go home?

Marie also had to wait at railroad tracks for a train to pass. How long did she wait?

6. Spencer and McKenna are on a long-distance bicycle ride. Spencer leaves one hour before McKenna. The graph below shows each rider's distance in miles from his or her house as a function of time since McKenna left to catch up with Spencer.

    a) Which function represents Spencer's distance? Which function represents McKenna's distance?

    b) One rider is speeding up as time passes and the other one is slowing down. Which one is which, and how can you tell from the graphs?

    c) Estimate when McKenna catches up to Spencer. How far have they traveled at that point in time?

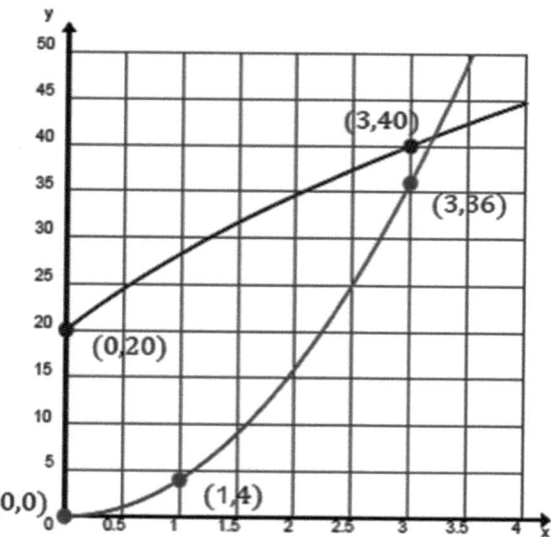

7.   The graph below shows distance versus time for a race between runners *A* and *B*. The race is already in progress, and the graph shows only the portion of the race that occurred after 11 A.M. The table below lists several characteristics of the graph. Interpret these characteristics in terms of what happened during this portion of the race. Include times and distances to support your interpretation. (*A sample response is given in the table.*)

DISTANCE *VS.* TIME

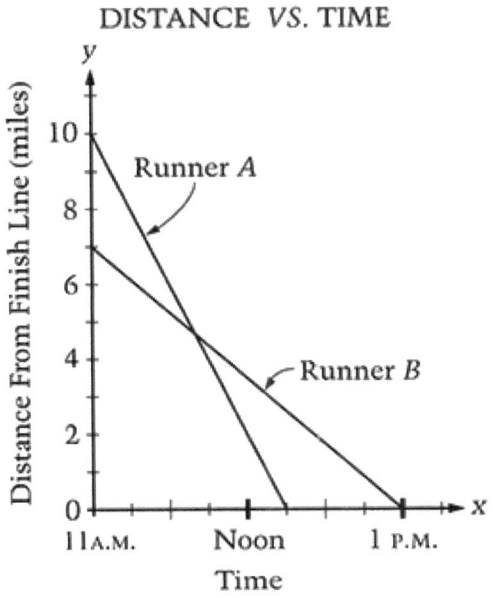

Characteristic of Graph	Interpretation in Terms of the Race
y-intercepts	At 11 A.M. Runner *A* is 10 miles from the finish line and Runner *B* is 7 miles from the finish line.
Slopes	
Point of intersection	
x-intercepts	

Regents Questions

MULTIPLE CHOICE

1. The graph below represents a jogger's speed during her 20-minute jog around her neighborhood.

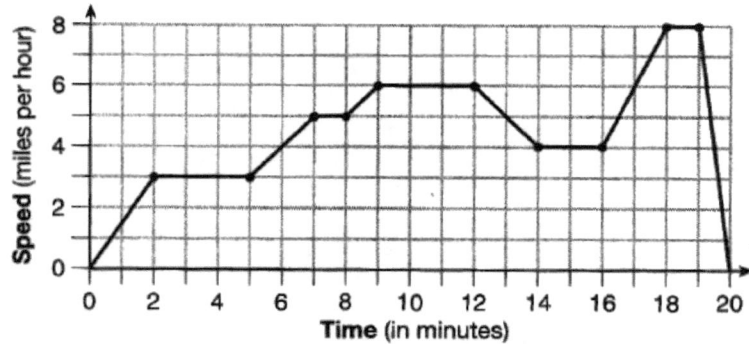

Which statement best describes what the jogger was doing during the 9–12 minute interval of her jog?

(1) She was standing still.

(2) She was increasing her speed.

(3) She was decreasing her speed.

(4) She was jogging at a constant rate.

2. A child is playing outside. The graph below shows the child's distance, $d(t)$, in yards from home over a period of time, t, in seconds.

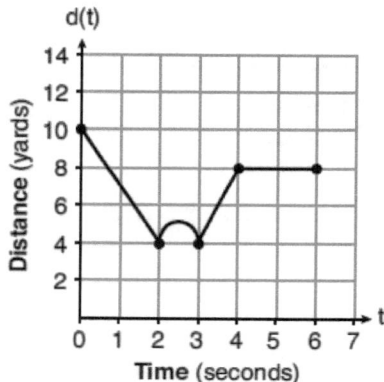

Which interval represents the child constantly moving closer to home?

(1) $0 \le t \le 2$

(2) $2 \le t \le 3$

(3) $3 \le t \le 4$

(4) $4 \le t \le 6$

3.   A café owner tracks the number of customers during business hours. The graph below models the data.

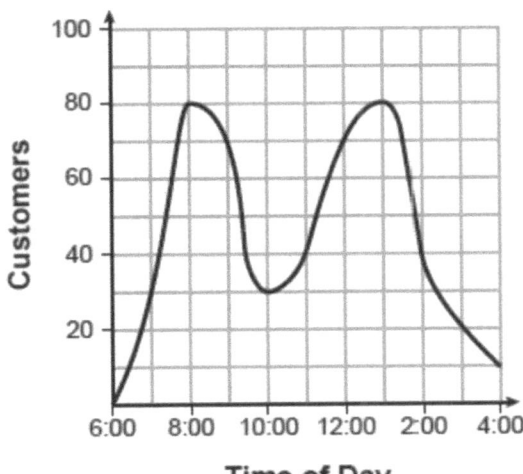

Based on the graph, the café owner saw a continual

  (1)   increase in customers from 6:00 to 11:00

  (2)   increase in customers from 12:00 to 3:00

  (3)   decrease in customers from 1:00 to 4:00

  (4)   decrease in customers from 11:00 to 2:00

4.   The graph below represents a dog walker's speed during his 30-minute walk around the neighborhood.

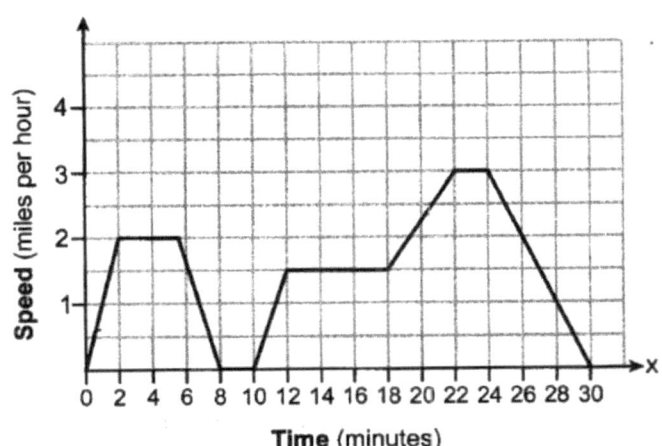

Which statement best describes what the dog walker was doing during the 12-18 minute interval of his walk?

  (1)   He was walking at a constant rate.   (3)   He was decreasing his speed.

  (2)   He was increasing his speed.   (4)   He was standing still.

301

CONSTRUCTED RESPONSE

5.   During a snowstorm, a meteorologist tracks the amount of accumulating snow. For the first three hours of the storm, the snow fell at a constant rate of one inch per hour. The storm then stopped for two hours and then started again at a constant rate of one-half inch per hour for the next four hours.

On the grid below, draw and label a graph that models the accumulation of snow over time using the data the meteorologist collected.

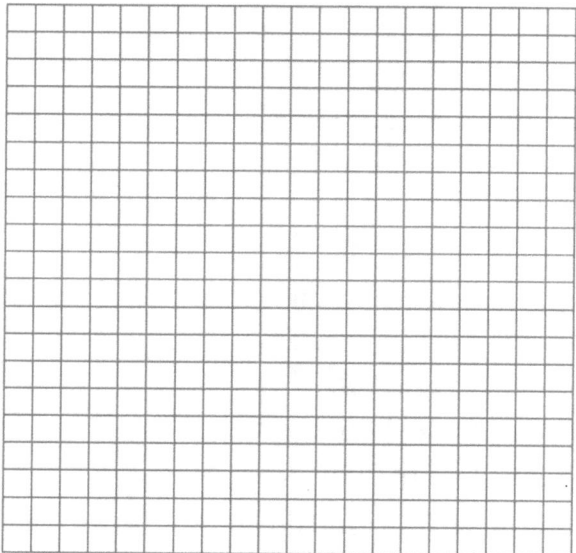

If the snowstorm started at 6 p.m., how many inches of snow had accumulated by midnight?

6.   A driver leaves home for a business trip and drives at a constant speed of 60 miles per hour for 2 hours. Her car gets a flat tire, and she spends 30 minutes changing the tire. She resumes driving and drives at 30 miles per hour for the remaining one hour until she reaches her destination. On the set of axes below, draw a graph that models the driver's distance from home.

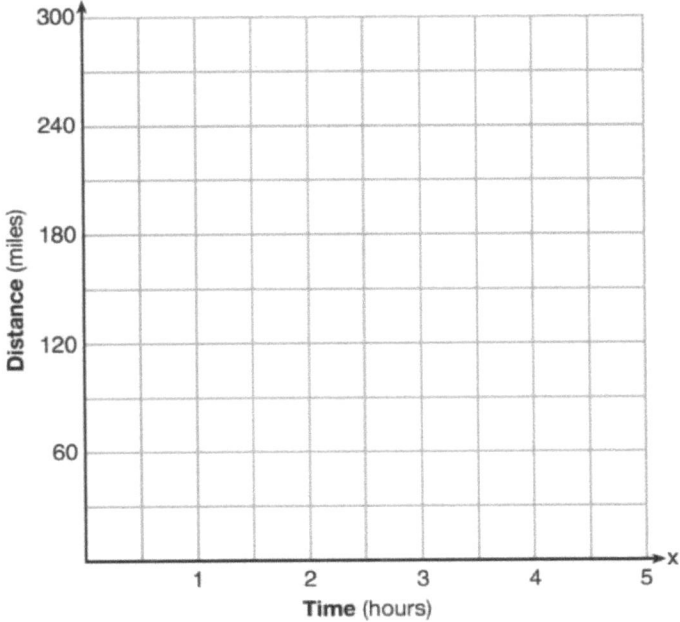

7.   An airplane leaves New York City and heads toward Los Angeles. As it climbs, the plane gradually increases its speed until it reaches cruising altitude, at which time it maintains a constant speed for several hours as long as it stays at cruising altitude. After flying for 32 minutes, the plane reaches cruising altitude and has flown 192 miles. After flying for a total of 92 minutes, the plane has flown a total of 762 miles.

Determine the speed of the plane, at cruising altitude, in miles per minute.

Write an equation to represent the number of miles the plane has flown, y, during x minutes at cruising altitude, only.

Assuming that the plane maintains its speed at cruising altitude, determine the total number of miles the plane has flown 2 hours into the flight.

8.   The graph below models Craig's trip to visit his friend in another state. In the course of his travels, he encountered both highway and city driving.

Based on the graph, during which interval did Craig most likely drive in the city? Explain your reasoning.

Explain what might have happened in the interval between *B* and *C*.

Determine Craig's average speed, to the *nearest tenth of a mile per hour*, for his entire trip.

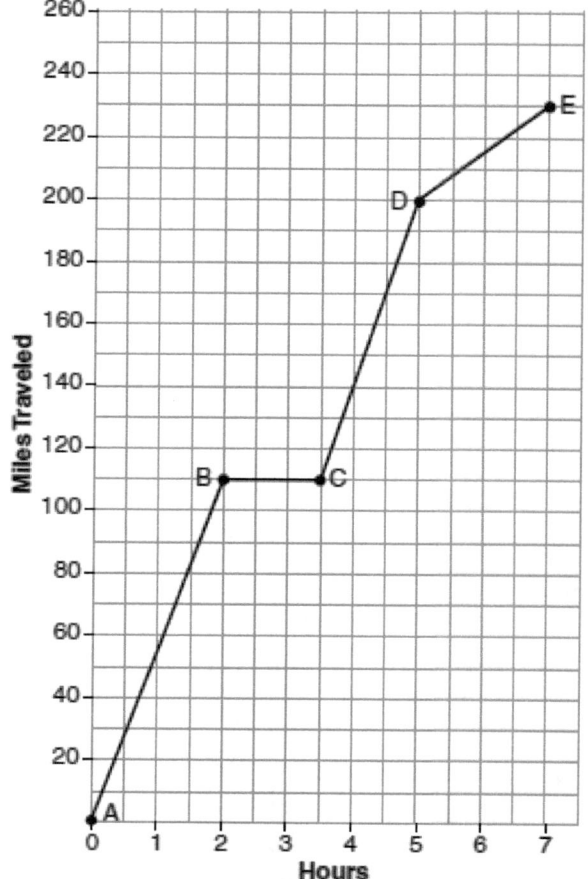

9.   The graph of $f(t)$ models the height, in feet, that a bee is flying above the ground with respect to the time it traveled in t seconds.

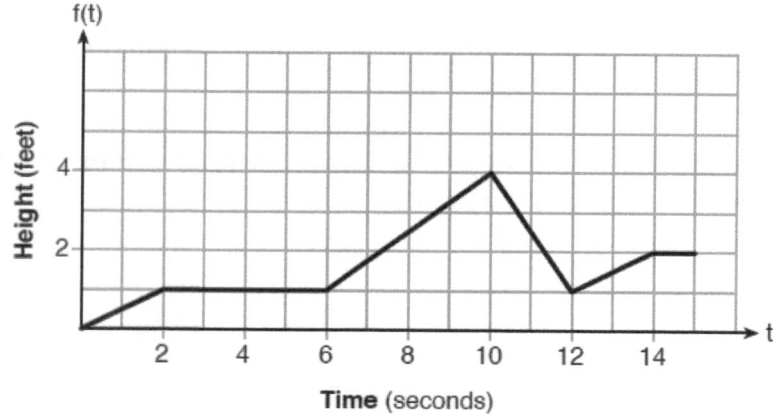

State all time intervals when the bee's rate of change is zero feet per second. Explain your reasoning.

10. One spring day, Elroy noted the time of day and the temperature, in degrees Fahrenheit. His findings are stated below.

At 6 a.m., the temperature was 50°F.
For the next 4 hours, the temperature rose 3° per hour.
The next 6 hours, it rose 2° per hour.
The temperature then stayed steady until 6 p.m.
For the next 2 hours, the temperature dropped 1° per hour.
The temperature then dropped steadily until the temperature was 56°F at midnight.

On the set of axes below, graph Elroy's data.

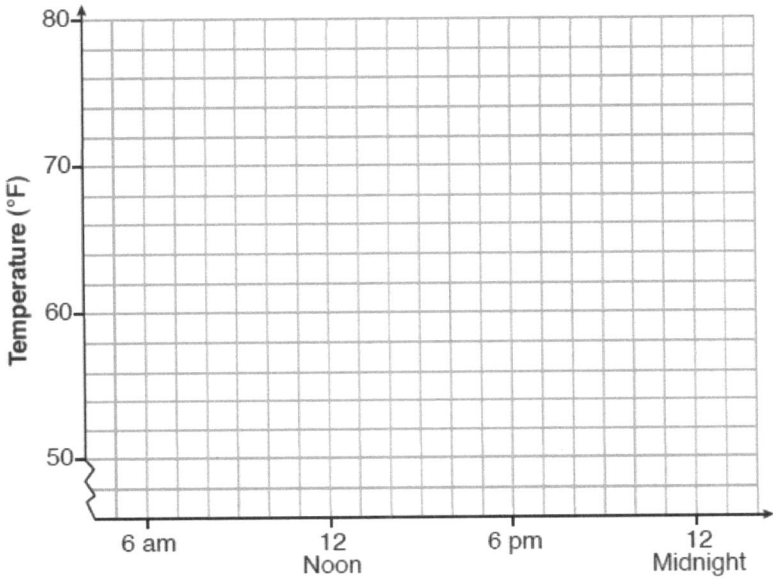

State the entire time interval for which the temperature was increasing.

Determine the average rate of change, in degrees per hour, from 6:00 p.m. to midnight.

11. A snowstorm started at midnight. For the first 4 hours, it snowed at an average rate of one-half inch per hour. The snow then started to fall at an average rate of one inch per hour for the next 6 hours. Then it stopped snowing for 3 hours. Then it started snowing again at an average rate of one-half inch per hour for the next 4 hours until the storm was over.

On the set of axes below, graph the amount of snow accumulated over the time interval of the storm.

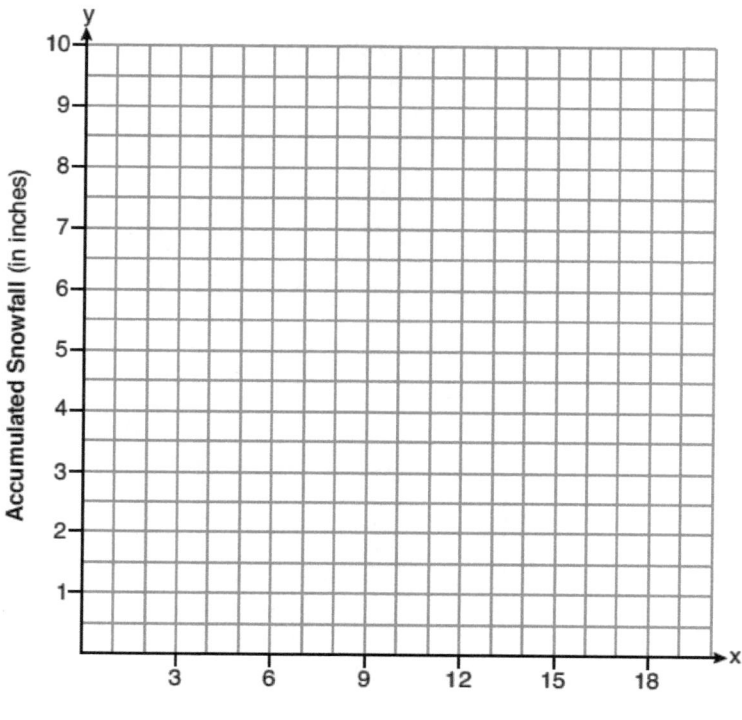

Elapsed Time (in hours)

Determine the average rate of snowfall over the length of the storm. State the rate, to the *nearest hundredth of an inch per hour.*

12. The graph below models the height of Sam's kite over a period of time.

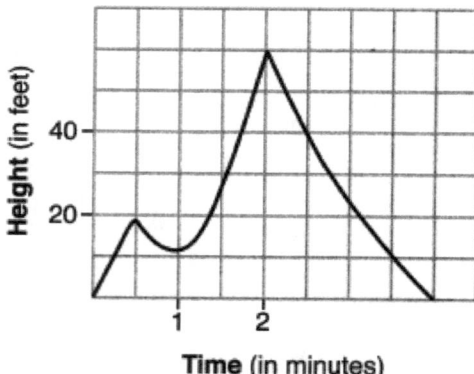

Time (in minutes)

Explain what the zeros of the graph represent in the context of the situation.
State the time intervals over which the height of the kite is increasing.
State the maximum height, in feet, that the kite reaches.

13. Thomas took a 140-mile bus trip to visit his grandparents. His trip is outlined on the graph below.

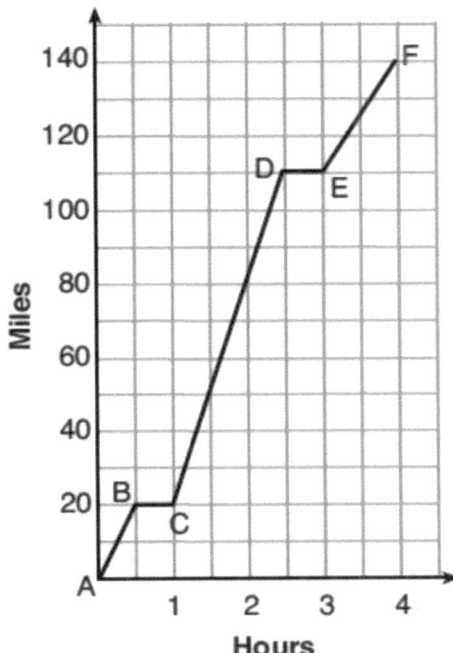

Explain what might have happened in the interval between *D* and *E*.
State the interval in which the bus traveled the fastest.
State how many miles per hour the bus was traveling during this interval.
What was the average rate of speed, in miles per hour, for Thomas' entire bus trip?

14. Anessa is studying the changes in population in a town. The graph below shows the population over 50 years.

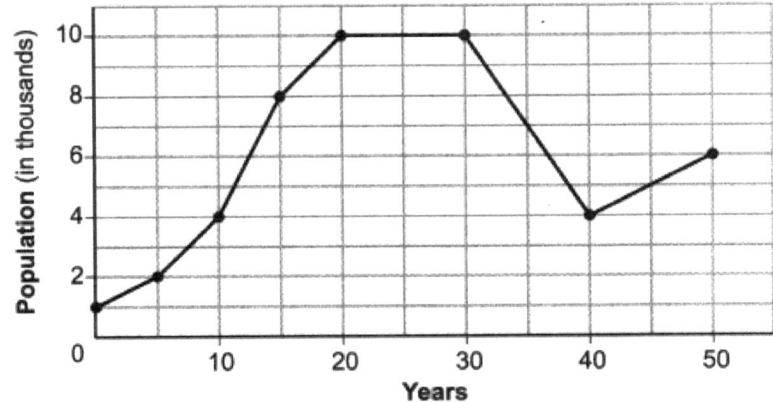

State the entire interval during which the population remained constant.
State the maximum population of the town over the 50-year period.
Determine the average rate of change from year 30 to year 40. Explain what your average rate of change means from year 30 to year 40 in the context of the problem.

15. Jean recorded temperatures over a 24-hour period one day in August in Syracuse, NY. Her results are shown in the table below.

Time (hour)	0	3	6	9	12	15	18	21	24
Temperature (°F)	80	75	70	78	92	89	85	80	74

Her data are modeled on the graph below.

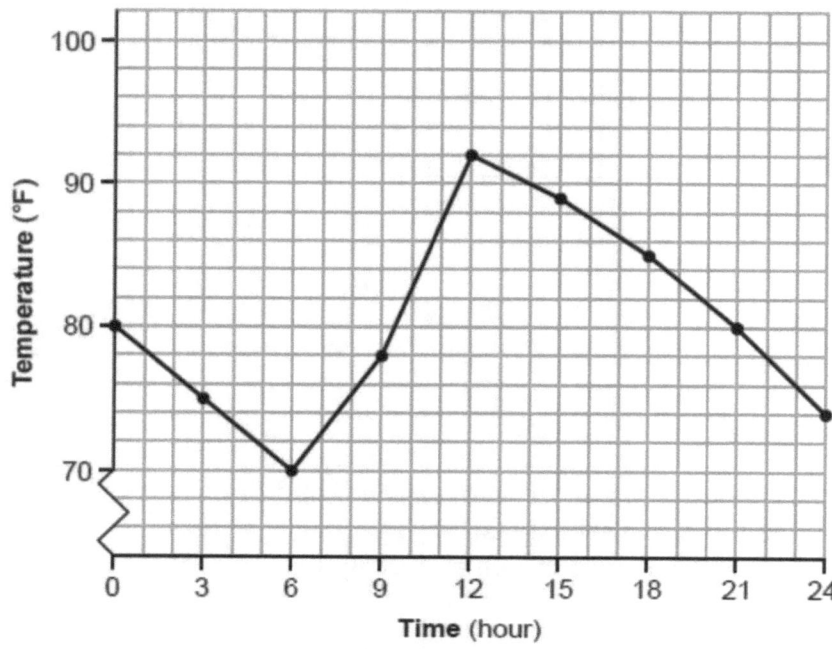

State the entire interval over which the temperature is increasing.

State the three-hour interval that has the greatest rate of change in temperature.

State the average rate of change from hour 12 to hour 24. Explain what this means in the context of the problem.

7.6   Systems of Functions

We can solve a system of functions graphically, by graphing both functions and finding their point of intersection. Or, we can solve the system algebraically, as follows.

We have seen that, if given two equations, such as $y = 3x - 5$ and $y = -x + 11$, we can solve the system of equations by *substitution*. We start by writing the new equation $3x - 5 = -x + 11$. We can then solve for x, and substitute this value of x into either of the original equations to find the corresponding value for y.

Note that in the example above, since y is already the isolated variable on the left sides of both equations, we can set the two expressions of x equal to each other, since they are both equal to y. This is known as the **transitive property of equality**: if $a = b$ and $b = c$, then $a = c$.

Similarly, if we have two functions of x, we can also determine when their values are the same by setting their expressions equal to each other and solving for the variable x.

Example: If $f(x) = 3x - 5$ and $g(x) = -x + 11$, then to determine when $f(x) = g(x)$, we set $3x - 5 = -x + 11$ and solve for x. Solving gives us $x = 4$.

Note that when we set two functions of x equal to each other, as in $f(x) = g(x)$, there are no y variables in the resulting equation. Therefore, the solutions to $f(x) = g(x)$ are the x-values at the points of intersection.

MODEL PROBLEM

Congress is considering two possible income tax plans for citizens. In the *flat* tax plan, citizens pay 30% of their entire income. In the *graduated* tax plan, citizens pay no taxes on the first $15,000 income but pay 35% of any income above $15,000. In function notation. write $f(x)$ to represent the taxes paid in the flat tax plan for an income of x, and write $g(x)$ to represent the taxes paid in the graduated tax plan for an income of x. Determine the amount of income x for which a citizen would pay the same amount in taxes under both plans.

Solution:

(A) $f(x) = 0.30x$
$g(x) = 0.35(x - 15000)$
$ = 0.35x - 5250$

(B) $0.30x = 0.35x - 5250$
$-0.05x = -5250$
$x = 105{,}000$
Income of $105,000

Explanation of steps:

(A)   Write the functions using the given information, simplifying if possible.

(B)   Set the function expressions equal to each other and solve for the variable.

PRACTICE PROBLEMS

1. Two health clubs, Club *A* and Club *B*, offer different membership plans. The cost for each plan includes a membership fee plus a monthly charge. The graph below represents the total membership costs of the two plans for one year.

   a) What is the membership fee for Club *A*?

   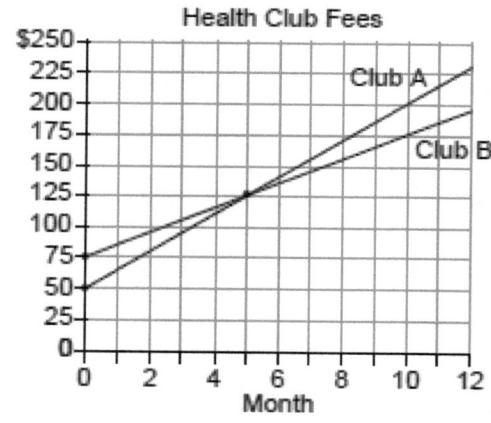

   b) What is the number of the month when the total cost is the same for both clubs, and what is the total cost for each club at that time?

   c) What is the monthly charge for Club *B*?

2. Both Tasha and her brother Tyson get an allowance of $5 each week. Tasha currently has $60 and decides to save her entire allowance each week. Tyson currently has $135 but spends all of his allowance plus an additional $10 each week. After how many weeks will they have the same amount of money?

3. A company produces and sells widgets. The company earns $25 for each widget sold, but it costs $20 per widget plus an annual fixed cost of $50,000 to produce them.

   a) In function notation, write $R(x)$ for the revenue earned by selling x widgets.

   b) In function notation, write $C(x)$ for the annual cost of producing x widgets.

   c) How many widgets need to be sold in a year to "break even" (that is, to have the cost equal to the revenue)?

4. Michael is trying to decide between two plumbing companies to fix his sink. The first company, Flow-Rite, charges $50 for a service call, plus an additional $36 per hour of labor. The second company, Gunk-Gone, charges $35 for a service call, plus an additional $39 per hour of labor. Let h represent the number of hours of labor.

   a) Write a rule for the function $f(h)$, which gives the cost of using Flow-Rite.

   b) Write a rule for the function $g(h)$, which gives the cost of using Gunk-Gone.

   c) At how many hours will the two companies charge the same amount of money?

Regents Questions

MULTIPLE CHOICE

1. Given the functions $h(x) = \frac{1}{2}x + 3$ and $j(x) = |x|$, which value of x makes $h(x) = j(x)$?

    (1) -2 (3) 3
    (2) 2   (4) -6

2. The functions $f(x)$ and $g(x)$ are graphed below.

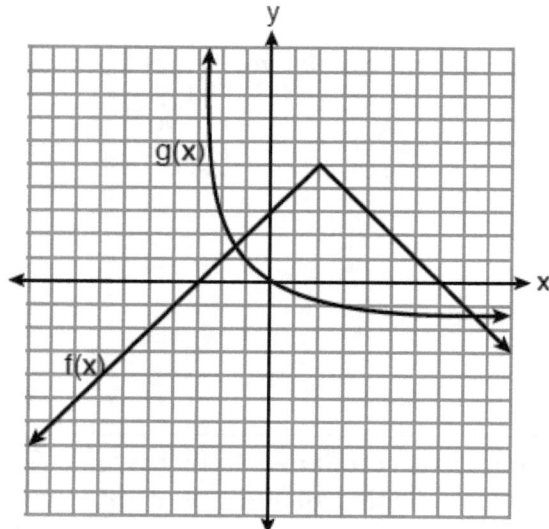

    Based on the graph, the solutions to the equation $f(x) = g(x)$ are
    (1) the x-intercepts
    (2) the y-intercepts
    (3) the x-values of the points of intersection
    (4) the y-values of the points of intersection

3. The functions $f(x)$ and $g(x)$ are graphed on the set of axes below.

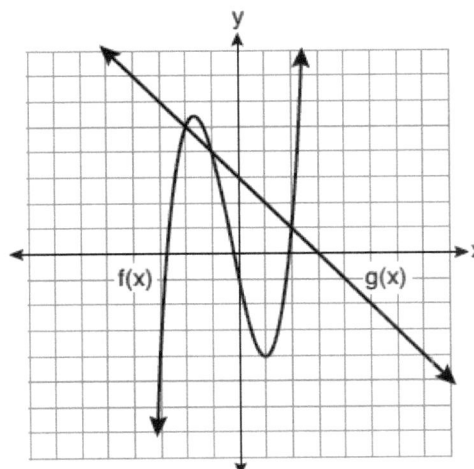

For which value of x is $f(x) \ne g(x)$?

(1) -1 (3) 3

(2) 2   (4) -2

4. If $f(x) = 2x + 6$ and $g(x) = |x|$ are graphed on the same coordinate plane, for which value of x is $f(x) = g(x)$?

(1) 6   (3) -2

(2) 2   (4) -6

CONSTRUCTED RESPONSE

5. A local business was looking to hire a landscaper to work on their property. They narrowed their choices to two companies. Flourish Landscaping Company charges a flat rate of $120 per hour. Green Thumb Landscapers charges $70 per hour plus a $1600 equipment fee.

Write a system of equations representing how much each company charges.

Determine and state the number of hours that must be worked for the cost of each company to be the same.

If it is estimated to take at least 35 hours to complete the job, which company will be less expensive? Justify your answer.

6.   Next weekend Marnie wants to attend either carnival *A* or carnival *B*. Carnival *A* charges $6 for admission and an additional $1.50 per ride. Carnival *B* charges $2.50 for admission and an additional $2 per ride.

In function notation, write $A(x)$ to represent the total cost of attending carnival *A* and going on *x* rides. In function notation, write $B(x)$ to represent the total cost of attending carnival *B* and going on *x* rides.

Determine the number of rides Marnie can go on such that the total cost of attending each carnival is the same.

Marnie wants to go on five rides. Determine which carnival would have the lower total cost. Justify your answer.

7.   Guy and Jim work at a furniture store. Guy is paid $185 per week plus 3% of his total sales in dollars, *x*, which can be represented by $g(x) = 185 + 0.03x$. Jim is paid $275 per week plus 2.5% of his total sales in dollars, *x*, which can be represented by $f(x) = 275 + 0.025x$. Determine the value of *x*, in dollars, that will make their weekly pay the same.

8.   A gardener is planting two types of trees:
        Type *A* is three feet tall and grows at a rate of 15 inches per year.
        Type *B* is four feet tall and grows at a rate of 10 inches per year.
    Algebraically determine exactly how many years it will take for these trees to be the same height.

9.   The graph below shows two functions, $f(x)$ and $g(x)$. State all the values of *x* for which $f(x) = g(x)$.

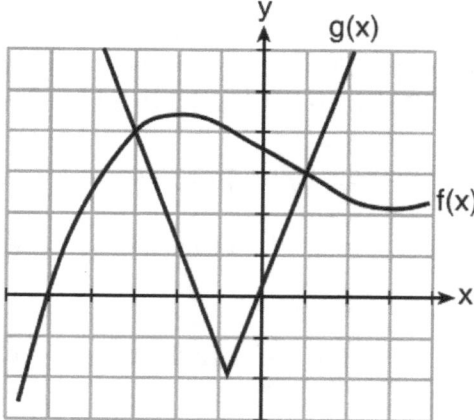

7.7   Combine Functions

We can perform **operations on functions**. For examples, we can define a function to be the sum, difference, product, or quotient of other functions.

Example: If $f(x) = x + 1$ and $g(x) = x - 2$, then we can find $h(x) = f(x) + g(x)$ by adding $(x + 1) + (x - 2) = 2x - 1$, so $h(x) = 2x - 1$.
We can check for any value of x. If $x = 5$, then $f(5) = 5 + 1 = 6$ and $g(5) = 5 - 2 = 3$, so $f(5) + g(5) = 6 + 3 = 9$. Also, $h(5) = 2(5) - 1 = 9$.

It is also common to use the following notation for combining functions:
$$(f + g)(x) = f(x) + g(x) \qquad (f - g)(x) = f(x) - g(x)$$

$$(fg)(x) = f(x)g(x) \qquad \left(\frac{f}{g}\right)(x) = \frac{f(x)}{g(x)} \text{ where } g(x) \neq 0$$

MODEL PROBLEM

If $f(x) = x^2 + 5$ and $g(x) = 2x - 1$, find $h(x) = f(x) - g(x)$. Find $h(-3)$.

Solution:

$h(x) = f(x) - g(x)$
(A) $= (x^2 + 5) - (2x - 1)$
(B) $= x^2 + 5 - 2x + 1$
$h(x) = x^2 - 2x + 6$
(C) $h(-3) = (-3)^2 - 2(-3) + 6$
$= 9 + 6 + 6 = 21$

Explanation of steps:

(A)   Substitute the rule for each function, using parentheses around each expression.
(B)   Simplify and express the resulting function rule in standard form.
(C)   To evaluate a function for a given value, substitute the value for the variable.
[To check, $h(-3)$ should equal
$f(-3) - g(-3) =$
$[(-3)^2 + 5] - [2(-3) - 1] =$
$14 - (-7) = 21.]$

315

PRACTICE PROBLEMS

1. If $f(x) = x^2 + x + 1$ and $g(x) = x - 5$, find $h(x) = f(x) + g(x)$.	2. If $f(x) = 2x + 1$ and $g(x) = x - 2$, find $h(x) = f(x) \cdot g(x)$.

3. To raise funds, a club is publishing and selling a calendar. The club has sold $500 in advertising and will sell copies of the calendar for $20 each. The cost of printing each calendar is $6. Let c be the number of calendars to be printed and sold.

   a) Write a rule for the function $R(c)$, which gives the revenue generated.

   b) Write a rule for the function $E(c)$, which gives the printing expenses.

   c) Describe how the function $P(c)$, which gives the club's profit, is related to $R(c)$ and $E(c)$, and write a rule for $P(c)$.

Regents Questions

MULTIPLE CHOICE

1. A company produces x units of a product per month, where $C(x)$ represents the total cost and $R(x)$ represents the total revenue for the month. The functions are modeled by $C(x) = 300x + 250$ and $R(x) = -0.5x^2 + 800x - 100$. The profit is the difference between revenue and cost where $P(x) = R(x) - C(x)$. What is the total profit, $P(x)$, for the month?

      (1) $P(x) = -0.5x^2 + 500x - 150$ (3) $P(x) = -0.5x^2 - 500x + 350$

      (2) $P(x) = -0.5x^2 + 500x - 350$ (4) $P(x) = -0.5x^2 + 500x + 350$

CONSTRUCTED RESPONSE

2. Given that $f(x) = 2x + 1$, find $g(x)$ if $g(x) = 2[f(x)]^2 - 1$.

CHAPTER 8.   EXPONENTIAL FUNCTIONS

8.1   Exponential Growth and Decay

When an amount is **increased** by a certain percent, r (for *rate*), we can calculate the new value by multiplying the original amount by $1 + r$, since $1 = 100\%$.
Example: A \$50 investment increases by 4%. To calculate the new value,
           $50(1 + 0.04) = 50(1.04) = \$52$.

Similarly, if an amount is **decreased** by a rate of r, we can multiply the amount by $1 - r$ to find the new value, since $1 = 100\%$.
Example: A \$50 investment loses 6% of its value. It is now worth
           $50(1 - 0.06) = 50(0.94) = \$47$.
           A 6% loss results in a new amount that is 94% of the original amount.

Often, an increase or decrease is calculated in regular intervals of time. This represents **exponential growth** (increase) or **exponential decay** (decrease).

An example of exponential growth is **compound interest**.
Example: Suppose a \$2,000 investment (principal) is invested in an account which earns 5%
           interest compounded annually over 4 years. This is represented by
           $2000(1.05)(1.05)(1.05)(1.05) = 2000(1.05)^4$.

Exponential decay can be used to calculate **depreciation**, which is a decrease in an asset's value over time.
Example: Suppose an industrial machine originally purchased for \$50,000 loses 10% of its
           value every year for 5 years. This is represented by
           $50000(0.90)(0.90)(0.90)(0.90)(0.90) = 50000(0.90)^5$.

The opposite of depreciation is **appreciation**, or the increase in an asset's value over time.

The **formula for exponential growth** is $y = a(1 + r)^x$, where a is the original amount, r is the constant rate of *increase*, x is the number of times the rate is applied, and y is the final amount. The **formula for exponential decay** is $y = a(1 - r)^x$ where r is the constant rate of *decrease*.

If we set b to the growth $(1 + r)$ or decay $(1 - r)$ factor, we can build a **general formula for exponential growth or decay**: $y = ab^x$ where $a > 0, b > 0, b \neq 1$. The formula represents *growth* when $b > 1$ or *decay* when $0 < b < 1$. The variable x is usually given in units of time. This can also be written in *function notation*; for example, $f(x) = ab^x$.

MODEL PROBLEM

The principal of a school predicts that enrollment at her school will increase by 15% each year for the next 4 years. If the current enrollment is 400 students, what would the enrollment be after the fourth year, *to the nearest whole number*, if the principal's predictions are true?

Solution:

    (A)   (B)

    $y = 400(1.15)^4 = 699.6025 \approx 700$

Explanation of steps:

    (A) Use the formula $y = ab^x$, where a is the original amount *[400]*, b is the growth factor *[115% or 1.15]*, and x is the number of times the rate is applied *[4, once each year]*.

    (B) Use your calculator to find the result, rounding as directed.

PRACTICE PROBLEMS

1. If x is the starting value before an exponential growth of 10% per day, write an expression for the value after 20 days.	2. If x is the starting value before an exponential decay of 2% per day, write an expression for the value after n days.
3. $2,500 is invested in a savings account that earns 3% interest compounded annually. Write an expression for the number of dollars in this account at the end of 4 years.	4. The current population of a town is 10,000. If the population increases by 20% each year, write an expression for population after t years?

5. If you invest $1500 in an account that pays 5% interest compounded annually, and you make no deposits or withdrawals on the account in 6 years, how much money do you have, to the *nearest cent*, at the end of 6 years?

6. If you invest $1000 in an account that pays 3% interest compounded annually, and you make no deposits or withdrawals on the account, how much money do you have, to the *nearest cent*, at the end of 5 years?

7. $2000 is invested in an account at a 3.5% interest rate compounded annually. No deposits or withdrawals are made on the account for 4 years. Determine, to the *nearest dollar*, the balance in the account after the 4 years.

8. A school raised $30,000 for an athletics fund. Each year the fund will decrease by 5%. Determine the amount of money, to the *nearest cent*, that will be left in the fund after 4 years.

9. A mouse population is 25,000 and is decreasing at a rate of 20% per year. What is the population after 3 years?

10. The population of Jacksonville is 3,810 and is growing at an annual rate of 3.5%. If this growth rate continues, what will be the approximate population in five years?

11. The value of a car purchased for $20,000 decreases at a rate of 12% per year. What will be the value of the car after 3 years?

12. A used car is purchased in July 2019 for $11,900. If the car depreciates (loses) 13% of its value each year, what is the value of the car, to the nearest hundred dollars, in July 2022?

13. On January 1, 2000, the price of gasoline was $1.39 per gallon. If the price increased by 0.5% per month, what was the cost of one gallon of gasoline, to the nearest cent, on January 1, 2001?

14. In a certain game tournament, 75% of the players are eliminated each round. If the tournament starts with 256 players, how many remain after three rounds?

Regents Questions

Multiple Choice

1. Krystal was given $3000 when she turned 2 years old. Her parents invested it at a 2% interest rate compounded annually. No deposits or withdrawals were made. Which expression can be used to determine how much money Krystal had in the account when she turned 18?

    (1) $3000(1 + 0.02)^{16}$

    (2) $3000(1 - 0.02)^{16}$

    (3) $3000(1 + 0.02)^{18}$

    (4) $3000(1 - 0.02)^{18}$

2. The value in dollars, $v(x)$, of a certain car after x years is represented by the equation $v(x) = 25,000(0.86)^x$. To the *nearest dollar*, how much more is the car worth after 2 years than after 3 years?

    (1) 2589

    (2) 6510

    (3) 15,901

    (4) 18,490

3. The function $V(t) = 1350(1.017)^t$ represents the value $V(t)$, in dollars, of a comic book t years after its purchase. The yearly rate of appreciation of the comic book is

    (1) 17%

    (2) 1.7%

    (3) 1.017%

    (4) 0.017%

4. The country of Benin in West Africa has a population of 9.05 million people. The population is growing at a rate of 3.1% each year. Which function can be used to find the population 7 years from now?

    (1) $f(t) = (9.05 \times 10^6)(1 - 0.31)^7$

    (2) $f(t) = (9.05 \times 10^6)(1 + 0.31)^7$

    (3) $f(t) = (9.05 \times 10^6)(1 + 0.031)^7$

    (4) $f(t) = (9.05 \times 10^6)(1 - 0.031)^7$

5. For a recently released movie, the function $y = 119.67(0.61)^x$ models the revenue earned, y, in millions of dollars each week, x, for several weeks after its release. Based on the equation, how much more money, in millions of dollars, was earned in revenue for week 3 than for week 5?

    (1) 37.27

    (2) 27.16

    (3) 17.06

    (4) 10.11

6. The equation $A = 1300(1.02)^7$ is being used to calculate the amount of money in a savings account. What does 1.02 represent in this equation?

    (1) 0.02% decay

    (2) 0.02% growth

    (3) 2% decay

    (4) 2% growth

7. A student invests $500 for 3 years in a savings account that earns 4% interest per year. No further deposits or withdrawals are made during this time. Which statement does *not* yield the correct balance in the account at the end of 3 years?

    (1) $500(1.04)^3$

    (2) $500(1-.04)^3$

    (3) $500(1+.04)(1+.04)(1+.04)$

    (4) $500 + 500(.04) + 520(.04) + 540.8(.04)$

8. The table below shows the temperature, $T(m)$, of a cup of hot chocolate that is allowed to chill over several minutes, m.

Time, m (minutes)	0	2	4	6	8
Temperature, T(m) (°F)	150	108	78	56	41

    Which expression best fits the data for $T(m)$?

    (1) $150(0.85)^m$ (3) $150(0.85)^{m-1}$

    (2) $150(1.15)^m$ (4) $150(1.15)^{m-1}$

9. Milton has his money invested in a stock portfolio. The value, $v(x)$, of his portfolio can be modeled with the function $v(x) = 30,000(0.78)^x$, where x is the number of years since he made his investment. Which statement describes the rate of change of the value of his portfolio?

    (1) It decreases 78% per year.   (3) It increases 78% per year.

    (2) It decreases 22% per year.   (4) It increases 22% per year.

10. The 2014 winner of the Boston Marathon runs as many as 120 miles per week. During the last few weeks of his training for an event, his mileage can be modeled by $M(w) = 120(.90)^{w-1}$, where w represents the number of weeks since training began. Which statement is true about the model $M(w)$?

    (1) The number of miles he runs will increase by 90% each week.

    (2) The number of miles he runs will be 10% of the previous week.

    (3) $M(w)$ represents the total mileage run in a given week.

    (4) w represents the number of weeks left until his marathon.

11. Anne invested $1000 in an account with a 1.3% annual interest rate. She made no deposits or withdrawals on the account for 2 years. If interest was compounded annually, which equation represents the balance in the account after the 2 years?

    (1) $A = 1000(1 - 0.013)^2$ (3) $A = 1000(1 - 1.3)^2$

    (2) $A = 1000(1 + 0.013)^2$ (4) $A = 1000(1 + 1.3)^2$

12. If a population of 100 cells triples every hour, which function represents $p(t)$, the population after t hours?

   (1) $p(t) = 3(100)^t$ (3) $p(t) = 3t + 100$

   (2) $p(t) = 100(3)^t$ (4) $p(t) = 100t + 3$

13. Mario's $15,000 car depreciates in value at a rate of 19% per year. The value, V, after t years can be modeled by the function $V = 15,000(0.81)^t$. Which function is equivalent to the original function?

   (1) $V = 15,000(0.9)^{9t}$ (3) $V = 15,000(0.9)^{\frac{t}{9}}$

   (2) $V = 15,000(0.9)^{2t}$ (4) $V = 15,000(0.9)^{\frac{t}{2}}$

14. The Ebola virus has an infection rate of 11% per day as compared to the SARS virus, which has a rate of 4% per day. If there were one case of Ebola and 30 cases of SARS initially reported to authorities and cases are reported each day, which statement is true?

   (1) At day 10 and day 53 there are more Ebola cases.

   (2) At day 10 and day 53 there are more SARS cases.

   (3) At day 10 there are more SARS cases, but at day 53 there are more Ebola cases.

   (4) At day 10 there are more Ebola cases, but at day 53 there are more SARS cases.

15. Jill invests $400 in a savings bond. The value of the bond, $V(x)$, in hundreds of dollars after x years is illustrated in the table below.

x	V(x)
0	4
1	5.4
2	7.29
3	9.84

Which equation and statement illustrate the approximate value of the bond in hundreds of dollars over time in years?

   (1) $V(x) = 4(0.65)^x$, and it grows.   (3) $V(x) = 4(1.35)^x$, and it grows.

   (2) $V(x) = 4(0.65)^x$, and it decays.   (4) $V(x) = 4(1.35)^x$, and it decays.

16. Marc bought a new laptop for $1250. He kept track of the value of the laptop over the next three years, as shown in the table below.

Years After Purchase	Value in Dollars
1	1000
2	800
3	640

Which function can be used to determine the value of the laptop for x years after the purchase?

(1) $f(x) = 1000(1.2)^x$ (3) $f(x) = 1250(1.2)^x$

(2) $f(x) = 1000(0.8)^x$ (4) $f(x) = 1250(0.8)^x$

17. A population of bacteria can be modeled by the function $f(x) = 1000(0.98)^t$, where t represents the time since the population started decaying, and $f(t)$ represents the population of the remaining bacteria at time t. What is the rate of decay for this population?

(1) 98%   (3) 0.98%

(2) 2%   (4) 0.02%

18. A high school sponsored a badminton tournament. After each round, one-half of the players were eliminated. If there were 64 players at the start of the tournament, which equation models the number of players left after 3 rounds?

(1) $y = 64(1 - .5)^3$ (3) $y = 64(1 - .3)^{0.5}$

(2) $y = 64(1 + .5)^3$ (4) $y = 64(1 + .3)^{0.5}$

19. The equation $V(t) = 12,000(0.75)^t$ represents the value of a motorcycle t years after it was purchased. Which statement is true?

(1) The motorcycle cost $9000 when purchased.

(2) The motorcycle cost $12,000 when purchased.

(3) The motorcycle's value is decreasing at a rate of 75% each year.

(4) The motorcycle's value is decreasing at a rate of 0.25% each year.

20. In the equation $A = P(1 \pm r)^t$, A is the total amount, P is the principal amount, r is the annual interest rate, and t is the time in years. Which statement correctly relates information regarding the annual interest rate for each given equation?

(1) For $A = P(1.025)^t$, the principal amount of money is increasing at a 25% interest rate.

(2) For $A = P(1.0052)^t$, the principal amount of money is increasing at a 52% interest rate.

(3) For $A = P(0.86)^t$, the principal amount of money is decreasing at a 14% interest rate.

(4) For $A = P(0.68)^t$, the principal amount of money is decreasing at a 68% interest rate.

21. Emily was given $600 for her high school graduation. She invested it in an account that earns 2.4% interest per year. If she does *not* make any deposits or withdrawals, which expression can be used to determine the amount of money that will be in the account after 4 years?

    (1) $600(1 + 0.24)^4$ (3) $600(1 + 0.024)^4$

    (2) $600(1 - 0.24)^4$ (4) $600(1 - 0.024)^4$

22. Sunny purchases a new car for $29,873. The car depreciates 20% annually. Which expression can be used to determine the value of the car after t years?

    (1) $29,873(.20)^t$ (3) $29,873(1 - .20)^t$

    (2) $29,873(20)^t$ (4) $29,873(1 + .20)^t$

23. Mike uses the equation $b = 1300(2.65)^x$ to determine the growth of bacteria in a laboratory setting. The exponent represents

    (1) the total number of bacteria currently present

    (2) the percent at which the bacteria are growing

    (3) the initial amount of bacteria

    (4) the number of time periods

24. Jim uses the equation $A = P(1 + 0.05)^t$ to find the amount of money in an account, A, of an investment, P, after t years. For this equation, which phrase describes the yearly rate of change?

    (1) decreasing by 5%   (3) increasing by 5%

    (2) decreasing by 0.05%   (4) increasing by 0.05%

25. Which situation represents exponential growth?

    (1) Aidan adds $10 to a jar each week.

    (2) A pine tree grows 1.5 feet per year.

    (3) Ella earns $20 per hour babysitting.

    (4) The number of people majoring in computer science doubles every 5 years.

26. Joe deposits $4000 into a certificate of deposit (CD) at his local bank. The CD earns 3% interest, compounded annually. The value of the CD in x years can be found using the function

    (1) $f(x) = 4000 + 0.3x$ (3) $f(x) = 4000(1.3)^x$

    (2) $f(x) = 4000 + 0.03x$ (4) $f(x) = 4000(1.03)^x$

CONSTRUCTED RESPONSE

27. The breakdown of a sample of a chemical compound is represented by the function $p(t) = 300(0.5)^t$, where $p(t)$ represents the number of milligrams of the substance and t represents the time, in years. In the function $p(t)$, explain what 0.5 and 300 represent.

28. Rhonda deposited $3000 in an account in the Merrick National Bank, earning 4.2% interest, compounded annually. She made no deposits or withdrawals. Write an equation that can be used to find B, her account balance after t years.

29. Dylan invested $600 in a savings account at a 1.6% annual interest rate. He made no deposits or withdrawals on the account for 2 years. The interest was compounded annually. Find, to the *nearest cent*, the balance in the account after 2 years.

30. The number of carbon atoms in a fossil is given by the function $y = 5100(0.95)^x$, where x represents the number of years since being discovered.

    What is the percent of change each year? Explain how you arrived at your answer.

31. The value, $v(t)$, of a car depreciates according to the function $v(t) = P(.85)^t$, where P is the purchase price of the car and t is the time, in years, since the car was purchased. State the percent that the value of the car *decreases* by each year. Justify your answer.

32. A population of rabbits in a lab, $p(x)$, can be modeled by the function $p(x) = 20(1.014)^x$, where x represents the number of days since the population was first counted. Explain what 20 and 1.014 represent in the context of the problem.

    Determine, to the *nearest tenth*, the average rate of change from day 50 to day 100.

33. A car was purchased for $25,000. Research shows that the car has an average yearly depreciation rate of 18.5%. Create a function that will determine the value, $V(t)$, of the car t years after purchase. Determine, to the *nearest cent*, how much the car will depreciate from year 3 to year 4.

34. Marilyn collects old dolls. She purchases a doll for $450. Research shows this doll's value will increase by 2.5% each year. Write an equation that determines the value, V, of the doll t years after purchase.

    Assuming the doll's rate of appreciation remains the same, will the doll's value be doubled in 20 years? Justify your reasoning.

35. On the day Alexander was born, his father invested $5000 in an account with a 1.2% annual growth rate. Write a function, $A(t)$, that represents the value of this investment t years after Alexander's birth.

    Determine, to the *nearest dollar*, how much more the investment will be worth when Alexander turns 32 than when he turns 17.

8.2   Graphs of Exponential Functions

An **exponential function** is a function in which x appears as an exponent in the equation. For this chapter, the general form can be written as $f(x) = ab^x$ where $a \neq 0, b > 0$, and $b \neq 1$.
Examples: $f(x) = 5^x$ or $g(x) = -3(0.5)^x$

An exponential function can be graphed using a table or a graphing calculator.
Example: We can graph $f(x) = 2^x$ as follows, using $y = f(x)$.

x	2^x	$y = f(x)$	(x, y)
-1	2^{-1}	$\frac{1}{2}$	$\left(-1, \frac{1}{2}\right)$
0	2^0	1	$(0,1)$
1	2^1	2	$(1,2)$
2	2^2	4	$(2,4)$
3	2^3	8	$(3,8)$

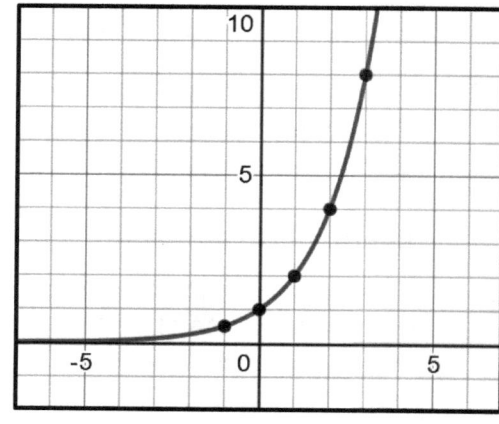

CALCULATOR TIP

To graph $y = 2^x$ on the calculator:

Press $\boxed{Y=}\boxed{2}\boxed{\wedge}\boxed{X,T,\Theta,n}\boxed{GRAPH}$

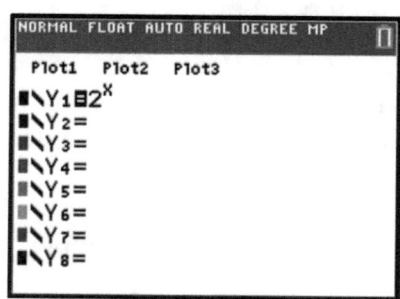

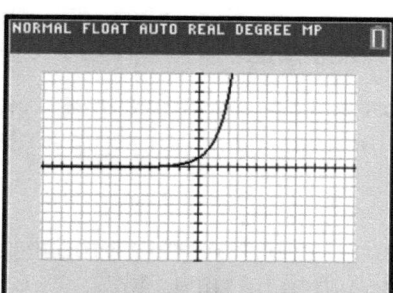

The behavior of the graph of an exponential function $f(x) = ab^x$ will depend on the values of a and b:

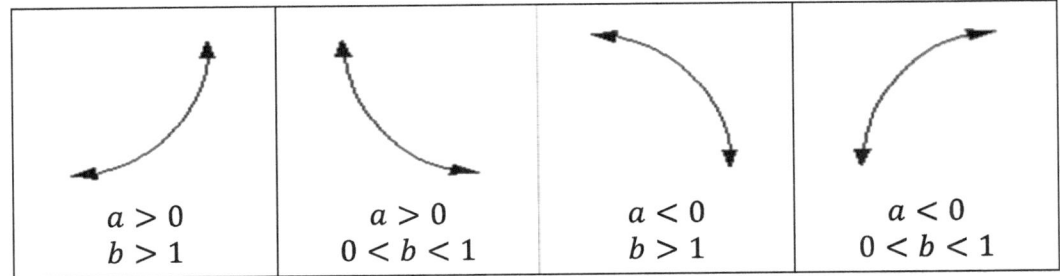

$a > 0$ $b > 1$	$a > 0$ $0 < b < 1$	$a < 0$ $b > 1$	$a < 0$ $0 < b < 1$

Example: The graph of the function $y = \left(\frac{1}{2}\right)^x$ looks like this.

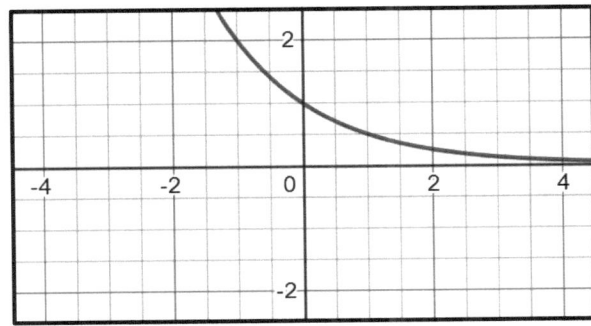

In general, $y = b^x$ will **increase** over the entire domain when $b > 1$ or **decrease** over the entire domain when $0 < b < 1$.

In the function $y = ab^x$, the **constant a** will tell us where the curve crosses the y-axis. Therefore, a is the y-intercept. (To find the y-intercept, substitute 0 for x; $b^0 = 1$.)

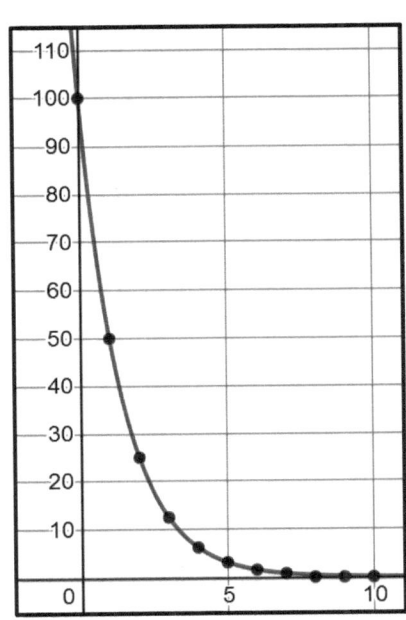

Example: For the function $y = 100\left(\frac{1}{2}\right)^x$, where $a = 100$ and $b = \frac{1}{2}$, the graph would intersect the y-axis at (0,100), as shown to the right.

An exponential function $y = ab^x$ will **never actually touch** (intersect) the x-axis. Neither a nor b is equal to zero, so ab^x (and therefore y) will never equal zero.

When a **is negative**, the graph of $y = ab^x$ is negative over the entire domain.

Example: The graph of $y = -3^x$ will look like the graph of $y = 3^x$ but **reflected** (flipped) over the x-axis.

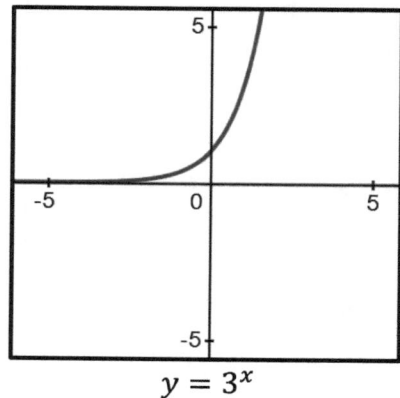

$$y = 3^x$$

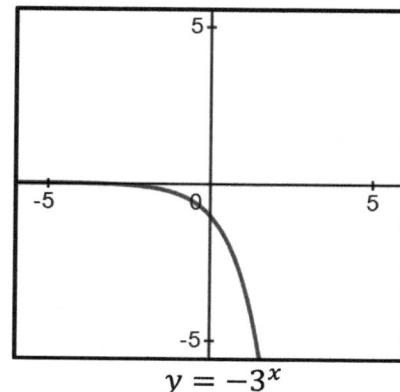

$$y = -3^x$$

Exponential functions may be used to represent **exponential growth** and **exponential decay**. In these special cases, the equation $y = ab^x$ is limited to $a > 0$.

- When $b > 1$, the function shows **exponential growth**.
- When $0 < b < 1$, the function shows **exponential decay**.

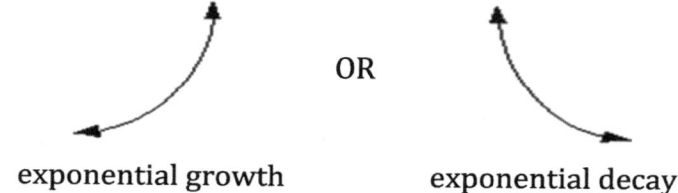

OR

exponential growth   exponential decay

In these cases, the y-intercept, a, represents the starting value at time $x = 0$. The value of b equals 1 plus or minus the percent of change at each of the x intervals of time.

Example: The function $y = 1000(1.05)^x$ represents the value of a bank account that is compounded annually for x years. The y-intercept (or starting value, a) is $1000, while $b = 1.05$ shows an exponential growth (interest) of 5% per year.

Finding the equation from a set of points:

If given a table or graph of a exponential function, you can use the calculator to find the equation. After entering the coordinates of the points, calculate the **exponential regression** using the calculator's ExpReg function, as described below.

Note that the exponential regression can only be calculated for a function that is always *positive* (that is, when $a > 0$). However, for a function with all negative y values, you can enter their additive inverses (positive values of y that produce a reflection of the function over the x-axis) and then negate the a that is found by the regression.

 CALCULATOR TIP

To find the equation for an exponential function:

1. Press STAT 1 to select Edit.
2. If values appear in the L1 or L2 columns, select the column heading and press CLEAR ENTER.
3. Enter the x values into the L1 column and the corresponding y values into the L2 column.
4. Press STAT <CALC> 0 for ExpReg.
5. On the next screen prompt, make sure L1 and L2 are selected for Xlist and Ylist. Next to Store RegEQ, enter ALPHA [F4] 1 to store the equation in Y1.

    [On the TI-83, you'll see an ExpReg prompt instead. Enter VARS <Y-VARS> 1 1.]
6. The screen will show the equation y=a*b^x along with the values of a and b.
7. To view the graph, press GRAPH. To see the equation, press Y=.

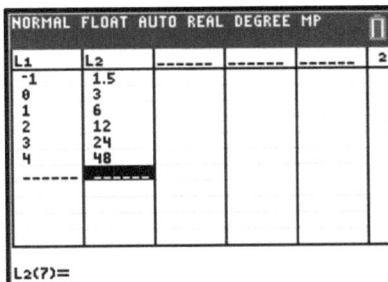 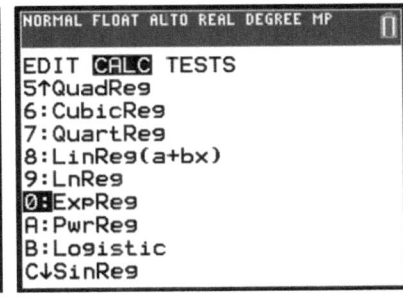 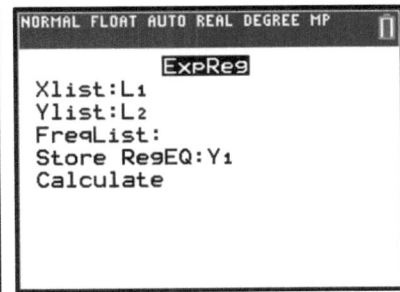

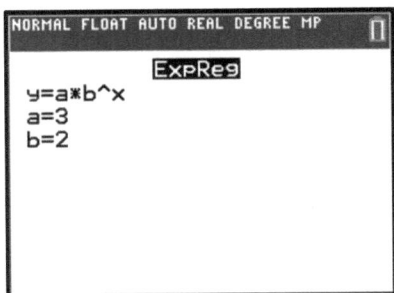 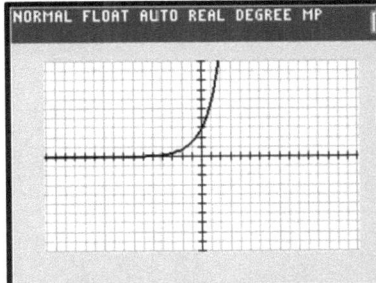 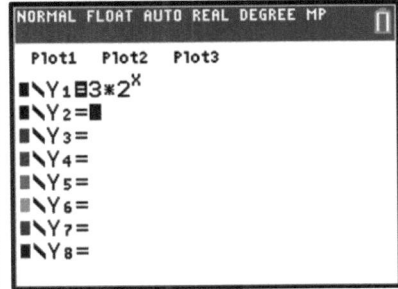

MODEL PROBLEM 1: *GRAPH AN EXPONENTIAL FUNCTION*

Graph the function $y = \frac{3}{2} \cdot 2^x$.

Solution:

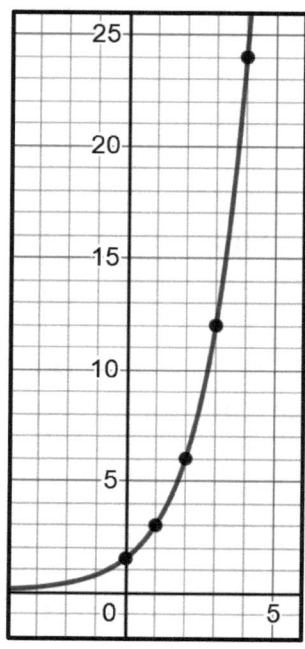

(A) x	(B) $\frac{3}{2} \cdot 2^x$	(C) y	(D) (x, y)
0	$\frac{3}{2} \cdot 2^0 = \frac{3}{2} \cdot 1$	1.5	(0,1.5)
1	$\frac{3}{2} \cdot 2^1 = \frac{3}{2} \cdot 2$	3	(1,3)
2	$\frac{3}{2} \cdot 2^2 = \frac{3}{2} \cdot 4$	6	(2,6)
3	$\frac{3}{2} \cdot 2^3 = \frac{3}{2} \cdot 8$	12	(3,12)
4	$\frac{3}{2} \cdot 2^4 = \frac{3}{2} \cdot 16$	24	(4,24)

Explanation of steps:
(A) Pick values of x.
(B) Substitute the values of x into the expression.
(C) Evaluate for y.
(D) Plot the resulting points on the graph.

PRACTICE PROBLEMS

1. Which graph may represent the exponential decay of a radioactive element?

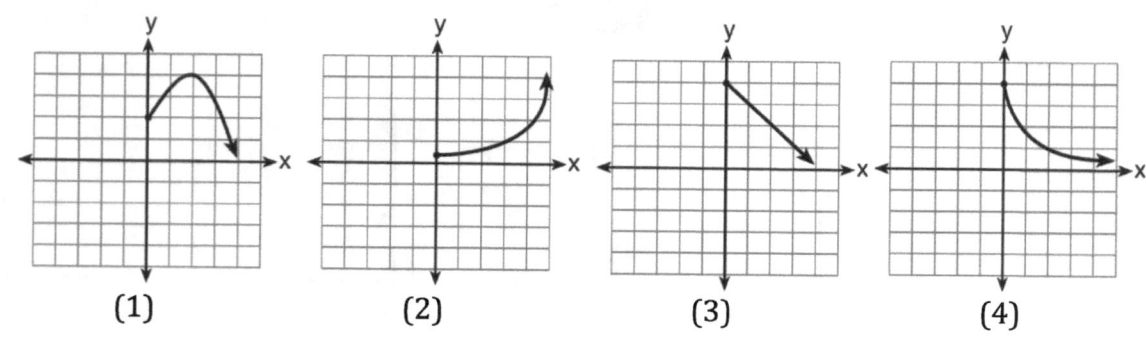

(1)   (2)   (3)   (4)

332

2. On the grid below, graph $y = 2^x$ over the interval $-1 \leq x \leq 3$.

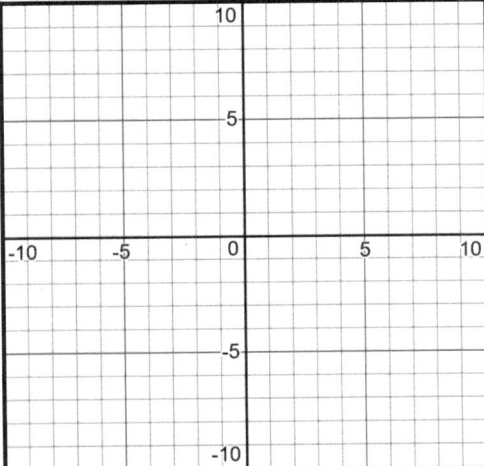

3. On the grid below, graph $y = 3^x$ over the interval $-1 \leq x \leq 2$.

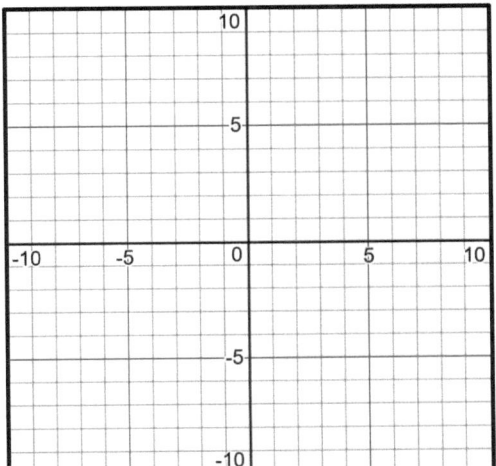

4. On the calculator, graph $y = \frac{1}{3} \cdot 2^x$. Sketch the graph below.

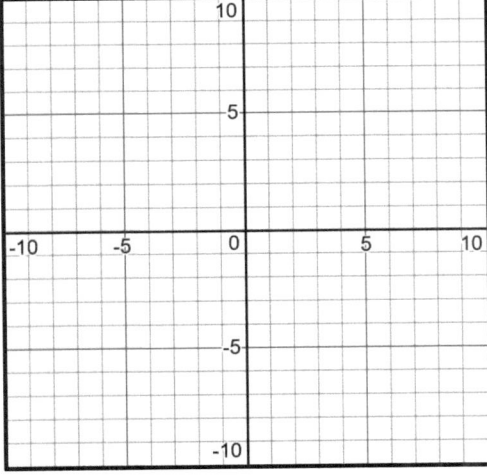

5. On the calculator, graph $y = 3 \cdot \left(\frac{1}{2}\right)^x$. Sketch the graph below.

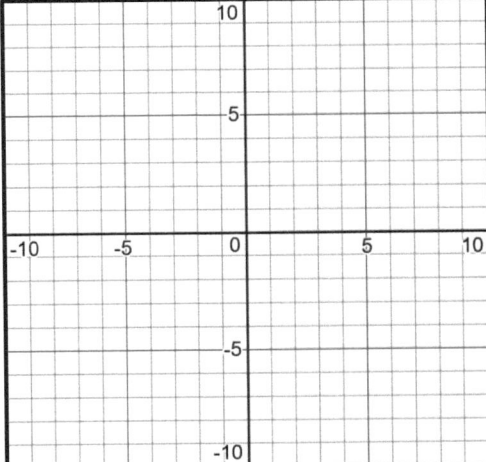

6. On the calculator, graph $y = 12(1.5)^x$. Sketch the graph below.

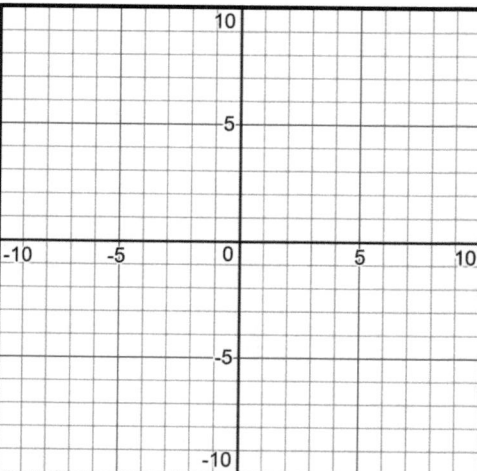

7. On the calculator, graph $y = 12(0.5)^x$. Sketch the graph below.

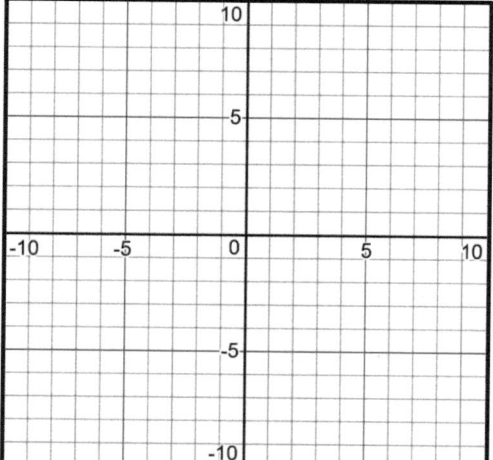

8. On the grid below, graph $y = -2^x$.

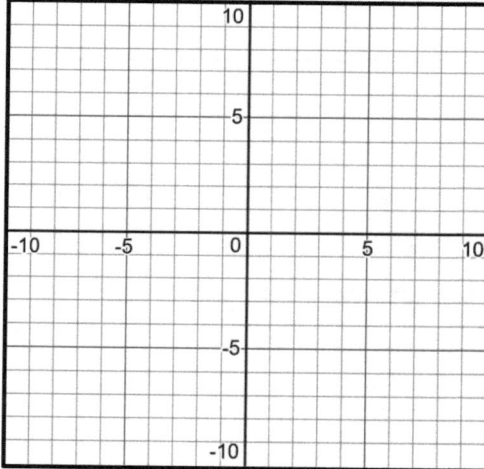

9. On the grid below, graph $y = 2^x - 5$ and explain how this graph differs from the exponential function $y = 2^x$.

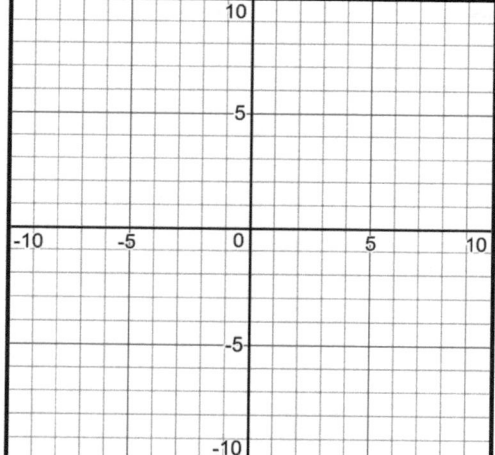

MODEL PROBLEM 2:   *DETERMINE THE EQUATION*

Write an equation for the exponential function using the values given in the table below.

x	1	2	3	4	5
y	0.75	2.25	6.75	20.25	60.75

Solution:

$$y = 0.25(3)^x$$

Explanation of steps:

On the calculator, press $\boxed{\text{STAT}}\boxed{1}$ to enter the values as L1 and L2, then use $\boxed{\text{STAT}}$<CALC>$\boxed{0}$ for ExpReg to determine the equation. *[a = 0.25 and b = 3]*

PRACTICE PROBLEMS

10. Write an equation for the exponential function using values given in the table below.

x	0	1	2	3	4
y	0.1	0.4	1.6	6.4	25.6

11. Write an equation for the exponential function graphed below.

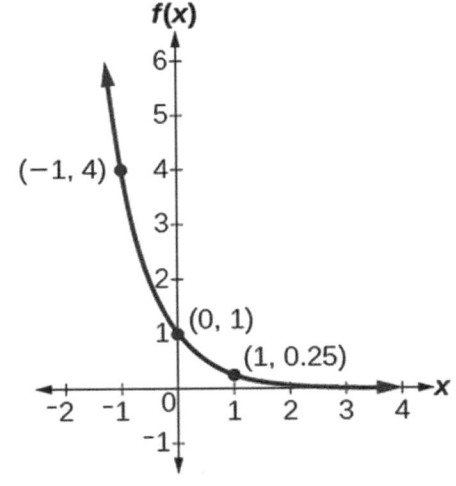

12. A spectrophotometer measures the concentration of ozone in the atmosphere at different altitudes, as shown by the table below. Write the exponential regression equation that models these data, rounding all values to the *nearest thousandth*.

Concentration of Ozone

Altitude (x)	Ozone Units (y)
0	0.7
5	0.6
10	1.1
15	3.0
20	4.9

13. The amount of radioactive substance decreases over times as a result of radioactive decay. The table below shows the amount of a certain radiactive substance remaining for selected years after 1990.

Years After 1990 (x)	0	2	5	9	14	17	19
Amount (y)	750	451	219	84	25	12	8

a) Write an exponential regression equation for this set of data, rounding all values to the *nearest thousandth*.

b) Using this equation, determine the amount of the substance that remained in 2002, to the *nearest integer*.

14. When a piece of paper is folded in half, the total thickness doubles. An unfolded piece of paper is 0.1 millimeter thick.

   a) Write an equation for the total thickness, t, as an exponential function of the number of folds, n.

   b) Graph the function for a discrete domain of whole numbers, $0 \leq n \leq 8$.

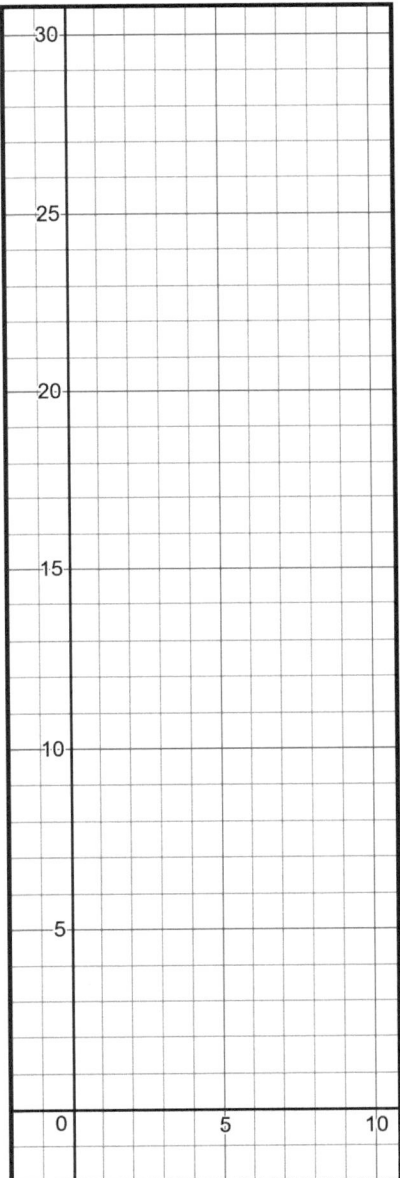

15. A ball is dropped from an initial height of 30 inches. On its first two bounces, the ball reaches heights of 15 and 7.5 inches.

   a) Write an equation for the height of the ball, h, as an exponential function of the number of bounces, n.

   b) Graph the function for a discrete domain of whole numbers, $0 \leq n \leq 6$.

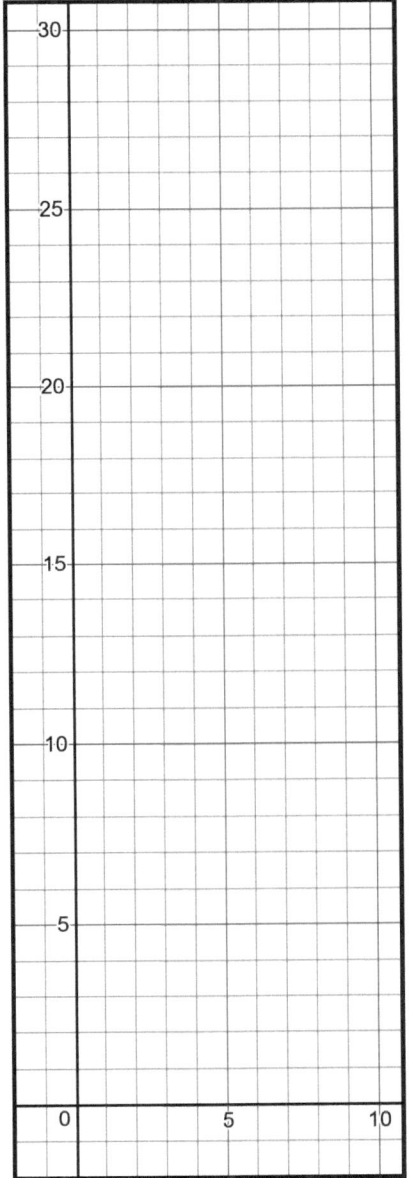

16. The average salary of baseball players in 1984 was $290,000. The table below shows the players' average salary over the following nine years.

Baseball Players' Salaries

Numbers of Years Since 1984	Average Salary (thousands of dollars)
0	290
1	320
2	400
3	495
4	600
5	700
6	820
7	1,000
8	1,250
9	1,580

a) Using this data, create a scatter plot on the grid below.

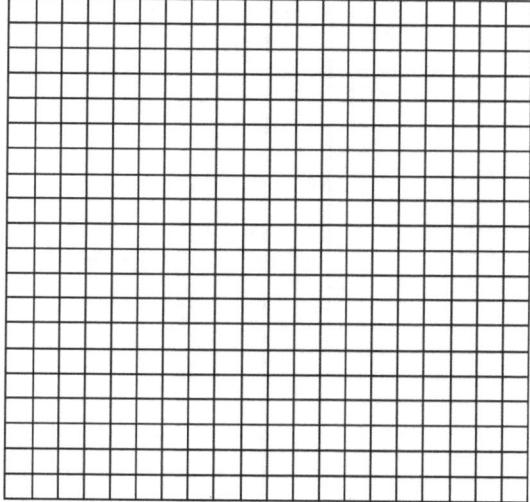

b) State the exponential regression equation with all values rounded to the *nearest hundredth*.

c) Using this equation, estimate the salary of a baseball player in the year 2005, to the *nearest thousand dollars*.

Regents Questions

MULTIPLE CHOICE

1. A population that initially has 20 birds approximately doubles every 10 years. Which graph represents this population growth?

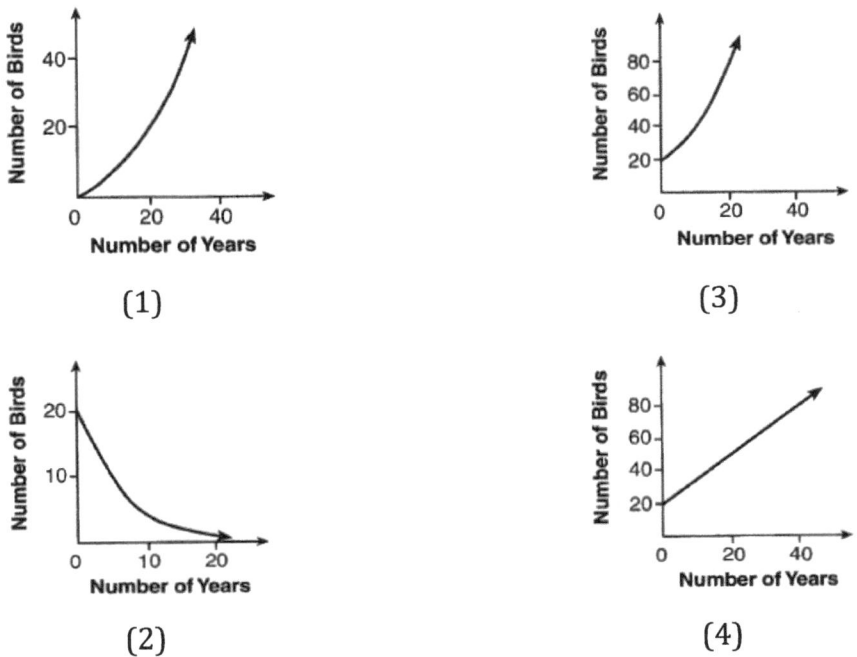

(1)   (3)

(2)   (4)

2. Some banks charge a fee on savings accounts that are left inactive for an extended period of time. The equation $y = 5000(0.98)^x$ represents the value, y, of one account that was left inactive for a period of x years.

   What is the y-intercept of this equation and what does it represent?
   (1) 0.98, the percent of money in the account initially
   (2) 0.98, the percent of money in the account after x years
   (3) 5000, the amount of money in the account initially
   (4) 5000, the amount of money in the account after x years

3.   The population of a small town over four years is recorded in the chart below, where 2013 is represented by $x = 0$. [Population is rounded to the nearest person]

Year	2013	2014	2015	2016
Population	3810	3943	4081	4224

The population, $P(x)$, for these years can be modeled by the function $P(x) = ab^x$ where b is rounded to the nearest thousandth. Which statements about this function are true?

 I. $a = 3810$
 II. $a = 4224$
 III. $b = 0.035$
 IV. $b = 1.035$

(1) I and III   (3) II and III
(2) I and IV   (4) II and IV

CONSTRUCTED RESPONSE

4.   Write an exponential equation for the graph shown below.

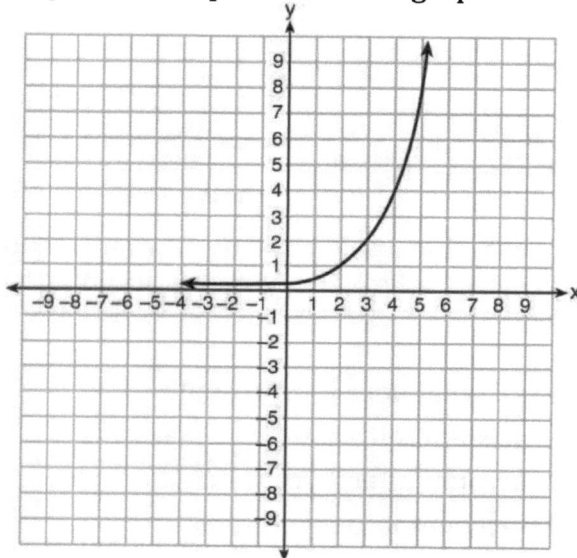

Explain how you determined the equation.

5. An application developer released a new app to be downloaded. The table below gives the number of downloads for the first four weeks after the launch of the app.

Number of Weeks	1	2	3	4
Number of Downloads	120	180	270	405

Write an exponential equation that models these data.

Use this model to predict how many downloads the developer would expect in the 26th week if this trend continues. Round your answer to the *nearest download*.

Would it be reasonable to use this model to predict the number of downloads past one year? Explain your reasoning.

6. Graph the function $f(x) = 2^x - 7$ on the set of axes below.

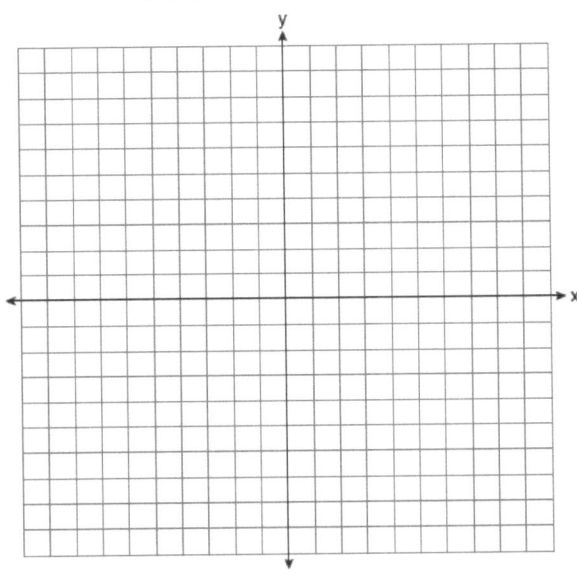

If $g(x) = 1.5x - 3$, determine if $f(x) > g(x)$ when $x = 4$. Justify your answer.

8.3   Rewrite Exponential Expressions

Recall the multiplication rule for exponents (from Section 5.3): when multiplying two factors with the same base, add the exponents.

Example: $x^5 \cdot x^3 = x^{5+3} = x^8$

In this section, we will also need to apply the **power rules for exponents**:

To find the power of a power, multiply the exponents.

Example: $(x^5)^3 = x^{5 \cdot 3} = x^{15}$

This works because $(x^5)^3 = (x^5)(x^5)(x^5) = x^{5+5+5} = x^{15}$

When raising a monomial to a power, raise each factor to that power.

Example: $(2x^5 y)^3 = 2^3 \cdot x^{5 \cdot 3} \cdot y^{1 \cdot 3} = 8x^{15} y^3$

When raising a fraction to a power, raise both the numerator and denominator to that power.

Example: $\left(\dfrac{2x^5}{3}\right)^2 = \dfrac{\left(2x^5\right)^2}{(3)^2} = \dfrac{4x^{10}}{9}$

These power rules can be used to rewrite expressions for exponential functions.

Examples: $f(x) = 2^{3x}$ can be rewritten as $f(x) = 8^x$ since $2^{3x} = (2^3)^x$.
             $g(x) = 3^{x+2}$ can be rewritten as $g(x) = 9(3)^x$ since $3^{x+2} = (3^x)(3^2)$.

▦ CALCULATOR TIP

We can use the calculator to check whether two expressions appear to be equivalent by using an arbitrary value of the variable and testing the equality of the two expressions. Although this method doesn't confirm that the expressions are equivalent for *all* values, it can help us recognize a likely error if they are not equal for the arbitrary value.

1. Store an arbitrary, preferably non-integer, value into x. For example, we can store 12.3 into x by pressing [1][2][.][3][STO▸][X,T,Θ,n].

2. Now, test whether the two expressions are equal using [2nd][TEST][1] for the equal sign. A result of 1 means they are equal, or a result of 0 means they are not.

Example: The screenshots below show how to test whether $3^{x+2} = 9(3^x)$.

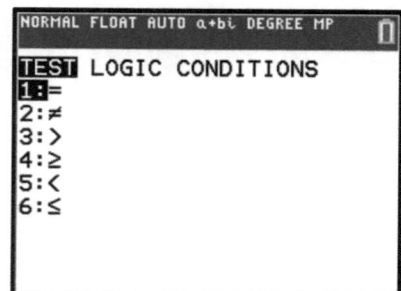

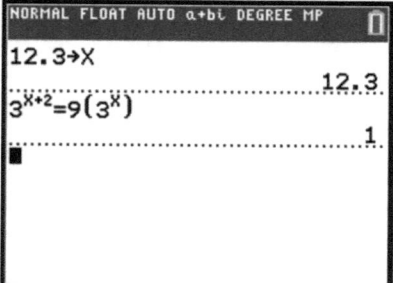

MODEL PROBLEM

If $f(x) = 2(0.5)^{3x}$, which of the following is an equivalent function?

       (1) $g(x) = 8(0.125)^x$ (3) $g(x) = 2(0.125)^x$

       (2) $g(x) = 6(1.5)^x$ (4) $g(x) = 2(1.5)^x$

Solution: (3)

Explanation of steps:

Only the quantity in parentheses *[(0.5)]* is being raised to a power. We can simplify by evaluating this quantity raised to the exponent's coefficient. *[$2(0.5)^{3x} = 2(0.5^3)^x = 2(0.125)^x$]*

PRACTICE PROBLEMS

1. Rewrite 5^{2x} as an equivalent expression with only x as the exponent.	2. Rewrite $10(1.1)^{5x}$ as an equivalent expression with only x as the exponent.
3. Rewrite 2^{3x+2} as an equivalent expression with only x as the exponent.	4. Rewrite $4(3)^{x+1}$ as an equivalent expression with only x as the exponent.

Regents Questions

MULTIPLE CHOICE

1.   Miriam and Jessica are growing bacteria in a laboratory. Miriam uses the growth function $f(t) = n^{2t}$ while Jessica uses the function $g(t) = n^{4t}$, where n represents the initial number of bacteria and t is the time, in hours. If Miriam starts with 16 bacteria, how many bacteria should Jessica start with to achieve the same growth over time?

       (1)   32   (3)   8

       (2)   16   (4)   4

2. A laboratory technician studied the population growth of a colony of bacteria. He recorded the number of bacteria every other day, as shown in the partial table below.

t (time, in days)	0	2	4
$f(t)$ (bacteria)	25	15,625	9,765,625

Which function would accurately model the technician's data?

(1) $f(t) = 25^t$ (3) $f(t) = 25t$

(2) $f(t) = 25^{t+1}$ (4) $f(t) = 25(t+1)$

3. The growth of a certain organism can be modeled by $C(t) = 10(1.029)^{24t}$, where $C(t)$ is the total number of cells after t hours. Which function is approximately equivalent to $C(t)$?

(1) $C(t) = 240(.083)^{24t}$ (3) $C(t) = 10(1.986)^t$

(2) $C(t) = 10(.083)^t$ (4) $C(t) = 240(1.986)^{\frac{t}{24}}$

4. A computer application generates a sequence of musical notes using the function $f(n) = 6(16)^n$, where n is the number of the note in the sequence and $f(n)$ is the note frequency in hertz. Which function will generate the same note sequence as $f(n)$?

(1) $g(n) = 12(2)^{4n}$ (3) $p(n) = 12(4)^{2n}$

(2) $h(n) = 6(2)^{4n}$ (4) $k(n) = 6(8)^{2n}$

5. Nora inherited a savings account that was started by her grandmother 25 years ago. This scenario is modeled by the function $A(t) = 5000(1.013)^{t+25}$, where $A(t)$ represents the value of the account, in dollars, t years after the inheritance. Which function below is equivalent to $A(t)$?

(1) $A(t) = 5000[(1.013)^t]^{25}$ (3) $A(t) = (5000)^t(1.013)^{25}$

(2) $A(t) = 5000[(1.013)^t + (1.013)^{25}]$ (4) $A(t) = 5000(1.013)^t(1.013)^{25}$

6. The number of bacteria grown in a lab can be modeled by $P(t) = 300 \cdot 2^{4t}$, where t is the number of hours. Which expression is equivalent to $P(t)$?

(1) $300 \cdot 8^t$ (3) $300^t \cdot 2^4$

(2) $300 \cdot 16^t$ (4) $300^{2t} \cdot 2^{2t}$

7. Materials A and B decay over time. The function for the amount of material A is $A(t) = 1000(0.5)^{2t}$ and for the amount of material B is $B(t) = 1000(0.25)^t$, where t represents time in days. On which day will the amounts of material be equal?

(1) initial day, only   (3) day 5, only

(2) day 2, only   (4) every day

8. The population of a city can be modeled by $P(t) = 3810(1.0005)^{7t}$, where $P(t)$ is the population after t years. Which function is approximately equivalent to $P(t)$?

(1) $P(t) = 3810(0.1427)^t$ (3) $P(t) = 26,670(0.1427)^t$

(2) $P(t) = 3810(1.0035)^t$ (4) $P(t) = 26,670(1.0035)^t$

9.   A laboratory technician used the function $t(m) = 2(3)^{2m+1}$ to model her research. Consider the following expressions:

   I. $6(3)^{2m}$ II. $6(6)^{2m}$ III. $6(9)^{m}$

   The function $t(m)$ is equivalent to
   - (1) I, only
   - (2) II, only
   - (3) I and III
   - (4) II and III

10. Which expression is equivalent to $(-4x^2)^3$?
   - (1) $-12x^6$
   - (2) $-12x^5$
   - (3) $-64x^6$
   - (4) $-64x^5$

11. Which expression is *not* equivalent to $(5^{2x})^3$?
   - (1) $(5^x)^6$
   - (2) $(5^{3x})^2$
   - (3) $(5^5)^x$
   - (4) $(5^2)^{3x}$

12. In an organism, the number of cells, $C(d)$, after d days can be represented by the function $C(d) = 120 \cdot 2^{3d}$. This function can also be expressed as
   - (1) $C(d) = 240^{3d}$
   - (2) $C(d) = 960 \cdot 2^d$
   - (3) $C(d) = 120 \cdot 6^d$
   - (4) $C(d) = 120 \cdot 8^d$

13. Which expression is equivalent to $(x + 4)^2 (x + 4)^3$?
   - (1) $(x + 4)^6$
   - (2) $(x + 4)^5$
   - (3) $(x^2 + 16)^6$
   - (4) $(x^2 + 16)^5$

14. The expression $300(4)^{x+3}$ is equivalent to
   - (1) $300(4)^x(4)^3$
   - (2) $300(4^x)^3$
   - (3) $300(4)^x + 300(4)^3$
   - (4) $300^x(4)^3$

15. Three expressions are shown below.
   I. $(x^3)^3$
   II. $x^4 \cdot x^5$
   III. $x^{10} \cdot x^{-1}$

   Which expressions are equivalent for all positive values of x?
   - (1) I and II, only
   - (2) I and III, only
   - (3) II and III, only
   - (4) I, II, and III

16. Three expressions are written below.
   A. $(2xy^2)^3$
   B. $(2x)^3 y^6$
   C. $(2x^2 y^2)(4xy^3)$

   Which expressions are equivalent to $8x^3 y^6$?
   - (1) A and B, only
   - (2) B and C, only
   - (3) A and C, only
   - (4) A, B, and C

8.4   Compare Linear and Exponential Functions

Using our knowledge about sequences, we can predict whether a continuous function is *linear* or *exponential* before we even graph the function. First, create a table using *equally spaced* values of the domain. Then look at the values of the function. If we can **add** a constant (a *common difference*) to each value to get the next value, the function is **linear**. If we can **multiply** each value by a constant (a *common ratio*) to get the next value, the function is **exponential**.

Example: Compare the tables of the two functions, $f(n) = 2n$ and $f(n) = 2^n$.
We will use values of n that are evenly spaced at 1 unit apart, from -2 to 3.

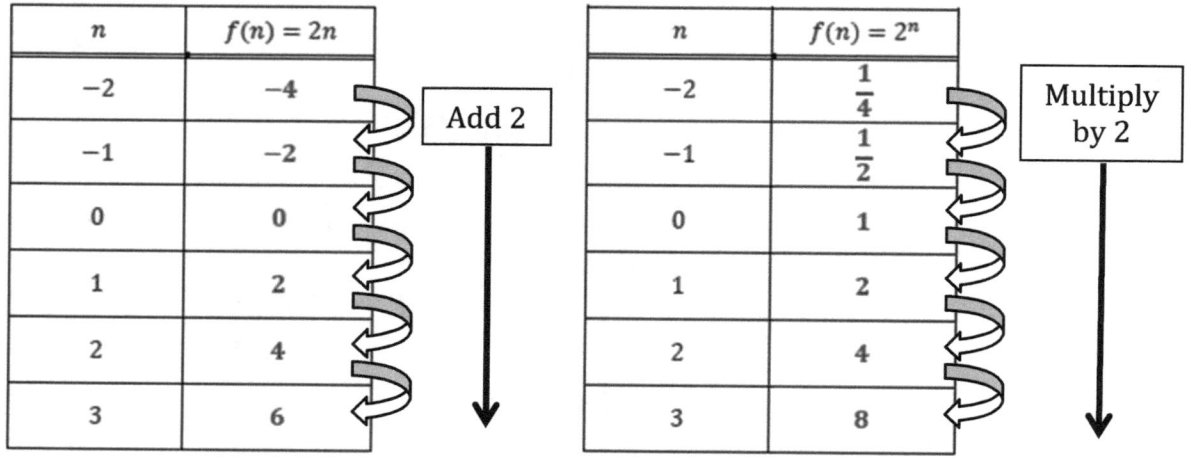

In the first table, we can add a constant, 2, to get the next function value, so this function is linear. In the second table, we can multiply each function value by 2 to get the next value, so the function is exponential.

An exponential growth function $g(x)$ will *always eventually exceed* any linear function $f(x)$. That is, x will eventually reach a value k for which $g(x) > f(x)$ for all $x > k$. We can find the value of k by graphing both functions on the calculator and determining the x-value at the rightmost point of intersection.

Example: Given the linear function $f(x) = 4x$ and the exponential function $g(x) = 2^x$, the graphs of the functions have two points of intersection, with the rightmost point of interection at (4,16). Therefore, for all values $x > 4$, we know $g(x) > f(x)$. This is shown in the Calculator Tip below.

In fact, an exponential growth function will *always eventually exceed* any polynomial function, such as quadratic functions (with an x^2 term), cubic functions (with an x^3 term), etc. We will learn more about quadratic and cubic functions later in this course.

CALCULATOR TIP

Use the calculator to find the *rightmost point of intersection* of two functions:

1. Press [Y=] and enter both equations.
2. Press [2nd][CALC][5] for intersect .
3. Press [ENTER] for the "First curve?" and "Second curve?" prompts.
4. For the "Guess?" prompt, use the arrow keys to move the cursor near the rightmost point of intersection. Then press [ENTER].
5. The coordinates of the point of intersection will be shown.

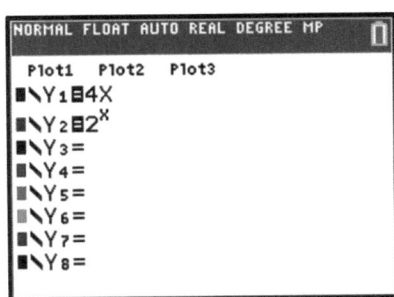 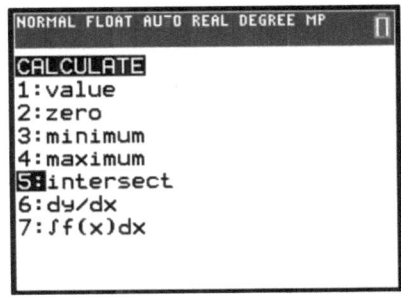 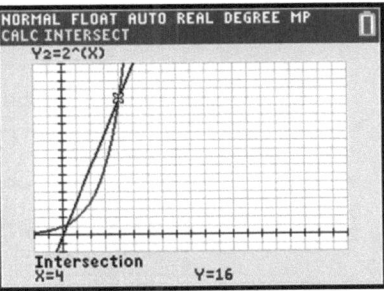

MODEL PROBLEM

Determine whether the function below is linear, exponential, or neither.

x	0	1	2	3	4
y	96	48	24	12	6

Solution:
    exponential

Explanation:
For a table with equally spaced values of x:
  a) if the same value is *added* to each y-value to get the next y, the function is linear
    [we would need to add −48, −24, −12, and −6, so it is not linear].
  b) if each y-value is *multiplied* by the same value to get the next y, the function is exponential
    [each y-value is multiplied by $\frac{1}{2}$ throughout, so it is exponential].
  c) otherwise, the function is neither.

PRACTICE PROBLEMS

1. Identify each function as linear, exponential, or neither.

a)

x	−3	−2	−1	0	1	2	3
y	14	10	6	2	−2	−6	−10

b)

x	−3	−2	−1	0	1	2	3
y	$\frac{1}{9}$	$\frac{1}{3}$	1	3	9	27	81

c)

x	−3	−2	−1	0	1	2	3
y	$\frac{1}{2}$	1	2	4	8	16	32

d)

x	−3	−2	−1	0	1	2	3
y	−27	−9	−3	0	3	9	27

2. A tile pattern is shown below. Create an explicit formula that could be used to determine the number of squares in the nth figure.

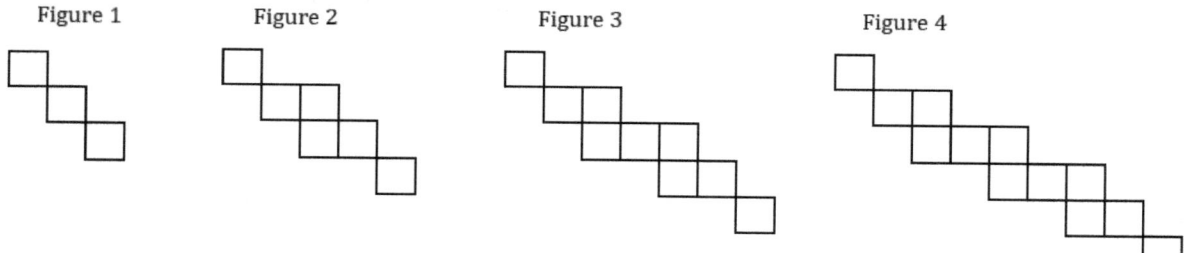

Figure 1   Figure 2   Figure 3   Figure 4

3. A tile pattern is shown below. Create an explicit formula that could be used to determine the number of black triangles in the nth figure.

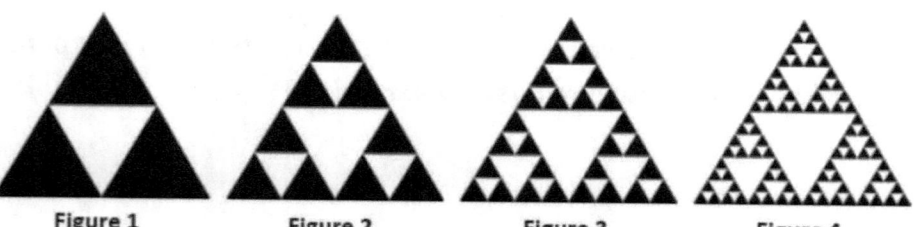

Figure 1   Figure 2   Figure 3   Figure 4

Regents Questions

MULTIPLE CHOICE

1. The table below shows the average yearly balance in a savings account where interest is compounded annually. No money is deposited or withdrawn after the initial amount is deposited.

Year	Balance, in Dollars
0	380.00
10	562.49
20	832.63
30	1232.49
40	1824.39
50	2700.54

Which type of function best models the given data?
- (1) linear function with a negative rate of change
- (2) linear function with a positive rate of change
- (3) exponential decay function
- (4) exponential growth function

2. The table below represents the function F.

x	3	4	6	7	8
$F(x)$	9	17	65	129	257

The equation that represents this function is
- (1) $F(x) = 3^x$
- (2) $F(x) = 3x$
- (3) $F(x) = 2^x + 1$
- (4) $F(x) = 2x + 3$

3. Which situation could be modeled by using a linear function?
- (1) a bank account balance that grows at a rate of 5% per year, compounded annually
- (2) a population of bacteria that doubles every 4.5 hours
- (3) the cost of cell phone service that charges a base amount plus 20 cents per minute
- (4) the concentration of medicine in a person's body that decays by a factor of one-third every hour

4.   Which table of values represents a linear relationship?

x	f(x)
−1	−3
0	−2
1	1
2	6
3	13

(1)

x	f(x)
−1	−3
0	−1
1	1
2	3
3	5

(3)

x	f(x)
−1	$\frac{1}{2}$
0	1
1	2
2	4
3	8

(2)

x	f(x)
−1	−1
0	0
1	1
2	8
3	27

(4)

5.   Alicia has invented a new app for smart phones that two companies are interested in purchasing for a 2-year contract.

Company _A_ is offering her $10,000 for the first month and will increase the amount each month by $5000. Company _B_ is offering $500 for the first month and will double their payment each month from the previous month.

Monthly payments are made at the end of each month. For which monthly payment will company _B_'s payment first exceed company _A_'s payment?

(1) 6   (3) 8
(2) 7   (4) 9

6.   Which function is shown in the table below?

x	f(x)
−2	$\dfrac{1}{9}$
−1	$\dfrac{1}{3}$
0	1
1	3
2	9
3	27

   (1) $f(x) = 3x$ (3) $f(x) = -x^3$
   (2) $f(x) = x + 3$ (4) $f(x) = 3^x$

7.   Grisham is considering the three situations below.
      I.   For the first 28 days, a sunflower grows at a rate of 3.5 cm per day.
      II.   The value of a car depreciates at a rate of 15% per year after it is purchased.
      III. The amount of bacteria in a culture triples every two days during an experiment.
      Which of the statements describes a situation with an equal difference over an equal interval?
      (1)   I, only   (3)   I and III
      (2)   II, only   (4)   II and III

8.   The tables below show the values of four different functions for given values of x.

x	f(x)		x	g(x)		x	h(x)		x	k(x)
1	12		1	−1		1	9		1	−2
2	19		2	1		2	12		2	4
3	26		3	5		3	17		3	14
4	33		4	13		4	24		4	28

      Which table represents a linear function?
      (1) $f(x)$ (3) $h(x)$
      (2) $g(x)$ (4) $k(x)$

9.   What is the largest integer, x, for which the value of $f(x) = 5x^4 + 30x^2 + 9$ will be greater than the value of $g(x) = 3^x$?
      (1)   7   (3)   9
      (2)   8   (4)   10

10. As x increases beyond 25, which function will have the largest value?
    (1) $f(x) = 1.5^x$ (3) $h(x) = 1.5x^2$
    (2) $g(x) = 1.5x + 3$ (4) $k(x) = 1.5x^3 + 1.5x^2$

11. Which scenario represents exponential growth?
    (1) A water tank is filled at a rate of 2 gallons/minute.
    (2) A vine grows 6 inches every week.
    (3) A species of fly doubles its population every month during the summer.
    (4) A car increases its distance from a garage as it travels at a constant speed of 25 miles per hour.

12. Vinny collects population data, $P(h)$, about a specific strain of bacteria over time in hours, h, as shown in the graph below.

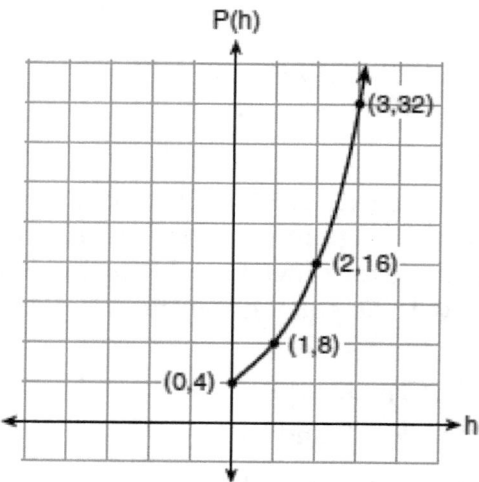

Which equation represents the graph of $P(h)$?
    (1) $P(h) = 4(2)^h$ (3) $P(h) = 3h^2 + 0.2h + 4.2$
    (2) $P(h) = \frac{46}{5}h + \frac{6}{5}$ (4) $P(h) = \frac{2}{3}h^3 - h^2 + 3h + 4$

13. One characteristic of all linear functions is that they change by
    (1) equal factors over equal intervals
    (2) unequal factors over equal intervals
    (3) equal differences over equal intervals
    (4) unequal differences over equal intervals

14. The highest possible grade for a book report is 100. The teacher deducts 10 points for each day the report is late. Which kind of function describes this situation?
    (1) linear   (3) exponential growth
    (2) quadratic   (4) exponential decay

15. Ian is saving up to buy a new baseball glove. Every month he puts $10 into a jar. Which type of function best models the total amount of money in the jar after a given number of months?
    (1) linear   (3) quadratic
    (2) exponential   (4) square root

16.   Which situation is *not* a linear function?

    (1)   A gym charges a membership fee of $10.00 down and $10.00 per month.

    (2)   A cab company charges $2.50 initially and $3.00 per mile.

    (3)   A restaurant employee earns $12.50 per hour.

    (4)   A $12,000 car depreciates 15% per year.

17.   During physical education class, Andrew recorded the exercise times in minutes and heart rates in beats per minute (bpm) of four of his classmates. Which table best represents a linear model of exercise time and heart rate?

Student 1

Exercise Time (in minutes)	Heart Rate (bpm)
0	60
1	65
2	70
3	75
4	80

(1)

Student 3

Exercise Time (in minutes)	Heart Rate (bpm)
0	58
1	65
2	70
3	75
4	79

(3)

Student 2

Exercise Time (in minutes)	Heart Rate (bpm)
0	62
1	70
2	83
3	88
4	90

(2)

Student 4

Exercise Time (in minutes)	Heart Rate (bpm)
0	62
1	65
2	66
3	73
4	75

(4)

18.   Which of the three situations given below is best modeled by an exponential function?

    I.   A bacteria culture doubles in size every day.

    II.   A plant grows by 1 inch every 4 days.

    III. The population of a town declines by 5% every 3 years.

    (1)   I, only   (3)   I and II

    (2)   II, only   (4)   I and III

19. The table below shows the weights of Liam's pumpkin, $l(w)$, and Patricia's pumpkin, $p(w)$, over a four-week period where w represents the number of weeks. Liam's pumpkin grows at a constant rate. Patricia's pumpkin grows at a weekly rate of approximately 52%.

Weeks w	Weight in Pounds l(w)	Weight in Pounds p(w)
6	2.4	2.5
7	5.5	3.8
8	8.6	5.8
9	11.7	8.8

Assume the pumpkins continue to grow at these rates through week 13. When comparing the weights of both Liam's and Patricia's pumpkins in week 10 and week 13, which statement is true?

(1) Liam's pumpkin will weigh more in week 10 and week 13.

(2) Patricia's pumpkin will weigh more in week 10 and week 13.

(3) Liam's pumpkin will weigh more in week 10, and Patricia's pumpkin will weigh more in week 13.

(4) Patricia's pumpkin will weigh more in week 10, and Liam's pumpkin will weigh more in week 13.

20. The function f is shown in the table below.

x	f(x)
0	1
1	3
2	9
3	27

Which type of function best models the given data?

(1) exponential growth function

(2) exponential decay function

(3) linear function with positive rate of change

(4) linear function with negative rate of change

21. Which situation can be modeled by a linear function?

(1) The population of bacteria triples every day.

(2) The value of a cell phone depreciates at a rate of 3.5% each year.

(3) An amusement park allows 50 people to enter every 30 minutes.

(4) A baseball tournament eliminates half of the teams after each round.

22. Which type of function is shown in the graph below?

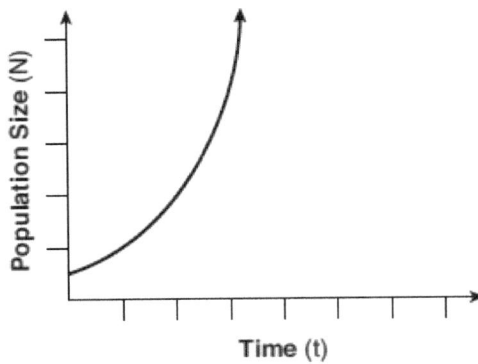

(1) linear   (3) square root
(2) exponential   (4) absolute value

23. Which situation could be modeled as a linear equation?
   (1) The value of a car decreases by 10% every year.
   (2) The number of fish in a lake doubles every 5 years.
   (3) Two liters of water evaporate from a pool every day.
   (4) The amount of caffeine in a person's body decreases by $\frac{1}{3}$ every 2 hours.

24. Eric deposits $500 in a bank account that pays 3.5% interest, compounded yearly. Which type of function should he use to determine how much money he will have in the account at the end of 10 years?
   (1) linear   (3) absolute value
   (2) quadratic   (4) exponential

25. Which table of values represents an exponential relationship?

x	f(x)
1	6
2	9
3	12
4	15
5	18

(1)

x	k(x)
1	4
2	16
3	64
4	256
5	1024

(3)

x	h(x)
1	2
2	7
3	12
4	17
5	22

(2)

x	p(x)
1	−9.5
2	−12
3	−14.5
4	−17
5	−19.5

(4)

26. Which situation could be modeled by a linear function?
   (1) The value of a car depreciates by 7% annually.
   (2) A gym charges a $50 initial fee and then $30 monthly.
   (3) The number of bacteria in a lab doubles weekly.
   (4) The amount of money in a bank account increases by 0.1 % monthly.

27. One Saturday afternoon, three friends decided to keep track of the number of text messages they received each hour from 8 a.m. to noon. The results are shown below.

   Emily said that the number of messages she received increased by 8 each hour.

   Jessica said that the number of messages she received doubled every hour.

   Chris said that he received 3 messages the first hour, 10 the second hour, none the third hour, and 15 the last hour.

Which of the friends' responses best classifies the number of messages they received each hour as a linear function?
   (1) Emily, only
   (2) Jessica, only
   (3) Emily and Chris
   (4) Jessica and Chris

28. Thirty-two teams are participating in a basketball tournament. Only the winning teams in each round advance to the next round, as shown in the table below.

Number of Rounds Completed, x	0	1	2	3	4	5
Number of Teams Remaining, f(x)	32	16	8	4	2	1

Which function type best models the relationship between the number of rounds completed and the number of teams remaining?

    (1) absolute value   (3) linear

    (2) exponential   (4) quadratic

29. Which function will have the greatest value when $x > 1$?

    (1) $g(x) = 2(5)^x$ (3) $h(x) = 2x^2 + 5$

    (2) $f(x) = 2x + 5$ (4) $k(x) = 2x^3 + 5$

30. Tables of values for four functions are shown below.

x	f(x)
0	6
1	7
2	10
3	15
4	22

x	h(x)
0	1
1	2
2	4
3	8
4	16

x	g(x)
0	0
1	−2
2	−2
3	0
4	4

x	j(x)
0	2
1	5
2	8
3	11
4	14

Which table best represents an exponential function?

    (1) $f(x)$ (3) $h(x)$

    (2) $g(x)$ (4) $j(x)$

CONSTRUCTED RESPONSE

31. About a year ago, Joey watched an online video of a band and noticed that it had been viewed only 843 times. One month later, Joey noticed that the band's video had 1708 views. Joey made the table below to keep track of the cumulative number of views the video was getting online.

Months Since First Viewing	Total Views
0	843
1	1708
2	forgot to record
3	7124
4	14,684
5	29,787
6	62,381

Write a regression equation that best models these data. Round all values to the *nearest hundredth*. Justify your choice of regression equation.

As shown in the table, Joey forgot to record the number of views after the second month. Use the equation from part *a* to estimate the number of full views of the online video that Joey forgot to record.

32. Caitlin has a movie rental card worth $175. After she rents the first movie, the card's value is $172.25. After she rents the second movie, its value is $169.50. After she rents the third movie, the card is worth $166.75.

Assuming the pattern continues, write an equation to define $A(n)$, the amount of money on the rental card after n rentals.

Caitlin rents a movie every Friday night. How many weeks in a row can she afford to rent a movie, using her rental card only? Explain how you arrived at your answer.

33. Rachel and Marc were given the information shown below about the bacteria growing in a Petri dish in their biology class.

Number of Hours, x	1	2	3	4	5	6	7	8	9	10
Number of Bacteria, $B(x)$	220	280	350	440	550	690	860	1070	1340	1680

Rachel wants to model this information with a linear function. Marc wants to use an exponential function. Which model is the better choice? Explain why you chose this model.

34.   The function, $t(x)$, is shown in the table below.

x	t(x)
−3	10
−1	7.5
1	5
3	2.5
5	0

Determine whether $t(x)$ is linear or exponential. Explain your answer.

35.   Consider the pattern of squares shown below:

Which type of model, linear or exponential, should be used to determine how many squares are in the *n*th pattern? Explain your answer.

36.   Michael has $10 in his savings account. Option 1 will add $100 to his account each week. Option 2 will double the amount in his account at the end of each week. Write a function in terms of x to model each option of saving. Michael wants to have at least $700 in his account at the end of 7 weeks to buy a mountain bike. Determine which option(s) will enable him to reach his goal. Justify your answer.

37.   Caleb claims that the ordered pairs shown in the table below are from a nonlinear function.

x	f(x)
0	2
1	4
2	8
3	16

State if Caleb is correct. Explain your reasoning.

38.   The number of people who attended a school's last six basketball games increased as the team neared the state sectional games. The table below shows the data.

Game	13	14	15	16	17	18
Attendance	348	435	522	609	696	783

State the type of function that best fits the given data. Justify your choice of a function type.

39.   Mike knows that $(3, 6.5)$ and $(4, 17.55)$ are points on the graph of an exponential function, $g(x)$, and he wants to find another point on the graph of this function. First, he subtracts 6.5 from 17.55 to get 11.05. Next, he adds 11.05 and 17.55 to get 28.6. He states that $(5, 28.6)$ is a point on $g(x)$. Is he correct? Explain your reasoning.

359

40.   The table below shows the value of a particular car over time.

Time (years)	Value (dollars)
0	20,000
5	10,550
10	5570
15	2940
20	1550

Determine whether a linear or exponential function is more appropriate for modeling this data. Explain your choice.

41.   Breanna creates the pattern of blocks below in her art class.

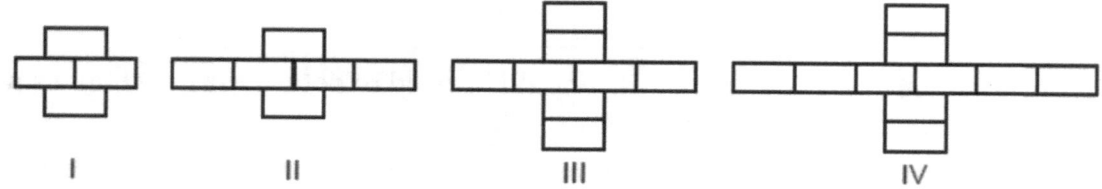

I   II   III   IV

A friend tells her that the number of blocks in the pattern is increasing exponentially. Is her friend correct? Explain your reasoning.

CHAPTER 9.   SEQUENCES

9.1   Arithmetic Sequences

A **sequence** is an ordered list of numbers, called terms. If we let the integer n represent the term number in the sequence, then a_n represents the nth term in the sequence. Unless otherwise specified, you can assume that the sequence is an **infinite sequence**, meaning that the values of n are the set of all counting numbers (or sometimes, whole numbers).

Examples: 9, 11, 13, 15, ...

$$a_1, a_2, a_3, a_4, ...$$

An **explicit formula** describes how to calculate each term (a_n) based on its term number (n). The formula allows us to determine the value of a specified term. By substituting the term number for n, we can evaluate the formula just as we would for any function.

Examples: The formula for the sequence above is $a_n = 2n + 7$. We can calculate the value of any term using this formula by substituting for n; for example, to calculate what the fourth term is, $a_4 = 2(4) + 7 = 15$.

An **arithmetic sequence** is one in which each term is obtained by *adding* the same number (called the **common difference**, represented by d) to the preceding term.

Example: 9, 11, 13, 15, ...

+2 +2 +2   so, $d = 2$

We can find the common difference of an arithmetic sequence by subtracting consecutive terms, such as $a_2 - a_1$ or $a_3 - a_2$. In general, the common difference $d = a_n - a_{n-1}$.

We can define an arithmetic sequence using the formula $a_n = a_1 + (n - 1)d$ where a_1 is the first term and d is the common difference.

Example: The sequence 9, 11, 13, 15, ... can be written as $a_n = 9 + (n - 1) \cdot 2$. By distributing the d and combining terms, we can simplify the formula:

$$a_n = 9 + (n - 1) \cdot 2$$
$$a_n = 9 + 2n - 2$$
$$a_n = 2n + 7$$

An *arithmetic sequence* is always a **linear** function. The common difference, d, is its *slope*.

Example: For the arithmetic sequence $9, 11, 13, 15, ...,$ $a_1 = 9$ and $d = 2$, so we can define the sequence by substituting for a_1 and d in $a_n = a_1 + (n-1)d$, giving us $a_n = 9 + (n-1) \cdot 2$. By simplifying, $a_n = 2n + 7$. The arithmetic sequence is graphed below. Since the function is discrete, the graph consists of isolated points, not a continuous line.

n	$a_n = 2n + 7$
1	9
2	11
3	13
4	15
5	17
6	19

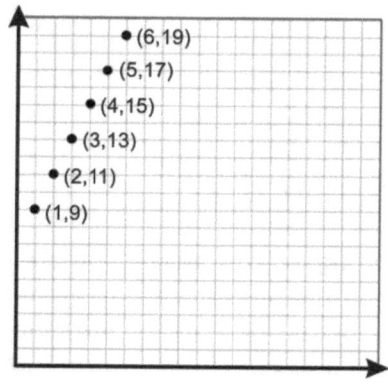

As can be seen by the table, we can **add** d (the slope) to each term to obtain the next term. In this graph, the set of points that make up the discrete function lie on a straight line.

A *linear regression* of these points, using the domain of real numbers x, would have an equation of $y = 2x + 7$, representing the continuous function, $f(x) = 2x + 7$.

To find a specific term of an arithmetic sequence:
   1. Determine a_1 (the first term) and d (the common difference) from the given terms.
   2. In the general formula for arithmetic sequences, substitute for n, a_1, and d, and evaluate.

Example: To find the 20th term of the arithmetic sequence $32, 37, 42, 47, ...$
            The general formula is $a_n = a_1 + (n-1)d$ and in this case, $a_1 = 32$, $d = 5$, and $n = 20$, so $a_{20} = 32 + (20-1)(5) = 127$.

CALCULATOR TIP

If the formula for an arithmetic sequence is known, the calculator can also be used to find the value of any terms.

1. Press 2nd|LIST|5 for the seq function.
2. On the TI-84 models with Stat Wizards turned on (in the MODE screen), you'll be prompted to enter the formula, the variable used in the formula, the starting and ending term numbers to display, and a step of 1 (see center screenshot below). Then press ENTER twice to Paste and display the result.
   [On the TI-83, this screen is skipped, so you will need to type these directly within the parentheses of the seq function, separated by commas, and then press ENTER.]

Example: To display the 25th term of the sequence $a_n = 2n + 7$, follow the above steps, as shown below. The variable X is used for convenience, but N could be used instead. Just be sure the variable used in the expression (Expr) is the same variable named on the next line. The display shows that 57 is the 25th term.

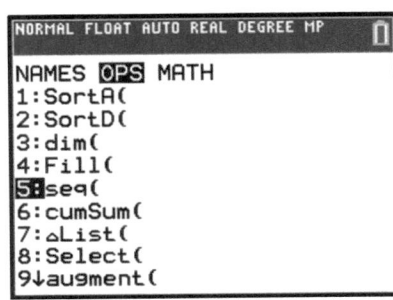
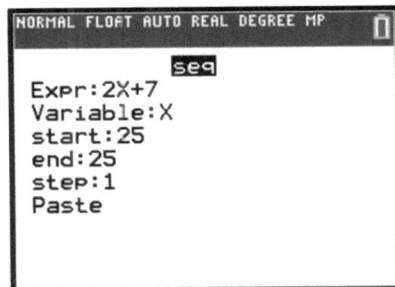
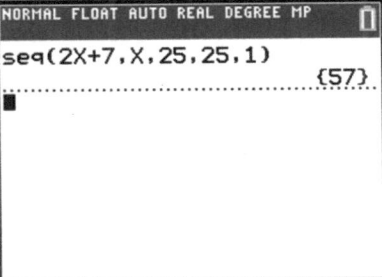

Because the common difference, d, is also the slope of the linear function, we can use the slope formula to calculate d given any two terms.

To write the formula for an arithmetic sequence given two terms:
1. Express the two terms as points. Find d, which is the slope of the line through these points, using the slope formula.
2. Substitute d, as well as n and a_n from one of the terms, into the general formula for an arithmetic sequence. Solve for a_1.
3. Write the general formula, substituting known values for a_1 and d, and simplify.

Example: To write a formula for the arithmetic sequence where $a_5 = 63$ and $a_8 = 99$, find d, which is the slope of the line through the points (5,63) and (8,99):
$$d = \frac{99 - 63}{8 - 5} = \frac{36}{3} = 12.$$ Then, substitute $d = 12$ and, using the first point, $n = 5$ and $a_n = 63$ into the general formula $a_n = a_1 + (n - 1)d$ to find a_1:
$$63 = a_1 + (5 - 1)(12)$$
$$63 = a_1 + 48$$
$$a_1 = 15$$
Now that we know $a_1 = 15$ and $d = 12$, we can write the formula:
$$a_n = a_1 + (n - 1)d$$
$$a_n = 15 + (n - 1)(12)$$
$$a_n = 12n + 3$$

MODEL PROBLEM 1:   *FIND THE NTH TERM*

Find the 15th term of the arithmetic sequence 5, 11, 17, 23, ...

Solution:	Explanation of steps:
(A) $a_1 = 5, d = 6$	(A) Find a_1 and d.
(B) $a_n = a_1 + (n - 1)d$	[a_1 is the first term, d is the common difference.]
(C) $a_{15} = 5 + (15 - 1) \cdot 6$	(B) Write the formula for arithmetic sequences.
(D) $a_{15} = 89$	(C) Substitute for a_1, d, and the term number, n.
	(D) Evaluate.

PRACTICE PROBLEMS

1. What is the common difference of the arithmetic sequence 5, 8, 11, 14, ... ?	2. What is the common difference in the sequence 8, 4, 0, −4, ... ?

3. Write a formula for the nth term of the arithmetic sequence, $15, 20, 25, 30, \dots$.

4. Write a formula for the nth term of arithmetic sequence, $10, 12, 14, 16, \dots$.

5. Find the eighth term of the arithmetic sequence for which $a_1 = 21$ and $d = 9$.

6. Find the 27th term of the arithmetic sequence, $5, 8, 11, 14, \dots$.

MODEL PROBLEM 2:   WRITE A FORMULA GIVEN TWO TERMS

Two terms of an arithmetic sequence are $a_8 = 21$ and $a_{27} = 97$. Write a rule for the nth term.

Solution:

(A) (8,21) and (27,97)

$$d = \frac{97 - 21}{27 - 8} = \frac{76}{19} = 4$$

(B) $a_n = a_1 + (n - 1)d$

$21 = a_1 + (8 - 1) \cdot 4$

$21 = a_1 + 32 - 4$

$-7 = a_1$

(C) $a_n = a_1 + (n - 1)d$

$a_n = -7 + (n - 1) \cdot 4$

$a_n = -7 + 4n - 4$

$a_n = 4n - 11$

Explanation of steps:

(A) Express the two terms as points. Find d, which is the slope of the line through these points, using the slope formula.

(B) Substitute d, as well as n and a_n from one of the terms [$a_8 = 21$, so use $n = 8$ and $a_n = 21$], into the general formula for an arithmetic sequence. Solve for a_1.

(C) Write the general formula, substituting known values for a_1 and d, and simplify.

PRACTICE PROBLEMS

7. Two terms of an arithmetic sequence are $a_6 = 10$ and $a_{21} = 55$. Write a formula for the nth term.	8. Two terms of an arithmetic sequence are $a_4 = -23$ and $a_{22} = 49$. Write a formula for the nth term.

Regents Questions

MULTIPLE CHOICE

1. The diagrams below represent the first three terms of a sequence.

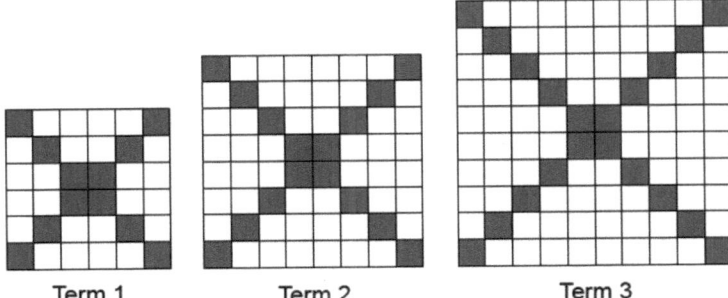

    Term 1   Term 2   Term 3

    Assuming the pattern continues, which formula determines a_n, the number of shaded squares in the nth term?

    (1) $a_n = 4n + 12$ (3) $a_n = 4n + 4$
    (2) $a_n = 4n + 8$ (4) $a_n = 4n + 2$

2. The third term in an arithmetic sequence is 10 and the fifth term is 26. If the first term is a_1, which is an equation for the nth term of this sequence?

    (1) $a_n = 8n + 10$ (3) $a_n = 16n + 10$
    (2) $a_n = 8n - 14$ (4) $a_n = 16n - 38$

3. In a sequence, the first term is 4 and the common difference is 3. The fifth term of this sequence is

    (1) -11 (3) 16
    (2) -8 (4) 19

4. On the main floor of the Kodak Hall at the Eastman Theater, the number of seats per row increases at a constant rate. Steven counts 31 seats in row 3 and 37 seats in row 6. How many seats are there in row 20?

    (1) 65   (3) 69
    (2) 67   (4) 71

5. For the sequence $-27, -12, 3, 18, ...,$ the expression that defines the nth term where $a_1 = -27$ is

    (1) $15 - 27n$ (3) $-27 + 15n$
    (2) $15 - 27(n - 1)$ (4) $-27 + 15(n - 1)$

6.   Given the following three sequences:
         I. $2, 4, 6, 8, 10$ ...
         II. $2, 4, 8, 16, 32$ ...
         III. $a, a + 2, a + 4, a + 6, a + 8$ ...
     Which ones are arithmetic sequences?
     (1)   I and II, only   (3)   II and III, only
     (2)   I and III, only   (4)   I, II, and III

7.   The shaded boxes in the figures below represent a sequence.

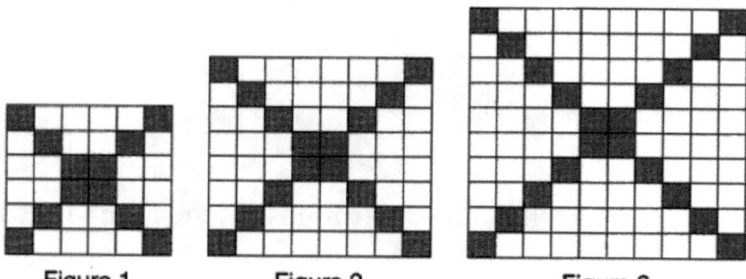

         Figure 1   Figure 2   Figure 3

     If figure 1 represents the first term and this pattern continues, how many shaded blocks will
     be in figure 35?
     (1)   55   (3)   420
     (2)   148   (4)   805

8.   Given: the sequence 4, 7, 10, 13, ...
     When using the arithmetic sequence formula $a_n = a_1 + (n - 1)d$ to determine the 10th term,
     which variable would be replaced with the number 3?
     (1) a_1 (3) a_n
     (2) n (4) d

9.   The first term in a sequence is 5 and the fifth term is 17. What is the common difference?
     (1)   2.4   (3)   3
     (2)   12   (4)   4

10. The 24th term of the sequence –5, –11, –17, –23, ... is
     (1)   –149   (3)   133
     (2)   –143   (4)   139

CONSTRUCTED RESPONSE

11. Determine the common difference of the arithmetic sequence in which $a_1 = 3$ and $a_4 = 15$.

12. Determine the common difference of the arithmetic sequence in which $a_1 = 5$ and $a_5 = 17$.
     Determine the 21st term of this sequence.

9.2   Geometric Sequences

A **geometric sequence** is one in which each term is obtained by *multiplying* the same number (called the **common ratio**, represented by r) to the preceding term.

Example: 5, 10, 20, 40, ...

so, $r = 2$

We can find the common ratio by dividing consecutive terms, such as $\dfrac{a_2}{a_1}$ or $\dfrac{a_3}{a_2}$. In general, the common ratio $r = \dfrac{a_n}{a_{n-1}}$.

We can define a geometric sequence using the formula $a_n = a_1 r^{n-1}$ where a_1 is the first term and r is the common ratio ($r \neq 0$).

Example: The formula for the above sequence can be written as $a_n = 5 \cdot 2^{n-1}$.

A *geometric sequence* is always an **exponential** function.

Example: For the geometric sequence $5, 10, 20, 40, \ldots$, $a_1 = 5$ and $r = 2$, so we can define the sequence by substituting for a_1 and r in $a_n = a_1 r^{n-1}$, giving us the formula $a_n = 5 \cdot 2^{n-1}$. This is graphed below.

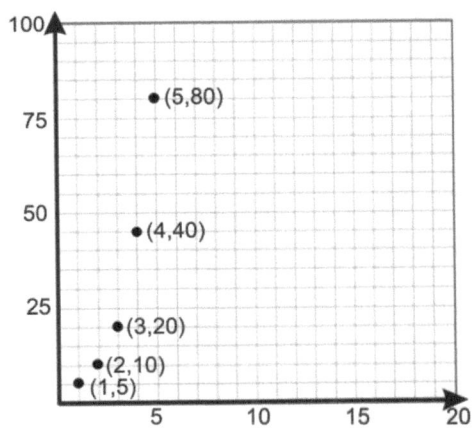

n	$a_n = 5 \cdot 2^{n-1}$
1	5
2	10
3	20
4	40
5	80
6	160

As can be seen by the table, we can **multiply** each term by r to obtain the next term. In this graph, the set of points that make up the discrete function lie on an exponential curve.

Since $r^{n-1} = \dfrac{r^n}{r}$, an alternate way of writing the geometric sequence formula is $a_n = \dfrac{a_1 r^n}{r}$. This will allow us to simplify our formulas so that they don't include binomial exponents.

Example: $a_n = 5 \cdot 2^{n-1}$ can be rewritten as $a_n = \dfrac{5 \cdot 2^n}{2} = \dfrac{5}{2}(2)^n$ or $a_n = 2.5(2)^n$.

To find a specific term of a geometric sequence:
1. Determine a_1 (the first term) and r (the common ratio) from the given terms.
2. In the general formula for geometric sequences, substitute for n, a_1, and r, and evaluate.

Example: To find the sixth term of the geometric sequence $8, 32, 128, \ldots$,
we know $a_1 = 8$, $r = 4$, and $n = 6$, so substitute into the general formula
$a_n = a_1 r^{n-1}$ to get $a_6 = 8(4)^{6-1} = 8(4)^5 = 8{,}192$.

 **CALCULATOR TIP**

If the formula for a geometric sequence is known, the calculator can also be used to find the value of any terms, just like we saw with arithmetic sequences.

1. Press [2nd][LIST][5] for the seq function.
2. On the TI-84 models, enter the formula, the variable used in the formula, the starting and ending term numbers to display, and a step of 1 next to the prompts on the next screen.

    Then press [ENTER] twice to Paste and display the result.
    [On the TI-83, this screen is skipped, so you will need to type these directly within the
    parentheses of the seq function, separated by commas, and then press [ENTER].]

Example: To display the 6th through 8th terms of the sequence $a_n = 5(3)^n$, follow the above
steps, as shown below. The terms are 3645, 10935, and 32805.

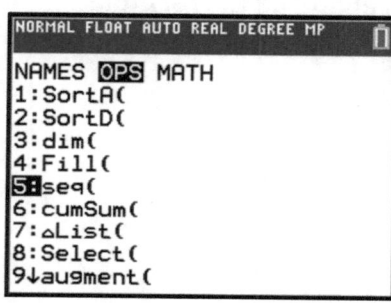 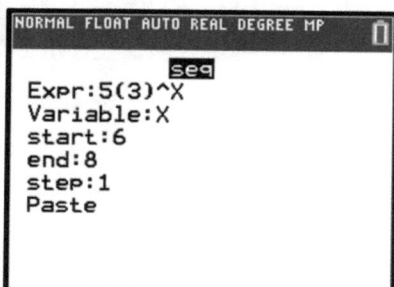 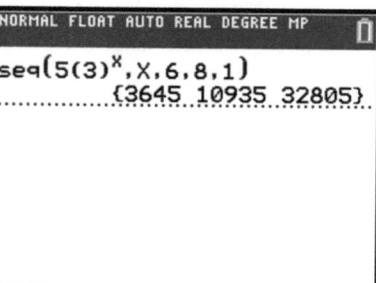

MODEL PROBLEM

Find the seventh term of the sequence: $3, -6, 12, -24, \ldots$.

Solution:

(A) $a_1 = 3; r = \frac{-6}{3} = -2$

(B) $a_n = a_1 r^{n-1}$

    $a_7 = 3(-2)^{7-1} = 192$

Explanation of steps:

(A) Determine a_1 (the first term) and r (the common ratio) from the given terms.

(B) In the general formula for geometric sequences, substitute for n [7], a_1, and r, and simplify.

PRACTICE PROBLEMS

1. What is the common ratio of the geometric sequence $12, 6, 3, 1.5, \ldots$?	2. What is the common ratio of the geometric sequence $-2, 4, -8, 16, \ldots$?
3. What is the common ratio of the geometric sequence $2, -8, 32, -128, \ldots$?	4. Write a formula for the nth term of the geometric sequence, $4, 10, 25, 62.5, \ldots$.

5. Write a formula for the nth term of the geometric sequence, $-1, 2, -4, 8, \ldots$.	6. What is the fifteenth term of the geometric sequence $5, -10, 20, -40, \ldots$?
7. Find the seventh term of the geometric sequence for which $a_1 = 6$ and $r = -\frac{1}{2}$.	8. Find the thirtieth term of the geometric sequence for which the tenth term is 512 and the fifteenth term is 16,384.

Regents Questions

MULTIPLE CHOICE

1. What is a common ratio of the geometric sequence whose first term is 5 and third term is 245?

    (1) 7   (3) 120
    (2) 49   (4) 240

2. If $x \neq 0$, then the common ratio of the sequence $x, 2x^2, 4x^3, 8x^4, 16x^5, \dots$ is

    (1) $2x$ (3) x
    (2) 2   (4) $\frac{1}{2}x$

3. In a geometric sequence, the first term is 4 and the common ratio is –3. The fifth term of this sequence is

    (1) 324   (3) –108
    (2) 108   (4) –324

4. The eleventh term of the sequence 3, –6, 12, –24, ..., is

    (1) –3072   (3) 3072
    (2) –6144   (4) 6144

CONSTRUCTED RESPONSE

5. Determine and state whether the sequence $1, 3, 9, 27, \dots$ displays exponential behavior. Explain how you arrived at your decision.

CHAPTER 10.   IRRATIONAL NUMBERS

10.1   Simplify Radicals

$\sqrt{x}$ represents the principal square root of x, or the non-negative value that, when multiplied by itself, is equal to x. The $\sqrt{}$ symbol is called a **radical sign**, and the quantity under the radical sign is called the **radicand**. A **radical** is a term containing a radical sign.

Example: $\sqrt{36} = 6$ (36 is the radicand.)

▦▢ CALCULATOR TIP

To find the square root of a number on the calculator, press [2nd][√], then type the radicand, then press [▸][ENTER].

[On the TI-83, the symbols "√(" appear; after typing the radicand, press [)][ENTER] instead.]

Example: Find $\sqrt{36}$ by entering [2nd][√][3][6][▸][ENTER]

   [or [2nd][√][3][6][)][ENTER] on the TI-83].

Since a positive number has a **negative square root** as well, we represent the negative square root by placing a negative sign before the radical. To indicate both the positive and negative roots, we use the $\pm$ **symbol** before the radical.

Examples: $\sqrt{36} = 6$ $-\sqrt{36} = -6$ since $(-6)(-6) = 36$ $\pm\sqrt{36} = \{-6, 6\}$

It is important to recognize that **squaring** and taking a **square root** are reverse operations. Therefore, $\sqrt{x^2} = x$ and $\left(\sqrt{y}\right)^2 = y$.

Examples: $\sqrt{3^2} = \sqrt{9} = 3$ and $\left(\sqrt{25}\right)^2 = 5^2 = 25$.

▦▢ CALCULATOR TIP

You can square a number on the calculator using the [x^2] key. To raise a number to a power other than 2, you can use the [^] key.

Examples: Find 5^2 by entering [5][x^2][ENTER], or find 5^4 by entering [5][^][4][ENTER].

Two more important rules are: $\sqrt{ab} = \sqrt{a}\sqrt{b}$ and $\sqrt{\dfrac{a}{b}} = \dfrac{\sqrt{a}}{\sqrt{b}}$ (for $b \neq 0$).

Examples: $\sqrt{36} = \sqrt{9 \cdot 4} = \sqrt{9}\sqrt{4} = 3 \cdot 2 = 6$ and $\sqrt{\dfrac{4}{9}} = \dfrac{\sqrt{4}}{\sqrt{9}} = \dfrac{2}{3}$.

To simplify a radical into simplest radical form:
1. Write the prime factorization of the radicand.
2. Group all pairs of factors, representing squares.
3. Remove the squares (pairs of factors) from the radicand by replacing them with their square roots (single factors outside the radical sign).
4. Multiply all factors outside the radicand, and all factors remaining inside the radicand.

Examples: $\sqrt{75} = \sqrt{3 \cdot \boxed{5 \cdot 5}} = 5\sqrt{3}$

$\sqrt{288} = \sqrt{\boxed{2 \cdot 2} \cdot \boxed{2 \cdot 2} \cdot 2 \cdot \boxed{3 \cdot 3}} = 2 \cdot 2 \cdot 3 \cdot \sqrt{2} = 12\sqrt{2}$

For larger radicands, you may wish to create a **prime factorization** tree.

Example: The prime factorization for 1,050 is $2 \cdot 3 \cdot 5 \cdot 5 \cdot 7$, or $2 \cdot 3 \cdot 5^2 \cdot 7$.

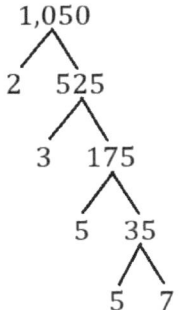

[last digit is 0 (even), so 1,050 is divisible by 2]

[5 + 2 + 5 = 12 is divisible by 3, so 525 is divisible by 3]

[last digit is 5, so 175 is divisible by 5]

[last digit is 5, so 35 is divisible by 5]

As described in the notes above, determining whether a number is divisible by 2, 3, or 5 can be done by inspection, using the following rules. A number is divisible by:

2	if the last digit is divisible by 2 (even)
3	if the *sum of the digits* is divisible by 3
5	if the last digit is divisible by 5 (0 or 5)

There are divisibility rules for prime numbers greater than 5, but they are a bit more complicated, so you can just check these on the calculator. For example, simply divide a number by 7 to see if the result is a whole number; if it is, then the number is divisible by that prime. Repeat for larger primes if necessary: 11, 13, 17, 19, 23, 29, 31, etc.

MODEL PROBLEM

Simplify $8\sqrt{90}$

Solution:

$$8\sqrt{90} = 8\sqrt{2\cdot\boxed{3\cdot3}\cdot5} = 8\cdot3\sqrt{2\cdot5} = 24\sqrt{10}$$

with labels (B), (C), (D) above, and (A) pointing to the factor tree.

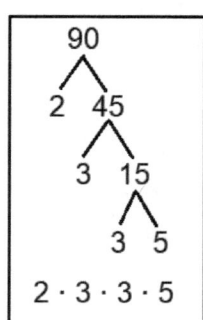

Explanation of steps:
- (A) Write the prime factorization of the radicand.
- (B) Group all pairs of factors, representing squares.
- (C) Remove the squares (pairs of factors) from the radicand by replacing them with their square roots (single factors) outside the radical sign.
- (D) Multiply all factors outside the radicand, and all factors remaining inside the radicand.

PRACTICE PROBLEMS

1. Simplify $\sqrt{12}$.	2. Simplify $\sqrt{50}$.
3. Simplify $\sqrt{32}$.	4. Express $4\sqrt{75}$ in simplest radical form.
5. Simplify $5\sqrt{20}$.	6. Simplify $3\sqrt{45}$.

7. Simplify $5\sqrt{72}$.	8. Simplify $2\sqrt{128}$.
9. Simplify $-3\sqrt{48}$.	10. Simplify $-\sqrt{98}$.
11. Express $2\sqrt{108}$ in simplest radical form.	12. Express $3\sqrt{250}$ in simplest radical form.
13. Simplify $\dfrac{\sqrt{32}}{4}$.	14. Simplify $\dfrac{7\sqrt{18}}{3}$.

Regents Questions

There are no Regents exam questions on this topic.

10.2   Operations with Radicals

Radicals may be combined by addition or subtraction only if, when expressed in simplest radical form, they are like radicals. **Like radicals** have the **same radicand**. (*Note:* they must also have the same index, but we are only concerned with square roots here.)

Sometimes unlike radicals may be simplified into like radicals.
Example: $\sqrt{12}$ and $\sqrt{75}$ can be simplified into $2\sqrt{3}$ and $5\sqrt{3}$, which are like radicals.

Combine like radicals just as you would combine like terms: add or subtract the coefficients and keep the radicand unchanged.
Example: $2\sqrt{3} + 5\sqrt{3} - \sqrt{3} = (2 + 5 - 1)\sqrt{3} = 6\sqrt{3}$

To multiply radicals, separately find the product of their coefficients and the product of their radicands, then simplify if possible.
Example: $\left(5\sqrt{3}\right)\left(2\sqrt{7}\right) = 10\sqrt{21}$

To divide radicals, separately find the quotient of their coefficients and the quotient of their radicands, then simplify if possible.
Example: $\dfrac{6\sqrt{72}}{3\sqrt{8}} = 2\sqrt{9} = 2 \cdot 3 = 6$

Sometimes, multiple operations involving radicals need to be performed.
Example: Simplify $\sqrt{2}\left(\sqrt{10} + 4\right)$ using the distributive property.
             $\sqrt{2}\left(\sqrt{10} + 4\right) = \sqrt{20} + 4\sqrt{2} = 2\sqrt{5} + 4\sqrt{2}$

MODEL PROBLEM 1: *ADDING OR SUBTRACTING RADICALS*

Express the sum $3\sqrt{8} + 2\sqrt{2}$ in simplest radical form.

Solution:

(A) $3\sqrt{8}$ can be simplified as follows: $3\sqrt{\boxed{2\cdot2}\cdot2} = 3\cdot2\sqrt{2} = 6\sqrt{2}$.

(B) So, $3\sqrt{8} + 2\sqrt{2} = 6\sqrt{2} + 2\sqrt{2} = 8\sqrt{2}$.

Explanation of steps:

(A) Express each term in simplest radical form.
(B) Combine like radicals by adding or subtracting their coefficients.

PRACTICE PROBLEMS

1. Add $\sqrt{75} + \sqrt{3}$.	2. Add $\sqrt{27} + \sqrt{12}$.
3. Find the sum of $\sqrt{50}$ and $\sqrt{32}$ in simplest radical form.	4. Find the sum of $\sqrt{27}$ and $\sqrt{108}$ in simplest radical form.
5. Write the sum $\sqrt{28} + \sqrt{63}$ in simplest radical form.	6. Write the sum $\sqrt{150} + \sqrt{24}$ in simplest radical form.

7. What is $3\sqrt{2} + \sqrt{8}$ expressed in simplest radical form?	8. What is $\sqrt{72} - 3\sqrt{2}$ expressed in simplest radical form?
9. Add $5\sqrt{7} + 3\sqrt{28}$.	10. Subtract $2\sqrt{50} - \sqrt{2}$.
11. Write the expression $6\sqrt{50} + 6\sqrt{2}$ in simplest radical form.	12. Express $\sqrt{25} - 2\sqrt{3} + \sqrt{27} + 2\sqrt{9}$ in simplest radical form.

MODEL PROBLEM 2:   *MULTIPLYING RADICALS*

Express the product $(5\sqrt{8})(7\sqrt{3})$ in simplest radical form.

Solution:

(A) $(5\sqrt{8})(7\sqrt{3}) = 35\sqrt{24}$

(B)   Simplifying, we get $35\sqrt{24} = 35\sqrt{\boxed{2 \cdot 2} \cdot 2 \cdot 3} = 35 \cdot 2\sqrt{2 \cdot 3} = 70\sqrt{6}$

Explanation of steps:

(A)   Find the product of the coefficients $[5 \times 7 = 35]$.
       Find the product of the radicands $[8 \times 3 = 24]$.

(B)   Express in simplest radical form.

PRACTICE PROBLEMS

13. Express $\sqrt{6} \cdot \sqrt{15}$ in simplest form.	14. What is the product of $4\sqrt{2}$ and $2\sqrt{6}$?
15. Express in simplest form: $\sqrt{90} \cdot \sqrt{40} - \sqrt{8} \cdot \sqrt{18}$	16. Express the product $3\sqrt{20}(2\sqrt{5} - 7)$ in simplest radical form.

17. Express $3\sqrt{7}\left(\sqrt{14} + 4\sqrt{56}\right)$ in simplest radical form.

18. Express the product of $\left(3 + \sqrt{5}\right)$ and $\left(3 - \sqrt{5}\right)$ in simplest form.

19. Express $y\sqrt{3} - \left(\sqrt{32} + y\sqrt{27}\right)$ in simplest radical form.

20. The length of a rectangle is $\left(3\sqrt{8} + 2\right)$ and the width is $\left(2\sqrt{2} + 1\right)$.

a) Express the perimeter of the rectangle in simplest radical form.

b) Express the area of the rectangle in simplest radical form.

MODEL PROBLEM 3:   *DIVIDING RADICALS*

Express $\dfrac{9\sqrt{20}}{3\sqrt{5}}$ in simplest radical form.

Solution:

$$\dfrac{9\sqrt{20}}{3\sqrt{5}} = 3\sqrt{4} = 3 \cdot 2 = 6$$

Explanation of steps:
(A) Find the quotient of the coefficients *[9 ÷ 3 = 3]*.
    Find the quotient of the radicands *[20 ÷ 5 = 4]*.
(B) Express in simplest radical form.

PRACTICE PROBLEMS

21. Express $\dfrac{\sqrt{65}}{\sqrt{5}}$ in simplest form.	22. Express in simplest form: $\dfrac{20\sqrt{100}}{4\sqrt{2}}$
23. Express $\dfrac{\sqrt{84}}{\sqrt{3}}$ in simplest radical form.	24. Express $\dfrac{6\sqrt{20}}{3\sqrt{5}}$ in simplest radical form.

25. Express $\dfrac{3\sqrt{75} + \sqrt{27}}{3}$ in simplest radical form.	26. Express $\dfrac{16\sqrt{21}}{2\sqrt{7}} - 5\sqrt{12}$ in simplest radical form.
27. Express in simplest form: $\dfrac{\sqrt{48} - 5\sqrt{27} + 2\sqrt{75}}{\sqrt{3}}$	28. Express in simplest form: $\dfrac{\sqrt{27} + \sqrt{75}}{\sqrt{12}}$

Regents Questions

MULTIPLE CHOICE

1. What is the sum of $3x\sqrt{7}$ and $2x\sqrt{7}$?

   (1) $5x\sqrt{7}$ (3) $5x\sqrt{14}$

   (2) $5x^2\sqrt{7}$ (4) $5x^2\sqrt{14}$

10.3 Rationalize Denominators

When working with algebraic fractions involving square roots, we prefer to change it to an equivalent fraction that does not include any square roots in the denominator. Eliminating radicals from the denominator is called **rationalizing the denominator**. Whenever we simplify fractions, we should rationalize their denominators.

Example: $\dfrac{3}{\sqrt{2}}$ has an irrational denominator.

If a fraction's denominator has a *monomial* containing a square root, we will multiply both the numerator and denominator by that square root. This will work because multiplying a square root by itself will eliminate the radical sign: $\sqrt{x} \cdot \sqrt{x} = x$.

Example: We can rationalize the denominator of $\dfrac{3}{\sqrt{2}}$ by multiplying by $\dfrac{\sqrt{2}}{\sqrt{2}}$, as in:

$$\frac{3}{\sqrt{2}} \cdot \frac{\sqrt{2}}{\sqrt{2}} = \frac{3 \cdot \sqrt{2}}{\sqrt{2} \cdot \sqrt{2}} = \frac{3\sqrt{2}}{2}$$

To rationalize a monomial denominator of a fraction:
1. Find the radical in the denominator, and create a new fraction using this radical in the numerator and denominator (a form of 1). Multiply the two fractions.
2. Simplify.

Example: To simplify $\dfrac{4}{3\sqrt{6}}$, multiply the fraction by $\dfrac{\sqrt{6}}{\sqrt{6}}$, which gives us

$$\frac{4}{3\sqrt{6}} \cdot \frac{\sqrt{6}}{\sqrt{6}} = \frac{4\sqrt{6}}{18} = \frac{2\sqrt{6}}{9}.$$

When adding or subtracting fractions, it is helpful to rationalize the denominators first.

MODEL PROBLEM 1:   *RATIONALIZING THE DENOMINATOR*

Rationalize the denominator of $\dfrac{10}{\sqrt{5}}$ and write the equivalent expression in simplest form.

Solution:

(A)   (B)   (C)

$$\frac{10}{\sqrt{5}} \cdot \left(\frac{\sqrt{5}}{\sqrt{5}}\right) = \frac{10\sqrt{5}}{5} = 2\sqrt{5}$$

Explanation of steps:

(A) If the fraction has a single irrational square root term in the denominator $[\sqrt{5}]$, create a new fraction with this radical as its numerator and denominator.

(B) Multiply the two fractions. The denominator of the product is now rational.

(C) Simplify.

PRACTICE PROBLEMS

1. Rationalize the denominator of $\dfrac{1}{\sqrt{7}}$.	2. Rationalize the denominator of $\dfrac{6}{\sqrt{2}}$ and simplify.
3. Rationalize the denominator of $\dfrac{5}{\sqrt{10}}$ and simplify.	4. Rationalize the denominator of $\dfrac{6}{\sqrt{21}}$ and simplify.
5. Rationalize the denominator of $\dfrac{8}{3\sqrt{6}}$ and simplify.	6. Rationalize the denominator of $\dfrac{10\sqrt{2}}{\sqrt{5}}$ and simplify.

7. Multiply $\dfrac{2}{\sqrt{3}} \times \dfrac{\sqrt{2}}{5}$ and express the product as a fraction with a rational denominator.

8. Simplify $\sqrt{\dfrac{16}{3}}$. Express as a fraction with a rational denominator.

MODEL PROBLEM 2:   *ADD FRACTIONS WITH IRRATIONAL DENOMINATORS*

Add $\dfrac{2}{\sqrt{3}} + \dfrac{1}{\sqrt{5}}$

Solution:

(A) $\dfrac{2}{\sqrt{3}} \cdot \left(\dfrac{\sqrt{3}}{\sqrt{3}}\right) = \dfrac{2\sqrt{3}}{3}$ $\dfrac{1}{\sqrt{5}} \cdot \left(\dfrac{\sqrt{5}}{\sqrt{5}}\right) = \dfrac{\sqrt{5}}{5}$

(B) $\dfrac{2\sqrt{3}}{3} + \dfrac{\sqrt{5}}{5} = \dfrac{2\sqrt{3}}{3} \cdot \left(\dfrac{5}{5}\right) + \dfrac{\sqrt{5}}{5} \cdot \left(\dfrac{3}{3}\right) = \dfrac{10\sqrt{3}}{15} + \dfrac{3\sqrt{5}}{15}$

(C) $= \dfrac{10\sqrt{3} + 3\sqrt{5}}{15}$

Explanation of steps:

(A) When adding fractions, first rationalize the denominators of both fractions, and simplify each fraction if possible.

(B) Now, add the fractions by finding the least common denominator *[the LCD is 15]* and converting the fractions into equivalent fractions with the same LCD.

(C) Write the sum as a single fraction by adding the numerators and keeping the LCD as the denominator.

PRACTICE PROBLEMS

9. Add $\frac{1}{\sqrt{3}} + \frac{1}{\sqrt{2}}$.	10. Add $\frac{1}{\sqrt{2}} + \frac{3}{\sqrt{5}}$.
11. Add $\frac{3}{\sqrt{5}} + \frac{4}{\sqrt{6}}$.	12. What is $\frac{3 - \sqrt{8}}{\sqrt{3}}$ expressed in simplest form?

13. What is $\sqrt{\frac{4}{3}} - \sqrt{\frac{3}{4}}$ expressed in simplest form?

Regents Questions

CONSTRUCTED RESPONSE

1. Rationalize: $\frac{3}{2\sqrt{6}}$

10.4   Closure

The set of real numbers is made up of **rational numbers** and **irrational numbers**. (Using the language of sets, the set of reals is the union of the two disjoint sets: the set of rationals and the set of irrationals.)

A rational number is any number that can be expressed as a fraction $\frac{a}{b}$ where a is an integer and b is a non-zero integer. Expressed in decimal form, rational numbers are terminating or repeating decimals, such as –100, 1.75. or $2.\overline{6}$. So, an irrational number is any real number that *cannot* be expressed as a fraction of integers. Irrational numbers, such as π and $\sqrt{2}$, are non-terminating, non-repeating decimals.

In fact, the *square roots of all whole numbers that are not perfect squares are irrational.*
Examples: $\sqrt{9}$ and $\sqrt{49}$ are rational, but $\sqrt{3}$ and $\sqrt{50}$ are irrational.

We have already seen (*in Section 1.1*) that that the set of *non-zero* rational numbers is *closed* under each of the four basic operations.

In contrast, the set of irrational numbers is **not closed** under any of the four basic operations.
Examples: We can add, subtract, multiply, or divide two irrational numbers and the result may be rational, as shown here.

$$\sqrt{2} + \left(-\sqrt{2}\right) = 0 \qquad \sqrt{2} - \sqrt{2} = 0 \qquad \sqrt{2} \cdot \sqrt{2} = 2 \qquad \frac{\sqrt{2}}{\sqrt{2}} = 1$$

When any of the four basic operations are performed on *non-zero* real numbers:
   a)   if both operands are **rational**, the result is **rational**;
   b)   if one of the operands is **rational** and the other operand is **irrational**, the result is *always* **irrational**;
   c)   if both operands are **irrational**, the result may be **rational** (e.g., $\sqrt{2} \cdot \sqrt{8} = \sqrt{16} = 4$) or **irrational** (e.g., $\sqrt{2} \cdot \sqrt{3} = \sqrt{6}$).

How do we know that statement b) above is true?
Remember, an irrational number is a non-repeating, non-terminating decimal. If we were to, say, add such a number to a rational number (that is, a terminating or repeating decimal), the result would still be a non-repeating, non-terminating decimal.
Examples: (a) $2 + \sqrt{2} = 2 + 1.414213 \ldots = 3.414213 \ldots$
                    (b) $\frac{1}{3} + \pi = 0.33333 \ldots + 3.14159 \ldots = 3.47492 \ldots$

Another way to look at it is by using our knowledge of closure.

For example: Is it possible to have $a + b = c$ where a is rational, b is irrational, and c is rational?

The answer is no, by the following proof:

1) If we solve the equation for b, we get $b = c - a$.
2) If both c and a are rational, and we know the set of rational numbers is closed under subtraction, then the difference $c - a$ would have to be rational.
3) This would mean that b would have to be rational, since $b = c - a$.
4) Since we originally said b is irrational, we've come to a contradiction.
5) So, it's not possible to have $a + b = c$ where a is rational, b is irrational, and c is rational. Therefore the sum c would have to be irrational.

This is known as a *proof by contradiction*. Similar explanations can be used to justify or prove this for the other basic operations.

MODEL PROBLEM

The famous "golden ratio" is defined as $\dfrac{1 + \sqrt{5}}{2}$. Is the golden ratio rational or irrational?

Solution:

$1 + \sqrt{5}$ is the sum of a rational and an irrational number, so it is irrational.

$\dfrac{1 + \sqrt{5}}{2}$ is the division of an irrational numerator and a rational denominator, so it is irrational.

Explanation of steps:

Look at each operation. If both operands are rational, then the result is rational. If one operand is rational and the other is irrational [as is the case with both operations performed here], then the result is irrational. An operation between two irrational numbers may have either a rational or irrational result.

PRACTICE PROBLEMS

1. Which of the following square roots is an irrational number?	2. State whether $2\sqrt{3}$ is rational or irrational.
(1) $-\sqrt{16}$ (3) $\sqrt{64}$ (2) $\sqrt{8}$ (4) $\sqrt{\dfrac{1}{64}}$	

3. State whether $\frac{\pi}{2}$ is rational or irrational. Justify your answer.

4. State whether $\frac{2 - \sqrt{29}}{4}$ is rational or irrational. Justify your answer.

5. Name at least three possible values of x that would make $x\sqrt{3}$ a rational number.

6. $\frac{22}{7}$ and 3.14 are often used as a rational approximations of π. Which of these is closer to the actual value of π?

Regents Questions

Multiple Choice

1. Given: $L = \sqrt{2}$
   $M = 3\sqrt{3}$
   $N = \sqrt{16}$
   $P = \sqrt{9}$

   Which expression results in a rational number?
   (1) $L + M$
   (2) $M + N$
   (3) $N + P$
   (4) $P + L$

2.   Which statement is *not* always true?
    (1)   The product of two irrational numbers is irrational.
    (2)   The product of two rational numbers is rational.
    (3)   The sum of two rational numbers is rational.
    (4)   The sum of a rational number and an irrational number is irrational.

3.   Which statement is *not* always true?
    (1)   The sum of two rational numbers is rational.
    (2)   The product of two irrational numbers is rational.
    (3)   The sum of a rational number and an irrational number is irrational.
    (4)   The product of a nonzero rational number and an irrational number is irrational.

4.   For which value of P and W is $P + W$ a rational number?
    (1) $P = \frac{1}{\sqrt{3}}$ and $W = \frac{1}{\sqrt{6}}$ (3) $P = \frac{1}{\sqrt{6}}$ and $W = \frac{1}{\sqrt{10}}$
    (2) $P = \frac{1}{\sqrt{4}}$ and $W = \frac{1}{\sqrt{9}}$ (4) $P = \frac{1}{\sqrt{25}}$ and $W = \frac{1}{\sqrt{2}}$

5.   Given the following expressions:
         I. $-\frac{5}{8} + \frac{3}{5}$ III. $(\sqrt{5}) \cdot (\sqrt{5})$
         II. $\frac{1}{2} + \sqrt{2}$ IV. $3 \cdot (\sqrt{49})$
     Which expression(s) result in an irrational number?
    (1)   II, only   (3)   I, III, IV
    (2)   III, only   (4)   II, III, IV

6.   The product of $\sqrt{576}$ and $\sqrt{684}$ is
    (1)   irrational because both factors are irrational
    (2)   rational because both factors are rational
    (3)   irrational because one factor is irrational
    (4)   rational because one factor is rational

7.   Which expression results in a rational number?
    (1) $\sqrt{121} - \sqrt{21}$ (3) $\sqrt{36} \div \sqrt{225}$
    (2) $\sqrt{25} \cdot \sqrt{50}$ (4) $3\sqrt{5} + 2\sqrt{5}$

8.   Which expression results in a rational number?
    (1) $\sqrt{2} \cdot \sqrt{18}$ (3) $\sqrt{2} + \sqrt{2}$
    (2) $5 \cdot \sqrt{5}$ (4) $3\sqrt{2} + 2\sqrt{3}$

9.   If $x = 2$, $y = 3\sqrt{2}$, and $w = 2\sqrt{8}$, which expression results in a rational number?
    (1) $x + y$ (3) $(w)(y)$
    (2) $y - w$ (4) $y \div x$

10. Which expression represents an irrational number?

     (1) $\sqrt{16} + \sqrt{1}$ (3) $\sqrt{36} + \sqrt{7}$

     (2) $\sqrt{25} + \sqrt{4}$ (4) $\sqrt{49} + \sqrt{9}$

CONSTRUCTED RESPONSE

11. Ms. Fox asked her class "Is the sum of 4.2 and $\sqrt{2}$ rational or irrational?" Patrick answered that the sum would be irrational. State whether Patrick is correct or incorrect. Justify your reasoning.

12. Determine if the product of $3\sqrt{2}$ and $8\sqrt{18}$ is rational or irrational. Explain your answer.

13. Is the sum of $3\sqrt{2}$ and $4\sqrt{2}$ rational or irrational? Explain your answer.

14. Jakob is working on his math homework. He decides that the sum of the expression $\frac{1}{3} + \frac{6\sqrt{5}}{7}$ must be rational because it is a fraction. Is Jakob correct? Explain your reasoning.

15. State whether $7 - \sqrt{2}$ is rational or irrational. Explain your answer.

16. A teacher wrote the following set of numbers on the board:
     $$a = \sqrt{20} \qquad\qquad b = 2.5 \qquad\qquad c = \sqrt{225}$$
     Explain why $a + b$ is irrational, but $b + c$ is rational.

17. Is the product of $\sqrt{16}$ and $\frac{4}{7}$ rational or irrational? Explain your reasoning.

18. Is the product of two irrational numbers always irrational? Justify your answer.

19. State whether the product of $\sqrt{3}$ and $\sqrt{9}$ is rational or irrational. Explain your answer.

20. Is the product of $\sqrt{1024}$ and –3.4 rational or irrational? Explain your reasoning.

21. Is the product of $\sqrt{8}$ and $\sqrt{98}$ rational or irrational? Justify your answer.

22. Given: $A = \sqrt{363}$ and $B = \sqrt{27}$
     Explain why $A + B$ is irrational. Explain why $A \cdot B$ is rational.

23. Classify the expression $\frac{2}{\sqrt{144}} + \frac{\sqrt{169}}{3}$ as rational or irrational. Explain your reasoning.

24. State whether $2\sqrt{3} + 6$ is rational or irrational. Explain your answer.

CHAPTER 11.   FACTORING

11.1   Factor Out the Greatest Common Factor

The **distributive property** allows us to express the product of a monomial and polynomial as a single polynomial. There are times when we would like to **factor** a polynomial, which means to break it down into a product of its factors. Factoring is a process that reverses multiplying.

As part of this process, we may need to find the **greatest common factor (GCF)** of the terms. Remember that we found the GCF of whole numbers by listing their prime factorizations and then determining which factors they had in common.

Example: GCF of 60 and 75. $60 = 2 \cdot 2 \cdot \underline{3} \cdot \underline{5}$ and $75 = \underline{3} \cdot \underline{5} \cdot 5$.
They have $3 \cdot 5 = 15$ in common, so 15 is the GCF.

If the terms of a polynomial have any factors in common, we can use the distributive property (in reverse) to break the polynomial down into factors.

To factor out the GCF:
1. Identifying the greatest common factor (GCF) of its terms.
2. Find the other factor by either
   a. gathering up the remaining factors from each term, or
   b. dividing the original polynomial by the GCF
3. Write the result as a product of the two factors.

Example: $3x^2 + 6x$ can be rewritten as $\underline{3} \cdot x \cdot x + 2 \cdot \underline{3} \cdot \underline{x}$, giving us a GCF of $3x$.
To find the other factor, we can either:
   a) gather up the remaining (*not underlined*) factors from the prime factorization $\underline{3} \cdot \underline{x} \cdot x + 2 \cdot \underline{3} \cdot \underline{x}$ to get $x + 2$, or

   b) we can divide the original polynomial by the GCF, as in $\dfrac{3x^2 + 6x}{3x} = x + 2$.

Write the result as the product of the two factors: $3x(x + 2)$.
Note: You can check your answer by applying the distributive property.

If all of the factors in a term's prime factorization are underlined as part of the GCF, you are left with a "hidden" factor of 1. So, you'll need to write the term 1 as part of the other factor. Remember, the other factor should have as many terms as the original polynomial.

Example: $3x^2 + 3x$ can be rewritten as $\underline{3} \cdot \underline{x} \cdot x + \underline{3} \cdot \underline{x}$, giving us a GCF of $3x$. When gathering up the remaining factors, we see that the all of the second term's factors are underlined, leaving us with just a "hidden" factor of 1. So, our other factor is $x + 1$. The result in factored form is $3x(x + 1)$.
The need to use a term of 1 is more easily seen when we divide by the GCF:
$$\frac{3x^2 + 3x}{3x} = \frac{3x^2}{3x} + \frac{3x}{3x} = x + 1$$

MODEL PROBLEM

Factor $8x^2y - 12xy + 20y^2$.

Solution:
(A) $8x^2y - 12xy + 20y^2 = \underline{2} \cdot \underline{2} \cdot 2 \cdot x \cdot x \cdot \underline{y} - \underline{2} \cdot \underline{2} \cdot 3 \cdot x \cdot \underline{y} + \underline{2} \cdot \underline{2} \cdot 5 \cdot \underline{y} \cdot y$
(B) $= 2 \cdot 2 \cdot y \, (2 \cdot x \cdot x - 3 \cdot x + 5 \cdot y)$
(C) $= 4y(2x^2 - 3x + 5y)$

Explanation of steps:
(A) Find the GCF of the terms. One way to do this is to expand each term using the prime factorization methods. Then underline any factors that are common to <u>all terms</u>.
(B) The underlined factors represent the GCF $[2 \cdot 2 \cdot y = 4y]$.
The factors that remain (not underlined) should be written in parentheses as the other factor. (If all of a term's factors are underlined, write 1.)
You can also find the second factor by dividing the original polynomial by the GCF
$\left[\dfrac{8x^2y - 12xy + 20y^2}{4y} = 2x^2 - 3x + 5y\right].$
(C) Write the result in unexpanded form.

PRACTICE PROBLEMS

1. Factor: $4x^2 - 6x$	2. Factor: $5a^2 - 10a$
3. Factor: $14x^3 + 7x$	4. Factor: $x^3 + x^2 - x$

5. Factor: $12x^3y + 18xy^2$

6. Factor: $2y^3 - 4y^2 + 2y$

7. Factor: $3x^3 - 6x^2 + 6x$

8. Factor: $-2x - 2y$

9. Factor: $3m^2n + 12mn^2$

10. Factor: $6x^3y^2z - 4x^2y^2$

Regents Questions

There are no Regents exam questions on this topic.

11.2   Factor a Trinomial

When we multiply two binomials (by FOIL), the result is often a trinomial.
Example: $(x - 3)(x + 2) = x^2 + 2x - 3x - 6 = x^2 - x - 6$

Therefore, we can often **factor a trinomial** into the product of two binomial factors.

If the trinomial is a second-degree polynomial in one variable in standard form with a lead coefficient of 1 (that is, if x is the variable, then the polynomial is of the form $x^2 + bx + c$, where b and c are integers), then we can use the following method to factor the trinomial.

To factor $x^2 + bx + c$ by the product-sum method:
1. Find two integers (if any) that *multiply* to give us c and *add* to give us b.
2. Factor the trinomial into two binomials in which the variable is written as each first term and the two integers are written as the last terms.

Remember that if c is positive, the two integers must have the same signs, but if c is negative, the two integers must have different signs.
Example: To factor $x^2 - x - 6$ [$b = -1$ *and* $c = -6$], we need two integers whose product is -6 and whose sum is -1. Those two integers are -3 and 2.
                 So, write the result as $(x - 3)(x + 2)$.
                 Note: You can check your answer by multiplying the two binomials.

Essentially, we are using the Box Method for multiplying binomials (from Section 5.3) in reverse:
Example: To factor $x^2 - x - 6$, place x^2 in the top left box and –6 in the bottom right box.

	x	
x	x^2	
		-6

                 We now need to fill the remaning headers with two integers whose product is –6 and whose sum is –1, namely –3 and 2. This will give us two terms in the diagonal that add up to the middle term: $-3x + 2x = -x$. The headers now give us the two binomial factors: $(x - 3)(x + 2)$.

	x	-3
x	x^2	$-3x$
2	$2x$	-6

Note that not all trinomials can be factored. For example, $x^2 + x + 1$ cannot be factored because there are no two integers whose product is 1 and whose sum is 1. A trinomial that cannot be factored is called a **prime trinomial**.

MODEL PROBLEM

Factor $x^2 - 8x + 12$.

Solution:
$(x - 6)(x - 2)$

Explanation of steps:
Find two integers that multiply to give us c [+12] and add to give us b [−8].
[The factors of 12 are: 12×1, 6×2, and 4×3, as well as $(-12) \times (-1)$, $(-6) \times (-2)$, and $(-4) \times (-3)$. The only pair that adds to −8 is −6 and −2. So, the answer is $(x - 6)(x - 2)$.]

PRACTICE PROBLEMS

1. Factor $x^2 + 9x + 14$	2. Factor $x^2 - 11x + 18$
3. Factor $x^2 - 6x - 27$	4. Factor $a^2 - a - 210$
5. Factor: $x^2 + 5x - 24$	6. Factor: $x^2 + 2x - 15$

7.   Factor: $x^2 - 10x - 24$

8.   Factor: $x^2 - 5x + 6$

9.   Determine whether the following trinomial is prime: $x^2 - 3x + 15$

10. If $x + 2$ is a factor of $x^2 + bx + 10$, what is the value of b?

11. Add the trinomials. Then, factor the result: $(-3x^2 + x - 2) + (4x^2 + 3x - 10)$

Regents Questions

MULTIPLE CHOICE

1. Which expression is equivalent to $x^4 - 12x^2 + 36$?
    (1) $(x^2 - 6)(x^2 - 6)$ (3) $(6 - x^2)(6 + x^2)$
    (2) $(x^2 + 6)(x^2 + 6)$ (4) $(x^2 + 6)(x^2 - 6)$

2. The trinomial $x^2 - 14x + 49$ can be expressed as
    (1) $(x - 7)^2$ (3) $(x - 7)(x + 7)$
    (2) $(x + 7)^2$ (4) $(x - 7)(x + 2)$

3. David correctly factored the expression $m^2 - 12m - 64$. Which expression did he write?
    (1) $(m - 8)(m - 8)$ (3) $(m - 16)(m + 4)$
    (2) $(m - 8)(m + 8)$ (4) $(m + 16)(m - 4)$

4. The expression $x^2 - 10x + 24$ is equivalent to
    (1) $(x + 12)(x - 2)$ (3) $(x + 6)(x + 4)$
    (2) $(x - 12)(x + 2)$ (4) $(x - 6)(x - 4)$

5. Which expression is equivalent to $x^2 + 5x - 6$?
    (1) $(x + 3)(x - 2)$ (3) $(x - 6)(x + 1)$
    (2) $(x + 2)(x - 3)$ (4) $(x + 6)(x - 1)$

CONSTRUCTED RESPONSE

6. In the equation $x^2 + 10x + 24 = (x + a)(x + b)$, b is an integer. Find algebraically all possible values of b.

11.3   Factor the Difference of Perfect Squares

When multiplying binomials, you may have noticed an interesting result when the binomials are a sum and difference of the same two terms. The middle terms of the product cancel out, leaving us with just the difference of the squares of the terms.
Example: $(x + 3)(x - 3) = x^2 - 3x + 3x - 9 = x^2 - 9$

So, if we want to factor a **difference of two perfect squares** into its two binomial factors, we can simply take the **sum and difference of the square roots** of the two terms.
Example: $x^2 - 9$ is a difference of two perfect squares.
            By taking the square root of each term (x and 3), we can factor this expression into the product of two binomials: $(x + 3)(x - 3)$.

In order to use this method, *both terms must be perfect squares* and there must be a *subtraction* sign between them.

Be sure to write the square roots in the same order that their squares appear in the problem. For example, $(x + 3)(x - 3)$ is not the same as $(3 + x)(3 - x)$; subtraction is not commutative.

MODEL PROBLEM

Factor $9x^4 - 25$.

Solution:
$(3x^2 + 5)(3x^2 - 5)$

Explanation of steps:
If the expression is the difference (subtraction) of two perfect squares, then take the square root of each term $[\sqrt{9x^4} = 3x^2 \text{ and } \sqrt{25} = 5]$.
Write the binomial factors as the sum $[3x^2 + 5]$ and difference $[3x^2 - 5]$ of the square roots, and express the answer as a product of the binomial factors $[(3x^2 + 5)(3x^2 - 5)]$.

PRACTICE PROBLEMS

1. Factor $x^2 - 36$	2. Factor $4x^2 - 9$
3. Factor $9 - x^2$	4. Factor $a^2 - 1$
5. Factor $49x^2 - y^2$	6. Factor $4a^2 - 9b^2$

7. Factor $x^2y^2 - 16$

8. Factor $x^{10} - 100$

9. Factor $100n^2 - 1$

10. Factor $121 - x^2$

11. Factor $9a^2 - 64b^2$

12. When $9x^2 - 100$ is factored, it is equivalent to $(3x - b)(3x + b)$. What is a value for b?

Regents Questions

MULTIPLE CHOICE

1. If the area of a rectangle is expressed as $x^4 - 9y^2$, then the product of the length and the width of the rectangle could be expressed as
   (1) $(x - 3y)(x + 3y)$
   (2) $(x^2 - 3y)(x^2 + 3y)$
   (3) $(x^2 - 3y)(x^2 - 3y)$
   (4) $(x^4 + y)(x - 9y)$

2. The expression $x^4 - 16$ is equivalent to
   (1) $(x^2 + 8)(x^2 - 8)$
   (2) $(x^2 - 8)(x^2 - 8)$
   (3) $(x^2 + 4)(x^2 - 4)$
   (4) $(x^2 - 4)(x^2 - 4)$

3. Which expression is equivalent to $16x^4 - 64$?
   (1) $(4x^2 - 8)^2$
   (2) $(8x^2 - 32)^2$
   (3) $(4x^2 + 8)(4x^2 - 8)$
   (4) $(8x^2 + 32)(8x^2 - 32)$

4. The expression $49x^2 - 36$ is equivalent to
   (1) $(7x - 6)^2$
   (2) $(24.5x - 18)^2$
   (3) $(7x - 6)(7x + 6)$
   (4) $(24.5x - 18)(24.5x + 18)$

5. Which expression is equivalent to $y^4 - 100$?
   (1) $(y^2 - 10)^2$
   (2) $(y^2 - 50)^2$
   (3) $(y^2 + 10)(y^2 - 10)$
   (4) $(y^2 + 50)(y^2 - 50)$

6. The expression $4x^2 - 25$ is equivalent to
   (1) $(4x - 5)(x + 5)$
   (2) $(4x + 5)(x - 5)$
   (3) $(2x + 5)(2x - 5)$
   (4) $(2x - 5)(2x - 5)$

7. The expression $w^4 - 36$ is equivalent to
   (1) $(w^2 - 18)(w^2 - 18)$
   (2) $(w^2 + 18)(w^2 - 18)$
   (3) $(w^2 - 6)(w^2 - 6)$
   (4) $(w^2 + 6)(w^2 - 6)$

8. The expression $16x^2 - 81$ is equivalent to
   (1) $(8x - 9)(8x + 9)$
   (2) $(8x - 9)(8x - 9)$
   (3) $(4x - 9)(4x + 9)$
   (4) $(4x - 9)(4x - 9)$

9. The expression $36x^2 - 9$ is equivalent to
   (1) $(6x - 3)^2$
   (2) $(18x - 4.5)^2$
   (3) $(6x + 3)(6x - 3)$
   (4) $(18x + 4.5)(18x - 4.5)$

10. The expression $9m^2 - 100$ is equivalent to
    (1) $(3m - 10)(3m + 10)$
    (2) $(3m - 10)(3m - 10)$
    (3) $(3m - 50)(3m + 50)$
    (4) $(3m - 50)(3m - 50)$

11.4   Factor Completely

Sometimes, more than one method of factoring must be performed in order to completely factor an expression. An expression is **factored completely** if all of the factors are prime.

To factor completely:
1. Factor out a greatest common factor, if there is one.
2. Factor any trinomials, if possible, or any differences of two perfect squares.
3. Repeat these steps for each factor until every factor is prime.

CALCULATOR TIP

We can use the calculator to check our factoring by a method we saw earlier.
1. Store an arbitrary, preferably non-integer, value into x. For example, we can store 12.3 into x by pressing ⟦1⟧⟦2⟧⟦.⟧⟦3⟧⟦STO▸⟧⟦X,T,Θ,n⟧.
2. Now, test whether the two expressions are equal using ⟦2nd⟧⟦TEST⟧⟦1⟧ for the equal sign. A result of 1 means they are equal, or a result of 0 means they are not.

Example: The screenshots below show how to test whether $3x^2 - 12$ and its factored form, $3(x + 2)(x - 2)$, are equivalent.

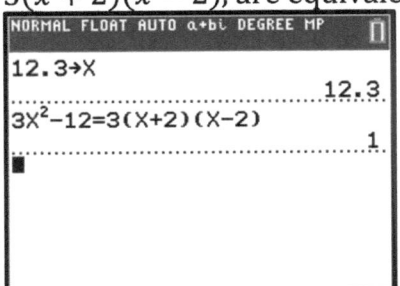

MODEL PROBLEM

Factor completely: $5x^4 - 5$

Solution:

$5x^4 - 5 =$
(A) $5(x^4 - 1) =$
(B) $5(x^2 + 1)(x^2 - 1) =$
(C) $5(x^2 + 1)(x + 1)(x - 1)$

Explanation of steps:
(A) Factor out the GCF, if any *[the GCF is 5]*.
(B) If any factor is a trinomial *[not here]* or a difference of two perfect squares *[x⁴ − 1]*, then factor it.
(C) Look at each factor to see if any of them can be factored further. *[The factor $x^2 - 1$ is a difference of two perfect squares and can be factored further.]*

PRACTICE PROBLEMS

1. Factor completely: $2y^2 + 12y - 54$	2. Factor completely: $3x^2 + 15x - 42$
3. Factor completely: $3x^2 - 27$	4. Factor completely: $2x^2 - 50$
5. Factor completely: $2a^2 - 10a - 28$	6. Factor completely: $x^3 + 8x^2 + 7x$
7. Factor completely: $2x^8 + 16x^7 + 32x^6$	8. Factor completely: $3ax^2 - 27a$

9. Factor completely: $5x^2y^3 - 180y$	10. Factor completely: $2x^5 - 32x$
11. Factor completely: $2x^2 + 10x - 12$	12. Factor completely: $a^3 - 4a$
13. Factor completely: $3x^3 - 33x^2 + 90x$	14. Factor completely: $36x^2 - 100y^6$
15. Factor completely: $4x^3y^3 - 36xy$	16. Factor completely: $6x - x^3 - x^2$

Regents Questions

MULTIPLE CHOICE

1. When factored completely, the expression $p^4 - 81$ is equivalent to
    (1) $(p^2 + 9)(p^2 - 9)$ (3) $(p^2 + 9)(p + 3)(p - 3)$
    (2) $(p^2 - 9)(p^2 - 9)$ (4) $(p + 3)(p - 3)(p + 3)(p - 3)$

2. When factored completely, $x^3 - 13x^2 - 30x$ is
    (1) $x(x + 3)(x - 10)$ (3) $x(x + 2)(x - 15)$
    (2) $x(x - 3)(x - 10)$ (4) $x(x - 2)(x + 15)$

3. Which expression is equivalent to $36x^2 - 100$?
    (1) $4(3x - 5)(3x - 5)$ (3) $2(9x - 25)(9x - 25)$
    (2) $4(3x + 5)(3x - 5)$ (4) $2(9x + 25)(9x - 25)$

4. Which expression is equivalent to $16x^2 - 36$?
    (1) $4(2x - 3)(2x - 3)$ (3) $(4x - 6)(4x - 6)$
    (2) $4(2x + 3)(2x - 3)$ (4) $(4x + 6)(4x + 6)$

5. Which expression is *not* equivalent to $2x^2 + 10x + 12$?
    (1) $(2x + 4)(x + 3)$ (3) $(2x + 3)(x + 4)$
    (2) $(2x + 6)(x + 2)$ (4) $2(x + 3)(x + 2)$

6. Which expression is equivalent to $18x^2 - 50$?
    (1) $2(3x + 5)^2$ (3) $2(3x - 5)(3x + 5)$
    (2) $2(3x - 5)^2$ (3) $2(3x - 25)(3x + 25)$

7. Which expression is *not* equivalent to $-4x^3 + x^2 - 6x + 8$?
    (1) $x^2(-4x + 1) - 2(3x - 4)$ (3) $-4x^3 + (x - 2)(x - 4)$
    (2) $x(-4x^2 - x + 6) + 8$ (4) $-4(x^3 - 2) + x(x - 6)$

8. Which expression is equivalent to $2x^2 + 8x - 10$?
    (1) $2(x - 1)(x + 5)$ (3) $2(x - 1)(x - 5)$
    (2) $2(x + 1)(x - 5)$ (4) $2(x + 1)(x + 5)$

9. When factored completely, $-x^3 + 10x^2 + 24x$ is
    (1) $-x(x + 4)(x - 6)$ (3) $-x(x + 2)(x - 12)$
    (2) $-x(x - 4)(x - 6)$ (4) $-x(x - 2)(x + 12)$

CONSTRUCTED RESPONSE

10. Factor the expression $x^4 + 6x^2 - 7$ completely.

11. Factor $x^4 - 16$ completely.

12. Factor the expression $x^4 - 36x^2$ completely.

13. Factor completely: $3y^2 - 12y - 288$

14. Factor completely: $4x^3 - 49x$

15. Factor $2x^2 + 16x - 18$ completely.

16. Factor $18x^2 - 2$ completely.

17. Factor $36 - 4x^2$ completely.

CHAPTER 12   QUADRATIC FUNCTIONS

12.1   Solve Simple Quadratic Equations

In a **linear equation**, the highest power for the variable x is 1. It is also called a first-degree equation. If an equation includes an x^2 term (and no term with x to a power higher than 2), it is called a **quadratic equation**, or a second-degree equation.

Example: $y = x^2$ is a quadratic equation; therefore, $f(x) = x^2$ is a quadratic function.

We have seen that tables of *linear* functions show a *common difference* in its values (as long as the x values change by constant amounts) and tables of *exponential* functions show a *common ratio*. To recognize quadratic functions, we will not find a common difference like we do for linear functions, but we'll find that the differences among their differences (called **second differences**) increase by a constant amount, as shown below.

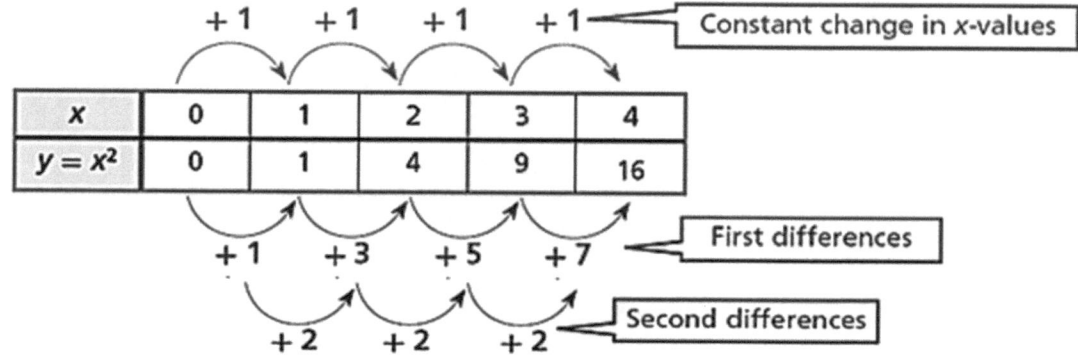

In this section, we will solve *simple* quadratic equations that have an x^2 term but no x term. The solutions to a quadratic equation are called its **roots** or **zeros**. Unlike linear equations, there are usually two solutions to a quadratic equation.

Example: For the equation $x^2 = 25$, the roots (*or zeros, or solutions*) are 5 and -5.
             After all, both $5^2 = 25$ and $(-5)^2 = 25$.

Note that when $x^2 = 0$, then zero is the only solution, and for $x^2 = -1$ (or any negative constant), there are no solutions among the real numbers. And so, it's possible for quadratic equations to have no real solutions, one solution, or two distinct solutions.

Also, based on the context, we may be able to discard one of the solutions as unreasonable.

Example: If $x^2 = 25$ represents the relationship between the length of a side of a square, x, and the area of the square, 25, a negative value for x (that is, -5) would not be a reasonable solution, so it can be discarded. The solution is 5.

One-step equations: To solve a simple quadratic equation like $x^2 = 25$, we can isolate the variable by performing the opposite of squaring – taking the **square root** – of both sides. We use the symbol, $\pm\sqrt{}$, to represent both the positive and negative square roots.

In general, for any non-negative constant c, we solve $x^2 = c$ by writing $x = \pm\sqrt{c}$. The solutions to a quadratic equation can also be written in set notation.

Example: (a) We solve $x^2 = 25$ by writing $x = \pm\sqrt{25} = \pm5$, or in set notation, $\{5, -5\}$.
(b) We solve $x^2 = 2$ by writing $x = \pm\sqrt{2}$, meaning that both $\sqrt{2}$ and $-\sqrt{2}$ are solutions to this equation. Since $\sqrt{2}$ is irrational, we cannot eliminate the radical sign, as we were able to do in the previous example. In set notation, the solutions are $\{\sqrt{2}, -\sqrt{2}\}$.

Generally, solutions are written in simplest radical form.
Example: If we find $x = \pm\sqrt{12}$, the solutions should be written as $\{2\sqrt{3}, -2\sqrt{3}\}$.

For this course, the constant c in the equation $x^2 = c$ will be non-negative.
Example: We cannot solve $x^2 = -1$ using the set of real numbers. This is because $\sqrt{-1}$, or the square root of any negative number, is not a real number.

Solving multi-step equations: It may take more than one step to solve a quadratic equation; we may need to isolate x^2 before taking the square root of both sides.
Example: $2x^2 + 1 = 33$ can be solved by subtracting 1 from both sides to get $2x^2 = 32$, then dividing both sides by 2 to get $x^2 = 16$. Finally, we can take the square root of both sides to give us $x = \pm\sqrt{16}$, or $x = \pm4$, so the roots are $\{4, -4\}$.

Finding the zeros of a function: The roots of a function are called its "zeros" because they represent the x-values at the points on its graph where $y = 0$, and therefore $f(x) = 0$. So, we can find the zeros of a quadratic function by *setting its expression equal to zero*.
Example: To find the zeros of $f(x) = x^2 - 9$, we solve the equation $x^2 - 9 = 0$. The zeros (or roots) are $\{3, -3\}$.

Literal equations: In an earlier section (see p. 39), we solved literal equations for one variable in terms of the other variables. We can solve literal equations for a squared variable by this same method: isolate the squared variable, then take the square root of both sides.
Example: Given $a = x^2 + b$, we can solve for x in terms of a and b.
First, subtract b from both sides to isolate the x^2, giving us $a - b = x^2$.
Then, take the square root of both sides, so $x = \pm\sqrt{a - b}$.

The standard form for a quadratic equation is $ax^2 + bx + c = 0$, where $a \neq 0$. The method in this section works only when $b = 0$; that is, when there is only an x^2 term in the equation. When $b \neq 0$ and the quadratic equation includes both an x^2 term and an x term, we would solve by one of the other methods in this unit (factoring, completing the square, or the quadratic formula).

MODEL PROBLEM 1:   *ONE-STEP EQUATION*

Solve $a^2 = 5$.

Solution:

   $a = \pm\sqrt{5}$ Solutions are $\{\sqrt{5}, -\sqrt{5}\}$.

Explanation:

Take the square root of both sides, making sure to include both the positive and negative root.

PRACTICE PROBLEMS

1. Solve for x: $x^2 = 81$	2. Solve for y: $y^2 = 20$

MODEL PROBLEM 2:   *MULTI-STEP EQUATION*

Solve $3x^2 - 2 = 46$.

Solution:

       $3x^2 - 2 = 46$
(A) $3x^2 = 48$
       $x^2 = 16$
(B) $x = \pm\sqrt{16}$
       $\{4, -4\}$

Explanation of steps:

(A) Isolate x^2. *[Add 2 to both sides, then divide both sides by 3]*
(B) Take the square root of both sides and write the solutions.

PRACTICE PROBLEMS

3. What is the solution set of the equation $3x^2 = 75$?	4. What is the solution set of the equation $4x^2 - 36 = 0$?
5. Solve: $2x^2 - 1 = 11$	6. Solve: $9x^2 + 5 = 9$

MODEL PROBLEM 3: *ZEROS OF A FUNCTION*

Find the zeros of the function $f(x) = 4x^2 - 1$.

Solution:

(A) $4x^2 - 1 = 0$
(B) $4x^2 = 1$
    $x^2 = \frac{1}{4}$
(C) $x = \pm\sqrt{\frac{1}{4}}$
    $\left\{-\frac{1}{2}, \frac{1}{2}\right\}$

Explanation of steps:

(A) Set $f(x) = 0$.
(B) Isolate x^2. *[Add 1 to both sides and divide both sides by 4]*
(C) Take the square root of both sides and write the solutions. *[The square root of ¼ is ½.]*

PRACTICE PROBLEMS

7. Find the zeros of the function, $f(x) = 5x^2 - 5$.	8. Find the zeros of the function, $g(m) = 3m^2$.

MODEL PROBLEM 4:　*LITERAL EQUATION*

The formula for the area of a circle is $A = \pi r^2$, where r is the radius. Express the radius in terms of the area.

Solution:

$$A = \pi r^2$$

(A) $\dfrac{A}{\pi} = r^2$

(B) $r = \sqrt{\dfrac{A}{\pi}}$

Explanation of steps:

(A) Isolate the squared variable for which we are solving. *[Divide both sides by π.]*

(B) Take the square root of both sides.
[We discard the negative square root in this case because the radius of a circle must be positive.]

PRACTICE PROBLEMS

9. Solve for h in terms of s:　$s = \dfrac{1}{2}h^2$	10. Solve for the positive value of a in terms of b and c: $$a^2 + b^2 = c^2$$

Regents Questions

MULTIPLE CHOICE

1. The formula for the volume of a cone is $V = \frac{1}{3}\pi r^2 h$. The radius, r, of the cone may be expressed as

   (1) $\sqrt{\frac{3V}{\pi h}}$

   (2) $\sqrt{\frac{V}{3\pi h}}$

   (3) $3\sqrt{\frac{V}{\pi h}}$

   (4) $\frac{1}{3}\sqrt{\frac{V}{\pi h}}$

2. If $4x^2 - 100 = 0$, the roots of the equation are

   (1) -25 and 25

   (2) -25, only

   (3) -5 and 5

   (4) -5, only

3. The equation for the volume of a cylinder is $V = \pi r^2 h$. The positive value of r, in terms of h and V, is

   (1) $r = \sqrt{\frac{V}{\pi h}}$

   (2) $r = \sqrt{V\pi h}$

   (3) $r = 2V\pi h$

   (4) $r = \frac{V}{2\pi}$

4. The distance a free falling object has traveled can be modeled by the equation $d = \frac{1}{2}at^2$, where a is acceleration due to gravity and t is the amount of time the object has fallen. What is t in terms of a and d?

   (1) $t = \sqrt{\frac{da}{2}}$

   (2) $t = \sqrt{\frac{2d}{a}}$

   (3) $t = \left(\frac{da}{d}\right)^2$

   (4) $t = \left(\frac{2d}{a}\right)^2$

5. Which value of x is a solution to the equation $13 - 36x^2 = -12$?

   (1) $\frac{36}{25}$

   (2) $\frac{25}{36}$

   (3) $-\frac{6}{5}$

   (4) $-\frac{5}{6}$

6. The solution to $2x^2 = 72$ is

   (1) $\{9, 4\}$

   (2) $\{-4, 9\}$

   (3) $\{6\}$

   (4) $\{\pm 6\}$

CONSTRUCTED RESPONSE

7. Ryker is given the graph of the function $y = \frac{1}{2}x^2 - 4$. He wants to find the zeros of the function, but is unable to read them exactly from the graph.

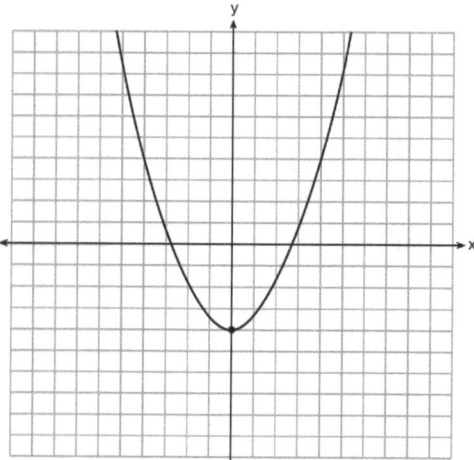

Find the zeros in simplest radical form.

8. The volume of a large can of tuna fish can be calculated using the formula $V = \pi r^2 h$. Write an equation to find the radius, r, in terms of V and h.

    Determine the diameter, to the *nearest inch*, of a large can of tuna fish that has a volume of 66 cubic inches and a height of 3.3 inches.

9. The height, H, in feet, of an object dropped from the top of a building after t seconds is given by $H(t) = -16t^2 + 144$. How many feet did the object fall between one and two seconds after it was dropped? Determine, algebraically, how many seconds it will take for the object to reach the ground.

10. Using the formula for the volume of a cone, express r in terms of V, h, and π.

11. The formula $F_g = \frac{GM_1M_2}{r^2}$ calculates the gravitational force between two objects where G is the gravitational constant, M_1 is the mass of one object, M_2 is the mass of the other object, and r is the distance between them. Solve for the positive value of r in terms of F_g, G, M_1, and M_2.

12. Solve the quadratic equation below for the exact values of x.
    $$4x^2 - 5 = 75$$

13. Solve $5x^2 = 180$ algebraically.

14. Solve $6x^2 - 42 = 0$ for the exact values of x.

12.2   Solve Quadratic Equations by Factoring

The **standard form** for a quadratic equation is $ax^2 + bx + c = 0$, where a, b, and c are real numbers $(a \neq 0)$ and zero is on one side of the equation.

Example: $x^2 - 3x + 5 = 0$ is a quadratic equation in standard form, where $a = 1$, $b = -3$, and $c = 5$.

If both a and b are not equal to 0 (that is, if there are both an x^2 and an x term), we need a new method of solving. Here, we look at solving quadratic equations by factoring.

To solve by factoring, we can use the following fact about a product of two (or more) factors equal to 0. If $jk = 0$, then at least one of the following must be true: $j = 0$ or $k = 0$. This is known as the **zero product property**.

If we can factor a quadratic equation into **factored form**, $g(x - m)(x - n) = 0$ where $g \neq 0$, then the zero product property tells us that either $x - m = 0$ or $x - n = 0$, so $x = m$ or $x = n$. Therefore, the roots are m and n.

To solve a quadratic equation by factoring:
1. transform the equation into standard form, with zero on one side
2. factor the polynomial completely
3. set each factor that contains a variable equal to zero
4. solve each resulting equation

Model Problem

Find the roots of the equation $x^2 + 6 = 5x$.

Solution:

(A) $x^2 + 6 = 5x$

$\underline{\quad -5x \quad -5x \quad}$

$x^2 - 5x + 6 = 0$

(B) $(x - 3)(x - 2) = 0$

(C) $x - 3 = 0$ or $x - 2 = 0$

(D) $x = 3$ or $x = 2$

The roots are 2 and 3.

Explanation of steps:

(A)   transform the equation into standard form, with zero on one side *[by subtracting the 5x from both sides]*.

(B)   factor the polynomial completely

(C)   set each factor equal to zero

(D)   solve each resulting equation

It is also a good idea to check each solution on your calculator. Check the original equation using $\boxed{2}\boxed{\text{STO}\blacktriangleright}\boxed{\text{X,T,}\Theta,n}$ then check it again using $\boxed{3}\boxed{\text{STO}\blacktriangleright}\boxed{\text{X,T,}\Theta,n}$.

PRACTICE PROBLEMS

1. Solve for x: $x^2 - 5x = 0$	2. Solve for x: $x^2 + 3x - 18 = 0$
3. Solve for x (by factoring): $4x^2 - 36 = 0$	4. Solve for x: $x^2 - 4x - 32 = 0$
5. Solve for x: $x^2 - 5x = 6$	6. Solve for x: $x^2 - 3 = 2x$
7. Find the roots of the equation $x^2 - x = 6$ algebraically.	8. Find the roots of the equation $x^2 = 30 - 13x$ algebraically.

9. Solve for x: $x^2 - 4x = x + 24$	10. Solve for x: $2x^2 + 10x = 12$
11. Solve for x: $x(x + 2) = 3$	12. Solve for x: $(x + 2)(x + 3) = 12$

Regents Questions

MULTIPLE CHOICE

1. What are the zeros of the function $f(x) = x^2 - 13x - 30$?
   - (1) -10 and 3
   - (2) 10 and -3
   - (3) -15 and 2
   - (4) 15 and -2

2. The zeros of the function $f(x) = 3x^2 - 3x - 6$ are
   - (1) -1 and -2
   - (2) 1 and -2
   - (3) 1 and 2
   - (4) -1 and 2

3. The zeros of the function $f(x) = 2x^2 - 4x - 6$ are
   - (1) 3 and -1
   - (2) 3 and 1
   - (3) -3 and 1
   - (4) -3 and -1

4. The zeros of the function $f(x) = x^2 - 5x - 6$ are
   - (1) -1 and 6
   - (2) 1 and -6
   - (3) 2 and -3
   - (4) -2 and 3

5.   What is the solution set of the equation $(x - 2)(x - a) = 0$?

        (1) -2 and a (3)   2 and a

        (2) -2 and $-a$ (4)   2 and $-a$

6.   The zeros of the function $p(x) = x^2 - 2x - 24$ are

        (1) -8 and 3   (3) -4 and 6

        (2) -6 and 4   (4) -3 and 8

CONSTRUCTED RESPONSE

7.   Write an equation that defines $m(x)$ as a trinomial where $m(x) = (3x - 1)(3 - x) + 4x^2 + 19$.

       Solve for x when $m(x) = 0$.

8.   Solve the equation for y. $(y - 3)^2 = 4y - 12$

9.   Janice is asked to solve $0 = 64x^2 + 16x - 3$.

       She begins the problem by writing the following steps:

                Line 1 $0 = 64x^2 + 16x - 3$

                Line 2 $0 = B^2 + 2B - 3$

                Line 3 $0 = (B + 3)(B - 1)$

       Use Janice's procedure to solve the equation for x.

       Explain the method Janice used to solve the quadratic equation.

10.   Determine all the zeros of $m(x) = x^2 - 4x + 3$, algebraically.

11.   Solve $x^2 - 8x - 9 = 0$ algebraically. Explain the first step you used to solve the given equation.

12.   Solve $x^2 - 9x = 36$ algebraically for all values of x.

12.3   **Find Quadratic Equations from Given Roots**

If given roots (zeros), then working backwards we can find an equation with those roots.

To find an equation given its roots:
1. Set x equal to each root. For a double root, the same root is used twice.
2. Get zero to one side of each equation.
3. Write as factors.
4. Multiply factors.

MODEL PROBLEM

Find a quadratic equation with roots of 5 and -1.

Solution:
(A) $x = 5 \ or \ x = -1$
(B) $x - 5 = 0 \qquad x + 1 = 0$
(C) $(x - 5)(x + 1) = 0$
(D) $x^2 - 4x - 5 = 0$

Explanation of steps:
(A)   Set x equal to each root.
(B)   Get zero to one side of each equation.
(C)   Write as factors.
(D)   Multiply factors.

PRACTICE PROBLEMS

1. Write a quadratic equation in standard form with roots of 10 and -2.	2. Write a quadratic equation in standard form with roots of 0 and 3.
3. Write a quadratic equation in standard form that has roots of -12 and 2.	4. Write a quadratic equation in standard form that has roots of -3 and 5.

5. Write a quadratic equation in standard form that has the solution set $\{-5, 2\}$.	6. Write a quadratic equation in standard form that has the solution set $\{1, 3\}$.
7. Write a quadratic equation in standard form with roots of $\frac{3}{2}$ and 2.	8. Write a quadratic equation in standard form with the double root of 1.

9. Write a quadratic equation in standard form with roots of 4 and -4.	10. When a 3rd degree polynomial is set equal to zero, it is a cubic equation. A cubic equation can have 3 roots. Find a cubic equation with roots of 0,1, and -1. *(We'll look at cubic equations in a later unit.)*

Regents Questions

MULTIPLE CHOICE

1. For which function defined by a polynomial are the zeros of the polynomial -4 and -6?

   (1) $y = x^2 - 10x - 24$ (3) $y = x^2 + 10x - 24$

   (2) $y = x^2 + 10x + 24$ (4) $y = x^2 - 10x + 24$

2. Keith determines the zeros of the function $f(x)$ to be -6 and 5. What could be Keith's function?

   (1) $f(x) = (x + 5)(x + 6)$ (3) $f(x) = (x - 5)(x + 6)$

   (2) $f(x) = (x + 5)(x - 6)$ (4) $f(x) = (x - 5)(x - 6)$

12.4   Equations with the Square of a Binomial

In an earlier section in this unit, we saw that we can solve an equation of the form $x^2 = c$, where c is a real number, by taking the square root of both sides: $x = \pm\sqrt{c}$.

In some equations, something more than just x may be squared. We may have a variable expression in parentheses as the base of the exponent (and no other variable terms outside the parentheses). In a quadratic equation, this expression may be a monomial or binomial. We can still follow the same steps to solve these types of equations. Once we isolate the expression being squared, we eliminate the exponent and parentheses by taking the square root of both sides. Since we need to use the $\pm$ symbol, this results in two linear equations to be solved, one for the positive square root and one for the negative square root.

Examples: To solve $(2x - 6)^2 = 4$, take the square root of both sides to get $2x - 6 = \pm\sqrt{4}$. This results in two linear equations: $2x - 6 = 2$ and $2x - 6 = -2$. Solving both equations, we find the two roots of the original quadratic equation, 4 and 2.

Irrational Roots

In some cases, the roots may be irrational. Since we cannot combine a rational number with an irrational number by addition or subtraction, the roots may be written as a sum and difference.

Example: If we solve $(x + 1)^2 = 2$ by taking the square root of both sides, we get $x + 1 = \pm\sqrt{2}$. This represents the two equations, $x + 1 = \sqrt{2}$ and $x + 1 = -\sqrt{2}$. In each case, subtracting 1 from both sides gives us $x = -1 + \sqrt{2}$ and $x = -1 - \sqrt{2}$, so our solutions, or roots, would be written as $\{-1 + \sqrt{2}, -1 - \sqrt{2}\}$.

Note: You do not need to write out both equations. It is common practice to go directly from $x + 1 = \pm\sqrt{2}$ to writing $x = -1 \pm \sqrt{2}$ before stating the two roots.

MODEL PROBLEM 1:   *IRRATIONAL ROOTS*

Solve $3(x-2)^2 = 15$.

Solution: **Explanation of steps:**

$3(x-2)^2 = 15$

(A) $(x-2)^2 = 5$

(B) $x - 2 = \pm\sqrt{5}$

(C) $x = 2 \pm \sqrt{5}$

(D) $\{2 - \sqrt{5}, 2 + \sqrt{5}\}$

(A) Isolate the binomial squared (ie, isolate the parentheses). *[Divide both sides by 3]*

(B) Take the square root of both sides.

(C) Isolate x. *[Add 2 to both sides]*

(D) Write the solution (roots) in set notation.

PRACTICE PROBLEMS

1. Solve $(x+5)^2 = 16$.	2. Solve $(x-4)^2 = 10$.
3. Solve $6(b-1)^2 = 48$.	4. Solve $5 - (m+1)^2 = -25$.

MODEL PROBLEM 2:   *ZEROS OF A FUNCTION*

Find the zeros of the function $f(x) = (x + 1)^2 - 1$.

Solution:	**Explanation of steps:**
(A) $(x + 1)^2 - 1 = 0$	(A) Set $f(x) = 0$.
(B) $(x + 1)^2 = 1$	(B) Solve the equation.
$\quad x + 1 = \pm\sqrt{1}$	*[Isolate the parentheses by adding 1 to both sides; take*
$\quad x + 1 = \pm 1$	*the square root of both sides; and isolate x by subtracting*
$\quad x = -1 \pm 1$	*1 from both sides]*
(C) $\{-2, 0\}$	(C) Write the zeros. *[$-1 + 1 = 0$ and $-1 - 1 = -2$]*

PRACTICE PROBLEMS

5. Find the zeros of the function $f(x) = (x - 2)^2$.	6. Find the zeros of the function $g(x) = 2(x + 5)^2 - 50$.

426

Regents Questions

MULTIPLE CHOICE

1. The zeros of the function $f(x) = (x+2)^2 - 25$ are
   - (1) -2 and 5
   - (2) -3 and 7
   - (3) -5 and 2
   - (4) -7 and 3

2. A student is asked to solve the equation $4(3x-1)^2 - 17 = 83$. The student's solution to the problem starts as
$$4(3x-1)^2 = 100$$
$$(3x-1)^2 = 25$$
   A correct next step in the solution of the problem is
   - (1) $3x - 1 = \pm 5$
   - (2) $3x - 1 = \pm 25$
   - (3) $9x^2 - 1 = 25$
   - (4) $9x^2 - 6x + 1 = 5$

3. The solution of the equation $(x+3)^2 = 7$ is
   - (1) $3 \pm \sqrt{7}$
   - (2) $7 \pm \sqrt{3}$
   - (3) $-3 \pm \sqrt{7}$
   - (4) $-7 \pm \sqrt{3}$

4. What is the solution of the equation $2(x+2)^2 - 4 = 28$?
   - (1) 6, only
   - (2) 2, only
   - (3) 2 and -6
   - (4) 6 and -2

5. What are the solutions to the equation $3(x-4)^2 = 27$?
   - (1) 1 and 7
   - (2) -1 and -7
   - (3) $4 \pm \sqrt{24}$
   - (4) $-4 \pm \sqrt{24}$

6. The solutions to $(x+4)^2 - 2 = 7$ are
   - (1) $-4 \pm \sqrt{5}$
   - (2) $4 \pm \sqrt{5}$
   - (3) -1 and -7
   - (4) 1 and 7

CONSTRUCTED RESPONSE

7. Find the zeros of $f(x) = (x-3)^2 - 49$, algebraically.

12.5   Complete the Square

Another method for solving quadratic equations is called **completing the square**. This method converts a quadratic equation into one of the form, $(x + p)^2 = q$, where p and q are constants, allowing the resulting equation to be easily solved by taking the square root of both sides.

We do this by manipulating the equation so that one side is a **perfect square trinomial** that can be factored into the *square of a binomial*, $(x + p)^2$, for some p. For a perfect square trinomial of the form $x^2 + bx + c$, the value of p will be $\frac{b}{2}$.

[Note: For the purposes of this course, b will always be even in order to avoid fractions.]

A trinomial $x^2 + bx + c$ is a perfect square if $\left(\frac{b}{2}\right)^2 = c$.

Example: $x^2 - 10x + 25$ is a perfect square trinomial because $\left(-\frac{10}{2}\right)^2 = (-5)^2 = 25$.

Therefore, it can be written as the square of a binomial, $(x - 5)^2$.

To solve a quadratic equation $x^2 + bx + c = 0$ by completing the square:
   1. Add the opposite of c to both sides.
   2. Add $\left(\frac{b}{2}\right)^2$ to both sides.
   3. Factor the trinomial into a binomial squared.
   4. Take the square root of both sides. Use the ± symbol on the right side and simplify radicals.
   5. Solve for x, remembering that ± gives two possible solutions.

Example: We can solve $x^2 - 4x - 21 = 0$ as follows.

$$x^2 - 4x = 21 \qquad\qquad \left(\frac{b}{2}\right)^2 = \left(\frac{-4}{2}\right)^2 = 4$$
$$x^2 - 4x + 4 = 21 + 4$$
$$(x - 2)^2 = 25$$
$$x - 2 = \pm 5$$
$$x = 2 \pm 5 = \{-3, 7\}$$

Note that $x^2 - 4x - 21$ is factorable, so we *could* have solved this by factoring with the same result: $(x + 3)(x - 7) = 0$ also gives us $x = \{-3, 7\}$.

The advantage of this method is that it can be used when trinomials cannot be factored. When we use this method on quadratic equations in which *prime trinomials* are set equal to zero, the roots (if there are any real roots) will be *irrational* (they will necessarily include *radicals*).

Example: $x^2 - 4x - 2 = 0$ cannot be solved by factoring, but we can complete the square.

$$x^2 - 4x = 2 \qquad\qquad \left(\frac{b}{2}\right)^2 = \left(\frac{-4}{2}\right)^2 = 4$$
$$x^2 - 4x + 4 = 2 + 4$$
$$(x - 2)^2 = 6$$
$$x - 2 = \pm\sqrt{6}$$
$$x = 2 \pm 5 = \{2 - \sqrt{6}, 2 + \sqrt{6}\}$$

You may prefer to use the Box Method (from Section 5.3) to complete steps 2 and 3 above.

Example: When solving $x^2 - 4x = 2$,

      1.   Place x^2 in the top left square and write x and x as headers: $(x)(x) = x^2$.

$$
\begin{array}{c|c|c}
 & x & \\
\hline
x & x^2 & \\
\hline
 & & \\
\end{array}
$$

      2.   Split the second term, $-4x$, in half and place the two halves ($-2x$ and $-2x$) in the diagonal boxes. Then label the headers accordingly: $(-2)(x) = -2x$.

$$
\begin{array}{c|c|c}
 & x & -2 \\
\hline
x & x^2 & -2x \\
\hline
-2 & -2x & \\
\end{array}
$$

      3.   Fill in the last box as the product of the two headers, $(-2)(-2) = 4$. This is the value to be added to both sides of the equation: $x^2 - 4x + 4 = 2 + 4$.

$$
\begin{array}{c|c|c}
 & x & -2 \\
\hline
x & x^2 & -2x \\
\hline
-2 & -2x & 4 \\
\end{array}
$$

      4.   The headers also give you the binomial to be squared, $x - 2$.
So, $x^2 - 4x + 4 = 2 + 4$ becomes $(x - 2)^2 = 6$.

The images below give a graphical demonstration as to why adding $\left(\frac{b}{2}\right)^2$ to $x^2 + bx$ will always result in a perfect square trinomial that can be rewritten as $\left(x + \frac{b}{2}\right)^2$.

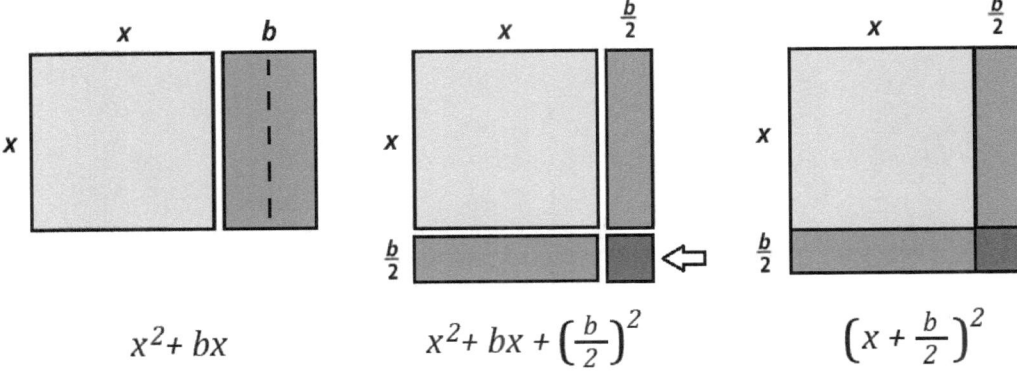

$$x^2 + bx \qquad\qquad x^2 + bx + \left(\frac{b}{2}\right)^2 \qquad\qquad \left(x + \frac{b}{2}\right)^2$$

MODEL PROBLEM 1:   *RATIONAL ROOTS*

Solve $x^2 + 6x - 7 = 0$ by completing the square.

Solution:

(A) $x^2 + 6x - 7 = 0$
$x^2 + 6x = 7$

(B) $\left(\frac{b}{2}\right)^2 = \left(\frac{6}{2}\right)^2 = 3^2 = 9$
$x^2 + 6x + 9 = 7 + 9$

(C) $(x + 3)^2 = 16$

(D) $x + 3 = \pm\sqrt{16}$
$x + 3 = \pm 4$

(E) $x = -3 \pm 4$
$\{-7, 1\}$

Explanation of steps:

(A) Add the opposite of c to both sides. *[Add 7 to both sides.]*

(B) Find $\left(\frac{b}{2}\right)^2$ and add it *[9]* to both sides of the equation.

(C) Factor the trinomial into a binomial squared.
[$x^2 + 6x + 9 = (x + 3)^2$].

(D) Take the square root of both sides. Use the $\pm$ symbol on the right side and simplify radicals. *[$\pm\sqrt{16} = \pm 4$]*

(E) Solve for x, remembering that $\pm$ gives two solutions *[$-3 - 4 = -7$ and $-3 + 4 = 1$].*

PRACTICE PROBLEMS

1. Solve $x^2 - 8x + 16 = 0$.	2. Solve $x^2 + 10x - 11 = 0$ by completing the square.

MODEL PROBLEM 2: *IRRATIONAL ROOTS*

Solve $x^2 - 2x - 1 = 0$ by completing the square.

Solution:

(A) $x^2 - 2x - 1 = 0$
$x^2 - 2x = 1$

(B) $\left(\frac{b}{2}\right)^2 = \left(\frac{-2}{2}\right)^2 = (-1)^2 = 1$
$x^2 - 2x + 1 = 1 + 1$

(C) $(x-1)^2 = 2$

(D) $x - 1 = \pm\sqrt{2}$

(E) $x = 1 \pm \sqrt{2}$
$\{1 + \sqrt{2}, 1 - \sqrt{2}\}$

Explanation of steps:

(A) Add the opposite of c to both sides.

(B) Add $\left(\frac{b}{2}\right)^2$ to both sides of the equation
[add 1 to both sides].

(C) Factor the left side into a binomial squared.
[$x^2 - 2x + 1 = (x-1)^2$].

(D) Take the square root of both sides. Use the $\pm$ symbol on the right side and simplify radicals.
[$\pm\sqrt{2}$ cannot be simplified.]

(E) Solve for x, remembering that $\pm$ gives two solutions *[we are left with two irrational roots].*

PRACTICE PROBLEMS

3. Solve $x^2 + 4x + 2 = 0$ by completing the square.	4. Solve $x^2 - 4x - 8 = 0$ by completing the square.

5. Solve $2x^2 - 12x + 4 = 0$ by completing the square.

6. Use completing the square to show that $x^2 - 2x + 3 = 0$ has no real solutions.

7. A rectangular pool has an area of 880 square feet. The length is 10 feet longer than the width. Find the dimensions of the pool, to the *nearest tenth of a foot*.

Regents Questions

MULTIPLE CHOICE

1. Which equation has the same solution as $x^2 - 6x - 12 = 0$?
   - (1) $(x + 3)^2 = 21$
   - (2) $(x - 3)^2 = 21$
   - (3) $(x + 3)^2 = 3$
   - (4) $(x - 3)^2 = 3$

2. What are the roots of the equation $x^2 + 4x - 16 = 0$?
   - (1) $2 \pm 2\sqrt{5}$
   - (2) $-2 \pm 2\sqrt{5}$
   - (3) $2 \pm 4\sqrt{5}$
   - (4) $-2 \pm 4\sqrt{5}$

3. Which equation has the same solutions as $x^2 + 6x - 7 = 0$?
   - (1) $(x + 3)^2 = 2$
   - (2) $(x - 3)^2 = 2$
   - (3) $(x - 3)^2 = 16$
   - (4) $(x + 3)^2 = 16$

4. What are the solutions to the equation $x^2 - 8x = 24$?
   - (1) $x = 4 \pm 2\sqrt{10}$
   - (2) $x = -4 \pm 2\sqrt{10}$
   - (3) $x = 4 \pm 2\sqrt{2}$
   - (4) $x = -4 \pm 2\sqrt{2}$

5. If Lylah completes the square for $f(x) = x^2 - 12x + 7$ in order to find the minimum, she must write $f(x)$ in the general form $f(x) = (x - a)^2 + b$. What is the value of a for $f(x)$?
   - (1) 6
   - (2) −6
   - (3) 12
   - (4) −12

6. When solving the equation $x^2 - 8x - 7 = 0$ by completing the square, which equation is a step in the process?
   - (1) $(x - 4)^2 = 9$
   - (2) $(x - 4)^2 = 23$
   - (3) $(x - 8)^2 = 9$
   - (4) $(x - 8)^2 = 23$

7. The method of completing the square was used to solve the equation $2x^2 - 12x + 6 = 0$. Which equation is a correct step when using this method?
   - (1) $(x - 3)^2 = 6$
   - (2) $(x - 3)^2 = -6$
   - (3) $(x - 3)^2 = 3$
   - (4) $(x - 3)^2 = -3$

8. What are the solutions to the equation $x^2 - 8x = 10$?
   - (1) $4 \pm \sqrt{10}$
   - (2) $4 \pm \sqrt{26}$
   - (3) $-4 \pm \sqrt{10}$
   - (4) $-4 \pm \sqrt{26}$

9. The quadratic equation $x^2 - 6x = 12$ is rewritten in the form $(x + p)^2 = q$, where q is a constant. What is the value of p?
   - (1) −12
   - (2) −9
   - (3) −3
   - (4) 9

10. Which equation has the same solution as $x^2 + 8x - 33 = 0$?
    - (1) $(x + 4)^2 = 49$
    - (2) $(x - 4)^2 = 49$
    - (3) $(x + 4)^2 = 17$
    - (4) $(x - 4)^2 = 17$

11. When solving $x^2 - 10x - 13 = 0$ by completing the square, which equation is a step in the process?

    (1) $(x - 5)^2 = 38$ (3) $(x - 10)^2 = 38$

    (2) $(x - 5)^2 = 12$ (4) $(x - 10)^2 = 12$

12. When using the method of completing the square, which equation is equivalent to $x^2 - 12x - 10 = 0$?

    (1) $(x + 6)^2 = -26$ (3) $(x - 6)^2 = -26$

    (2) $(x + 6)^2 = 46$ (4) $(x - 6)^2 = 46$

13. When completing the square for $x^2 - 18x + 77 = 0$, which equation is a correct step in this process?

    (1) $(x - 9)^2 = 4$ (3) $x = \pm 13$

    (2) $(x - 3)^2 = 2$ (4) $x - 9 = \pm 9$

14. Which equation is equivalent to $x^2 - 6x + 4 = 0$?

    (1) $(x - 3)^2 = -4$ (3) $(x - 3)^2 = 6$

    (2) $(x - 3)^2 = 5$ (4) $(x - 3)^2 = 9$

CONSTRUCTED RESPONSE

15. A student was given the equation $x^2 + 6x - 13 = 0$ to solve by completing the square. The first step that was written is shown below.
$$x^2 + 6x = 13$$
The next step in the student's process was
$$x^2 + 6x + c = 13 + c.$$
State the value of c that creates a perfect square trinomial. Explain how the value of c is determined.

16. Solve the equation $x^2 - 6x = 15$ by completing the square.

17. Solve the following equation by completing the square: $x^2 + 4x = 2$

18. Use the method of completing the square to determine the exact values of x for the equation $x^2 - 8x + 6 = 0$.

19. Determine the exact values of x for $x^2 - 8x - 5 = 0$ by completing the square.

20. Express the equation $x^2 - 8x = -41$ in the form $(x - p)^2 = q$.

12.6 Quadratic Formula and the Discriminant

Another method for solving quadratic equations of the form $ax^2 + bx + c = 0$ is by the use of the **quadratic formula**:

$$x = \frac{-b \pm \sqrt{b^2 - 4ac}}{2a}$$

Example: To solve $x^2 - 2x - 1 = 0$ by the quadratic formula, write the formula but substitute 1 for a, –2 for b, and –1 for c, then evaluate:

$$x = \frac{-(-2) \pm \sqrt{(-2)^2 - 4(1)(-1)}}{2(1)} = \frac{2 \pm \sqrt{8}}{2} = \frac{2 \pm 2\sqrt{2}}{2} = 1 \pm \sqrt{2}$$

For quadratic equations that cannot be solved by factoring, you may prefer this method over completing the square. For the purposes of this course, any time that $a \neq 1$ for the trinomial $ax^2 + bx + c$, then this is the method to use.

The formula is derived by completing the square on the equation, $ax^2 + bx + c = 0$. The $\pm$ symbol in the formula allows for the possibility of two solutions (*roots*).

435

CALCULATOR TIP

You can check your solutions using the calculator; however, be aware that the calculator will display irrational numbers as decimal approximations.

First, store the values of a, b, and c in variables, then type and evaluate each part of the formula. Since $\pm$ can't be entered as an operation, the formula needs to be entered twice, first with $+$ and then with $-$.

Example: To check that the roots of $x^2 - 2x - 1 = 0$ are $1 \pm \sqrt{2}$,

1) Store 1 into A, –2 into B, and –1 into C by pressing ①⟦STO▸⟧⟦ALPHA⟧[A] ⟦ALPHA⟧[:] ⟦(-)⟧②⟦STO▸⟧⟦ALPHA⟧[B] ⟦ALPHA⟧[:] ⟦(-)⟧①⟦STO▸⟧⟦ALPHA⟧[C] ⟦ENTER⟧. Multiple instructions, separated by colons, may be entered on one line.

2) Find $\dfrac{-b + \sqrt{b^2 - 4ac}}{2a}$ by typing ⟦(⟧⟦(-)⟧⟦ALPHA⟧[B]⟦+⟧⟦2nd⟧⟦√⟧
⟦ALPHA⟧[B]⟦x^2⟧⟦–⟧④⟦ALPHA⟧[A]⟦ALPHA⟧[C]⟦▸⟧⟦)⟧ ⟦÷⟧ ⟦(⟧②⟦ALPHA⟧[A]⟦)⟧ ⟦ENTER⟧.

3) To find $\dfrac{-b - \sqrt{b^2 - 4ac}}{2a}$, recall the formula by pressing ⟦2nd⟧[ENTRY], then use the arrows to move to the $+$ and change it to a $-$. Press ⟦ENTER⟧ again.

4) Check if the decimal approximations match the exact answers, $1 \pm \sqrt{2}$, by calculating ①⟦+⟧⟦2nd⟧⟦√⟧②⟦▸⟧⟦ENTER⟧ and ①⟦–⟧⟦2nd⟧⟦√⟧②⟦▸⟧⟦ENTER⟧.

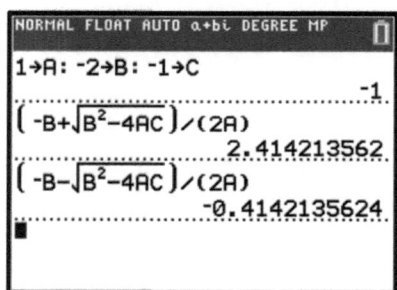
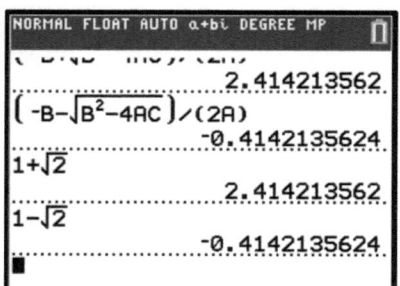

The part of the quadratic formula that is under the square root symbol, $b^2 - 4ac$, is called the _discriminant_.

The **discriminant** $b^2 - 4ac$ tells us about the **number and nature of the roots**.
- If the **discriminant is negative**, then the formula includes the square root of a negative number, so there are **no real roots**. _(You will learn about imaginary roots in Algebra II.)_
- If the **discriminant is zero**, the square root term "disappears" from the formula ($\sqrt{0} = 0$), leaving just $x = \dfrac{-b}{2a}$. Therefore, the equation has only **one distinct real root**. As long as a and b are both rational, the root will be rational.
  (Note: Since a quadratic generally has two roots, some people prefer to say that when the discriminant is zero, there are **two equal roots**.)
- If the **discriminant is positive**, the $\pm$ symbol before the radical causes **two different real roots** to be produced by the formula.
  - If the discriminant is a _perfect square_, the radical sign will be eliminated, meaning the _two roots are rational_ (assuming a and b are rational), which means the equation could have been factored over the integers.
  - If the discriminant is _not a perfect square_, the radical sign cannot be eliminated, so the _two roots are irrational_.

We can relate these situations to the graphs of the corresponding parabolas. The discriminant determines how many x-intercepts (_roots_) there are.

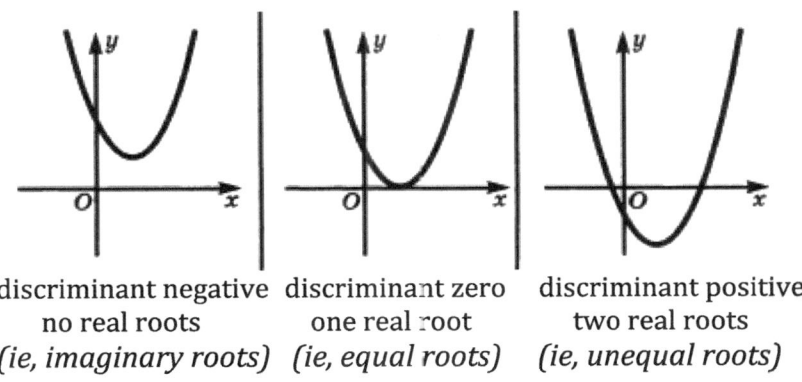

discriminant negative   discriminant zero   discriminant positive
no real roots   one real root   two real roots
(ie, imaginary roots)   _(ie, equal roots)_   _(ie, unequal roots)_

▦▢ CALCULATOR TIP

If you plan on using the quadratic formula more than once, it may be convenient to store the formula in your calculator as a reusable program. *Be aware, however, that if your calculator is placed into Test Mode prior to an exam, you will no longer have access to the program.*

To create a program to calculate the quadratic formula:

1. Press PRGM\<NEW\>1 for Create New.
2. Type a name, such as QUAD, using the letter keys, and press ENTER. Alpha lock is on, so you don't have to press the ALPHA key.
3. Press PRGM\<I/O\>2 for Prompt. Type ALPHA[A], ALPHA[B], ALPHA[C]ENTER.
4. To create the formula, $m = \dfrac{-b + \sqrt{b^2 - 4ac}}{2a}$, type ([(-)]ALPHA[B]+2nd[√]
   ALPHA[B]x^2-4ALPHA[A]ALPHA[C])) ÷ (2ALPHA[A])STO►ALPHA[M]ENTER.
5. To create the formula, $n = \dfrac{-b - \sqrt{b^2 - 4ac}}{2a}$, type ([(-)]ALPHA[B]-2nd[√]
   ALPHA[B]x^2-4ALPHA[A]ALPHA[C])) ÷ (2ALPHA[A])STO►ALPHA[N]ENTER.
6. Press PRGM\<I/O\>3 for Disp. Type the line "X=",M,N using the keys,
   ALPHA["]X,T,Θ,n2ndMATH1ALPHA["], ALPHA[M], ALPHA[N]ENTER.
7. Press 2nd[QUIT]. This will save your program.
8. Press PRGM, select the QUAD program, and press ENTERENTER.
9. Enter a value for each prompt and press ENTER. For example, to find the roots of $x^2 - 2x - 1 = 0$, enter 1, –2, and –1.
10. The roots are shown.

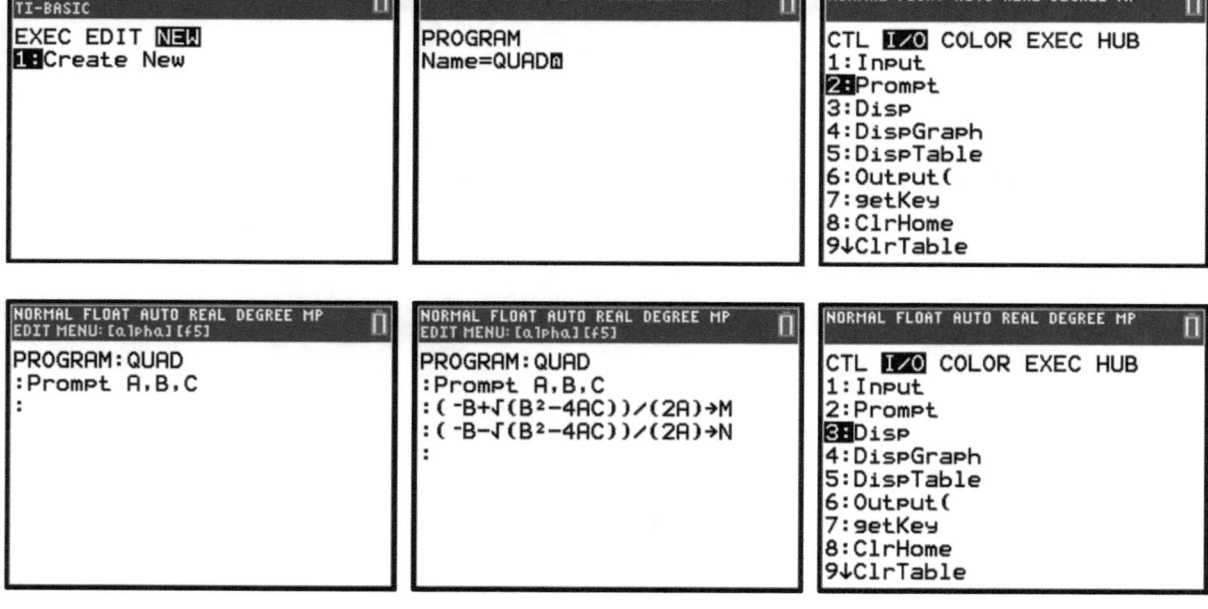

continued on the next page ...

MODEL PROBLEM

In the diagram to the right, a smaller rectangle with a length of x and width of 2 is enclosed inside a larger rectangle with a length of $2x + 3$ and a width of x. The shaded area is 21 square units. Find the length of the smaller rectangle.

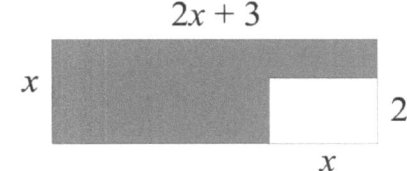

Solution:

(A)   area of large rectangle – area of small rectangle = shaded area

(B)

$$x(2x + 3) - 2x = 21$$
$$2x^2 + 3x - 2x = 21$$
$$2x^2 + x = 21$$
$$2x^2 + x - 21 = 0$$

(C)

$$x = \frac{-b \pm \sqrt{b^2 - 4ac}}{2a} = \frac{-1 \pm \sqrt{1^2 - 4(2)(-21)}}{2(2)}$$

$$= \frac{-1 \pm \sqrt{1 + 168}}{4} = \frac{-1 \pm \sqrt{169}}{4} = \frac{-1 \pm 13}{4}$$

$$x = \frac{12}{4} = 3 \quad \text{or} \quad x = \frac{-14}{4} = -3.5 \ (rejected)$$

The length of the smaller rectangle is 3 units.

Explanation of steps:

(A) Write an equation for the shaded area.

(B) Substitute expressions for the areas (products of the lengths and widths) of the rectangles. Simplify and express in standard form.

(C) Substitute a, b and c [2, 1, $and -21$] into the quadratic formula, and evaluate. Reject any negative roots, since the length cannot be negative.

PRACTICE PROBLEMS

1. A quadratic function has two real roots. How many times does the graph of the function cross the x-axis? (1) 1 (3) 0 (2) 2 (4) Cannot be determined	2. The roots of $x^2 + 4x + 7$ are (1) not real (2) equal (3) rational (4) irrational
3. The roots of the equation $9x^2 + 3x - 4 = 0$ are (1) not real (2) real, rational, and equal (3) real, rational, and unequal (4) real, irrational, and unequal	4. The roots of the equation $2x^2 + 4 = 9x$ are (1) real, rational, and equal (2) real, rational, and unequal (3) real, irrational, and unequal (4) not real
5. Solve the equation $6x^2 - 2x - 3 = 0$ and express the answer in simplest radical form.	6. Solve the equation $2x^2 + 7x - 3 = 0$ and express the answer in simplest radical form.
7. Solve $x^2 + 7x + 8 = 0$ by the quadratic formula.	8. Solve $2x^2 - 8x + 3 = 0$ by the quadratic formula.

Regents Questions

MULTIPLE CHOICE

1. If the quadratic formula is used to find the roots of the equation $x^2 - 6x - 19 = 0$, the correct roots are

    (1) $3 \pm 2\sqrt{7}$ (3) $3 \pm 4\sqrt{14}$

    (2) $-3 \pm 2\sqrt{7}$ (4) $-3 \pm 4\sqrt{14}$

2. How many real-number solutions does $4x^2 + 2x + 5 = 0$ have?

    (1) one   (3) zero

    (2) two   (4) infinitely many

3. The roots of $x^2 - 5x - 4 = 0$ are

    (1) 1 and 4   (3) −1 and −4

    (2) $\dfrac{5 \pm \sqrt{41}}{2}$ (4) $\dfrac{-5 \pm \sqrt{41}}{2}$

CONSTRUCTED RESPONSE

4. How many real solutions does the equation $x^2 - 2x + 5 = 0$ have? Justify your answer.

5. Fred's teacher gave the class the quadratic function $f(x) = 4x^2 + 16x + 9$.

    a) State two different methods Fred could use to solve the equation $f(x) = 0$.

    b) Using one of the methods stated in part a, solve $f(x) = 0$ for x, to the nearest tenth.

6. Given: $g(x) = 2x^2 + 3x + 10$
                 $k(x) = 2x + 16$
    Solve the equation $g(x) = 2k(x)$ algebraically for x, to the nearest tenth. Explain why you chose the method you used to solve this quadratic equation.

7. Solve for x to the *nearest tenth*: $x^2 + x - 5 = 0$.

8. Is the solution to the quadratic equation written below rational or irrational? Justify your answer.
    $$0 = 2x^2 + 3x - 10$$

9. Use the quadratic formula to solve $x^2 - 4x + 1 = 0$ for x. Round the solutions to the *nearest hundredth*.

10. Solve $3d^2 - 8d + 3 = 0$ algebraically for all values of d, rounding to the *nearest tenth*.

11. Solve $x^2 + 3x - 9 = 0$ algebraically for all values of x. Round your answer to the *nearest hundredth*.

12. Using the quadratic formula, solve $3x^2 - 2x - 6 = 0$ for all values of x. Round your answers to the *nearest hundredth*.

12.7   Word Problems – Quadratic Equations

Some verbal problems will require writing and solving quadratic equations. As with any type of verbal problem, try to represent the situation by writing an equation (or system of equations). If the equation is quadratic (in at least one term the variable is squared), then use the methods for solving a quadratic equation.

Geometric area problems often result in quadratic equations.

Example: If the length and width of a rectangle are expressions in terms of the same variable, say x, then the area, as the product of these expressions, would include that variable squared (x^2).

Important: If either root of the equation is not possible in the given situation, reject it.

Example: When finding the length of a side of a rectangle, reject any negative solutions.

MODEL PROBLEM

Find two consecutive whole numbers such that their product is 42.

Solution:

(A)   Let x represent the smaller number. The larger number is $x + 1$.

(B) $x(x + 1) = 42$

(C) $x^2 + x = 42$

(D) $x^2 + x - 42 = 0$

(E) $(x + 7)(x - 6) = 0$

(F) $x + 7 = 0$ *or* $x - 6 = 0$

(G) $x = -7$ *or* $x = 6$

(H)   Numbers are 6 and 7.

Explanation of steps:

(A)   Represent an unknown quantity as a variable and express other quantities in terms of this variable.

(B)   Write an equation for the given situation.

(C)   Simplify both sides of the equation.

(D)   Since the equation is quadratic *[there's a x^2 term]*, transform it into standard form by getting zero to one side *[by subtracting 42 from both sides]*.

(E)   Factor completely.

(F)   Set each factor containing a variable equal to 0.

(G)   Solve each resulting equation. Reject any impossible solutions. *[The problem asked for whole numbers, so –7 is not a possible solution]*.

(H)   Be sure to answer the problem. *[If x, the smaller number, is 6, then the larger, next consecutive whole number is 7.]*

PRACTICE PROBLEMS

1. The square of a positive number decreased by twice the number is 48. Find the number.	2. The larger of two positive numbers is 8 more than the smaller. The sum of their squares is 104. Find the two numbers.
3. When 36 is subtracted from the square of a number, the result is five times the number. What is the positive solution?	4. The square of a positive number is 24 more than 5 times the number. What is the value of the number?
5. The area of the rectangular playground enclosure at South School is 500 square meters. The length of the playground is 5 meters longer than the width. Find the dimensions of the playground, in meters.	6. Tamara has two sisters. One of the sisters is 7 years older than Tamara. The other sister is 3 years younger than Tamara. The product of Tamara's two sisters' ages is 24. How old is Tamara?

7. Find two negative consecutive odd integers such that their product is 63.	8. Find three consecutive odd integers such that the product of the first and the second exceeds the third by 8.
9. Find two consecutive whole numbers where the product of the larger and 10 more than the smaller is 90.	10. Three brothers have ages that are consecutive even integers. The product of the first and third boys' ages is 20 more than twice the second boy's age. Find the age of *each* of the three boys.
11. The hypotenuse of a right triangle is 6 meters long. One leg is 1 meter longer than the other. Find the lengths of *both* legs of the triangle, to the *nearest hundredth of a meter*. 	12. From ground level, a projectile is shot into the air. Its height, in feet, is modeled by $p(x) = -16x^2 + 32x$, where x is the number of elapsed seconds. Determine the total number of seconds that the projectile will be in the air.

13. A pump is used to drain water from a pool. The amount of water in the pool is modeled by the function $w(t) = -5t^2 - 8t + 120$, where $w(t)$ represents the number of gallons of water in the pool after the pump has operated for t minutes.

    a) How many gallons of water were in the pool when the pump was first turned on?

    b) Determine, to the *nearest tenth of a minute*, the amount of time it takes for all the water in the pool to drain.

14. In the accompanying diagram, the large rectangle $ABCD$ is made up of four smaller rectangles with dimensions as indicated.

    a)   Represent, in terms of x, the area of $ABCD$.

    b)   Find the area of *each* of rectangles I, II, III, and IV.

    c) Show that the area obtained in part (a) above is equal to the sum of the areas obtained in part (b) above.

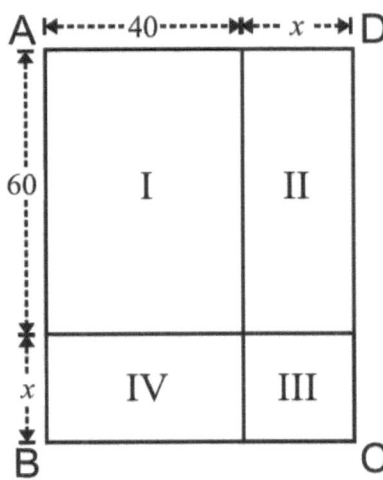

Regents Questions

MULTIPLE CHOICE

1. Sam and Jeremy have ages that are consecutive odd integers. The product of their ages is 783. Which equation could be used to find Jeremy's age, j, if he is the younger man?

   (1) $j^2 + 2 = 783$ (3) $j^2 + 2j = 783$

   (2) $j^2 - 2 = 783$ (4) $j^2 - 2j = 783$

2. Sara was asked to solve this word problem: "The product of two consecutive integers is 156. What are the integers?" What type of equation should she create to solve this problem?

   (1) linear   (3) exponential

   (2) quadratic   (4) absolute value

3. Abigail's and Gina's ages are consecutive integers. Abigail is younger than Gina and Gina's age is represented by x. If the difference of the square of Gina's age and eight times Abigail's age is 17, which equation could be used to find Gina's age?

   (1) $(x + 1)^2 - 8x = 17$ (3) $x^2 - 8(x + 1) = 17$

   (2) $(x - 1)^2 - 8x = 17$ (4) $x^2 - 8(x - 1) = 17$

4. A movie theater's popcorn box is a rectangular prism with a base that measures 6 inches by 4 inches and has a height of 8 inches. To create a larger box, both the length and the width will be increased by x inches. The height will remain the same. Which function represents the volume, $V(x)$, of the larger box?

   (1) $V(x) = (6 + x)(4 + x)(8 + x)$ (3) $V(x) = (6 + x) + (4 + x) + (8 + x)$

   (2) $V(x) = (6 + x)(4 + x)(8)$ (4) $V(x) = (6 + x) + (4 + x) + (8)$

5. The length of a rectangular flat-screen television is six inches less than twice its width, x. If the area of the television screen is 1100 square inches, which equation can be used to determine the width, in inches?

   (1) $x(2x - 6) = 1100$ (3) $2x + x(2x - 6) = 1100$

   (2) $x(6 - 2x) = 1100$ (4) $2x + x(6 - 2x) = 1100$

CONSTRUCTED RESPONSE

6.   A rectangular garden measuring 12 meters by 16 meters is to have a walkway installed around it with a width of x meters, as shown in the diagram below. Together, the walkway and the garden have an area of 396 square meters.

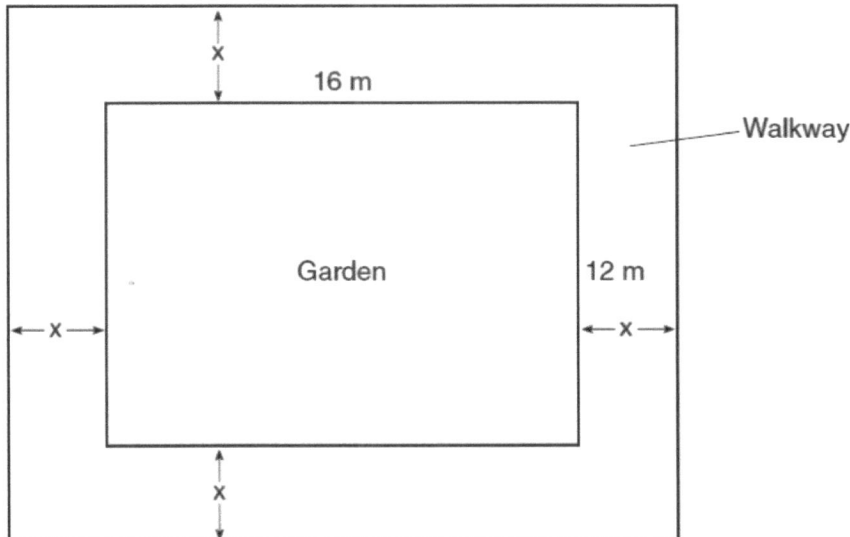

      Write an equation that can be used to find x, the width of the walkway. Describe how your equation models the situation.

      Determine and state the width of the walkway, in meters.

7.   A school is building a rectangular soccer field that has an area of 6000 square yards. The soccer field must be 40 yards longer than its width. Determine algebraically the dimensions of the soccer field, in yards.

8.   New Clarendon Park is undergoing renovations to its gardens. One garden that was originally a square is being adjusted so that one side is doubled in length, while the other side is decreased by three meters.

      The new rectangular garden will have an area that is 25% more than the original square garden. Write an equation that could be used to determine the length of a side of the original square garden.

      Explain how your equation models the situation.

      Determine the area, in square meters, of the new rectangular garden.

9.   John and Sarah are each saving money for a car. The total amount of money John will save is given by the function $f(x) = 60 + 5x$. The total amount of money Sarah will save is given by the function $g(x) = x^2 + 46$. After how many weeks, x, will they have the same amount of money saved? Explain how you arrived at your answer.

10.   A landscaper is creating a rectangular flower bed such that the width is half of the length. The area of the flower bed is 34 square feet. Write and solve an equation to determine the width of the flower bed, to the _nearest tenth of a foot_.

11. A toy rocket is launched from the ground straight upward. The height of the rocket above the ground, in feet, is given by the equation $h(t) = -16t^2 + 64t$, where t is the time in seconds. Determine the domain for this function in the given context. Explain your reasoning.

12. A rectangular picture measures 6 inches by 8 inches. Simon wants to build a wooden frame for the picture so that the framed picture takes up a maximum area of 100 square inches on his wall. The pieces of wood that he uses to build the frame all have the same width.

    Write an equation or inequality that could be used to determine the maximum width of the pieces of wood for the frame Simon could create.

    Explain how your equation or inequality models the situation.

    Solve the equation or inequality to determine the maximum width of the pieces of wood used for the frame to the *nearest tenth of an inch*.

13. A contractor has 48 meters of fencing that he is going to use as the perimeter of a rectangular garden. The length of one side of the garden is represented by x, and the area of the garden is 108 square meters.

    Determine, algebraically, the dimensions of the garden in meters.

14. When an apple is dropped from a tower 256 feet high, the function $h(t) = -16t^2 + 256$ models the height of the apple, in feet, after t seconds. Determine, algebraically, the number of seconds it takes the apple to hit the ground.

15. The length of a rectangular sign is 6 inches more than half its width. The area of this sign is 432 square inches. Write an equation in one variable that could be used to find the number of inches in the dimensions of this sign.

    Solve this equation algebraically to determine the dimensions of this sign, in inches.

16. Julia is 4 years older than twice Kelly's age, x. The product of their ages is 96. Write an equation that models this situation.

    Determine Kelly's age algebraically.

    State the difference between Julia's and Kelly's ages, in years.

CHAPTER 13.   PARABOLAS

13.1   Find Roots Given a Parabolic Graph

The **standard form** of a quadratic function is $y = ax^2 + bx + c$ where $a \neq 0$.
Example: $y = x^2 - 4x + 3$ or $f(x) = -2x^2 - x + 1$

The graph of a quadratic function is a U-shaped curve called a **parabola**.

The points on the parabola represent the (x, y) values for which the quadratic equation is true. As we saw with linear equations (page 92), we can **determine whether a point is on a graph** by substituting the x-value and y-value for the variables x and y in the equation of the graph and then checking if these values make the equation true. If a point is on a graph, it is a solution to the equation represented by the graph.
Example: (3,4) is on the graph represented by $y = x^2 - 2x + 1$ because by substituting
              3 for x and 4 for y, we get $4 = 3^2 - 2(3) + 1$, which is true.

In a quadratic function, if $a > 0$, then the parabola "**opens up**" like the letter U, but if $a < 0$, then the parabola "**opens down**" like an upside-down U.
Example: The graph of $y = x^2 - 4x + 3$ opens up, but $y = -2x^2 - x + 1$ opens down.

Just as we have for other functions, the **y-intercept** of $y = ax^2 + bx + c$ is the constant, c. We know this because the graph of the function crosses the y-axis when $x = 0$, which gives us $y = c$.

When $y = 0$, we can find the **roots** or **zeros** algebraically by solving for x.
Example: The roots of $y = x^2 - 4x + 3$ are the solutions to the equation $x^2 - 4x + 3 = 0$.
              Factoring the trinomial, we get $(x - 1)(x - 3) = 0$, so the roots are 1 and 3.

Graphically, $y = 0$ for all points along the x-axis. Therefore, the **roots** or **zeros** are the x-coordinates of the points **where the parabola crosses the x-axis**, also called the **x-intercepts**.
Example: Since the roots of the quadratic function $y = x^2 - 4x + 3$ are 1 and 3, the parabola
              will cross the x-axis at points (1,0) and (3,0).

MODEL PROBLEM 1:   *DETERMINING WHETHER A POINT IS ON THE GRAPH*

Does the graph whose equation is $y = -x^2 - x + 3$ contain the point $(2, -3)$?

Solution:
(A) $-3 = -2^2 - 2 + 3$?
(B) $-3 = -4 - 2 + 3 = -3$, so yes, $(2, -3)$ is in the solution set.

Explanation of steps:
(A) Substitute for x and y.
(B) Evaluate both sides of the equation to determine if the equation is true.

PRACTICE PROBLEMS

1. Which of the following points is *not* on the graph of the parabola whose equation is $y = 2x^2 + x - 1$?	2. If $(h, 11)$ is a point on the graph of $y = x^2 - 3x + 1$, which of the following can be the value of h?
(a) $(-1, 0)$ (c) $(-2, -9)$ (b) $(0, -1)$ (d) $(2, 9)$	(a) -2 (c) 1 (b) 0 (d) 2

MODEL PROBLEM 2:   *DETERMINING THE ROOTS*

What are the root(s) of the quadratic equation associated with this graph?

Solution:
The roots are 0 and 4.

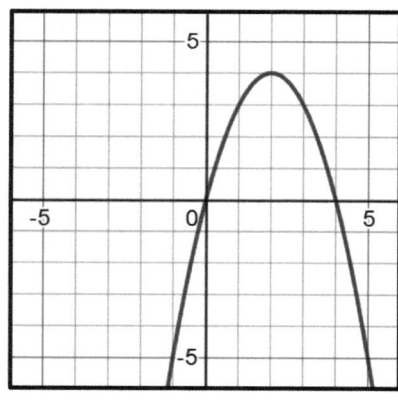

Explanation of steps:
The x-intercepts, or the x-coordinates of the points where the parabola crosses the x-axis, are the roots of the equation. *[Since the parabola crosses the x-axis at (0,0) and (4,0), the roots are 0 and 4.]*

PRACTICE PROBLEMS

3. What are the root(s) of the quadratic equation associated with this graph?

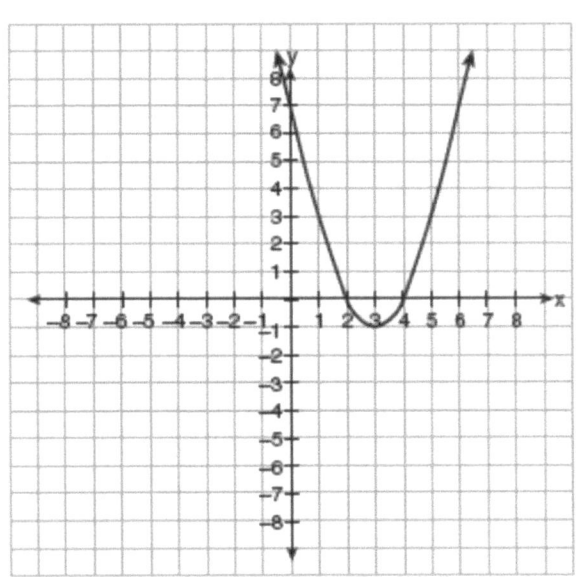

4. What are the root(s) of the quadratic equation associated with this graph?

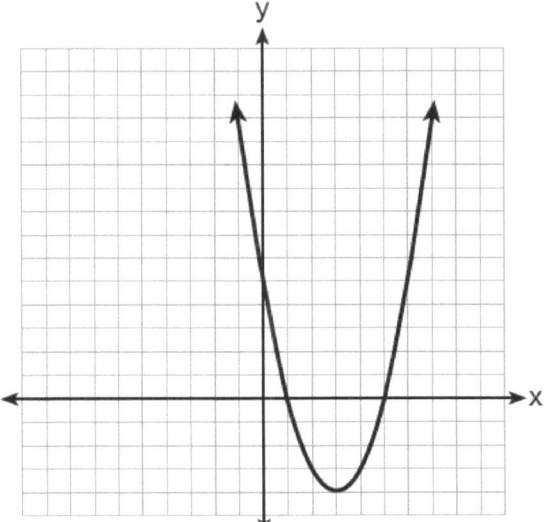

5. What are the root(s) of the quadratic equation associated with this graph?

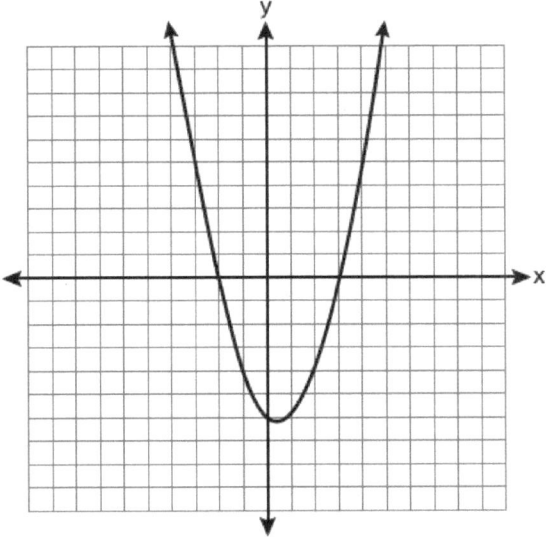

6. What are the root(s) of the quadratic equation associated with this graph?

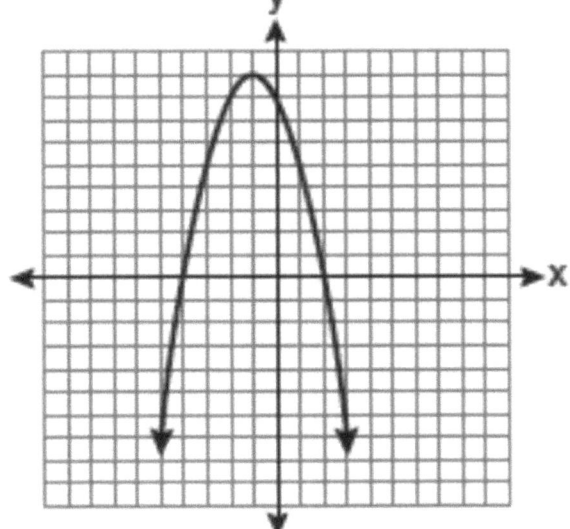

7. What are the root(s) of the quadratic equation associated with this graph?

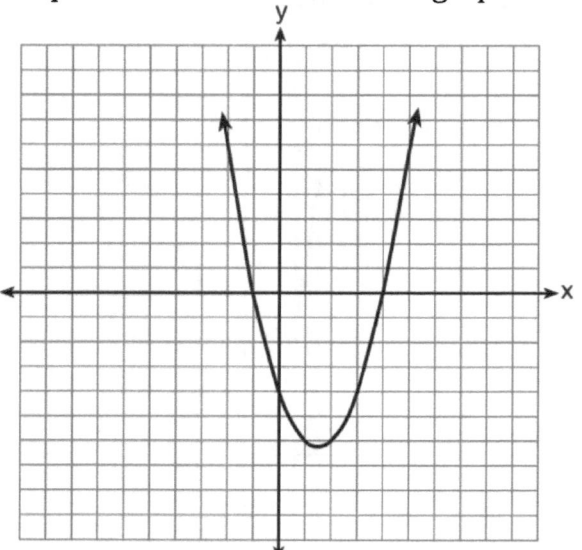

8. What are the root(s) of the quadratic equation associated with this graph?

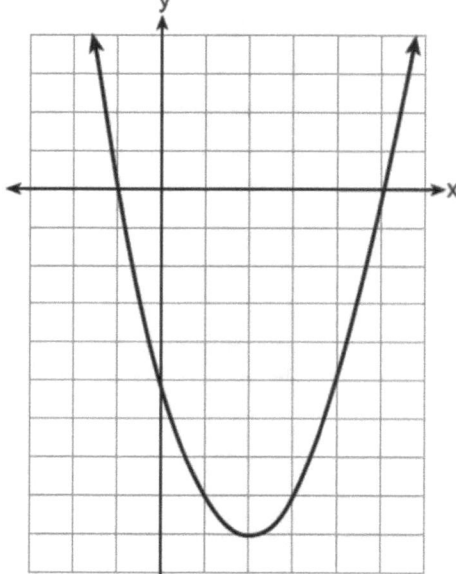

9. What are the root(s) of the quadratic equation associated with this graph?

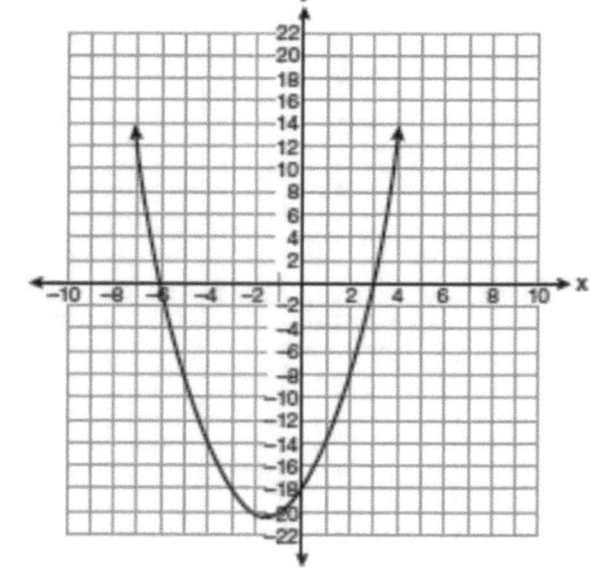

10. What are the root(s) of the quadratic equation associated with this graph?

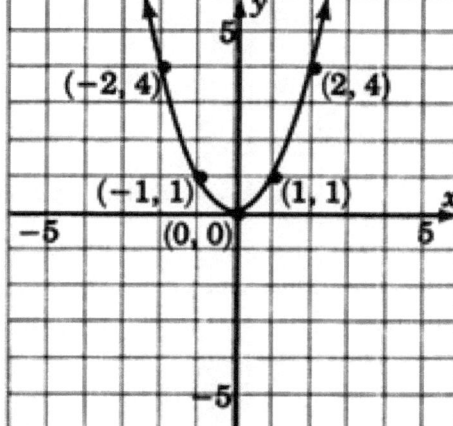

Regents Questions

MULTIPLE CHOICE

1. The graphs below represent functions defined by polynomials. For which function are the zeros of the polynomials 2 and −3?

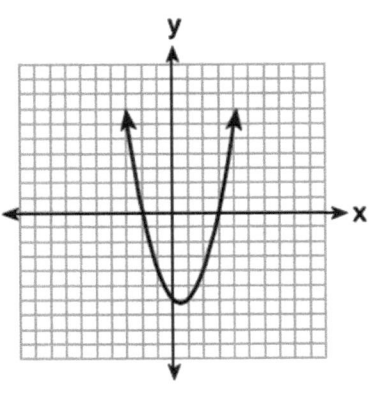

(1)

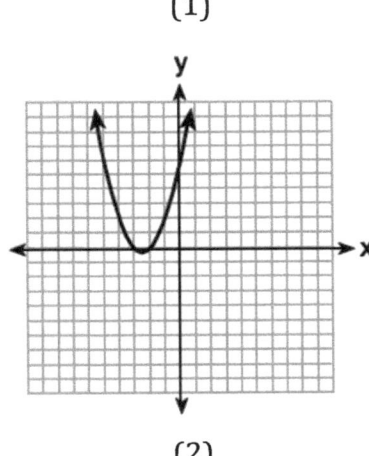

(2)

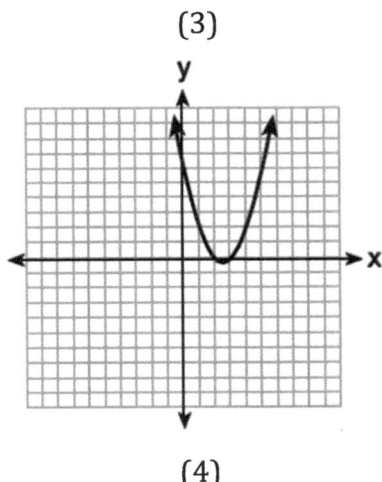

(3)

(4)

2. Which point is *not* on the graph represented by $y = x^2 + 3x − 6$?

    (1) $(−6,12)$ (3) $(2,4)$
    (2) $(−4,−2)$ (4) $(3,−6)$

3.   Which function has zeros of −4 and 2?

$$f(x) = x^2 + 7x - 8$$
(1)

$$g(x) = x^2 - 7x - 8$$
(3)

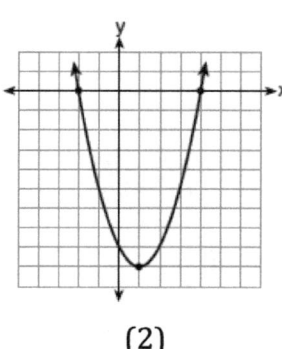   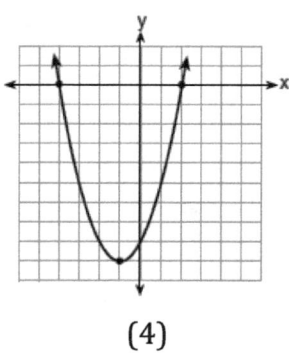

(2)   (4)

4.   The graph of $y = \frac{1}{2}x^2 - x - 4$ is shown below. The points $A(-2,0)$, $B(0,-4)$, and $C(4,0)$ lie on this graph.

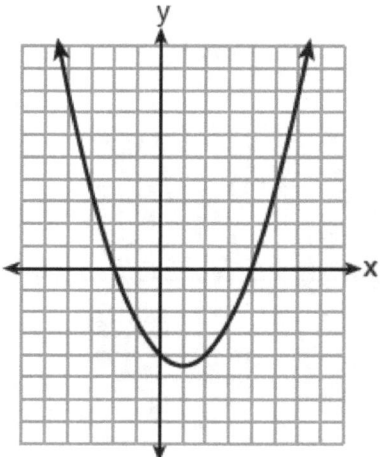

Which of these points can determine the zeros of the equation $y = \frac{1}{2}x^2 - x - 4$?

(1) A, only   (3) A and C, only
(2) B, only   (4) A, B, and C

5.   Which ordered pair does *not* represent a point on the graph of $y = 3x^2 - x + 7$?

(1) $(-1.5, 15.25)$ (3) $(1.25, 10.25)$
(2) $(0.5, 7.25)$ (4) $(2.5, 23.25)$

6.   The graphs below represent four polynomial functions. Which of these functions has zeros of 2 and –3?

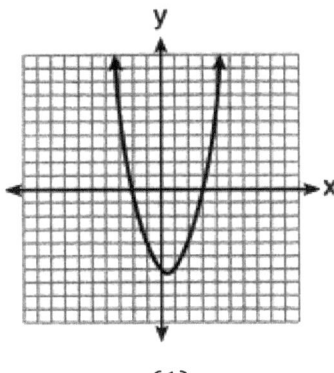

(1)

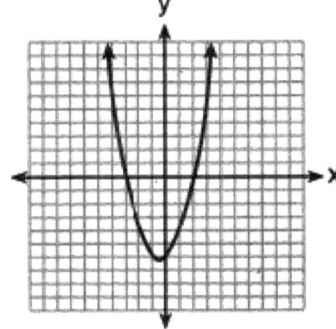

(3)

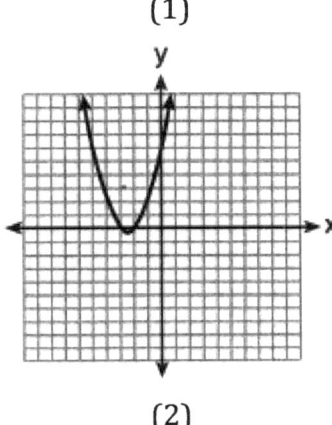

(2)

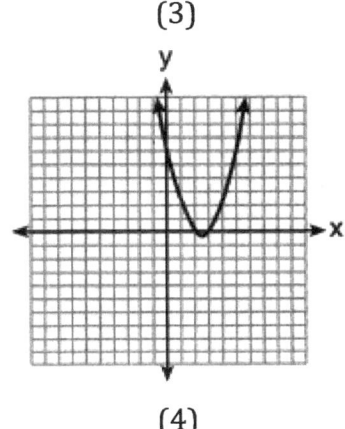

(4)

7.   The function *f* is graphed on the set of axes below.

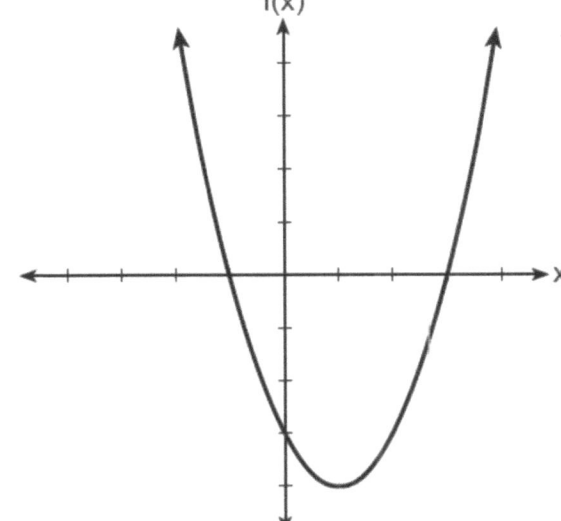

What is a possible factorization of this function?

(1) $f(x) = (x - 1)(x + 3)$ (3) $f(x) = (x + 1)(x - 4)$
(2) $f(x) = (x + 1)(x - 3)$ (4) $f(x) = (x - 1)(x + 4)$

CONSTRUCTED RESPONSE

8.   The function $r(x)$ is defined by the expression $x^2 + 3x - 18$. Use factoring to determine the zeros of $r(x)$. Explain what the zeros represent on the graph of $r(x)$.

9.   The graph of the function $f(x) = ax^2 + bx + c$ is given below.

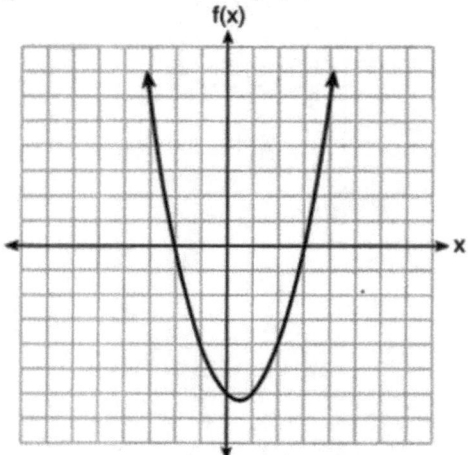

Could the factors of $f(x)$ be $(x + 2)$ and $(x - 3)$?

Based on the graph, explain why or why *not*.

13.2   Find Vertex and Axis Graphically

A **parabola** is a graph of a quadratic function of the form $y = ax^2 + bx + c$ where $a \neq 0$.

The **vertex** of a parabola is the minimum or maximum point, or turning point, on the graph. It is the lowest point on the curve if the parabola opens up ($a > 0$), or the highest point if the parabola opens down ($a < 0$).

The **axis of symmetry** is the vertical line that crosses through the vertex. Its equation is in the form $x =$ *[the x-coordinate of the vertex]*. The axis of symmetry divides the parabola into two parts that are mirror images of each other.

Example: If the vertex of a parabola is $(-2, 3)$, its x-coordinate is -2, so the axis of symmetry is a line whose equation is $x = -2$. In other words, a vertical line drawn from the vertex to the x-axis would cross the axis at -2.

MODEL PROBLEM

What are the vertex and axis of symmetry of the following parabola?

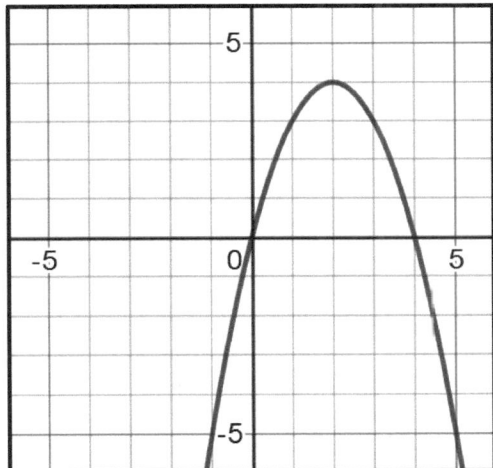

Solution:
  (A)   The vertex is $(2, 4)$.
  (B)   The axis of symmetry is $x = 2$.

Explanation of steps:
  (A)   The vertex is the turning point of the parabola. *[Since this parabola opens down, the vertex is the highest point, (2,4).]*
  (B)   The axis of symmetry is a vertical line through the vertex. *[Given that the vertex is (2,4), a vertical line through this point would include all points where x equals 2, and would cross the x-axis at 2, so the equation of the axis of symmetry is x = 2.]*

PRACTICE PROBLEMS

1. What are the vertex and the axis of symmetry of the parabola shown below?

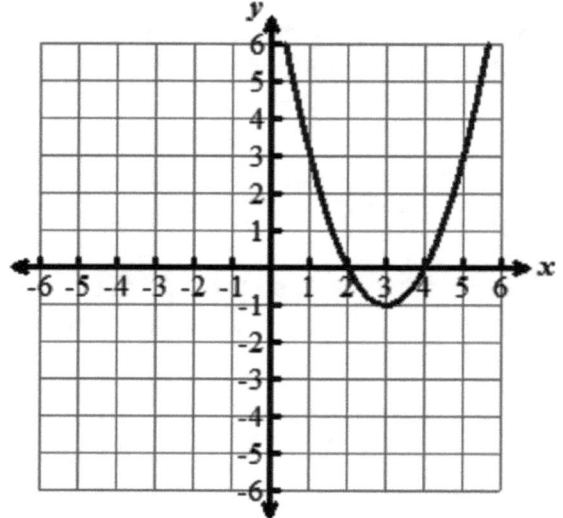

2. What are the vertex and the axis of symmetry of the parabola shown below?

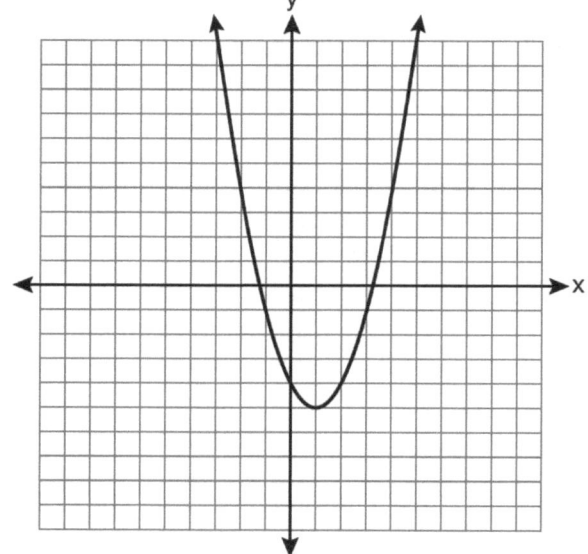

3. What are the vertex and the axis of symmetry of the parabola shown below?

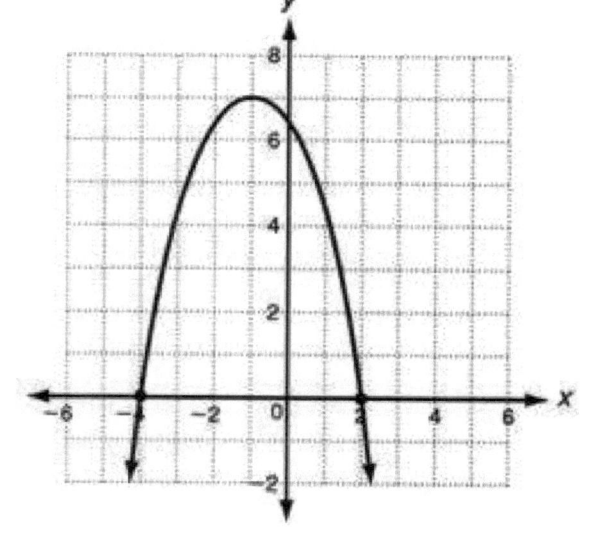

4. What are the vertex and the axis of symmetry of the parabola shown below?

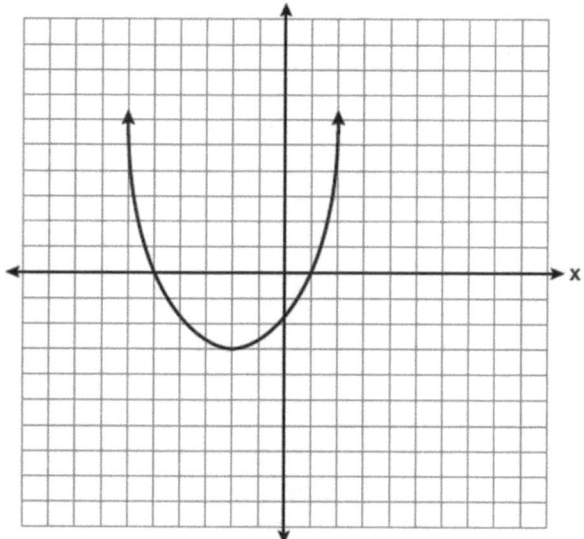

5. What are the vertex and the axis of symmetry of the parabola shown below?

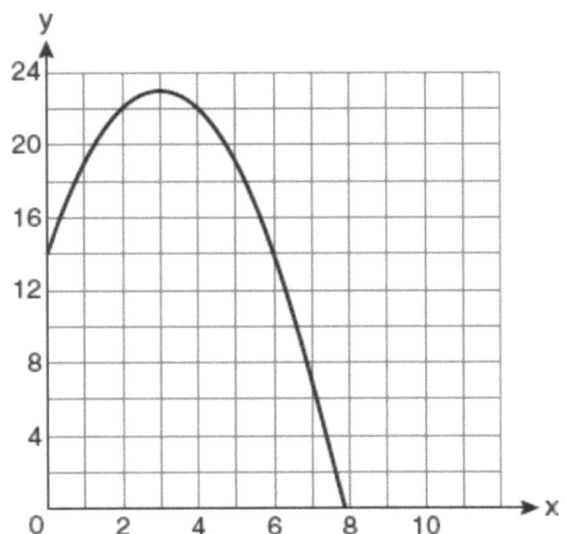

6. What are the vertex and the axis of symmetry of the parabola shown below?

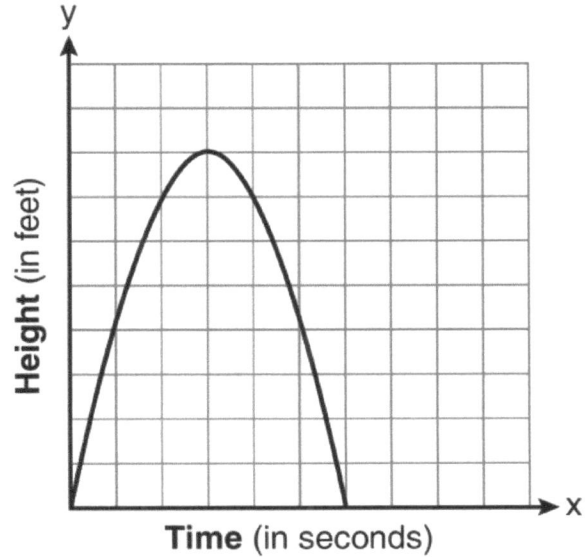

7. What are the vertex and the axis of symmetry of the parabola shown below?

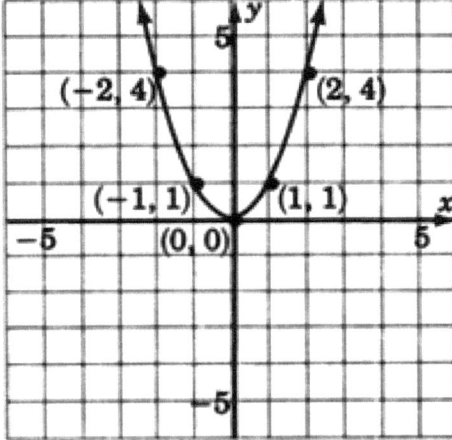

8. If the equation of the axis of symmetry of a parabola is $x = 2$, at which pair of points could the parabola intersect the x-axis?

(1)   (3,0) and (5,0)

(2)   (3,0) and (2,0)

(3)   (3,0) and (1,0)

(4)   (−3,0) and (−1,0)

Regents Questions

MULTIPLE CHOICE

1. The function $f(x)$ is graphed on the set of axes below.

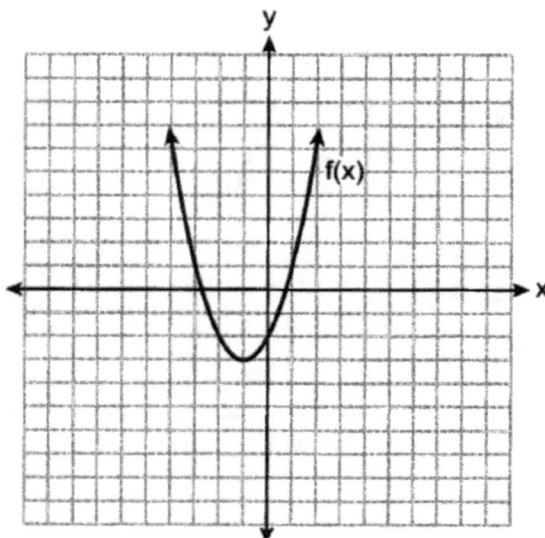

What is the equation of the axis of symmetry for $f(x)$?

(1) $x = -1$ (3) $y = -1$

(2) $x = -3$ (4) $y = -3$

13.3   Find Vertex and Axis Algebraically

A **parabola** is a graph of a quadratic function of the form $y = ax^2 + bx + c$ where $a \neq 0$.
Example: For $y = x^2 - 4x + 3$, $a = 1$, $b = -4$, and $c = 3$.

The **axis of symmetry** is a vertical line that divides the parabola into two parts that are mirror images of each other. The **equation for the axis of symmetry** is: $x = \dfrac{-b}{2a}$.

Example: The axis of symmetry for the parabola whose equation is $y = x^2 - 4x + 3$
is the line $x = \dfrac{-b}{2a} = \dfrac{-(-4)}{2(1)} = 2$, or simply $x = 2$.

An alternate method for finding the *equation for the axis of symmetry* is to find the *average of the roots*. This may be easier if the quadratic function is already factored. The **factored form** of a quadratic equation is $y = g(x - m)(x - n)$, where m and n are the roots.
Example: If we factor $y = x^2 - 4x + 3$, we get $y = (x - 3)(x - 1)$, so the roots are $\{3, 1\}$.
Therefore, the axis of symmetry is at $x = \dfrac{m + n}{2} = \dfrac{3 + 1}{2} = 2$.

The **vertex** (or turning point) of the parabola is the lowest point (*minimum*) on the curve if the parabola *opens up* ($a > 0$), or the highest point (*maximum*) if the parabola *opens down* ($a < 0$). Since the minimum or maximum point for a parabola is at its vertex, we will often need to find the vertex to determine the largest or smallest value of a real world quadratic function.

The vertex lies on the axis of symmetry. The **x-coordinate of the vertex** is determined by the equation for the axis of symmetry. The **y-coordinate of the vertex** can be found by substituting for x in the quadratic equation.
Example: For $y = x^2 - 4x + 3$, the axis of symmetry is $x = 2$, so substitute 2 for x.
$y = (2)^2 - 4(2) + 3 = -1$, so the vertex is the point $(2, -1)$.

MODEL PROBLEM 1:   *GIVEN AN EQUATION*

Find the axis of symmetry and vertex for the parabola with equation $y = x^2 + 12x + 32$.

Solution:

(A) $a = 1$ and $b = 12$

(B) $x = \dfrac{-b}{2a} = \dfrac{-(12)}{2(1)} = -6$

(C) $y = x^2 + 12x + 32$
$y = (-6)^2 + 12(-6) + 32$
$y = -4$

(D)   Axis of symmetry: $x = -6$.
Vertex: $(-6, -4)$.

Explanation of steps:

(A)   The values for a and b come from the coefficients of the x^2 and x terms of the equation.

(B)   Substitute for a and b in the axis of symmetry equation $x = \dfrac{-b}{2a}$ and evaluate.

(C)   Substitute the value of x found in the previous step into the original equation to find the value of y. You now have the coordinates of the vertex.

(D)   State your answers.

461

PRACTICE PROBLEMS

1. Find the axis of symmetry and vertex of the parabola whose equation is $y = -x^2 + 4x - 8$.	2. Find the axis of symmetry and vertex of the parabola whose equation is $y = x^2 - 6x + 10$.
3. What is the vertex of the parabola whose equation is $y = 3x^2 + 6x - 1$?	4. What is the minimum point of the parabola whose equation is $y = 2x^2 + 8x + 9$?
5. Find the axis of symmetry and vertex of the parabola whose equation is $y = x^2 + 2x$.	6. Find the axis of symmetry and vertex of the parabola whose equation is $y = 3x^2 + 1$.

7. Find the axis of symmetry and vertex of the parabola whose equation is $y = -2x^2 - 8x + 3$	8. Find the axis of symmetry and vertex of the parabola whose equation is $y = -x^2 - 2x + 1$

MODEL PROBLEM 2: *VERBAL PROBLEM*

You have a 500-ft. roll of chain link fencing and a large field. You want to fence in a rectangular playground area. What is the largest such playground area you can enclose?

Solution:
   (A)   The perimeter $P = 2l + 2w$, so $500 = 2l + 2w$, or $250 = l + w$.
   (B)   Therefore, $l = -w + 250$.
   (C)   The area $A = lw$, so $A = (-w + 250)(w)$, or $A = -w^2 + 250w$.
   (D)   The axis of symmetry is at $w = \dfrac{-b}{2a} = \dfrac{-250}{2(-1)} = 125$.
   (E)   When $w = 125$, $l = -w + 250 = -125 + 250 = 125$.
   (F)   So the maximum area is $A = lw = (125)(125) = 15{,}625$ sq. ft.

Explanation of steps:
   (A)   The amount of fencing tells us the perimeter *[500 ft.]*. Write the formula and substitute.
   (B)   From the perimeter equation, we can solve for *l* in terms of *w*.
   (C)   Now, we can substitute for *l* and *w* into the area formula, resulting in a quadratic function.
   (D)   Since this quadratic function graphs as a parabola that opens down *[a = −1]*, the maximum area is at its vertex. Finding the axis of symmetry will find us the value of *w* when the area is at its maximum. *[w = 125 ft.]*
   (E)   Once we have *w*, we can find *l*. *[Substitute into the formula from step B.]*
   (F)   With both dimensions known, calculate the area.

PRACTICE PROBLEMS

9. A manufacturer has daily production costs of $C(x) = 0.25x^2 - 8x + 800$ where C is the total cost, in dollars, and x is the number of units produced. What is the minimum daily cost, and how many units should be produced to yield the minimum cost?

10. A business owner estimates her weekly profits, p, in dollars, by the function $p(w) = -4w^2 + 160w$, where w represents the number of workers she hires. What is the number of workers she should hire in order to earn the greatest profit?

Regents Questions

Multiple Choice

1. Which quadratic function has the largest maximum?

   $h(x) = (3 - x)(2 + x)$ $k(x) = -5x^2 - 12x + 4$

        (1)   (3)

x	f(x)
−1	−3
0	5
1	9
2	9
3	5
4	−3

   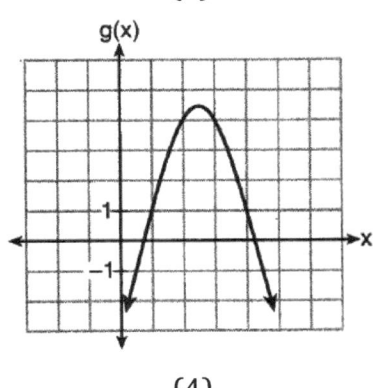

        (2)   (4)

2. Given the following quadratic functions:

   $$g(x) = -x^2 - x + 6$$

x	−3	−2	−1	0	1	2	3	4	5
n(x)	−7	0	5	8	9	8	5	0	−7

   Which statement about these functions is true?
   (1) Over the interval $-1 \le x \le 1$, the average rate of change for $n(x)$ is less than that for $g(x)$.
   (2) The y-intercept of $g(x)$ is greater than the y-intercept for $n(x)$.
   (3) The function $g(x)$ has a greater maximum value than $n(x)$.
   (4) The sum of the roots of $n(x) = 0$ is greater than the sum of the roots of $g(x) = 0$.

3.   The graph representing a function is shown below.

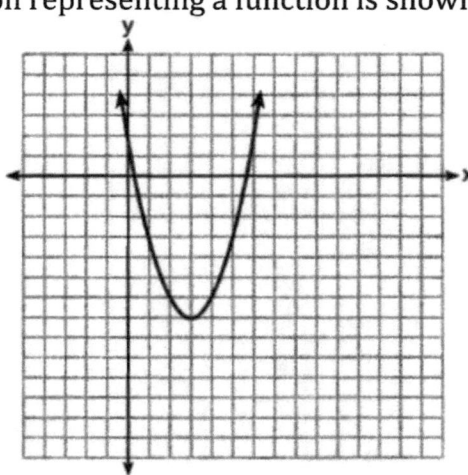

Which function has a minimum that is *less* than the one shown in the graph?
  (1) $y = x^2 - 6x + 7$ (3) $y = x^2 - 2x - 10$
  (2) $y = |x + 3| - 6$ (4) $y = |x - 8| + 2$

4.   The range of the function $f(x) = x^2 + 2x - 8$ is all real numbers
  (1) less than or equal to -9 (3) less than or equal to -1
  (2) greater than or equal to -9 (4) greater than or equal to -1

5.   Which quadratic function has the largest maximum over the set of real numbers?

       $f(x) = -x^2 + 2x + 4$ $g(x) = -(x - 5)^2 + 5$

              (1)   (3)

x	k(x)
−1	−1
0	3
1	5
2	5
3	3
4	−1

x	h(x)
−2	−9
−1	−3
0	1
1	3
2	3
3	1

              (2)   (4)

466

6.   Which of the quadratic functions below has the *smallest* minimum value?

$$h(x) = x^2 + 2x - 6 \qquad\qquad k(x) = (x + 5)(x + 2)$$

(1)   (3)

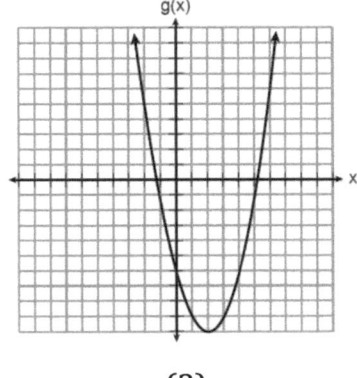

g(x)

x	f(x)
−1	−2
0	−5
1	−6
2	−5
3	−2

(2)   (4)

7.   The quadratic functions $r(x)$ and $q(x)$ are given below.

x	r(x)
−4	−12
−3	−15
−2	−16
−1	−15
0	−12
1	7

$$q(x) = x^2 + 2x - 8$$

The function with the *smaller* minimum value is

(1) $q(x)$, and the value is −9   (3) $r(x)$, and the value is −16
(2) $q(x)$, and the value is −1   (4) $r(x)$, and the value is −2

8.   Which interval represents the range of the function $h(x) = 2x^2 - 2x - 4$?

(1) $(0.5, \infty)$ (3) $[0.5, \infty)$
(2) $(-4.5, \infty)$ (4) $[-4.5, \infty)$

9.   Which quadratic function has the *smallest* minimum value?

$$f(x) = 6x^2 + 5x - 2$$ $$g(x) = 6(x - 2)^2 - 2$$

                 (1)   (3)

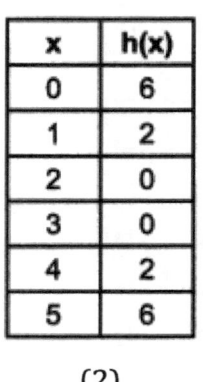

                 (2)

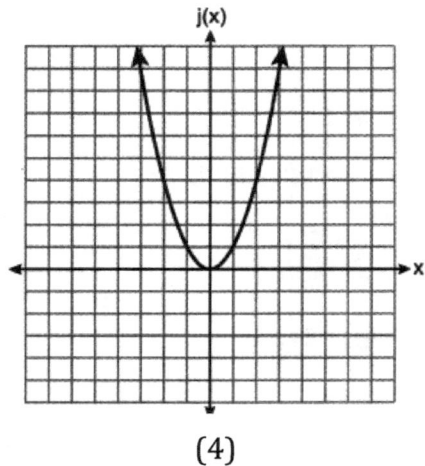

                 (4)

10.   The range of $f(x) = x^2 + 2x - 5$ is the set of all real numbers
       (1)   less than or equal to –6   (3)   less than or equal to –1
       (2)   greater than or equal to –6   (4)   greater than or equal to –1

CONSTRUCTED RESPONSE

11.   Let *f* be the function represented by the graph below.

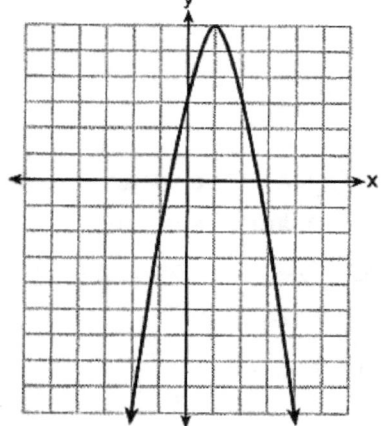

       Let *g* be a function such that $g(x) = -\frac{1}{2}x^2 + 4x + 3$.

       Determine which function has the larger maximum value. Justify your answer.

12.   Let $h(t) = -16t^2 + 64t + 80$ represent the height of an object above the ground after t
       seconds. Determine the number of seconds it takes to achieve its maximum height. Justify
       your answer.

       State the time interval, in seconds, during which the height of the object *decreases*. Explain
       your reasoning.

468

13. An Air Force pilot is flying at a cruising altitude of 9000 feet and is forced to eject from her aircraft. The function $h(t) = -16t^2 + 128t + 9000$ models the height, in feet, of the pilot above the ground, where t is the time, in seconds, after she is ejected from the aircraft.

    Determine and state the vertex of $h(t)$. Explain what the second coordinate of the vertex represents in the context of the problem.

    After the pilot was ejected, what is the maximum number of feet she was above the aircraft's cruising altitude? Justify your answer.

14. If the zeros of a quadratic function, F, are -3 and 5, what is the equation of the axis of symmetry of F? Justify your answer.

15. A ball is projected up into the air from the surface of a platform to the ground below. The height of the ball above the ground, in feet, is modeled by the function $f(t) = -16t^2 + 96t + 112$, where t is the time, in seconds, after the ball is projected.

    State the height of the platform, in feet.

    State the coordinates of the vertex. Explain what it means in the context of the problem. State the entire interval over which the ball's height is *decreasing*.

13.4 Graph Parabolas

The **standard form** of a quadratic function is $y = ax^2 + bx + c$ where $a \neq 0$.
Example: For $y = x^2 - 4x + 3$, $a = 1$, $b = -4$, and $c = 3$.

The graph of a quadratic function is called a **parabola**. If $a > 0$, then the parabola "**opens up**" like the letter U, but if $a < 0$, then the parabola "**opens down**" like an upside-down U.
Example: For $y = x^2 - 4x + 3$, $a = 1$, so the parabola opens up.

We can **graph the parabola** by:
1. drawing the axis of symmetry as a dashed line for reference only
2. plotting the vertex
3. plotting the y-intercept and its reflection
4. plotting any additional points
5. connecting the points with a solid curve

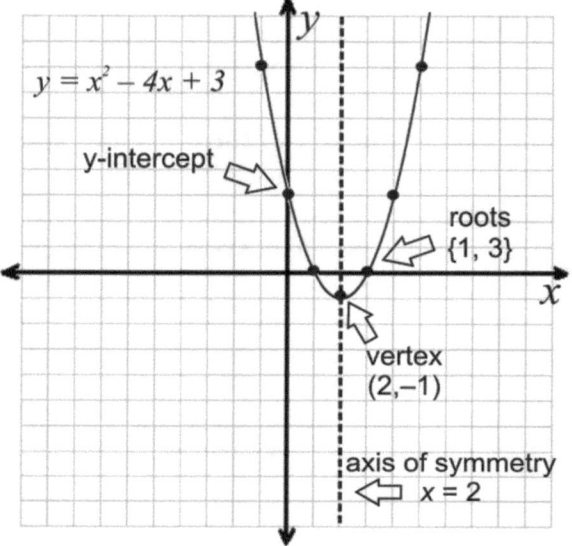

The **equation for the parabola's axis of symmetry**, a vertical line that divides the parabola into two parts that are mirror images of each other, is $x = \dfrac{-b}{2a}$.

Example: The axis of symmetry for the parabola whose equation is $y = x^2 - 4x + 3$ is the line $x = \dfrac{-b}{2a} = \dfrac{-(-4)}{2(1)} = 2$, or simply $x = 2$.

The axis of symmetry gives us the **x-coordinate of the vertex**. You can find **y-coordinate of the vertex** by substituting for x in the quadratic equation.
Example: For $y = x^2 - 4x + 3$, the axis of symmetry is $x = 2$, so substitute 2 for x.
$y = (2)^2 - 4(2) + 3 = -1$, so the vertex is the point $(2, -1)$.

The **y-intercept of the parabola** (the y coordinate of the point where the parabola crosses the y-axis), is c. The **reflection of the y-intercept** over the axis of symmetry (as long as the axis of symmetry is not $x = 0$) is another point on the parabola.
Example: For $y = x^2 - 4x + 3$, $c = 3$, so the y-intercept is 3.
The reflection of $(0,3)$ over the axis of symmetry is $(4,3)$.

We can **plot additional points** on the curve by substituting any integer value of x into the equation to find the corresponding value for y. We can also plot its reflection.
Example: For $y = x^2 - 4x + 3$, we substitute 5 for x to get $y = (5)^2 - 4(5) + 3 = 8$.
So, the point $(5,8)$ is on the parabola. Its reflection is the point $(-1,8)$.

For graphs of any non-linear functions, you are required to plot and label at least 3 points.

▦☐ CALCULATOR TIP

You can use the graphing calculator to **graph a parabola**.

Example: To graph $y = x^2 - 4x + 3$, press [Y=]. Then, next to "$Y_1 =$", type in the quadratic expression $x^2 - 4x + 3$ using [X,T,Θ,n] for the variable, x, and the [x^2] button for the exponent, 2. Then. press [GRAPH] to view the parabola.

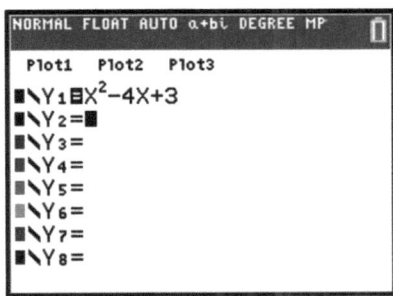 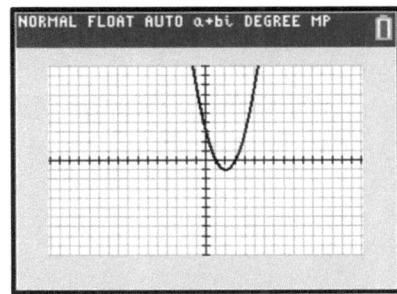

▦☐ CALCULATOR TIP

The calculator can help you find the **roots** of a graphed parabola:

- Using the [TABLE] feature:

  Press [2nd][TABLE] to see a table of (x, y) coordinates of points on the parabola. Press the [▾] key to scroll down for more points. You can find the roots by looking for the values of x when **y equals 0** *[for the equation $y = x^2 - 4x + 3$, the roots are 1 and 3]*.

- Using the [CALC] feature:

  Press [2nd][CALC][2] for zero. Look for the first point where the parabola crosses the x-axis. For the "Left Bound?" prompt, use the [◄][►] keys to move the cursor to left of this point and press [ENTER]. For the "Right Bound?" prompt, use the [►] key to move the cursor to right of this point (near the vertex) and press [ENTER] twice. The coordinates are shown; the value of x is the root. If there is a second point where the parabola crosses the x-axis, repeat these steps, moving the cursor to the left and right of the point to find its coordinates; the value of x is another root.

▦☐ CALCULATOR TIP

The calculator can also help you find the **vertex** of a graphed parabola:

Press [2nd][CALC] and then press either [3] for minimum if the parabola opens up $(a > 0)$ or [4] for maximum if the parabola opens down $(a < 0)$. For the "Left Bound?" prompt, use the [◄][►] keys to move the cursor to any point along the left side of the parabola and press [ENTER]. For the "Right Bound?" prompt, use the [►] key to move the cursor to any point along the right side of the parabola and press [ENTER] twice. The coordinates of the vertex (or a close approximation) will be shown. You may need to round to find the actual coordinates of the vertex.

We have already seen that the calculator's LinReg and ExpReg functions allow us to find the equations of linear and exponential functions, respectively, given some points. Likewise, given a graph of a parabola, it is also possible to find its equation using the calculator's QuadReg function, as described below. QuadReg is short for **quadratic regression**.

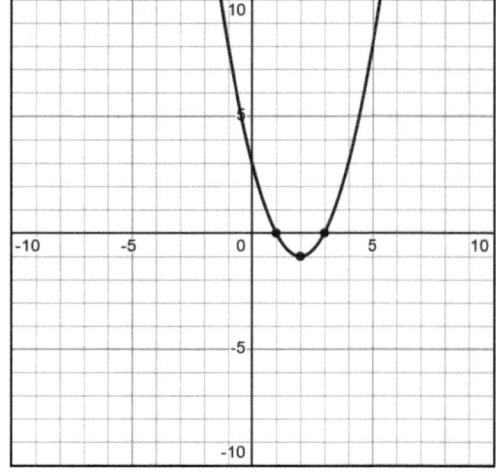

Example: Entering at least three points on this parabola, such as $(1,0)$, $(2,-1)$, and $(3,0)$, will allow the calculator to find the quadratic equation, $y = x^2 - 4x + 3$.

 **CALCULATOR TIP**

To find the equation for a quadratic function:

1. Press $\boxed{\text{STAT}}\boxed{1}$ to select Edit.
2. If values appear in the L1 or L2 columns, select the column heading and press $\boxed{\text{CLEAR}}\boxed{\text{ENTER}}$.
3. Enter the *x* values into the L1 column and the corresponding *y* values into the L2 column.
4. Press $\boxed{\text{STAT}}$<CALC>$\boxed{5}$ for QuadReg.
5. On the next screen prompt, make sure L1 and L2 are selected for Xlist and Ylist. Next to Store RegEQ, enter $\boxed{\text{ALPHA}}\boxed{\text{F4}}\boxed{1}$ to store the equation in Y1.

   [On the TI-83, you'll see an QuadReg prompt instead. Enter $\boxed{\text{VARS}}$<Y-VARS>$\boxed{1}\boxed{1}$.]
6. The screen will show the equation y=ax²+bx+c along with the values of *a*, *b*, and *c*.
7. To view the graph, press $\boxed{\text{GRAPH}}$. To see the equation, press $\boxed{\text{Y=}}$.

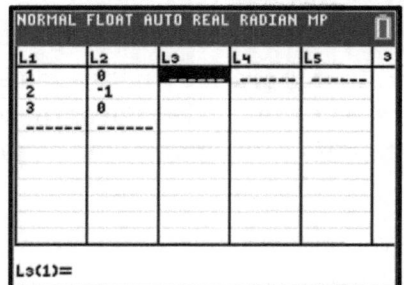   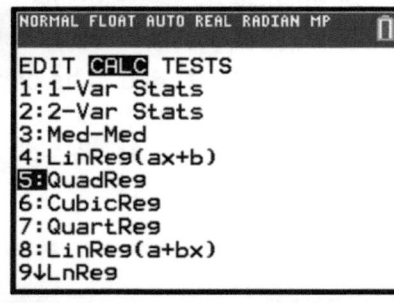   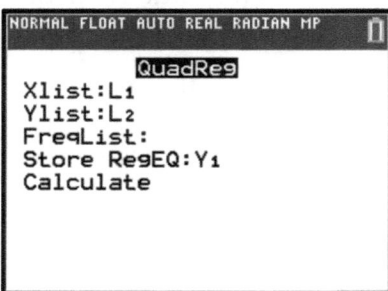

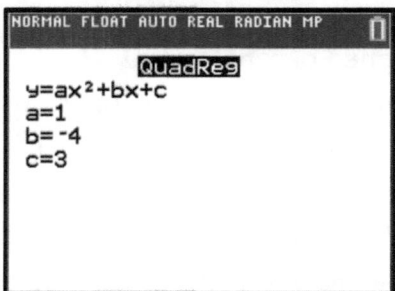   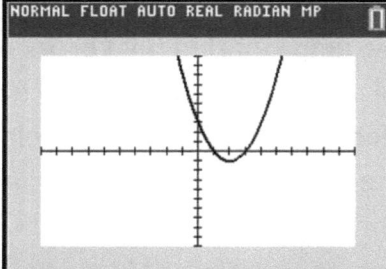

MODEL PROBLEM

Graph the equation $y = -x^2 + 2x + 3$.

Solution:

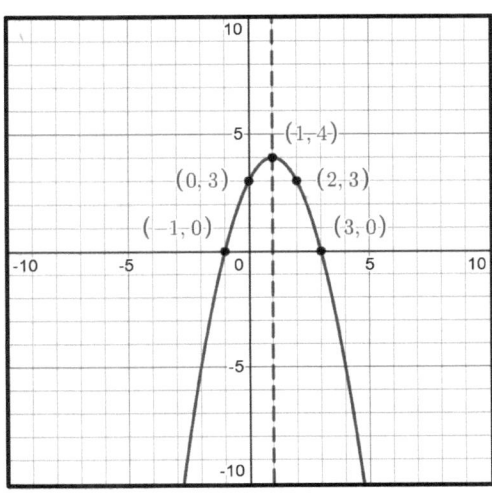

Explanation of steps:

(A) The axis of symmetry is $x = \dfrac{-b}{2a} = \dfrac{-(2)}{2(-1)} = 1$.
A dashed vertical line, $x = 1$, is drawn.

(B) Substituting $x = 1$, $y = -(1)^2 + 2(1) + 3 = 4$, so the vertex is $(1, 4)$.

(C) The y-intercept, c, is 3, so $(0,3)$ and its reflection $(2,3)$ are plotted.

(D) When $x = 3$, $y = -(3)^2 + 2(3) + 3 = 0$, so the point $(3,0)$ and its reflection $(-1,0)$ can be plotted as additional points. Connect the points to draw the parabola.

PRACTICE PROBLEMS

1. Graph the parabola whose equation is $y = 2x^2 - 8x + 4$.	2. Graph the parabola whose equation is $y = -x^2 + 6x - 5$.
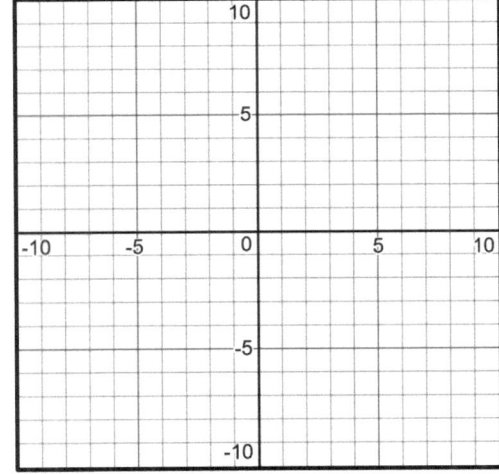	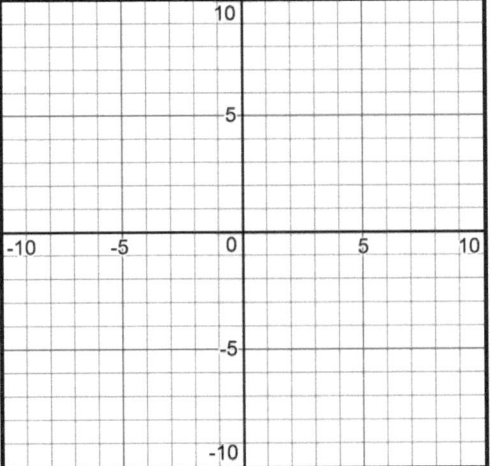

3. Graph the parabola whose equation is
    $y = -x^2 + 4x - 8.$

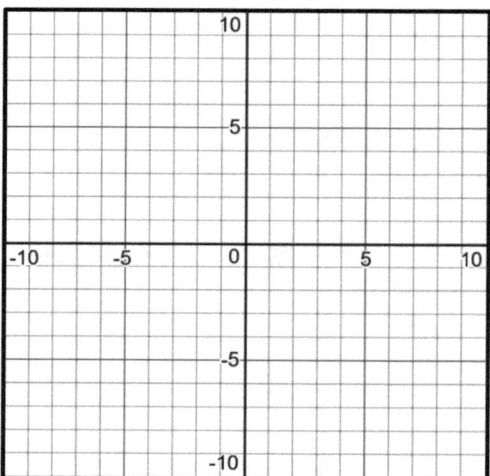

4. Graph the parabola whose equation is
    $y = x^2 - 6x + 10.$

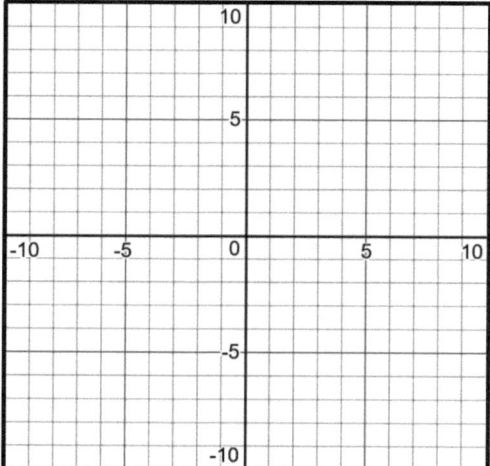

5. Graph the parabola whose equation is
    $y = 3x^2 + 6x - 1.$

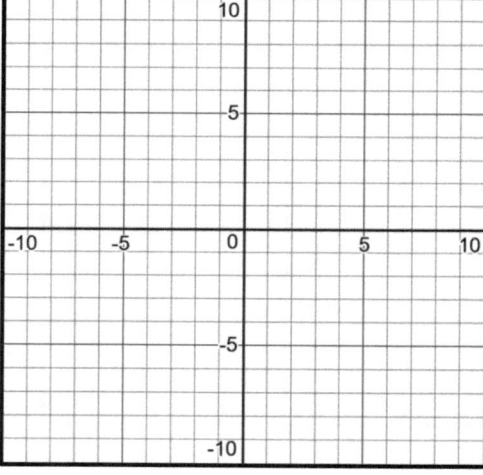

6. Graph the parabola whose equation is
    $y = 2x^2 + 8x + 9.$

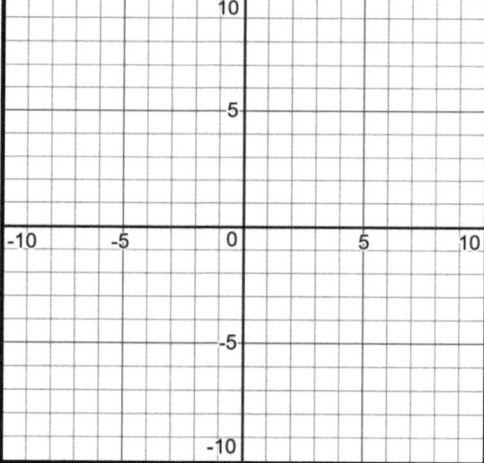

7. Graph the parabola whose equation is $y = x^2 + 2x$.

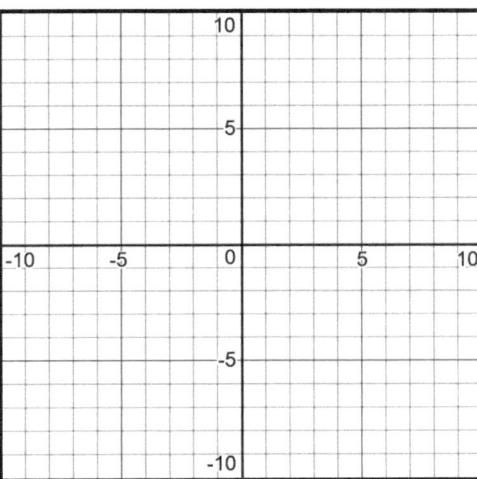

8. Graph the parabola whose equation is $y = 3x^2 + 1$.

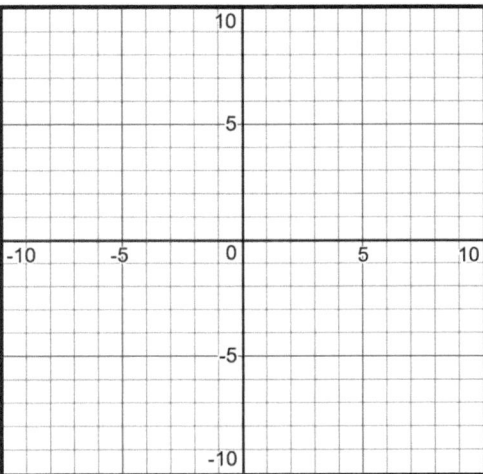

9. Graph the equation $y = x^2 - 2x - 3$. Using the graph, determine the roots of the equation $x^2 - 2x - 3 = 0$.

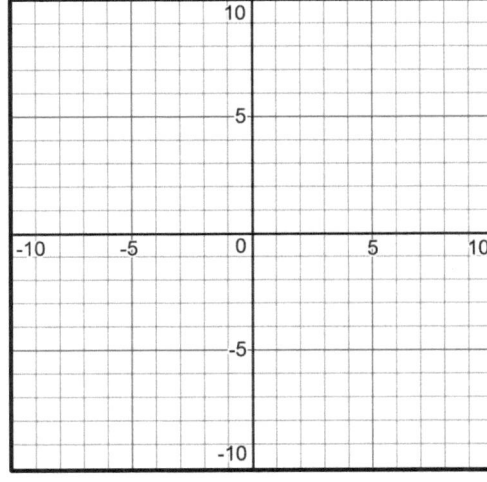

10. Graph the equation $y = x^2 + 2x - 8$. Using the graph, determine the roots of the equation $x^2 + 2x - 8 = 0$.

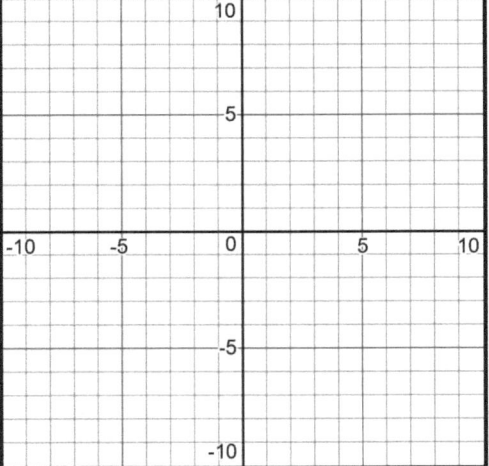

11. A ball is tossed in the air so that the path of the ball is modeled by the equation $y = -x^2 + 6x$, where y represents the height of the ball in feet and x is the time in seconds.

a) Graph $y = -x^2 + 6x$ for $0 \le x \le 6$ on the grid below.

b) At what time, x, is the ball at its highest point?

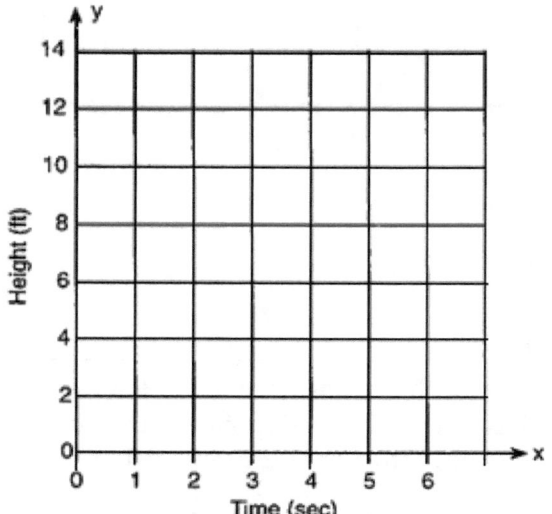

12. A ball is tossed in the air so that the path of the ball is modeled by the equation $h = -8t^2 + 40t$, where h represents the height of the ball in feet and t is the time in seconds.

a) Graph $h = -8t^2 + 40t$ for $0 \le t \le 5$ on the grid below.

b) At what time, t, is the ball at its highest point?

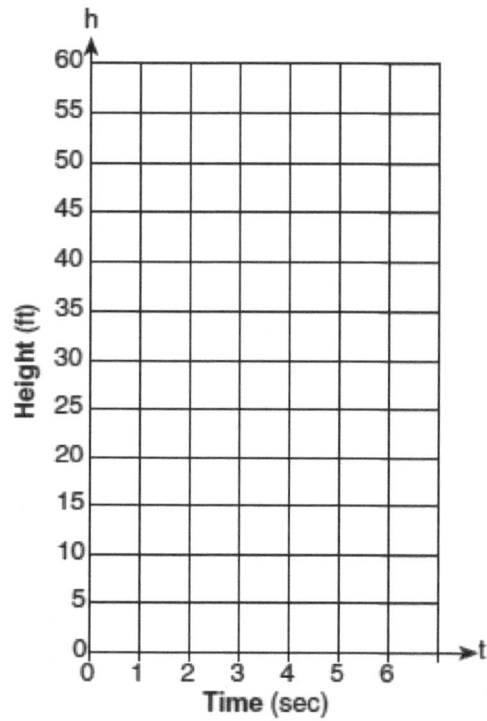

476

13. The shape of a parabolic arch is represented by the equation $y = -2x^2 + 12x$, where y is the height of the arch in feet. The arch is 6 feet wide at its base.

a) Graph the parabola below.

b) What is the maximum height of the arch?

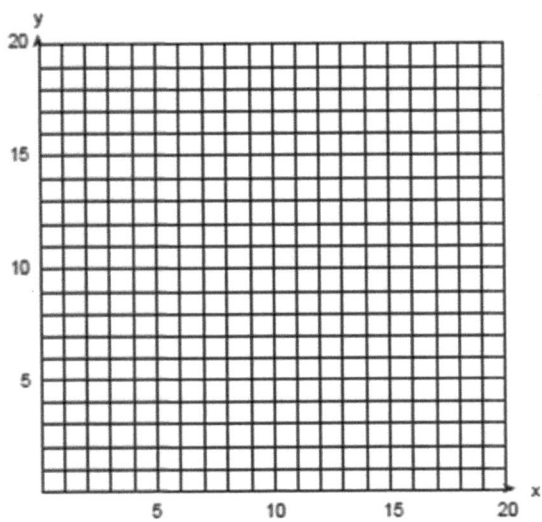

14. The shape of a parabolic arch is represented by the equation $y = -x^2 + 20x$, where y is the height of the arch in feet. The arch is 20 feet wide at its base.

a) Graph the parabola below.

b) What is the maximum height of the arch?

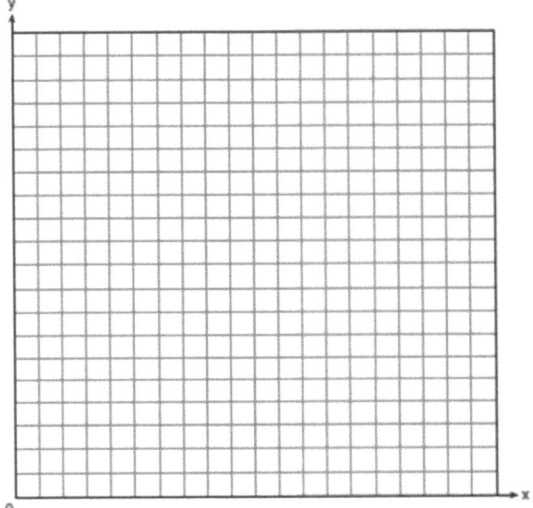

Regents Questions

MULTIPLE CHOICE

1. The height of a rocket, at selected times, is shown in the table below.

Time (sec)	0	1	2	3	4	5	6	7
Height (ft)	180	260	308	324	308	260	180	68

   Based on these data, which statement is *not* a valid conclusion?
   (1) The rocket was launched from a height of 180 feet.
   (2) The maximum height of the rocket occurred 3 seconds after launch.
   (3) The rocket was in the air approximately 6 seconds before hitting the ground.
   (4) The rocket was above 300 feet for approximately 2 seconds.

2. Morgan throws a ball up into the air. The height of the ball above the ground, in feet, is modeled by the function $h(t) = -16t^2 + 24t$, where t represents the time, in seconds, since the ball was thrown. What is the appropriate domain for this situation?
   (1) $0 \le t \le 1.5$ (3) $0 \le h(t) \le 1.5$
   (2) $0 \le t \le 9$ (4) $0 \le h(t) \le 9$

3. The expression $-4.9t^2 + 50t + 2$ represents the height, in meters, of a toy rocket t seconds after launch. The initial height of the rocket, in meters, is
   (1) 0   (3) 4.9
   (2) 2   (4) 50

4. The height of a ball Doreen tossed into the air can be modeled by the function $h(x) = -4.9x^2 + 6x + 5$, where x is the time elapsed in seconds, and $h(x)$ is the height in meters. The number 5 in the function represents
   (1) the initial height of the ball
   (2) the time at which the ball reaches the ground
   (3) the time at which the ball was at its highest point
   (4) the maximum height the ball attained when thrown in the air

5.   Three functions are shown below.

A: $g(x) = -\frac{3}{2}x + 4$

B: $f(x) = (x + 2)(x + 6)$

C:

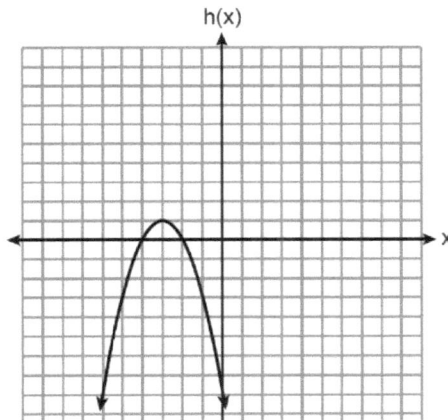

Which statement is true?

(1) B and C have the same zeros.

(3) B has a minimum and C has a maximum.

(2) A and B have the same y-intercept.

(4) C has a maximum and A has a minimum.

6.   A ball is thrown into the air from the top of a building. The height, $h(t)$, of the ball above the ground t seconds after it is thrown can be modeled by $h(t) = -16t^2 + 64t + 80$. How many seconds after being thrown will the ball hit the ground?

(1)   5   (3)   80

(2)   2   (4)   144

7. Which statement is true about the functions $f(x)$ and $g(x)$, given below?

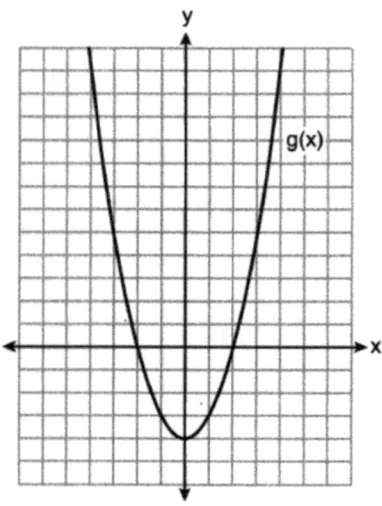

$$f(x) = -x^2 - 4x - 4$$

   (1) The minimum value of $g(x)$ is greater than the maximum value of $f(x)$.

   (2) $f(x)$ and $g(x)$ have the same y-intercept.

   (3) $f(x)$ and $g(x)$ have the same roots.

   (4) $f(x) = g(x)$ when $x = -4$.

8. Ian throws a ball up in the air and lets it fall to the ground. The height of the ball, $h(t)$, is modeled by the equation $h(t) = -16t^2 + 6t + 3$, with $h(t)$ measured in feet, and time, t, measured in seconds. The number 3 in $h(t)$ represents

   (1) the maximum height of the ball

   (2) the height from which the ball is thrown

   (3) the number of seconds it takes for the ball to reach the ground

   (4) the number of seconds it takes for the ball to reach its maximum height

CONSTRUCTED RESPONSE

9. A football player attempts to kick a football over a goal post. The path of the football can be modeled by the function $h(x) = -\frac{1}{225}x^2 + \frac{2}{3}x$, where x is the horizontal distance from the kick, and $h(x)$ is the height of the football above the ground, when both are measured in feet. On the set of axes below, graph the function $y = h(x)$ over the interval $0 \leq x \leq 150$.

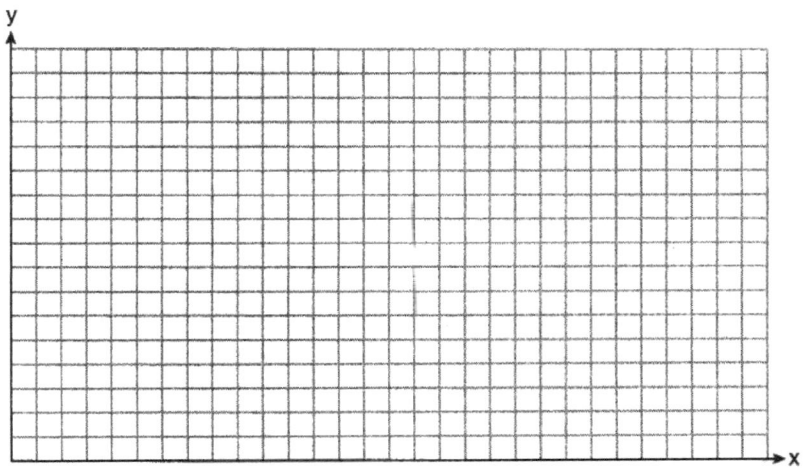

Determine the vertex of $y = h(x)$. Interpret the meaning of this vertex in the context of the problem.

The goal post is 10 feet high and 45 yards away from the kick. Will the ball be high enough to pass over the goal post? Justify your answer.

10. Graph $f(x) = x^2$ and $g(x) = 2^x$ for $x \geq 0$ on the set of axes below.

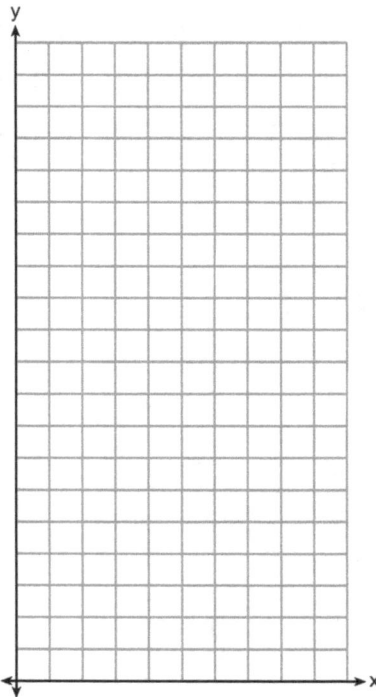

State which function, $f(x)$ or $g(x)$, has a greater value when $x = 20$. Justify your reasoning.

11. On the set of axes below, draw the graph of $y = x^2 - 4x - 1$.

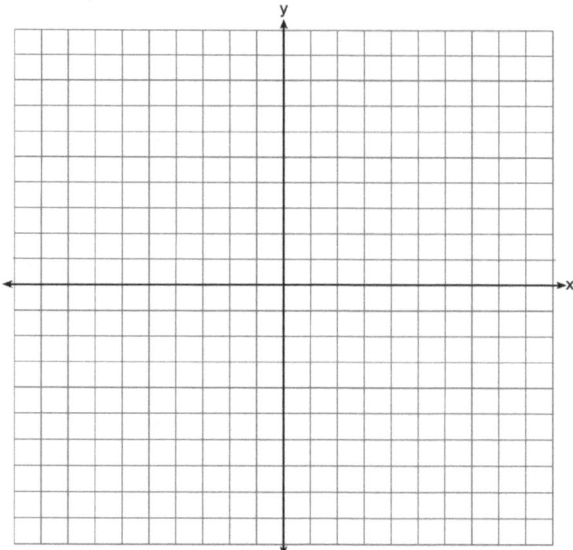

State the equation of the axis of symmetry.

12. Alex launched a ball into the air. The height of the ball can be represented by the equation $h = -8t^2 + 40t + 5$, where h is the height, in units, and t is the time, in seconds, after the ball was launched. Graph the equation from $t = 0$ to $t = 5$ seconds.

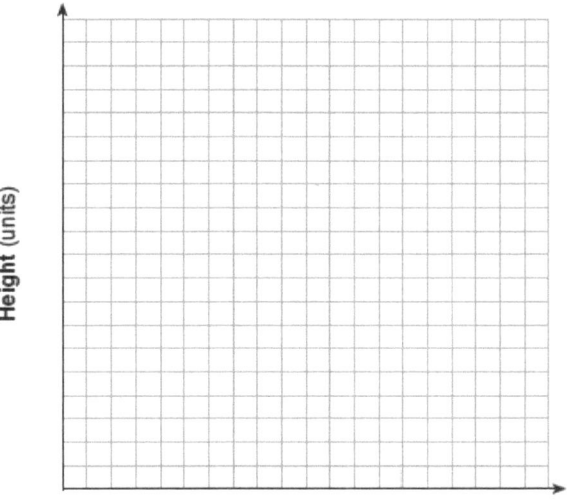

State the coordinates of the vertex and explain its meaning in the context of the problem.

13. Graph the function $f(x) = -x^2 - 6x$ on the set of axes below.

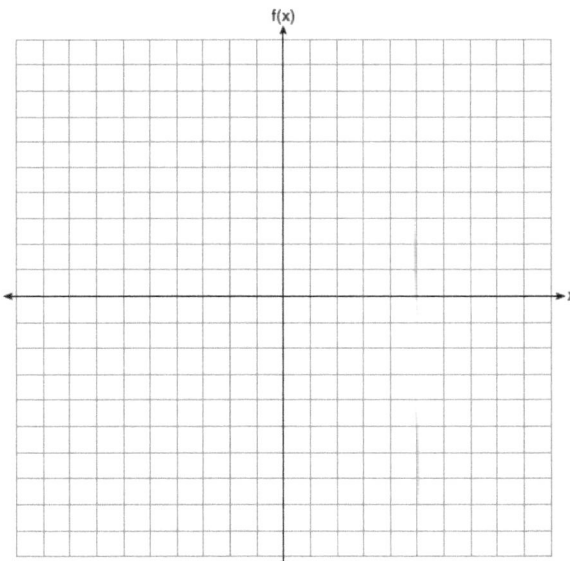

State the coordinates of the vertex of the graph.

14. Paul plans to have a rectangular garden adjacent to his garage. He will use 36 feet of fence to enclose three sides of the garden. The area of the garden, in square feet, can be modeled by $f(w) = w(36 - 2w)$, where w is the width in feet.

On the set of axes below, sketch the graph of $f(w)$.

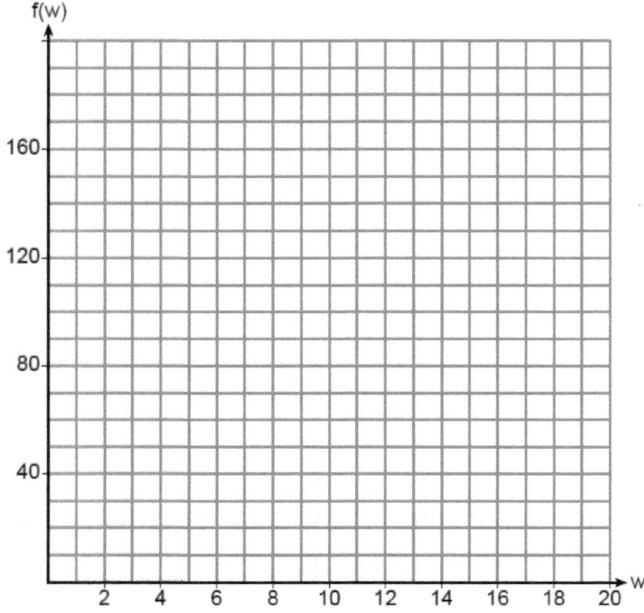

Explain the meaning of the vertex in the context of the problem.

15. Michael threw a ball into the air from the top of a building. The height of the ball, in feet, is modeled by the equation $h = -16t^2 + 64t + 60$, where t is the elapsed time, in seconds. Graph this equation on the set of axes below.

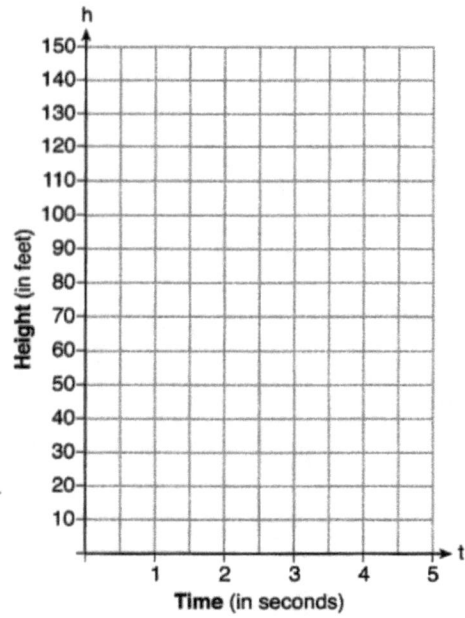

Determine the average rate of change, in feet per second, from when Michael released the ball to when the ball reached its maximum height.

16.   On the set of axes below, graph $f(x) = x^2 - 1$ and $g(x) = 3^x$.

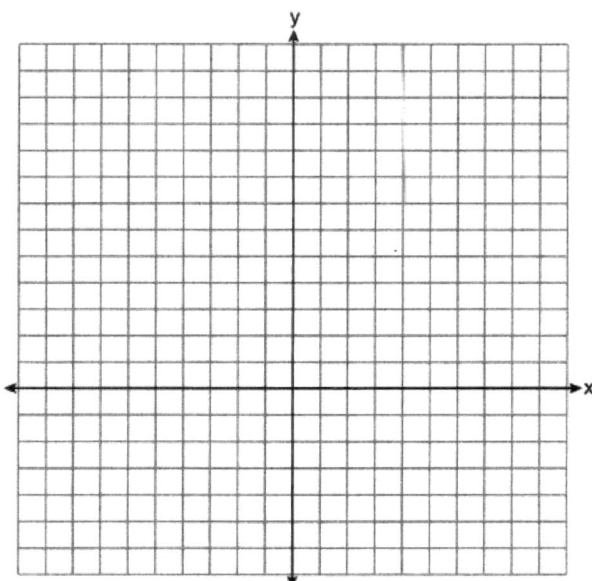

Based on your graph, for how many values of x does $f(x) = g(x)$. Explain your reasoning.

17. The path of a rocket is modeled by the function $h(t) = -4.9t^2 + 49t$, where h is the height, in meters, above the ground and t is the time, in seconds, after the rocket is launched. Sketch the graph on the set of axes below.

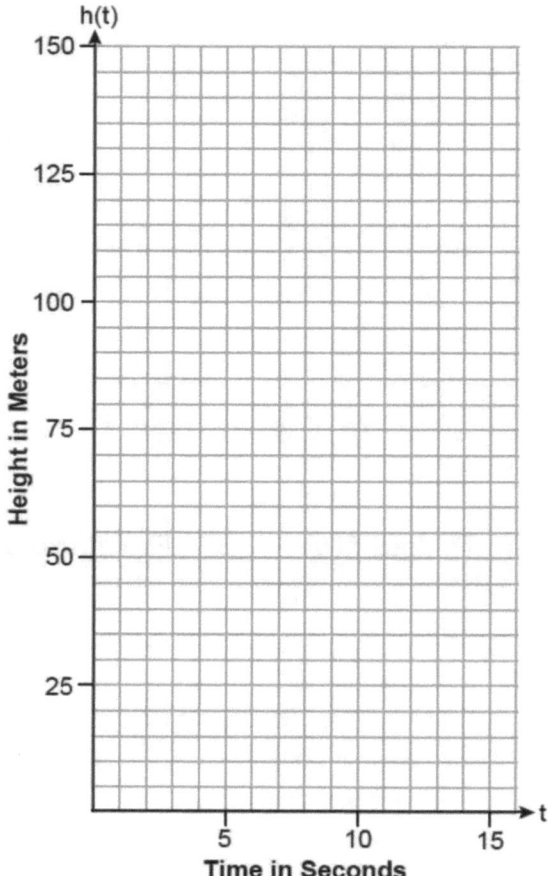

State the vertex of this function. Explain what the vertex means in the context of this situation.

18. While playing golf, Laura hit her ball from the ground. The height, in feet, of her golf ball can be modeled by $h(t) = -16t^2 + 48t$, where t is the time in seconds. Graph $h(t)$ on the set of axes below.

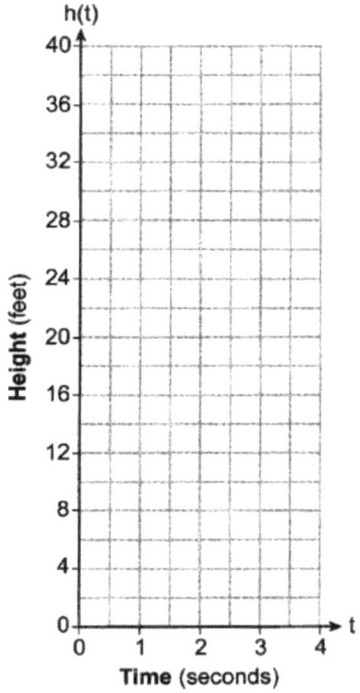

What is the maximum height, in feet, that the golf ball reaches on this hit?

How many seconds does it take the golf ball to hit the ground?

13.5   Vertex Form

It is sometimes helpful to transform an equation, $y = x^2 + bx + c$, into **vertex form**: $y = (x - h)^2 + k$. In vertex form, the values of h and k represent the coordinates of the vertex (h, k) of the parabola. This allows us to more easily determine the vertex from the equation.

To transform a quadratic equation into vertex form, we use the steps for **completing the square**, which we have learned previously.

To convert a quadratic function $y = x^2 + bx + c$ into vertex form:
1. Add the opposite of c to both sides.
2. Add $\left(\frac{b}{2}\right)^2$ to both sides.
3. Factor the trinomial into a binomial squared.
4. Isolate y. Write the equation in vertex form, as $y = (x - h)^2 + k$.
5. The vertex is the point (h, k).

Once we have an equation in vertex form, we can **state the vertex**. However, be careful about the signs of the coordinates. Since there's subtraction inside the parentheses of $y = (x - h)^2 + k$, we need to negate the sign of the second term in parentheses to find h. Outside parentheses, the sign of k can be taken as is.

Examples: For $y = (x - 3)^2 + 2$, the vertex is (3,2).
                 For $y = (x + 1)^2 - 1$, the vertex is $(-1, -1)$.

To graph a function in vertex form as a parabola, start by plotting the vertex, then use a table to plot points for values of x that are just above or below h (we usually choose consecutive integers).
Example: To graph $y = (x - 2)^2 - 1$, plot the vertex $(2, -1)$, then plot additional points using a table. As a parabola, the points should reflect over the axis of symmetry.

x	$(x - 2)^2 - 1$	y
-1	$(-1 - 2)^2 - 1$	8
0	$(0 - 2)^2 - 1$	3
1	$(1 - 2)^2 - 1$	0
2	$(2 - 2)^2 - 1$	-1
3	$(3 - 2)^2 - 1$	0
4	$(4 - 2)^2 - 1$	3
5	$(5 - 2)^2 - 1$	8

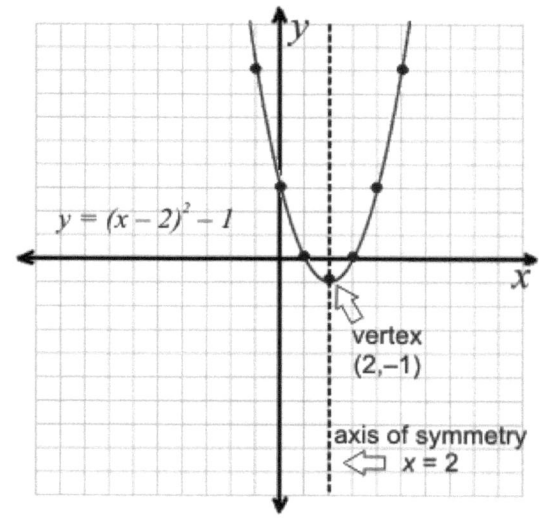

488

MODEL PROBLEM

Transform the equation $y = x^2 + 4x + 1$ into vertex form. State the coordinates of the vertex.

Solution:

$$y = x^2 + 4x + 1$$
(A) $y - 1 = x^2 + 4x$

(B) $\left(\dfrac{b}{2}\right)^2 = \left(\dfrac{4}{2}\right)^2 = 2^2 = 4$

$$y - 1 + 4 = x^2 + 4x + 4$$
(C) $y + 3 = (x + 2)^2$

(D) $y = (x + 2)^2 - 3$

(E)   Vertex is $(-2, -3)$.

Explanation of steps:

(A)   Add the opposite of c to both sides.

(B)   Add $\left(\dfrac{b}{2}\right)^2$ to both sides.

(C)   Factor the trinomial into a binomial squared.

(D)   Isolate y. Write the equation in vertex form, as $y = (x - h)^2 + k$.

(E)   The vertex is the point (h, k).

PRACTICE PROBLEMS

1. Transform the equation $y = x^2 + 6x + 10$ into vertex form and state the coordinates of the vertex.	2. Transform the equation of the function $f(x) = x^2 + 10x + 21$ into vertex form and state the coordinates of the vertex.
3. Graph the parabola $y = (x - 3)^2 + 1$. 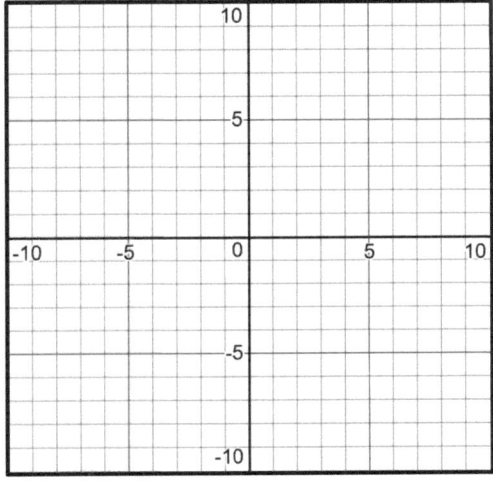	4. Graph the parabola $y = (x - 1)^2 - 3$. 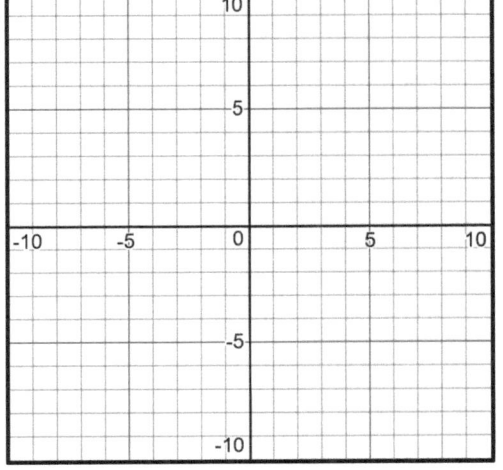

Regents Questions

MULTIPLE CHOICE

1. In the function $f(x) = (x - 2)^2 + 4$, the minimum value occurs when x is
   (1) -2 (3) -4
   (2) 2 (4) 4

2. Which equation is equivalent to $y - 34 = x(x - 12)$?
   (1) $y = (x - 17)(x + 2)$ (3) $y = (x - 6)^2 + 2$
   (2) $y = (x - 17)(x - 2)$ (4) $y = (x - 6)^2 - 2$

3. Which equation and ordered pair represent the correct vertex form and vertex for $f(x) = x^2 - 12x + 7$?
   (1) $f(x) = (x - 6)^2 + 43, (6,43)$ (3) $f(x) = (x - 6)^2 - 29, (6,-29)$
   (2) $f(x) = (x - 6)^2 + 43, (-6,43)$ (4) $f(x) = (x - 6)^2 - 29, (-6,-29)$

4. Which statement is true about the quadratic functions $g(x)$, shown in the table below, and $f(x) = (x - 3)^2 + 2$?

x	g(x)
0	4
1	-1
2	-4
3	-5
4	-4
5	-1
6	4

   (1) They have the same vertex.   (3) They have the same axis of symmetry.
   (2) They have the same zeros.   (4) They intersect at two points.

5. Which equation is equivalent to $y = x^2 + 24x - 18$?
   (1) $y = (x + 12)^2 - 162$ (3) $y = (x - 12)^2 - 162$
   (2) $y = (x + 12)^2 + 126$ (4) $y = (x - 12)^2 + 126$

6.   Three quadratic functions are given below.

      I.

$$f(x) = (x + 2)^2 + 5$$

      II.

x	−4	−3	−2	−1	0	1
$g(x)$	−3	2	5	5	2	−3

      III.

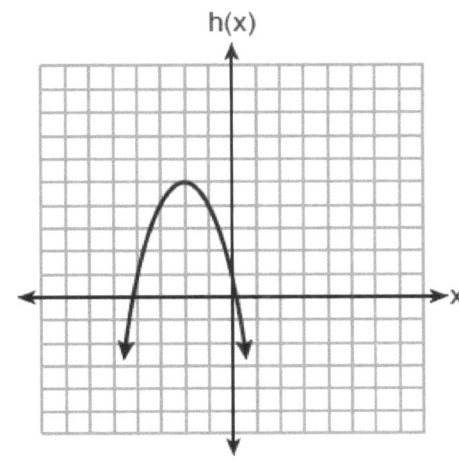

Which of these functions have the same vertex?
   (1)   I and II, only   (3)   I and III, only
   (2)   II and III, only   (4)   I, II, and III

CONSTRUCTED RESPONSE

7.   Determine and state the vertex of $f(x) = x^2 - 2x - 8$ using the method of completing the square.

8.   Use the method of completing the square to determine the vertex of $f(x) = x^2 - 14x - 15$. State the coordinates of the vertex.

CHAPTER 14.   QUADRATIC-LINEAR SYSTEMS

14.1   Solve Quadratic-Linear Systems Algebraically

A system of equations that includes one quadratic equation and one linear equation can be solved algebraically using the **substitution method.**

1. First solve each equation for y so that y is set equal to both a quadratic expression and a linear expression in terms of x.
2. Then, since both expressions are equal to y, set them equal to each other, and solve the resulting quadratic equation for x.
3. Once you have the value(s) of x, find the corresponding value(s) of y by substituting into one of the original equations.

MODEL PROBLEM

Solve the following system of equations algebraically:
$$y = x^2 - x - 6 \qquad y = 2x - 2$$

Solution:

(A) By substitution, $x^2 - x - 6 = 2x - 2$.

(B) $x^2 - x - 6 = 2x - 2$
$$\underline{-2x + 2 \quad - 2x + 2}$$
$$x^2 - 3x - 4 = 0$$
$$(x - 4)(x + 1) = 0$$
$$x = \{-1, 4\}$$

(C) When $x = -1, y = 2(-1) - 2 = -4$.
When $x = 4, y = 2(4) - 2 = 6$.

(D) Solutions: $(-1, -4)$ and $(4,6)$

Explanation of steps:

(A) Since both expressions are equal to y, set them equal to each other. That is, substitute the linear expression for y in the quadratic equation.

(B) Solve the quadratic equation by getting zero to one side, then factoring.

(C) For each root, substitute for x in the linear equation to find the corresponding value of y.

(D) Express the solutions as ordered pairs.

▦ CALCULATOR TIP

We can check each solution using the calculator. For example, to check whether $(-1, -4)$ is a solution to the system above, enter:

[(-)][1][STO▸][X,T,Θ,n] [ALPHA][:][(-)][4][STO▸][ALPHA][Y] [ENTER]

[ALPHA][Y] [2nd][TEST][1] [X,T,Θ,n][x²][−][X,T,Θ,n][−][6] [ENTER]   (1 means true)

[ALPHA][Y] [2nd][TEST][1] [2][X,T,Θ,n][−][2] [ENTER]   (1 means true)

Repeat these steps for $(4,6)$ by storing 4 in x and 6 in y and testing both equations again.

PRACTICE PROBLEMS

1. Solve the following system of equations algebraically: $$y = x^2 - 5$$ $$y = -4x$$	2. Solve the following system of equations algebraically: $$y = x^2 + 4x + 1$$ $$y = 5x + 3$$
3. Solve the following system of equations algebraically: $$y = x^2 + 2x - 1$$ $$y = 3x + 5$$	4. Solve the following system of equations algebraically: $$y = x^2 + 4x - 2$$ $$y = 2x + 1$$
5. Solve the following system of equations algebraically: $$y = x^2 + 7x + 22$$ $$y + 3x = 1$$	6. Solve the following system of equations algebraically: $$y + 3x = 6$$ $$x^2 = y + 2x + 6$$

7. Solve the following system of equations algebraically:
$$y = x^2 + 2x - 8$$
$$y = 2x + 1$$

8. Solve the following system of equations algebraically:
$$y = x^2 - 6x + 9$$
$$y = -9x + 19$$

9. Solve the following system of equations algebraically:
$$y = x^2 + 5x - 17$$
$$y = x - 5$$

10. Solve the following system of equations algebraically:
$$y = 3x - 6$$
$$y = x^2 - x - 6$$

Regents Questions

MULTIPLE CHOICE

1. If $f(x) = x^2 - 2x - 8$ and $g(x) = \frac{1}{4}x - 1$, for which value of x is $f(x) = g(x)$?

     (1) -1.75 and -1.438 (3) -1.438 and 0

     (2) -1.75 and 4   (4) 4 and 0

2. Which pair of equations would have $(-1, 2)$ as a solution?

     (1) $y = x + 3$ and $y = 2^x$ (3) $y = x^2 - 3x - 2$ and $y = 4x + 6$

     (2) $y = x - 1$ and $y = 2x$ (4) $2x + 3y = -4$ and $y = -\frac{1}{2}x - \frac{3}{2}$

3. If $f(x) = x^2 + 2x + 1$ and $g(x) = 7x - 5$, for which values of x is $f(x) = g(x)$?

     (1) -1 and 6   (3) -3 and -2

     (2) -6 and -1 (4) 2 and 3

CONSTRUCTED RESPONSE

4. If $f(x) = x^2$ and $g(x) = x$, determine the value(s) of x that satisfy the equation $f(x) = g(x)$.

5. Solve the following systems of equations algebraically for all values of x and y:

$$y = x^2 + 5x = 17$$
$$x - y = 5$$

14.2 Solve Quadratic-Linear Systems Graphically

To solve a system of equations graphically, **graph both equations** on the same set of axes and determine the **point(s) of intersection**. These points are the solutions to the system.

The graph of a quadratic-linear system consists of a **parabola** (from a quadratic equation) and a **line** (from a linear equation). The parabola and line may intersect at two points, so there may be two solutions to the system.

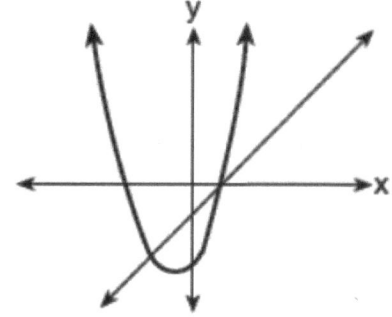

Example: The graph below shows that the solutions to the system, $y = x^2 - 4x + 3$ and $y = -2x + 6$, are the points of intersection, $(-1,8)$ and $(3,0)$.

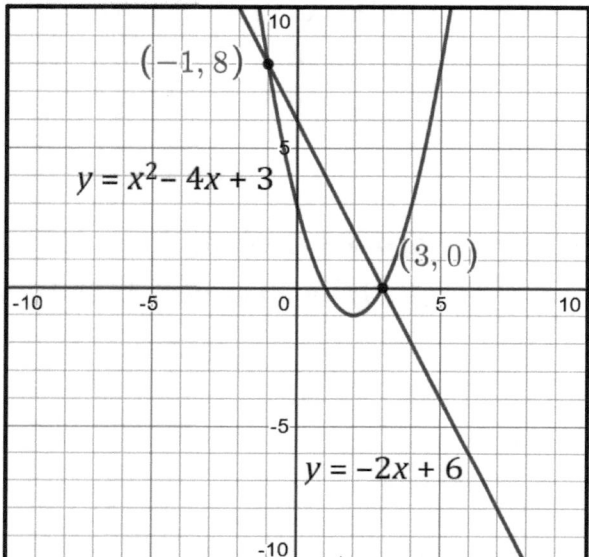

Although you are not required to label the two graphs when they have different degrees (such as one linear and one quadratic), it is still helpful to do so.

CALCULATOR TIP

You can also use the calculator to find the point(s) of intersection:

1. Press $\boxed{Y=}$ and enter both equations.
2. Press $\boxed{2nd}\boxed{CALC}\boxed{5}$ for intersect.
3. Press $\boxed{ENTER}$ for the "First curve?" and "Second curve?" prompts.
4. For the "Guess?" prompt, use the arrow keys to move the cursor near one of the points of intersection and press $\boxed{ENTER}$. The coordinates of the closest point of intersection will be shown.
5. If there appears to be a second point of intersection, repeat steps 2 to 4 but move the cursor near the second point in response to the "Guess?" prompt.

Example: The screenshots below show that the soltuions to the system of equations, $y = x^2 - 4x + 3$ and $y = -2x + 6$, are $(-1, 8)$ and $(3, 0)$.

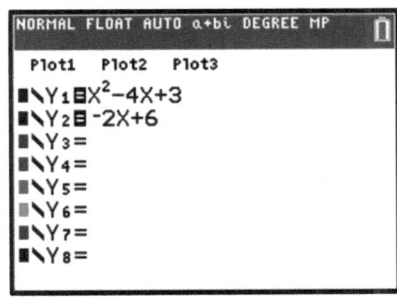 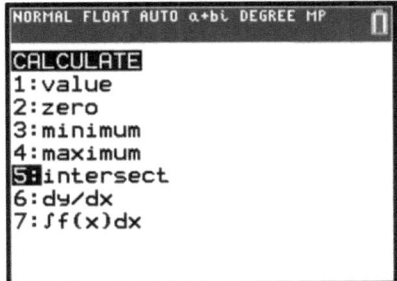

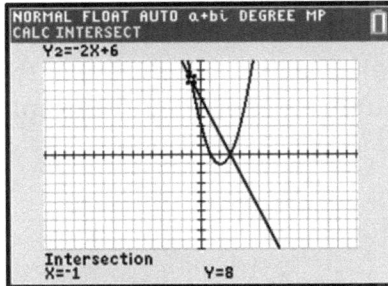 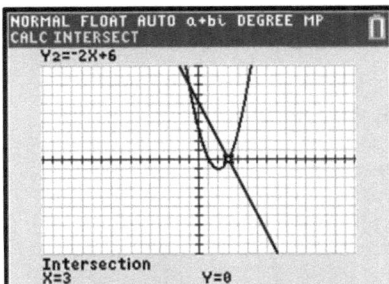

MODEL PROBLEM

Solve the following system of equations graphically:
$$y = x^2 - 4x + 3$$
$$y + 1 = x$$

Solution:

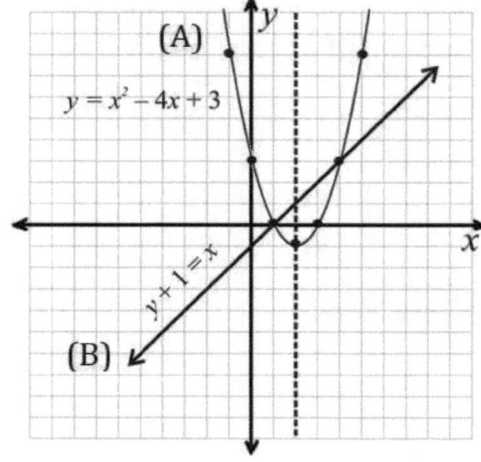

(C) Solutions: (1,0) and (4,3)

Explanation of steps:

(A) Graph the quadratic equation as a parabola. Include at least 3 points with integer values of x on each side of the axis of symmetry.
[The axis of symmetry is $x = 2$ and vertex is $(2, -1)$. Plot additional points with x-coordinates of 3, 4, and 5, plus their reflections.]

(B) Graph the linear equation as a line.
[First transform the equation into slope-intercept form, $y = x - 1$.]

(C) State the point(s) of intersection as the solutions

PRACTICE PROBLEMS

1. Which point is a solution of the system of equations shown on the graph?

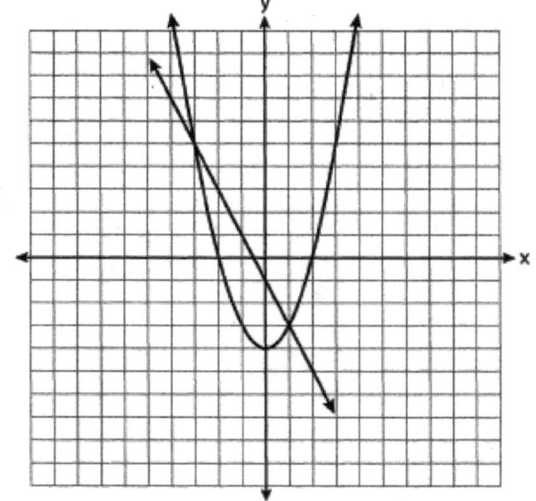

(1)	$(-3, 1)$	(3)	$(0, -1)$
(2)	$(-3, 5)$	(4)	$(0, -4)$

2. Which point is a solution of the system of equations shown on the graph?

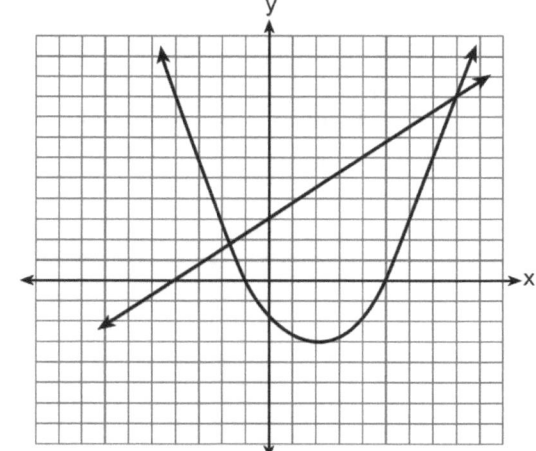

(1)	$(8, 9)$	(3)	$(0, 3)$
(2)	$(5, 0)$	(4)	$(2, -3)$

3. Solve the following system graphically:
$$y = x^2 + 4x - 2$$
$$y = 2x + 1$$

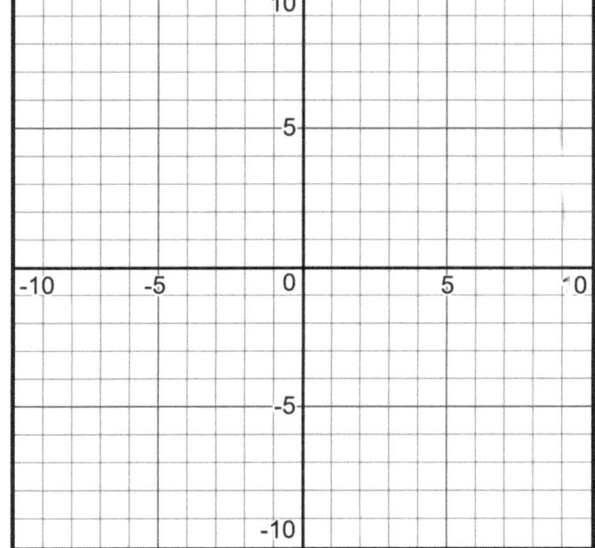

4. Solve the following system graphically:
$$y = x^2 + 2x - 1$$
$$y = 3x + 5$$

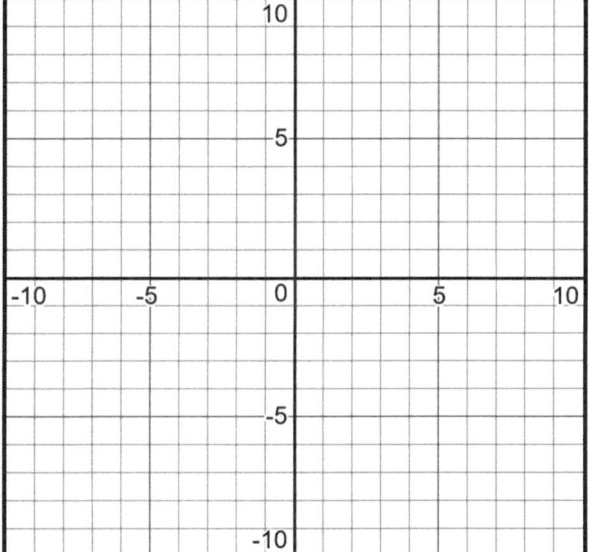

5. Solve the following system graphically:
$$y = x^2 + 4x + 1$$
$$y = 5x + 3$$

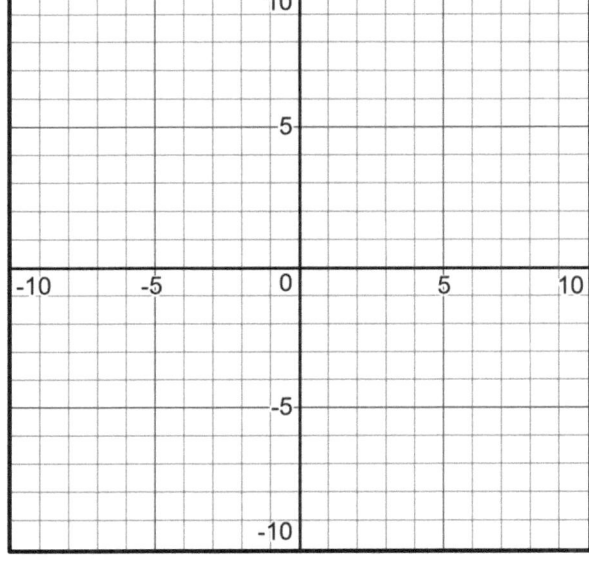

6. Solve the following system graphically:
$$y = x^2 + 4x - 1$$
$$y + 3 = x$$

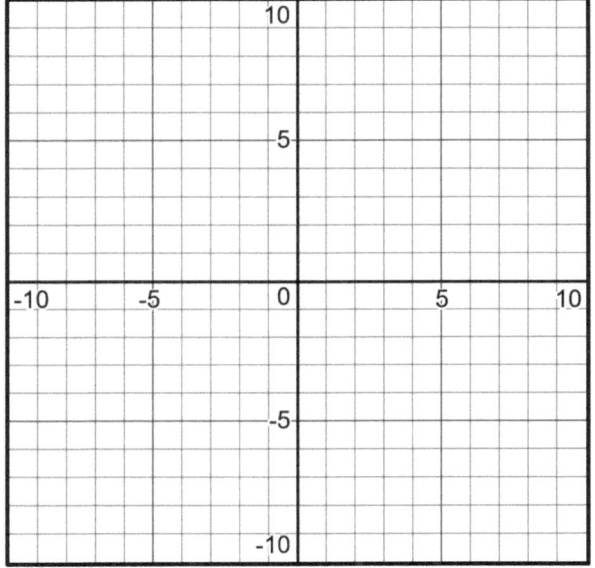

7. Solve the following system graphically:

$y = x^2 - 6x + 5$

$2x + y = 5$

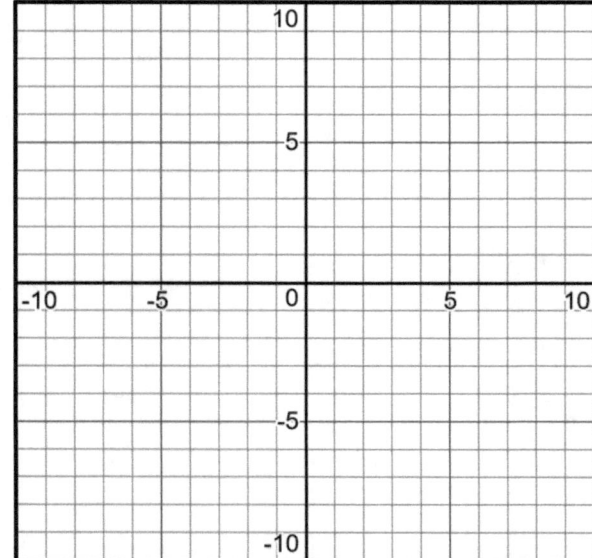

8. Solve the following system graphically:

$y = x^2 + 4x - 5$

$y = x - 1$

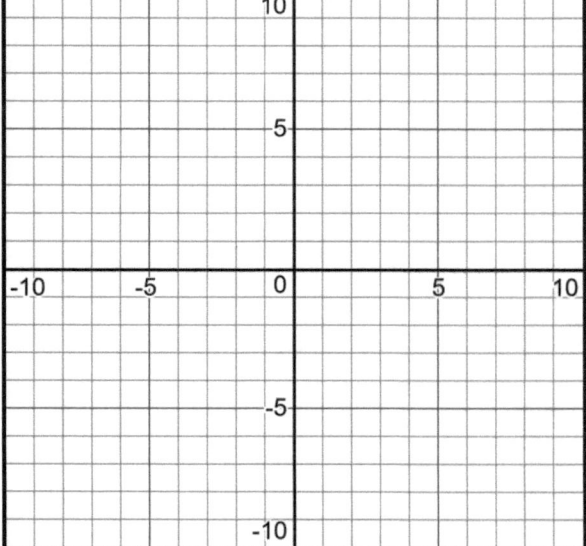

9. Solve the following system graphically:

$y = x^2 - 6x + 1$

$y + 2x = 6$

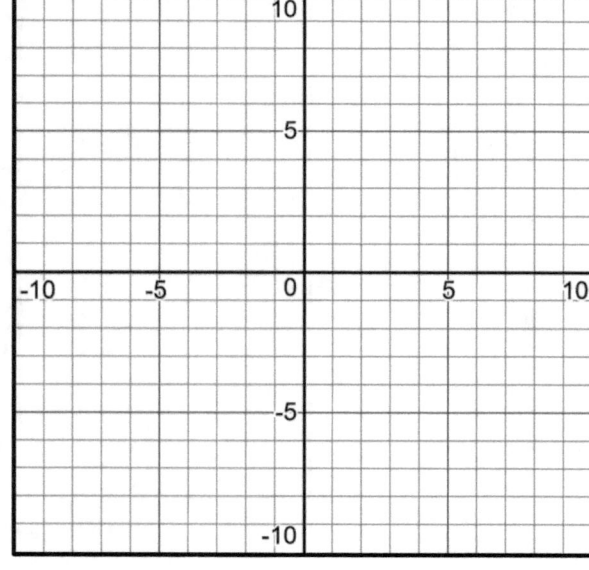

10. Solve the following system graphically:

$y = -x^2 + 6x - 3$

$x + y = 7$

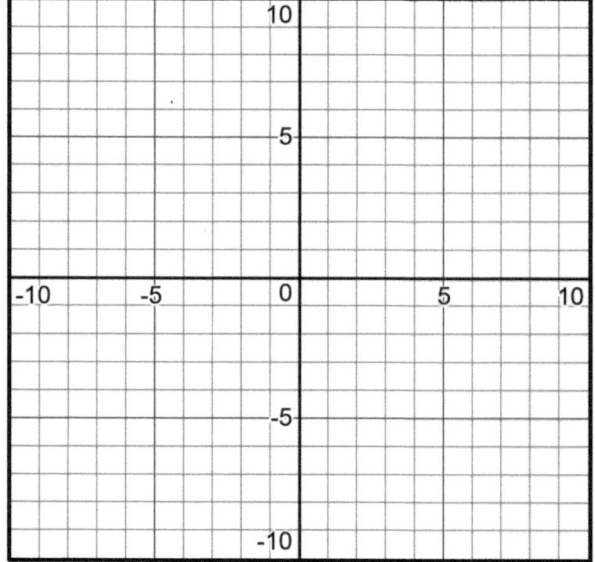

11. Solve the following system graphically:

$$y = -x^2 - 4x + 12$$
$$y = -2x + 4$$

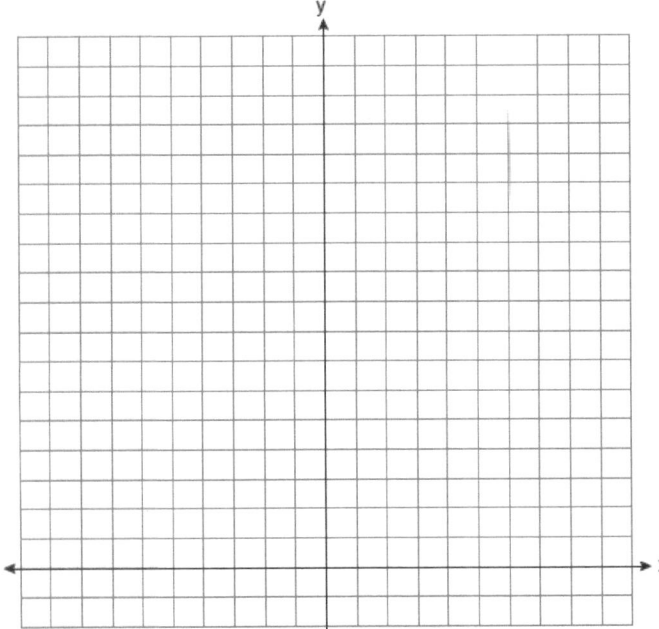

Regents Questions

MULTIPLE CHOICE

1. Nancy works for a company that offers two types of savings plans. Plan *A* is represented on the graph below.

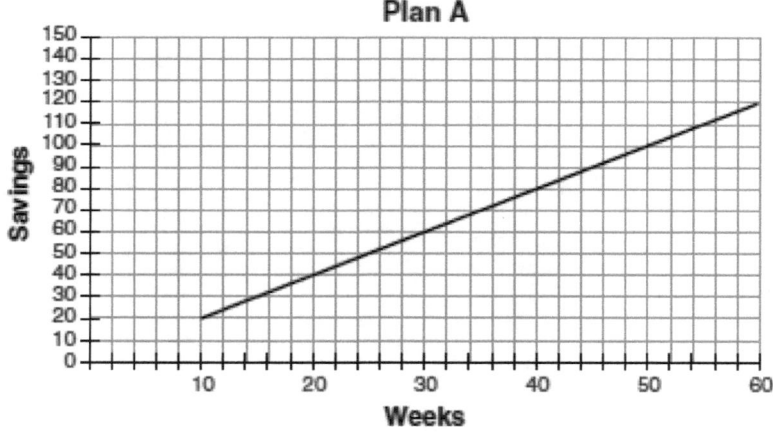

Plan *B* is represented by the function $f(x) = 0.01 + 0.05x^2$, where x is the number of weeks. Nancy wants to have the highest savings possible after a year. Nancy picks Plan *B*. Her decision is

    (1) correct, because Plan *B* is an exponential function and will increase at a faster rate

    (2) correct, because Plan *B* is a quadratic function and will increase at a faster rate

    (3) incorrect, because Plan *A* will have a higher value after 1 year

    (4) incorrect, because Plan *B* is a quadratic function and will increase at a slower rate

2. The graphs of $y = x^2 - 3$ and $y = 3x - 4$ intersect at approximately

    (1) $(0.38, -2.85)$, only   (3) $(0.38, -2.85)$ and $(2.62, 3.85)$

    (2) $(2.62, 3.85)$, only   (4) $(0.38, -2.85)$ and $(3.85, 2.62)$

3. A quadratic function and a linear function are graphed on the same set of axes. Which situation is *not* possible?

    (1) The graphs do not intersect.   (3) The graphs intersect in two points.

    (2) The graphs intersect in one point.   (4) The graphs intersect in three points.

CONSTRUCTED RESPONSE

4. A company is considering building a manufacturing plant. They determine the weekly production cost at site A to be $A(x) = 3x^2$ while the production cost at site B is $B(x) = 8x + 3$, where x represents the number of products, _in hundreds_, and $A(x)$ and $B(x)$ are the production costs, _in hundreds of dollars_.

Graph the production cost functions on the set of axes below and label them site A and site B.

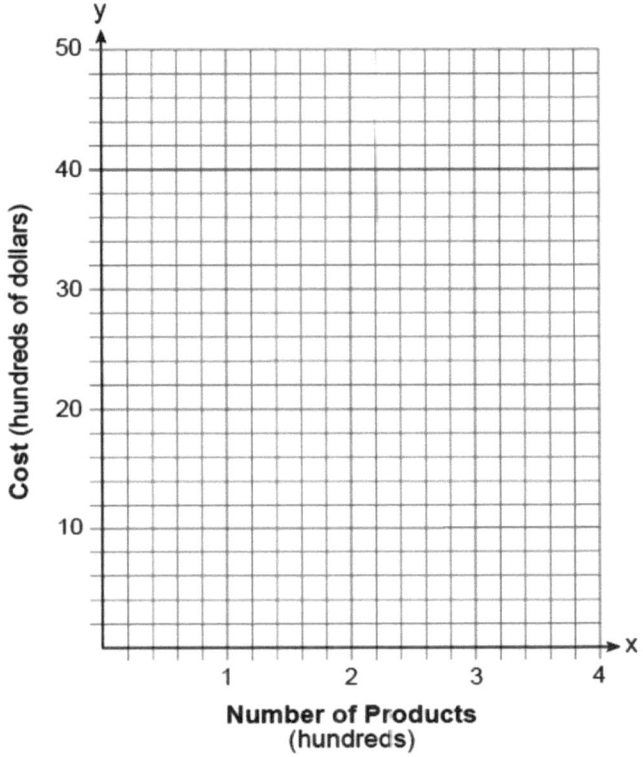

State the positive value(s) of x for which the production costs at the two sites are equal. Explain how you determined your answer.

If the company plans on manufacturing 200 products per week, which site should they use? Justify your answer.

5. Let $f(x) = -2x^2$ and $g(x) = 2x - 4$. On the set of axes below, draw the graphs of $y = f(x)$ and $y = g(x)$.

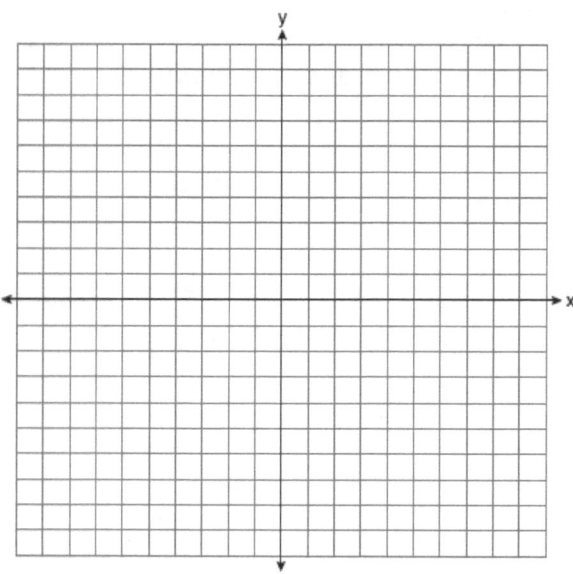

Using this graph, determine and state _all_ values of x for which $f(x) = g(x)$.

6. Graph $y = f(x)$ and $y = g(x)$ on the set of axes below.
$$f(x) = 2x^2 - 8x + 3$$
$$g(x) = -2x + 3$$

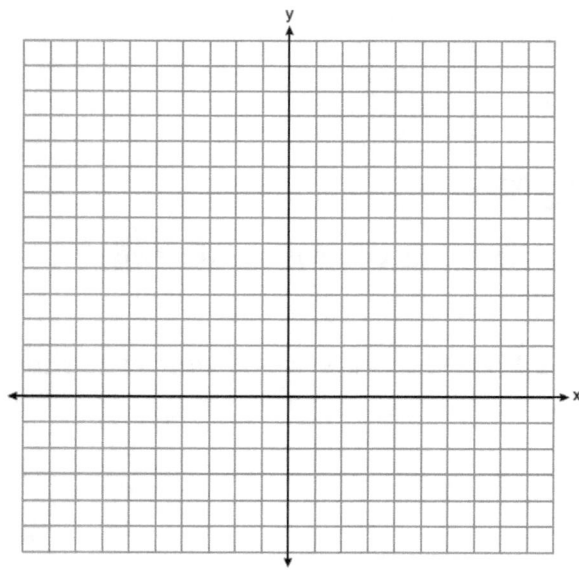

Determine and state all values of x for which $f(x) = g(x)$.

7.   Graph $f(x)$ and $g(x)$ on the set of axes below.

$$f(x) = x^2 - 4x + 3$$
$$g(x) = \frac{1}{2}x + 1$$

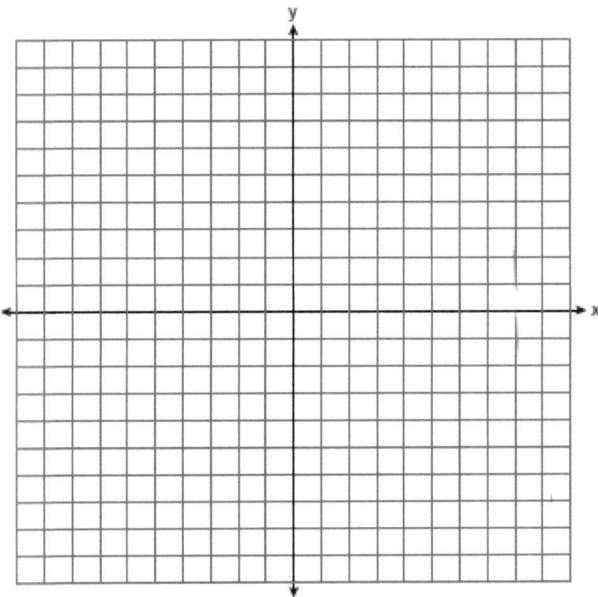

Based on your graph, state *one* value of x that satisfies $f(x) = g(x)$. Explain your reasoning.

8.   Graph $f(x) = |x| + 1$ and $g(x) = -x^2 - 6x + 1$ on the set of axes below.

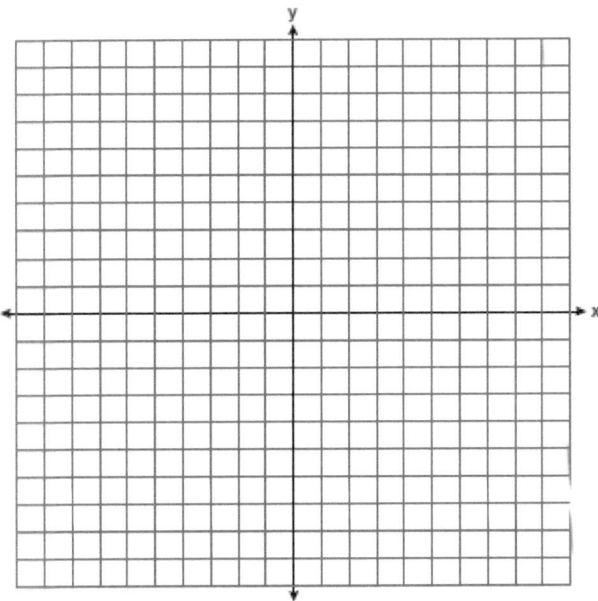

Based on your graph, determine all values of x for which $f(x) = g(x)$.

CHAPTER 15.   CUBIC AND RADICAL FUNCTIONS

15.1   Cubic Functions

A **cubic function** is one that is defined by a polynomial with a degree of three; that is, it includes an x^3 term. The graph of the simplest (*parent*) cubic function, $y = x^3$, is shown below.

x	y
-2	-8
-1	-1
0	0
1	1
2	8

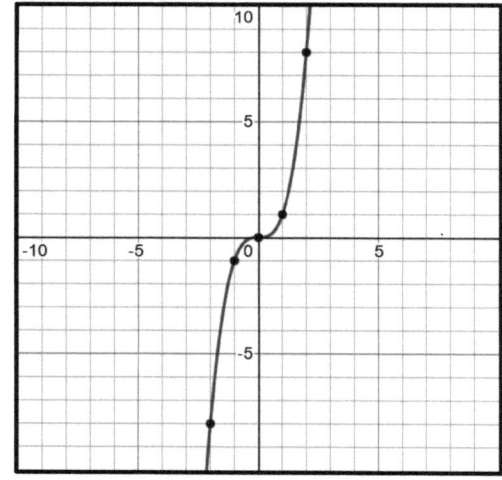

The general form of a cubic function is $f(x) = ax^3 + bx^2 + cx + d$.

It is sometimes possible to find the **roots** (zeros) of a cubic function by factoring.

Example: For $f(x) = x^3 - x^2 - 2x$, the roots can be found by factoring.

$$x^3 - x^2 - 2x = 0$$
$$x(x^2 - x - 2) = 0$$
$$x(x + 1)(x - 2) = 0$$

So, the roots are $\{-1, 0, 2\}$.

As with quadratic functions, the real roots of the function are the **x-intercepts** on the graph.

Example: The function graphed below has roots of -4, -1, and 2.

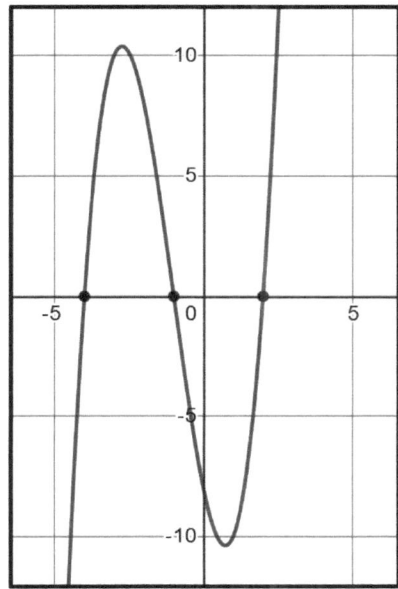

This means it has **factors** of $x + 4$, $x + 1$, and $x - 2$:

$$x = -4 \qquad x = -1 \qquad x = 2$$
$$x + 4 = 0 \qquad x + 1 = 0 \qquad x - 2 = 0$$
$$(x + 4)(x + 1)(x - 2) = 0$$

Multiplying the factors, we get the equation of this cubic function:
$$f(x) = x^3 + 3x^2 - 6x - 8.$$

However, this is not the only cubic function with these factors. The function to the right also has the same roots but an additional constant factor of $\frac{1}{4}$, which "flattens" the graph.

The equation of this cubic function is:
$$f(x) = \frac{x^3 + 3x^2 - 6x - 8}{4}$$

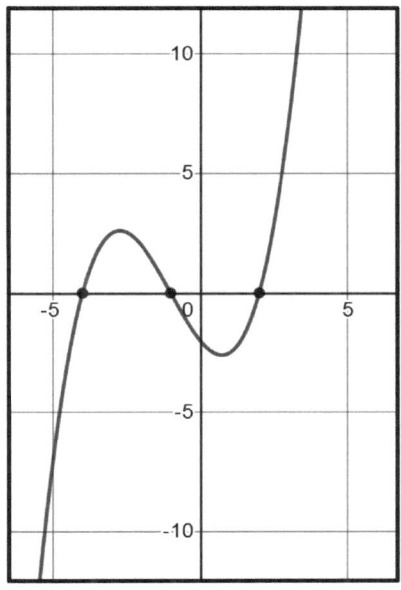

MODEL PROBLEM

Which of the following could represent the graph of $f(x)$ shown below?

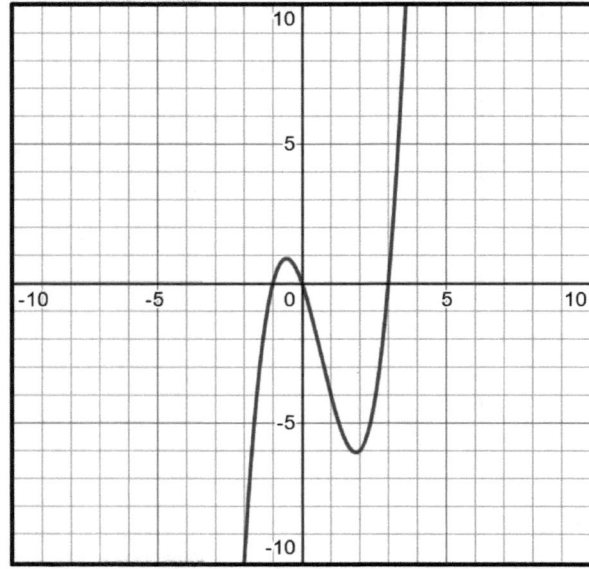

(1) $f(x) = (x - 1)(x + 3)$ (3) $f(x) = (x + 1)(x - 3)$

(2) $f(x) = x(x - 1)(x + 3)$ (4) $f(x) = x(x + 1)(x - 3)$

Solution: (4)

Explanation of steps:
   (A)   The x-intercepts *[−1, 0, and 3]* represent the real roots.
   (B)   If a is a root, then $(x - a)$ is a factor. *[If 3 is a root, then $(x - 3)$ is a factor; if 0 is a root, x is a factor; and if −1 is a root, $(x + 1)$ is a factor.]*

PRACTICE PROBLEMS

1. Graph $y = x^3 + 3$ by completing the table for integers $-2 \le x \le 2$.
    Based on the graph, how many real roots does the function appear to have?

x	y
-2	
-1	
0	
1	
2	

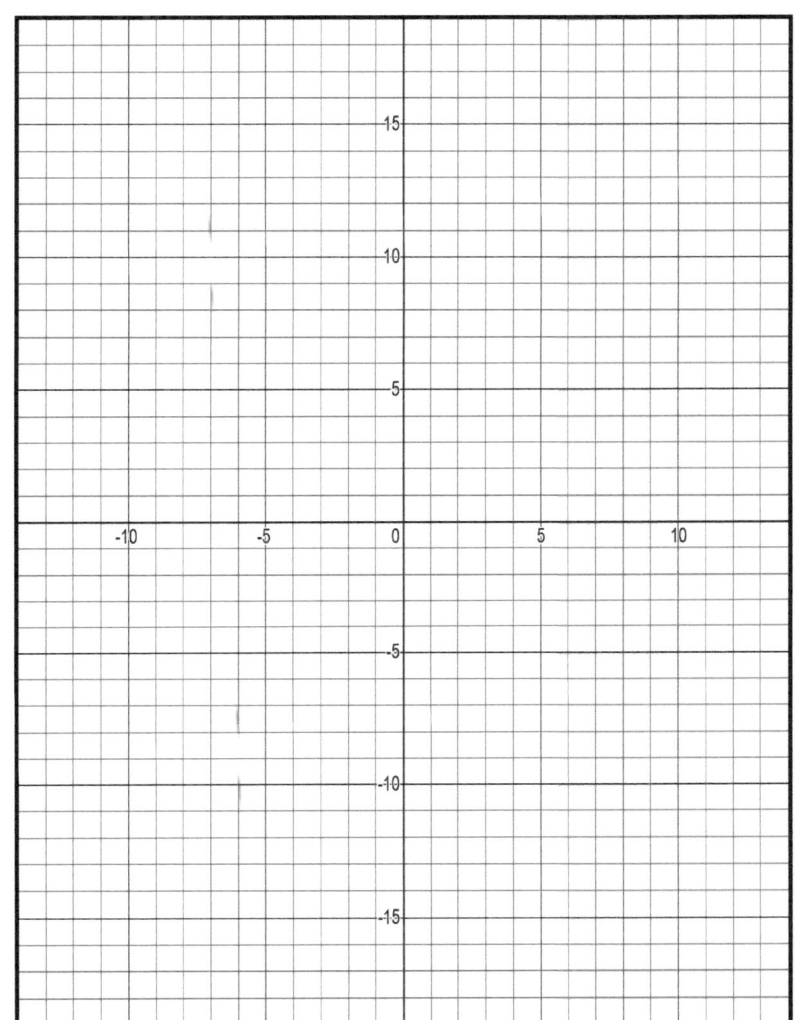

2. Graph $y = x^3 - 9x + 5$ by completing the table for integers $-3 \leq x \leq 3$.
Based on the graph, how many real roots does the function appear to have?

x	y
−3	
−2	
−1	
0	
1	
2	
3	

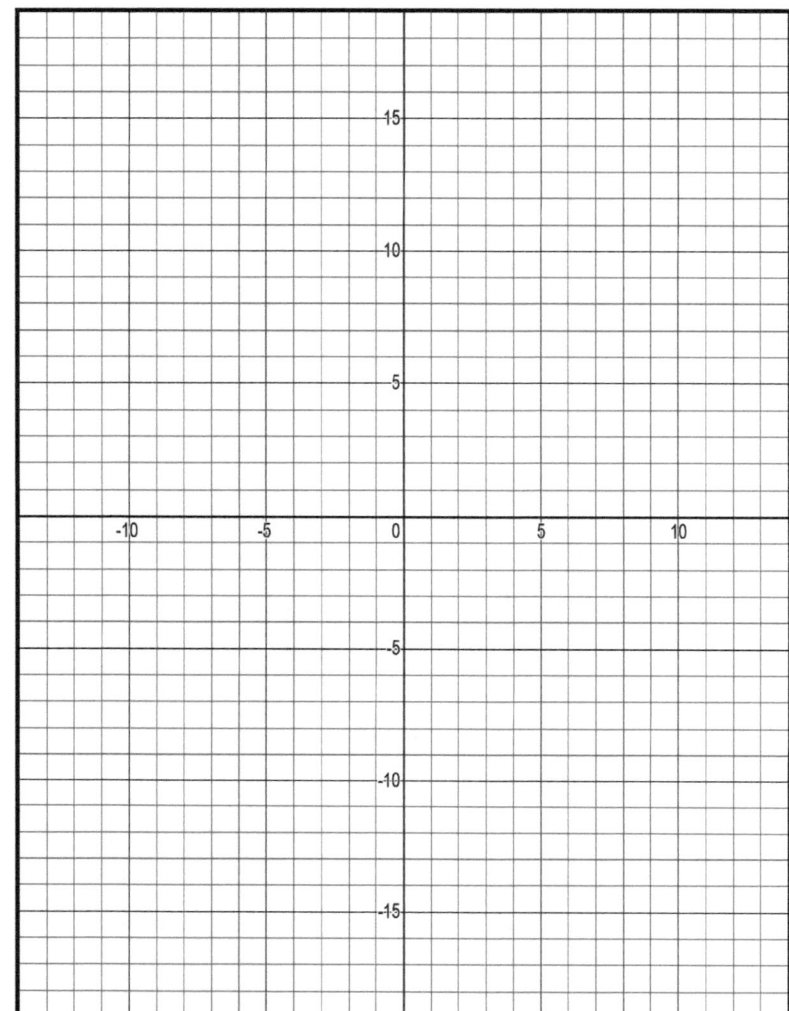

Regents Questions

MULTIPLE CHOICE

1. A polynomial function contains the factors x, $x - 2$, and $x + 5$. Which graph(s) below could represent the graph of this function?

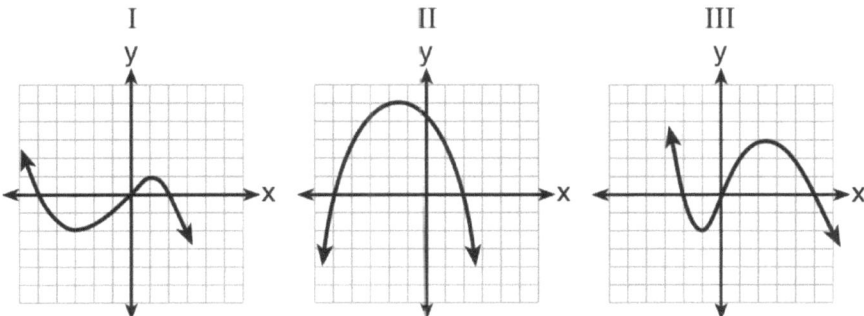

    I   II   III

    (1)   I, only   (3)   I and III

    (2)   II, only   (4)   I, II, and III

2. Which equation(s) represent the graph below?

    I $y = (x + 2)(x^2 - 4x - 12)$

    II $y = (x - 3)(x^2 + x - 2)$

    III $y = (x - 1)(x^2 - 5x - 6)$

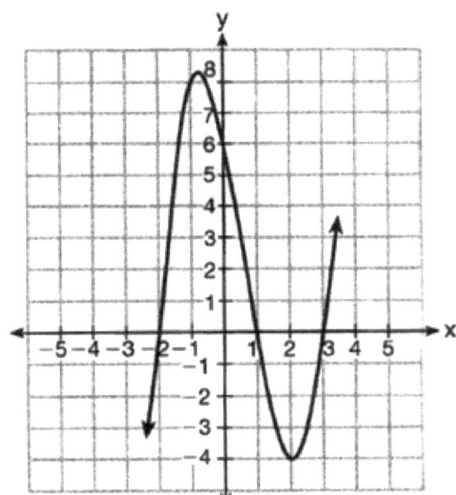

    (1)   I, only   (3)   I and II

    (2)   II, only   (4)   II and III

3.   The graph of $f(x)$ is shown below.

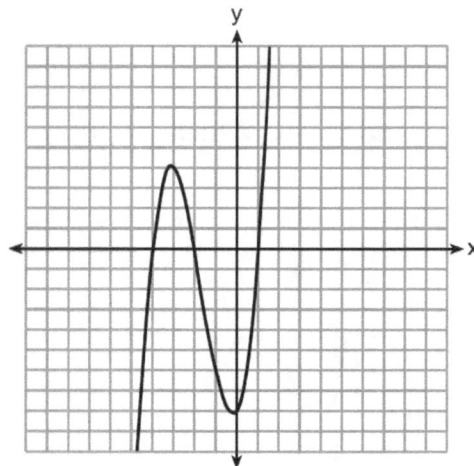

      Which function could represent the graph of $f(x)$?
      (1) $f(x) = (x + 2)(x^2 + 3x - 4)$ (3) $f(x) = (x + 2)(x^2 + 3x + 4)$
      (2) $f(x) = (x - 2)(x^2 + 3x - 4)$ (4) $f(x) = (x - 2)(x^2 + 3x + 4)$

4.   Based on the graph below, which expression is a possible factorization of $p(x)$?

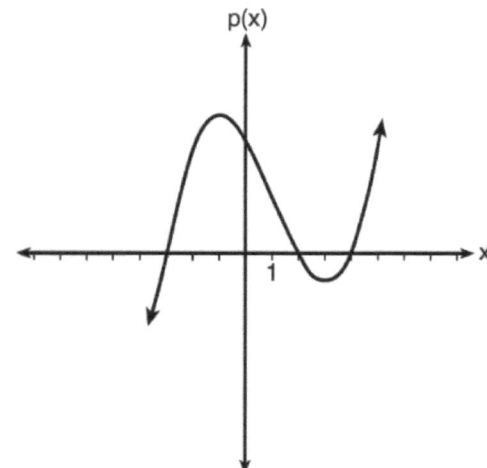

      (1) $(x + 3)(x - 2)(x - 4)$ (3) $(x + 3)(x - 5)(x - 2)(x - 4)$
      (2) $(x - 3)(x + 2)(x + 4)$ (4) $(x - 3)(x + 5)(x + 2)(x + 4)$

5.   Which polynomial function has zeros at -3, 0, and 4?
      (1) $f(x) = (x + 3)(x^2 + 4)$ (3) $f(x) = x(x + 3)(x - 4)$
      (2) $f(x) = (x^2 - 3)(x - 4)$ (4) $f(x) = x(x - 3)(x + 4)$

6.   Wenona sketched the polynomial $P(x)$ as shown on the axes below.

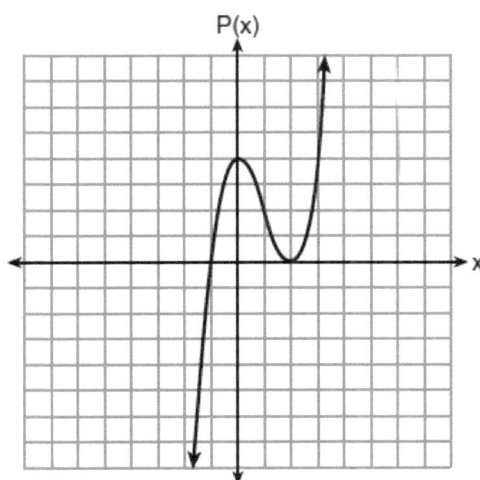

Which equation could represent $P(x)$?

(1) $P(x) = (x + 1)(x - 2)^2$ (3) $P(x) = (x + 1)(x - 2)$

(2) $P(x) = (x - 1)(x + 2)^2$ (4) $P(x) = (x - 1)(x + 2)$

7.   The zeros of the function $f(x) = 2x^3 + 12x - 10x^2$ are

(1) $\{2, 3\}$ (3) $\{0, 2, 3\}$

(2) $\{-1, 6\}$ (4) $\{0, -1, 6\}$

8.   Which ordered pair would *not* be a solution to $y = x^3 - x$?

(1) $(-4, -60)$ (3) $(-2, -6)$

(2) $(-3, -24)$ (4) $(-1, -2)$

9.   A cubic function is graphed on the set of axes below.

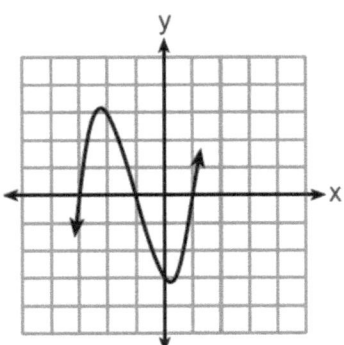

Which function could represent this graph?

(1) $f(x) = (x - 3)(x - 1)(x + 1)$ (3) $h(x) = (x - 3)(x - 1)(x + 3)$

(2) $g(x) = (x + 3)(x + 1)(x - 1)$ (4) $k(x) = (x + 3)(x + 1)(x - 1)$

513

10. A polynomial function is graphed below.

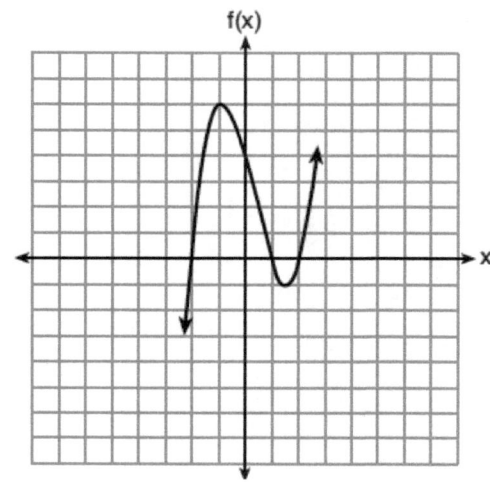

Which function could represent this graph?
- (1) $f(x) = (x + 1)(x^2 + 2)$
- (2) $f(x) = (x - 1)(x^2 - 2)$
- (3) $f(x) = (x - 1)(x^2 - 4)$
- (4) $f(x) = (x + 1)(x^2 + 4)$

11. The zeros of the function $f(x) = x^3 - 9x^2$ are
- (1) 9, only
- (2) 0 and 9
- (3) 0 and 3, only
- (4) $-3, 0,$ and 3

12. Which sketch represents the polynomial function $f(x) = x(x + 6)(x + 3)$?

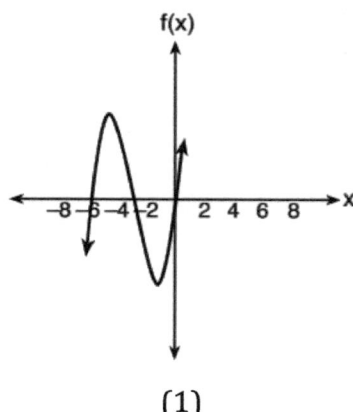

(1)

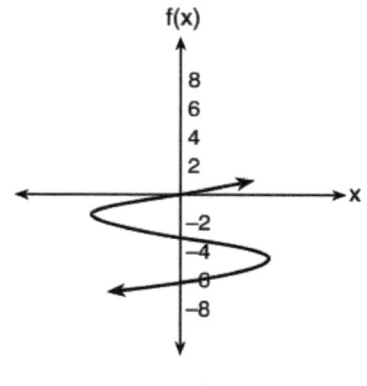

(3)

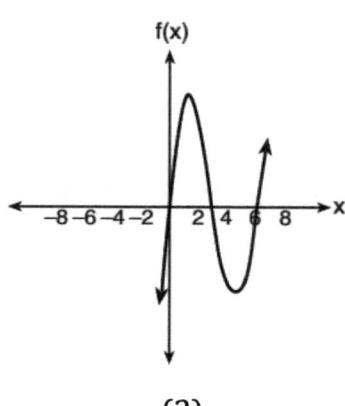

(2)

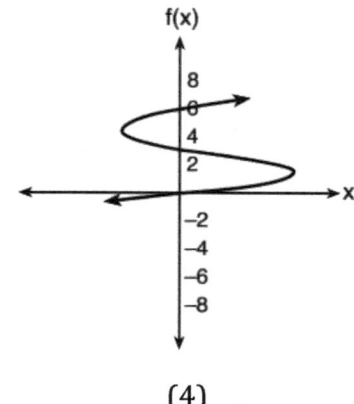

(4)

13.   If the zeros of the function $g(x)$ are $\{-3, 0, 4\}$, which function could represent $g(x)$?

    (1) $g(x) = (x + 3)(x - 4)$ (3) $g(x) = x(x + 3)(x - 4)$

    (2) $g(x) = (x - 3)(x + 4)$ (4) $g(x) = x(x - 3)(x + 4)$

14.   A function is graphed below.

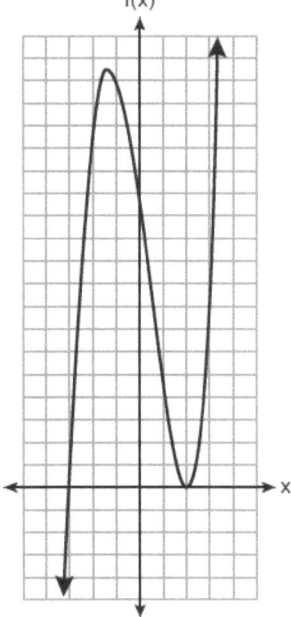

A possible equation for this function is

    (1) $f(x) = (x + 2)(x - 3)$ (3) $f(x) = (x - 2)^2(x + 3)$

    (2) $f(x) = (x - 2)(x + 3)$ (4) $f(x) = (x - 2)(x + 3)(x - 12)$

15.   Which point is a solution to $y = x^3 - 2x$?

    (1) $(-3, -21)$ (3) $(1,1)$

    (2) $(-2,10)$ (4) $(4,2)$

16.   What are the zeros of $m(x) = x(x^2 - 16)$?

    (1) -4 and 4, only   (3) $-4, 0$, and 4

    (2) -8 and 8, only   (4) $-8, 0$, and 8

17.   The zeros of a polynomial function are -2, 4, and 0. What are all the factors of this function?

    (1) $(x + 2)$ and $(x - 4)$ (3) $x, (x + 2)$, and $(x - 4)$

    (2) $(x - 2)$ and $(x + 4)$ (4) $x, (x - 2)$, and $(x + 4)$

CONSTRUCTED RESPONSE

18.   Explain how to determine the zeros of $f(x) = (x + 3)(x - 1)(x - 8)$. State the zeros of the function.

19.   Determine algebraically the zeros of $f(x) = 3x^3 + 21x^2 + 36x$.

20. The function $f(x)$ is graphed on the set of axes below.

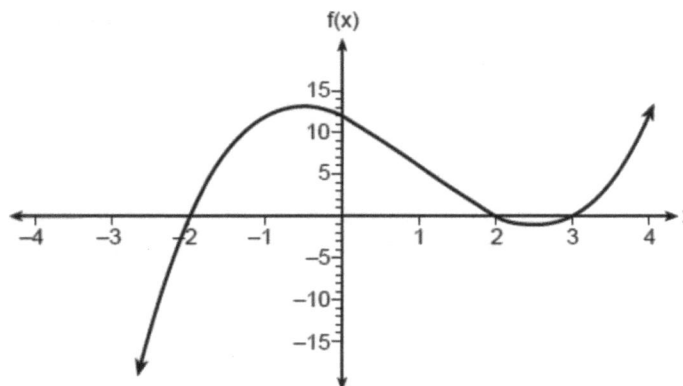

State the zeros of $f(x)$. Explain your reasoning.

15.2 Square Root Functions

A **square root function** is a function that has the independent variable, x, in the radicand.

Examples: $y = \sqrt{x} + 2$ or $y = \sqrt{x - 3}$

The simplest (*parent*) square root function, $y = \sqrt{x}$, can be graphed as follows:

x	y
0	0
1	1
4	2
9	3
16	4

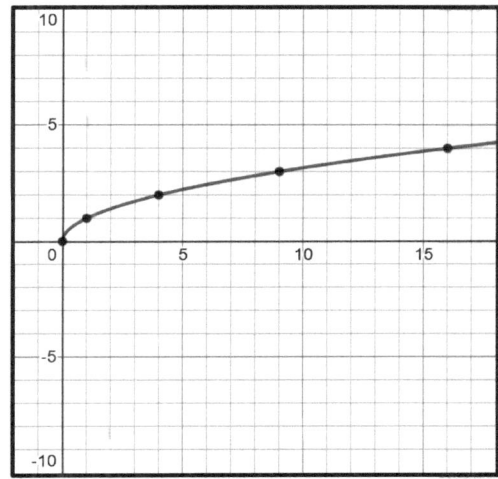

Note that for real numbers, the radicand cannot be negative. This is because there are no real numbers that have negative squares: the square of a positive number is positive and the square of a negative number is positive. So, the domain of a square root function is restricted to only values of x for which the radicand is at least zero.

CALCULATOR TIP

To enter a square root on the calculator, press [2nd][√], followed by the radicand.

On the TI-84 in MathPrint mode, exit the radicand by pressing the [▶] key.

On the TI-83, end the radicand by pressing [)] to close the parentheses.

MODEL PROBLEM

Which function is represented by the graph below?

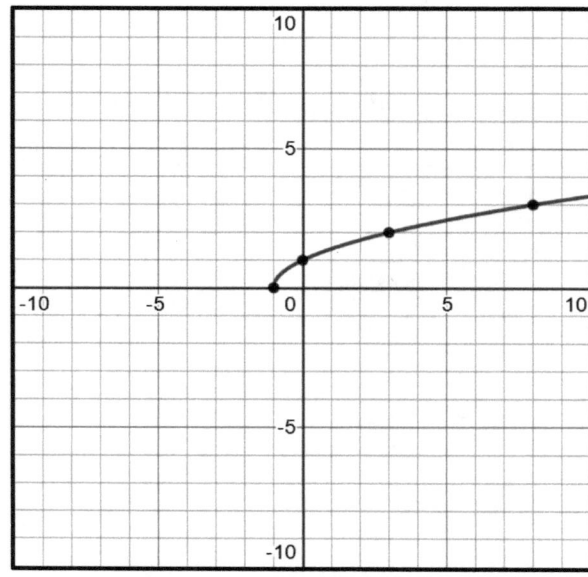

(1) $f(x) = \sqrt{x} + 1$ (3) $f(x) = \sqrt{x+1}$

(2) $f(x) = \sqrt{x} - 1$ (4) $f(x) = \sqrt{x-1}$

Solution: (3)

Explanation of steps:
Test the given points to determine which function they satisfy.
[The points $(-1,0), (0,1), (3,2),$ *and* $(8,3)$ *satisfy* $y = \sqrt{x+1}$ *by substituting each pair of x and y values into the equation. For example,* $(8,3)$ *satisfies the equation because* $3 = \sqrt{8+1}.]$

[Note: In Chapter 16, we will also see that the graph could be recognized as a translation of the parent function $y = \sqrt{x}$ *by one unit to the left.]*

PRACTICE PROBLEMS

1. Which of the following graphs represent the function $y = \sqrt{x} - 1$?

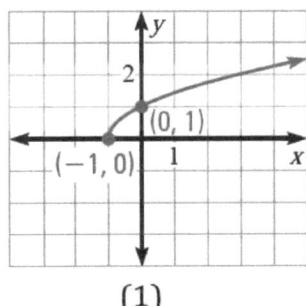 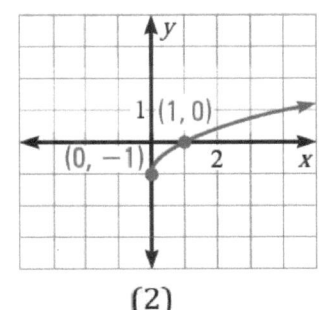 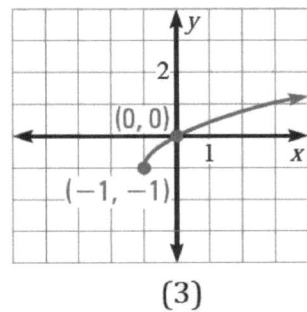

      (1)   (2)   (3)

2. Williams High School relocated into a larger building and immediately began recruiting students to increase enrollment. The number of students enrolled, y, is modeled by the function $y = 90\sqrt{3x} + 400$, where x is the number of months the new school building has been open.
   a) Construct a table of values.
   b) Sketch the function on the grid.
   c) Find the number of students enrolled exactly 3 months after the building opened.
   d) After how many months will 940 students be enrolled?

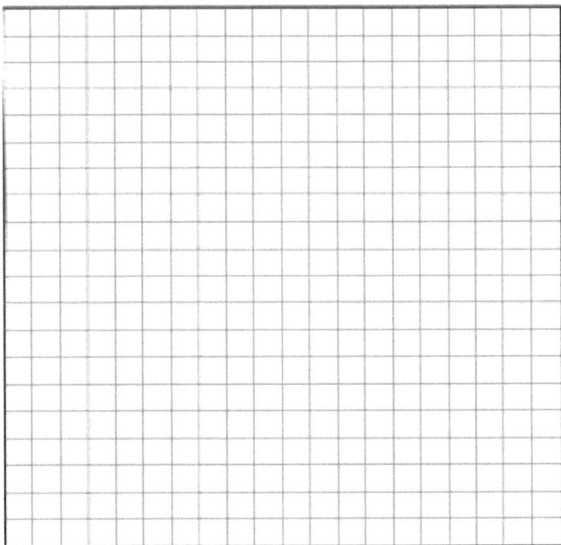

Regents Questions

MULTIPLE CHOICE

1.　Which graph represents $y = \sqrt{x - 2}$?

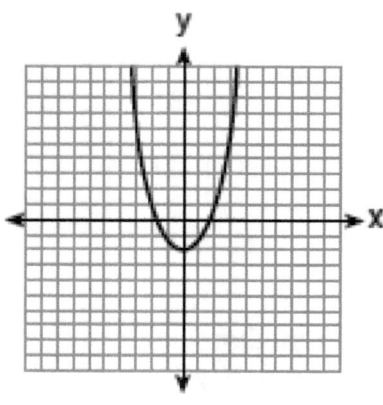

(1)

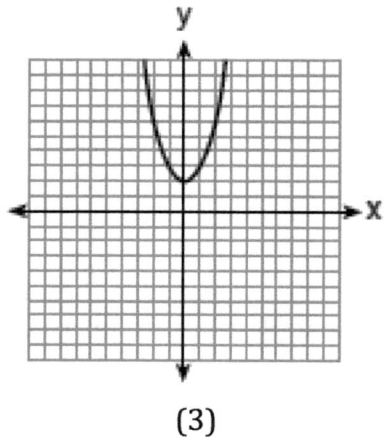

(3)

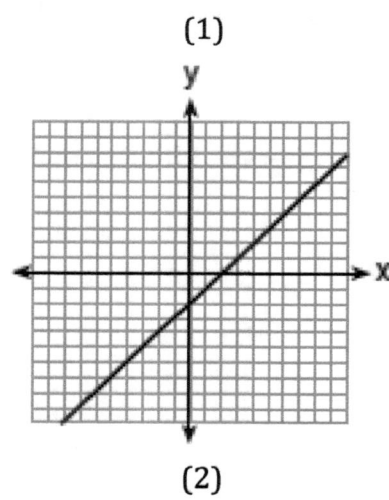

(2)

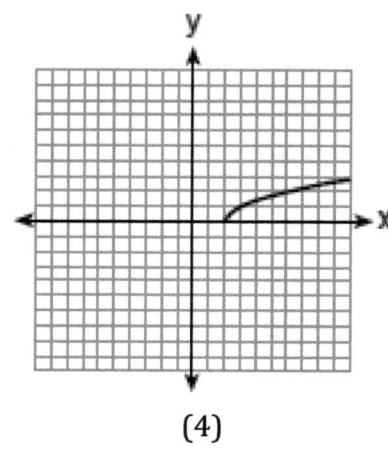

(4)

2.   Which function has the *smallest y*-intercept value?

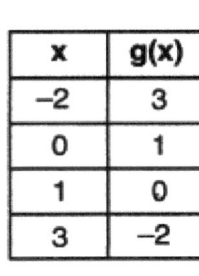

x	g(x)
−2	3
0	1
1	0
3	−2

(1)

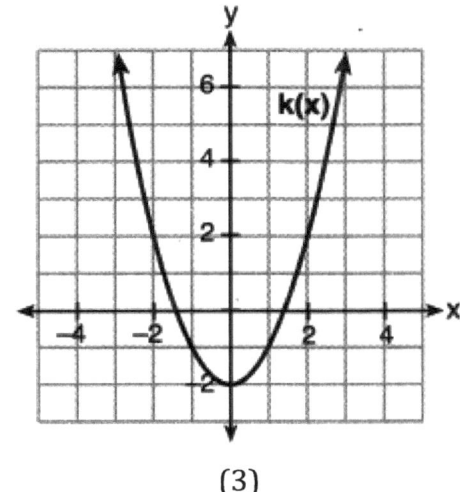

(3)

$$h(x) = \sqrt{x} - 3$$

(2)

$$f(x) = x^2 + 2x - 1$$

(4)

CONSTRUCTED RESPONSE

3.   Draw the graph of $y = \sqrt{x} - 1$ on the set of axes below.

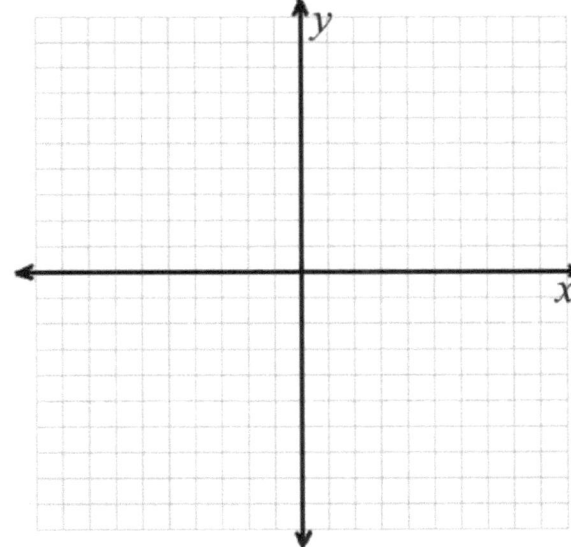

4.   Graph the function $y = -\sqrt{x + 3}$ on the set of axes below.

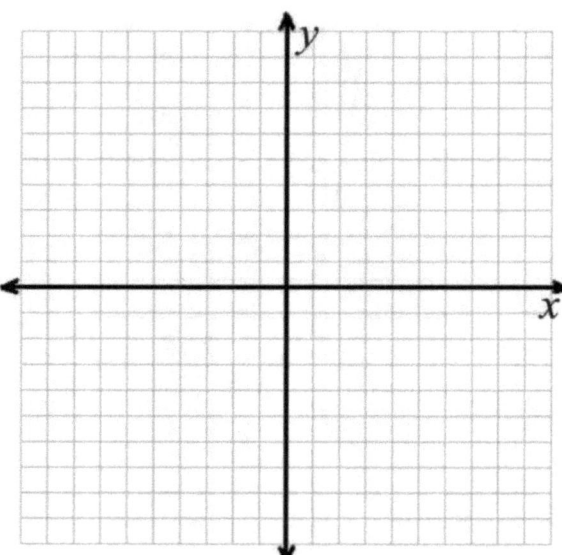

5.   Graph $f(x) = \sqrt{x + 2}$ over the domain $-2 \leq x \leq 7$.

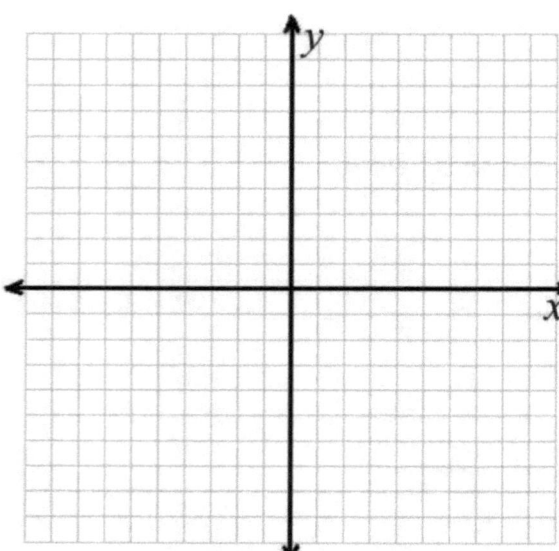

6.   Graph $f(x) = -\sqrt{x} + 1$ on the set of axes below.

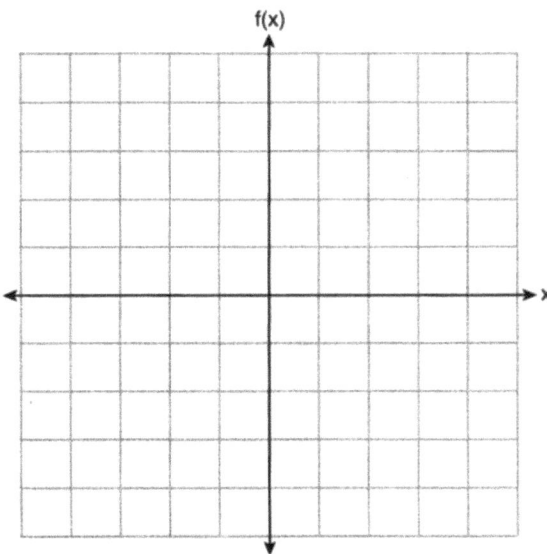

7.   Graph the function $g(x) = \sqrt{x + 3}$ on the set of axes below.

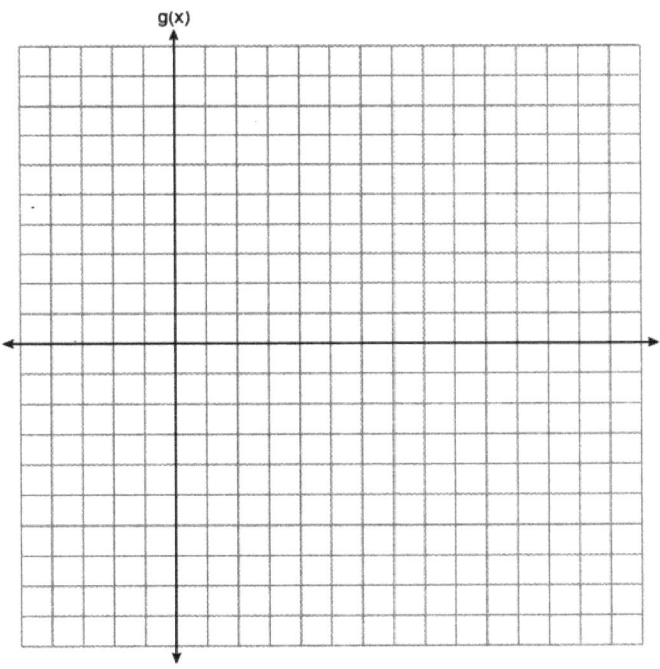

CHAPTER 16. TRANSFORMATIONS OF FUNCTIONS

16.1 Translations

The absolute value function $y = |x|$ and the quadratic function $y = x^2$ are graphed below.

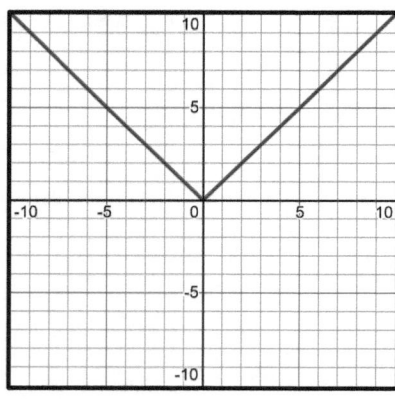 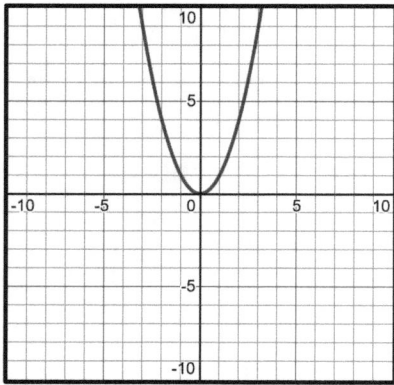

The functions above are called parent functions. A **parent function** is the simplest function of a family of functions that preserves the definition or shape of the entire family.

Example: $y = x^2$ is the parent function for the entire family of quadratic functions that have the general form $y = ax^2 + bx + c$ where $a \neq 0$, since these are all quadratic by definition and are all shaped as parabolas.

A function $f(x)$ may be:
- **translated** by adding/subtracting a constant, as in $f(x) + k$ or $f(x + k)$
- **reflected** by negation, as in $-f(x)$ or $f(-x)$
- **stretched** by multiplying/dividing by a constant, as in $k \cdot f(x)$ or $f(kx)$

A summary of the types of transformations we will cover in this chapter is given below.

$f(x) + k$	$(x, y) \rightarrow (x, y + k)$	vertically shifts the graph up $(k > 0)$ or down $(k < 0)$
$f(x + k)$	$(x, y) \rightarrow (x - k, y)$	horizontally shifts the graph left $(k > 0)$ or right $(k < 0)$
$-f(x)$	$(x, y) \rightarrow (x, -y)$	reflects the graph over the x-axis
$f(-x)$	$(x, y) \rightarrow (-x, y)$	reflects the graph over the y-axis
$k \cdot f(x)$	$(x, y) \rightarrow (x, ky)$	vertically stretches by a factor of k
$f(kx)$	$(x, y) \rightarrow (kx, y)$	horizontally stretches by a factor of $\frac{1}{k}$

In this section, we will look at translations. Reflections and stretches will be covered later in this chapter.

In a **translation**, the graph of a function is shifted up, down, left, or right.

Vertical shifts

If we add a constant k to a function's output expression, as in $f(x) + k$, its graph will shift **up** (if $k > o$) or **down** (if $k < 0$). In a vertical shift, each point (x, y) maps to $(x, y + k)$.

Example: The parent functions $y = |x|$ and $y = x^2$ may be translated as follows: $y = |x| + k$ or $y = x^2 + k$. Adding k to the function's output vertically shifts the function *up* (for a <u>positive k</u>) or *down* (for a <u>negative k</u>).

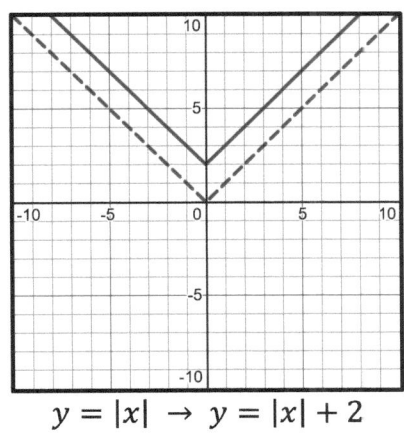 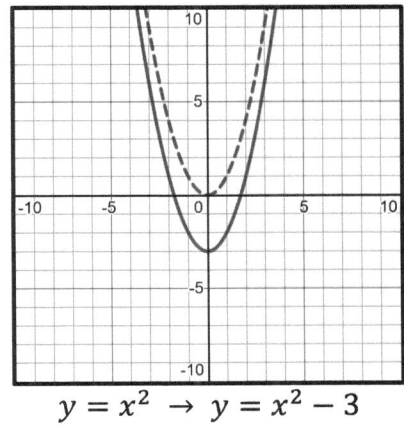

$$y = |x| \;\rightarrow\; y = |x| + 2$$ $$y = x^2 \;\rightarrow\; y = x^2 - 3$$

Horizontal shifts

If we add a constant k to a function's input expression, as in $f(x + k)$, its graph will shift **left** (if $k > o$) or **right** (if $k < 0$). In a horizontal shift, each point (x, y) maps to $(x - k, y)$.

Example: The parent functions $y = |x|$ and $y = x^2$ may be translated as follows: $y = |x + k|$ or $y = (x + k)^2$. Adding k to the function's input horizontally shifts the function **left** (for a <u>positive k</u>) or **right** (for a <u>negative k</u>).

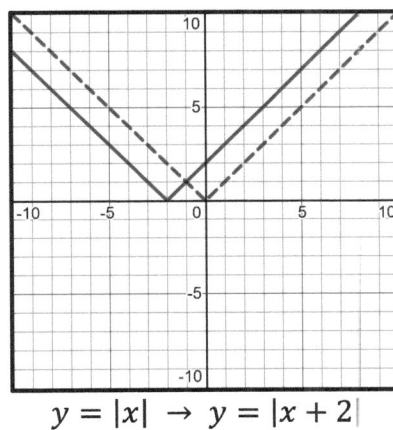 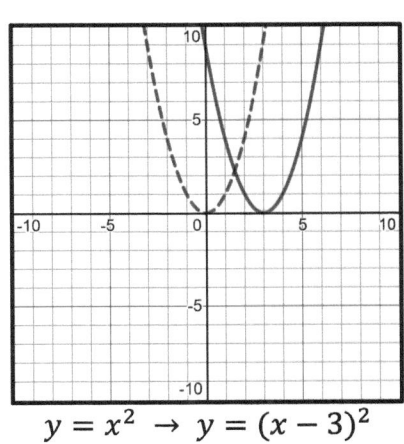

$$y = |x| \;\rightarrow\; y = |x + 2|$$ $$y = x^2 \;\rightarrow\; y = (x - 3)^2$$

MODEL PROBLEM

Below is a graph of the parent function $y = \sqrt{x}$. Draw a graph of $y = \sqrt{x-1} + 2$.

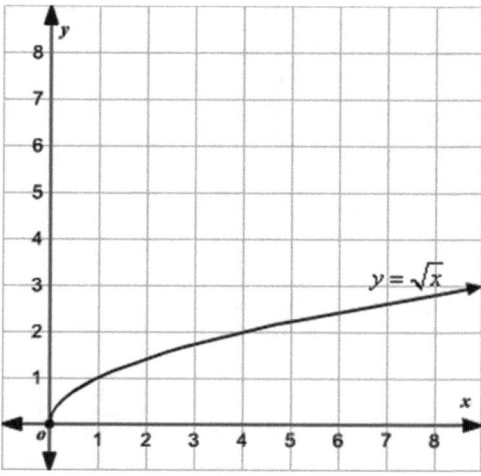

Solution:

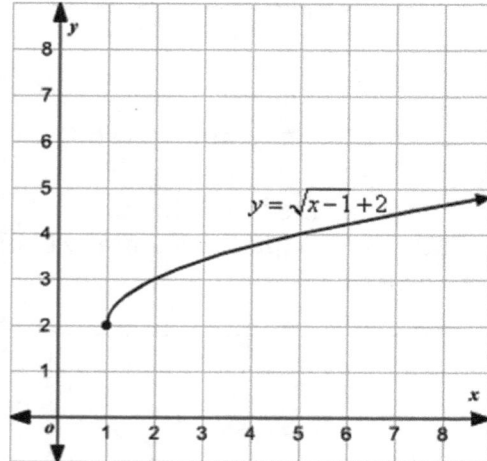

Explanation of steps:

The graph of $f(x) = \sqrt{x-a} + b$ can be obtained by translating the graph of the parent function $f(x) = \sqrt{x}$ by a units to the right and b units up. *[For $y = \sqrt{x-1} + 2$, this means 1 unit to the right and 2 units up.]* Note that this also changes the *domain* and *range*, from $x \geq 0$ and $y \geq 0$, to $x \geq a$ and $y \geq b$, respectively.

PRACTICE PROBLEMS

1. What is the equation of a graph translated up 3 units, if the original graph is $y = x^2$? (1) $y = (x - 3)^2$ (2) $y = (x + 3)^2$ (3) $y = x^2 - 3$ (4) $y = x^2 + 3$	2. Describe how the graph $g(x) = (x + 2)^2$ is related to the graph of $f(x) = x^2$. (1) a translation 2 units down of $f(x)$ (2) a translation 2 units left of $f(x)$ (3) a translation 2 units up of $f(x)$ (4) a translation 2 units right of $f(x)$
3. If the original graph is $y = |x|$, then the graph of $y = |x - 2|$ has been (1) shifted up 2 units (2) shifted down 2 units (3) shifted right 2 units (4) shifted left 2 units	4. If the graph of $y = |x| + 2$ is shifted 3 units down, what is the equation of the new graph?
5. Write the equation of the graph $y = |x|$ after it is shifted 4 units to the left.	6. Write the equation of the graph $y = x^2$ after it is shifted 5 units up and 2 units to the right.

Regents Questions

MULTIPLE CHOICE

1. Given the graph of the line represented by the equation $f(x) = -2x + b$, if b is increased by 4 units, the graph of the new line would be shifted 4 units
    - (1) right
    - (2) up
    - (3) left
    - (4) down

2. The graph of $y = f(x)$ is shown below.

    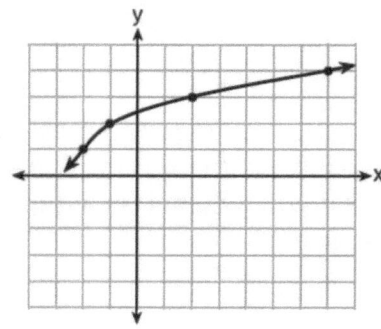

    What is the graph of $y = f(x + 1) - 2$?

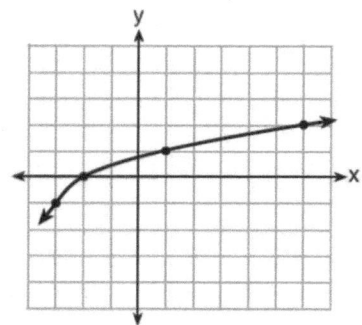

(1)

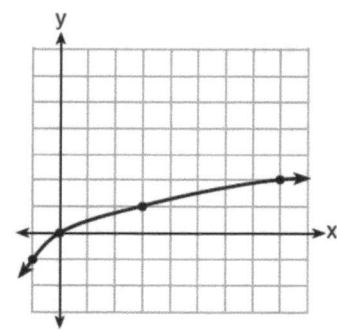

(3)

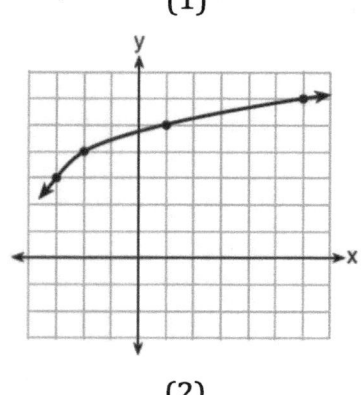

(2)

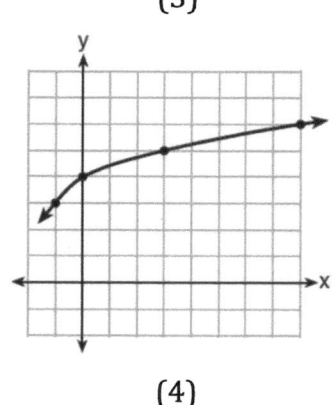

(4)

3. If the original function $f(x) = 2x^2 - 1$ is shifted to the left 3 units to make the function $g(x)$, which expression would represent $g(x)$?

    (1) $2(x - 3)^2 - 1$ (3) $2x^2 + 2$

    (2) $2(x + 3)^2 - 1$ (4) $2x^2 - 4$

4. Compared to the graph of $f(x) = x^2$, the graph of $g(x) = (x - 2)^2 + 3$ is the result of translating $f(x)$

    (1)   2 units up and 3 units right   (3)   2 units right and 3 units up

    (2)   2 units down and 3 units up   (4)   2 units left and 3 units right

5. Given the parent function $f(x) = x^3$, the function $g(x) = (x - 1)^3 - 2$ is the result of a shift of

    (1)   1 unit left and 2 units down   (3)   1 unit right and 2 units down

    (2)   1 unit left and 2 units up   (4)   1 unit right and 2 units up

6. Given: $f(x) = (x - 2)^2 + 4$
                 $g(x) = (x - 5)^2 + 4$

When compared to the graph of $f(x)$, the graph of $g(x)$ is

    (1)   shifted 3 units to the left   (3)   shifted 5 units to the left

    (2)   shifted 3 units to the right   (4)   shifted 5 units to the right

7. Josh graphed the function $f(x) = -3(x - 1)^2 + 2$. He then graphed the function $g(x) = -3(x - 1)^2 - 5$ on the same coordinate plane. The vertex of $g(x)$ is

    (1)   7 units below the vertex of $f(x)$ (3)   7 units to the right of the vertex of $f(x)$

    (2)   7 units above the vertex of $f(x)$ (4)   7 units to the left of the vertex of $f(x)$

8. The functions $f(x) = x^2 - 6x + 9$ and $g(x) = f(x) + k$ are graphed below.

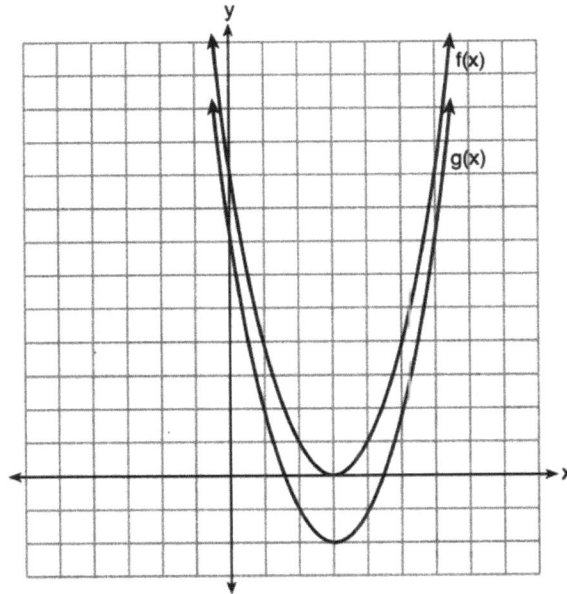

Which value of k would result in the graph of $g(x)$?

    (1)   0   (3) -3

    (2)   2   (4) -2

9.   If the parent function of $f(x)$ is $p(x) = x^2$, then the graph of the function
      $f(x) = (x - k)^2 + 5$, where $k > 0$, would be a shift of
      (1) k units to the left and a move of 5 units up
      (2) k units to the left and a move of 5 units down
      (3) k units to the right and a move of 5 units up
      (4) k units to the right and a move of 5 units down

10.   The graph of $y = f(x)$ is shown below.

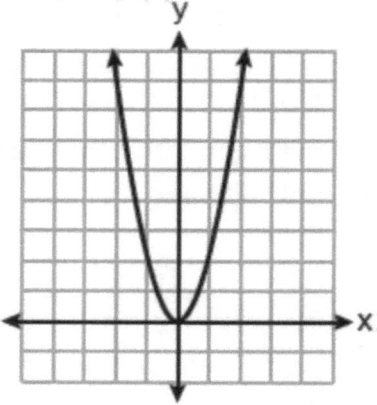

      Which graph represents $y = f(x - 2) + 1$?

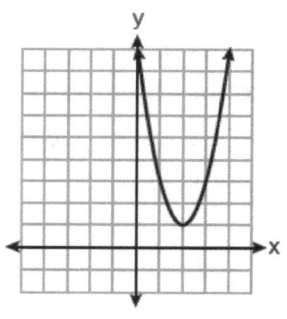

(1)

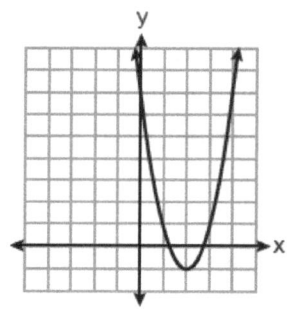

(3)

(2)

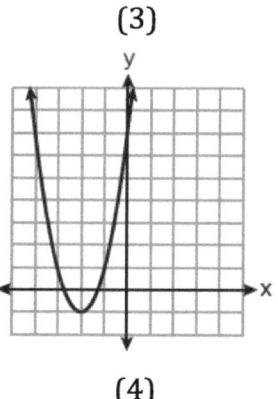

(4)

11.   If $f(x) = x^2$, which function is the result of shifting $f(x)$ 3 units left and 2 units down?
      (1) $g(x) = (x + 2)^2 - 3$ (3) $j(x) = (x + 3)^2 - 2$
      (2) $h(x) = (x - 2)^2 + 3$ (4) $k(x) = (x - 3)^2 + 2$

CONSTRUCTED RESPONSE

12. The vertex of the parabola represented by $f(x) = x^2 - 4x + 3$ has coordinates $(2, -1)$. Find the coordinates of the vertex of the parabola defined by $g(x) = f(x - 2)$. Explain how you arrived at your answer.

13. On the axes below, graph $f(x) = |3x|$.

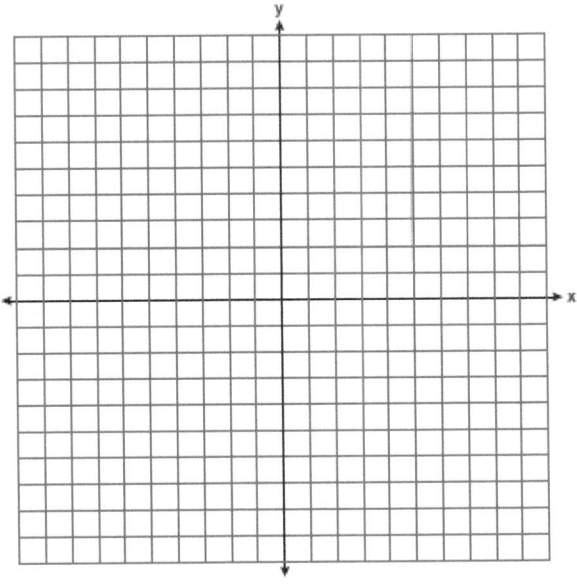

If $g(x) = f(x) - 2$, how is the graph of $f(x)$ translated to form the graph of $g(x)$?

If $h(x) = f(x - 4)$, how is the graph of $f(x)$ translated to form the graph of $h(x)$?

14. Graph the function $y = |x - 3|$ on the set of axes below.

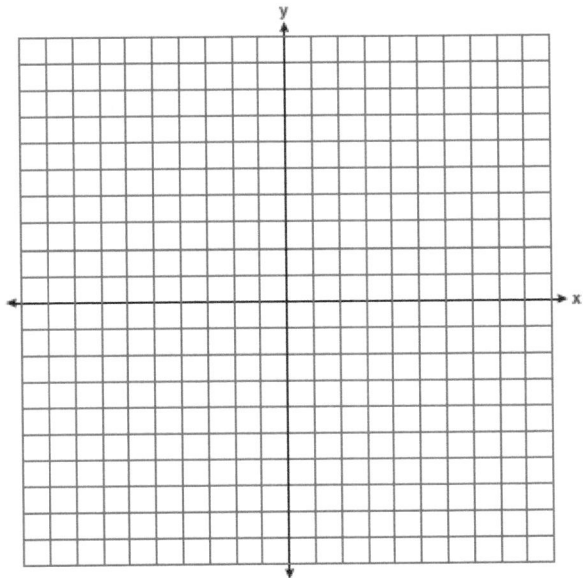

Explain how the graph of $y = |x - 3|$ has changed from the related graph $y = |x|$.

15. In the diagram below, $f(x) = x^3 + 2x^2$ is graphed. Also graphed is $g(x)$, the result of a translation of $f(x)$.

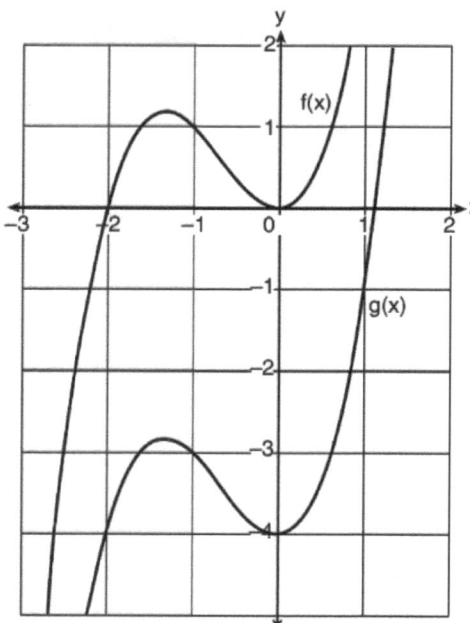

Determine an equation of $g(x)$. Explain your reasoning.

16. Richard is asked to transform the graph of $b(x)$ below.

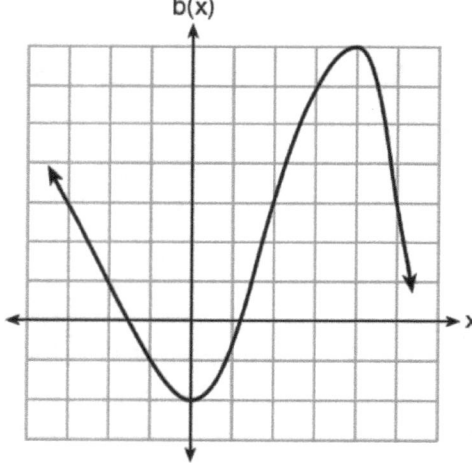

The graph of $b(x)$ is transformed using the equation $h(x) = b(x - 2) - 3$. Describe how the graph of $b(x)$ changed to form the graph of $h(x)$.

17. Describe the effect that each transformation below has on the function $f(x) = |x|$, where $a > 0$.

$$g(x) = |x - a|$$
$$h(x) = |x| - a$$

18. The graph of the function $p(x)$ is represented below. On the same set of axes, sketch the function $p(x + 2)$.

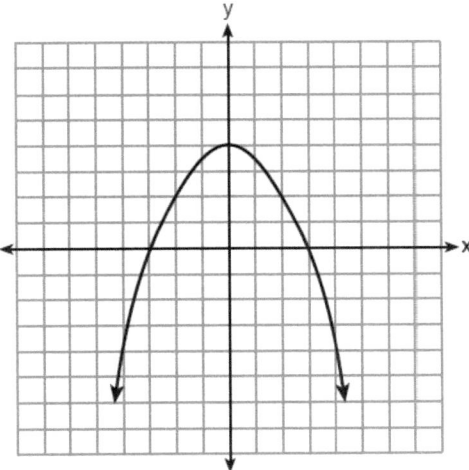

19. Describe the transformations performed on the graph of $f(x) = x^2$ to obtain the graph of $g(x)$ when $g(x) = (x - 3)^2 - 4$.

20. A student is given the functions $f(x) = (x + 1)^2$ and $g(x) = (x + 3)^2$. Describe the transformation that maps $f(x)$ onto $g(x)$.

16.2   Reflections

In a **reflection**, the graph of a function is reflected over the x-axis or the y-axis.

Vertical reflections

If we negate a function's output expression, as in $-f(x)$, its graph will reflect vertically over the x-axis. In a reflection over the x-axis, each point (x, y) maps to $(x, -y)$.

Example: The parent functions $y = |x|$ and $y = x^2$ may be reflected as follows:
$y = -|x|$ or $y = -x^2$ would flip the graph upside down over the x-axis.

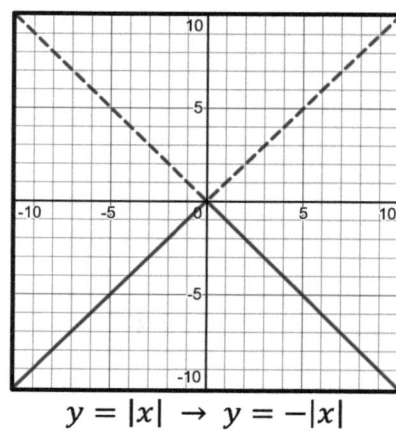 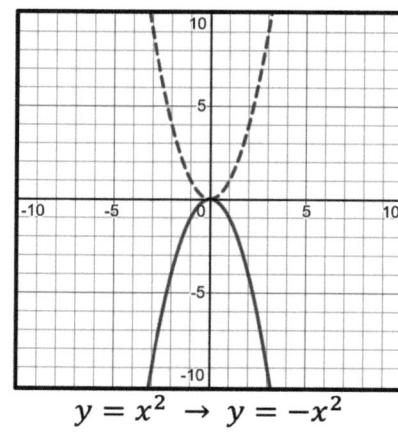

$$y = |x| \ \rightarrow \ y = -|x| \qquad\qquad y = x^2 \ \rightarrow \ y = -x^2$$

Horizontal reflections

If we negate a function's input expression, as in $f(-x)$, its graph will reflect horizontally over the y-axis. In a reflection over the y-axis, each point (x, y) maps to $(-x, y)$.

Example: Since the parent functions $y = |x|$ and $y = x^2$ are symmetric over the y-axis,
reflecting them by $y = |-x|$ or $y = (-x)^2$ merely maps the graphs over themselves.

For a **sequence of transformations**, perform the transformations in this order:
1. vertical reflections
2. translations
3. horizontal reflections

Example: Here is a graph of $y = (x + 2)^2 - 3$. It is the graph of $y = x^2$ after it is translated vertically left by 2 and horizontally down by 3.

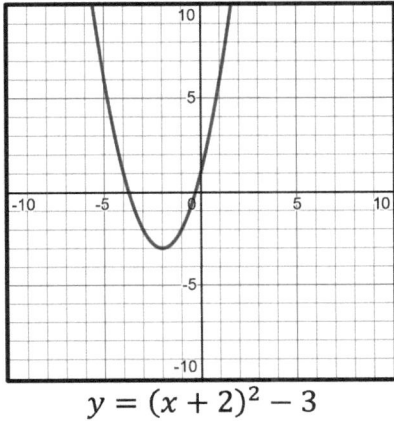

$$y = (x + 2)^2 - 3$$

Here are the graphs of $y = -(x + 2)^2 - 3$, which reflects $y = x^2$ over the x-axis (vertically) *before* the translations, and of $y = (-x + 2)^2 - 3$, which reflects $y = x^2$ over the y-axis (horizontally) *after* the translations.

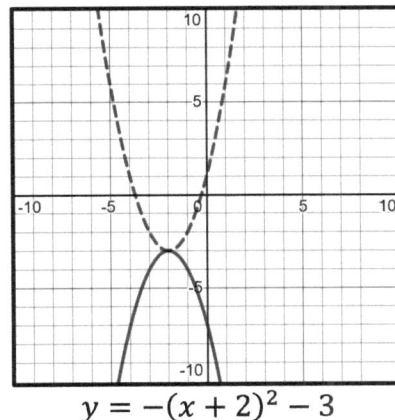

$$y = -(x + 2)^2 - 3$$

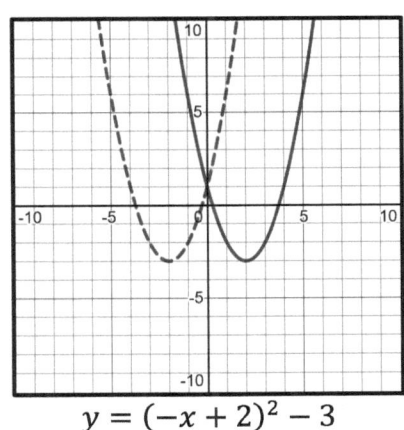

$$y = (-x + 2)^2 - 3$$

MODEL PROBLEM

Write the equation of the graph $y = \sqrt{x}$ after it is reflected over the x-axis and shifted 5 units to the left.

Solution:

$$y = -\sqrt{x + 5}$$

Explanation of steps:
To reflect vertically, negate the <u>output</u> *[place the negation sign <u>outside</u> the radical]*. To shift horizontally, add to (or subtract from) the <u>input</u> *[add 5 <u>inside</u> the radical]*.

PRACTICE PROBLEMS

1. If the original graph is $y = \sqrt{x}$ and the transformed graph is $y = -\sqrt{x} - 1$, then the graph has been

    (1)   reflected over the x-axis and shifted down 1

    (2)   reflected over the x-axis and shifted right 1

    (3)   reflected over the y-axis and shifted down 1

    (4)   reflected over the y-axis and shifted right 1

2. The graph of $y = |x + 2|$ is shown below.

    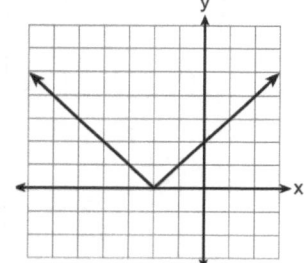

    Which graph represents $y = -|x + 2|$?

    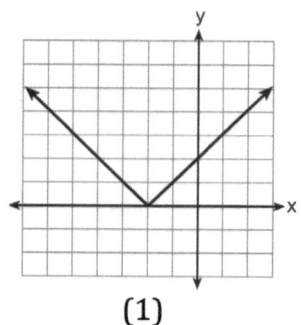   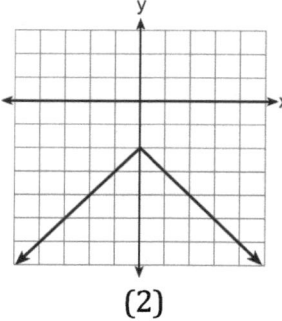   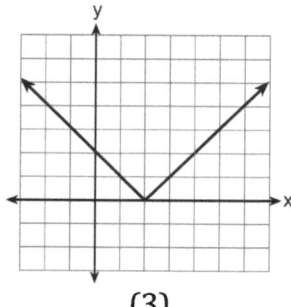   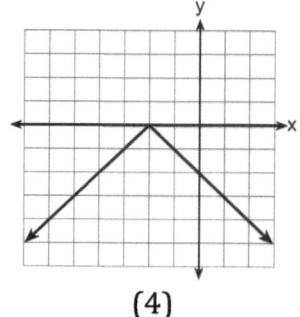

    (1)   (2)   (3)   (4)

3. The diagram below shows the graph of $y = -x^2 - c$.

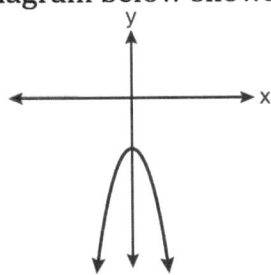

Which graph represents $y = x^2 - c$?

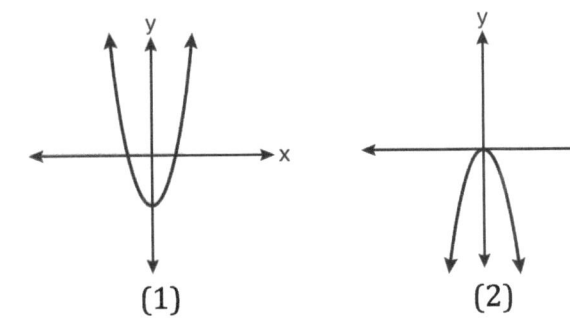 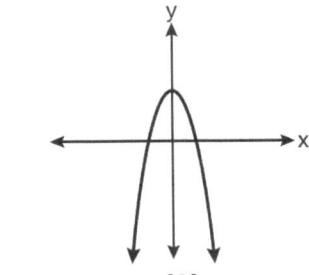 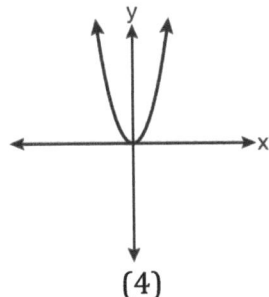

(1)   (2)   (3)   (4)

4. Write the equation of the graph $y = x^2$ after it is reflected over the x-axis and shifted 1 unit to the right.

5. Write the equation of the graph $y = \sqrt{x}$ after it is shifted 3 units left and then reflected over the y-axis.

6. Write the equation of the graph $y = x^3$ after it is shifted 2 units down and 1 unit to the right and then reflected over the y-axis.

Regents Questions

There are no Regents exam questions on this topic.

16.3   Stretches

The graph of a function may be stretched or compressed. When a graph is vertically stretched, it becomes **more narrow**. When a graph is vertically compressed, it becomes **wider**.

[Note: For the purposes of this course, the topic of horizontal stretches is intentionally omitted.]

If we multiply or divide a function's output expression by a constant k, as in $k \cdot f(x)$, its graph will **vertically stretch** (if $k > 1$) or **vertically compress** (if $0 < k < 1$) by a factor of k. In either case, each point (x, y) maps to (x, ky).

Example: The parent functions $y = |x|$ and $y = x^2$ may be stretched or compressed as follows: $y = k|x|$ or $y = kx^2$ would vertically stretch the graph by k and make it more narrow (for $k > 1$) or vertically compress the graph and make it flatter and wider (for $0 < k < 1$).

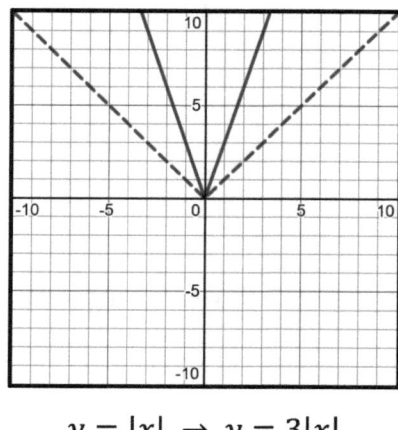

$$y = |x| \ \rightarrow \ y = 3|x|$$

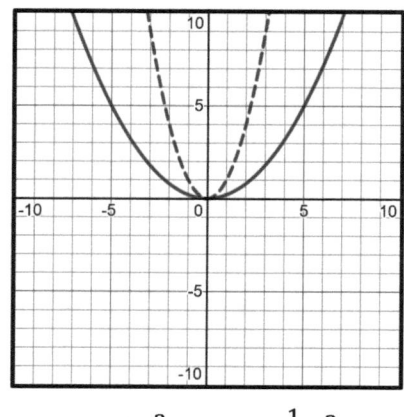

$$y = x^2 \ \rightarrow \ y = \tfrac{1}{5}x^2$$

For a **sequence of transformations**, perform the transformations in this order:
1. <u>vertical</u> reflections and/or stretches or compressions
2. translations
3. <u>horizontal</u> reflections

Order of transformations:
If $g(x) = a \cdot f(dx + c) + b$, then $g(x)$ is a transformation of $f(x)$ *in this order*:
1. vertically stretched or compressed by a (and also reflected over the x-axis if a is negative)
2. vertically translated by b and/or horizontally translated by c
3. horizontally reflected over the y-axis if $d = -1$

Example: If $f(x) = \sqrt{x}$ and $g(x) = 2\sqrt{x + 1} + 4$, then g is a transformation of f which is first vertically stretched by a factor of $a = 2$, then translated up by $b = 4$ units and left by $c = 1$ units.

These rules apply only if $f(x)$ is written using a single x-variable term.
Example: We could *not* apply these transformation rules to get $g(x) = 3x^2 + 2x - 1$.

MODEL PROBLEM

The graph of the function $y = \sqrt{x}$ is vertically stretched by a factor of 4 and flipped (reflected) over the x-axis. What is the equation of the resulting graph?

Solution:

$$
\begin{array}{ccc}
\text{(A)} & \text{(B)} & \text{(C)} \\
y = \sqrt{x} & \rightarrow \quad y = 4\sqrt{x} & \rightarrow \quad y = -4\sqrt{x}
\end{array}
$$

Explanation of steps:
(A) Start with the original equation.
(B) Perform the first transformation *[stretch by multiplying by the given factor]*.
(C) Perform the next transformation on the result *[negating will flip it over the x-axis]*.

PRACTICE PROBLEMS

1. The graph of $y = x^2$ is shown below.

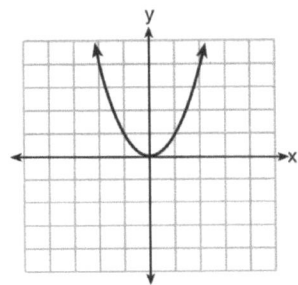

Which graph represents $y = 2x^2$?

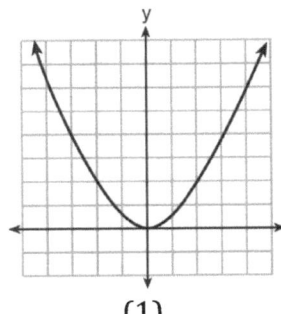 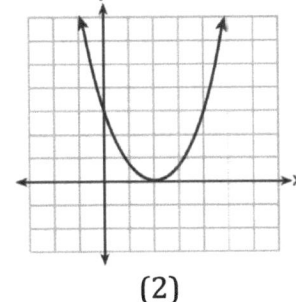 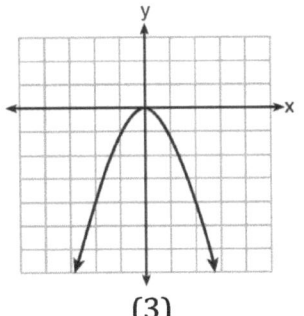 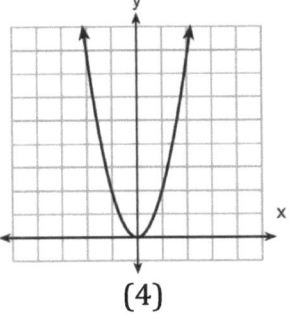

(1)   (2)   (3)   (4)

2. The graph of the equation $y = |x|$ is shown in the diagram below.

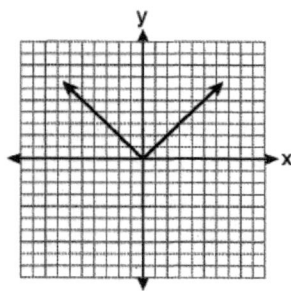

Which diagram could represent a graph of the equation $y = a|x|$ when $-1 < a < 0$?

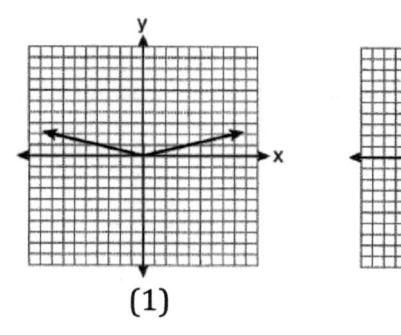   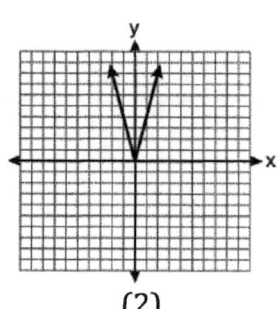   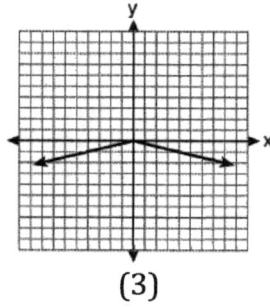   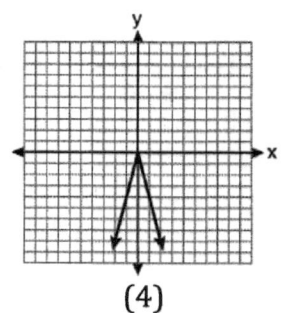

   (1)   (2)   (3)   (4)

3. What is the relationship between the graphs of $f(x) = x^2$ and $g(x) = -3x^2$?

    (1) The graph of $g(x)$ is wider and opens in the opposite direction from the graph of $f(x)$.

    (2) The graph of $g(x)$ is narrower and opens in the opposite direction from the graph of $f(x)$.

    (3) The graph of $g(x)$ is wider and is three units below the graph of $f(x)$.

    (4) The graph of $g(x)$ is narrower and is three units to the left of the graph of $f(x)$.

4. To vertically compress the graph of $y = x^2$ by a factor of $\frac{1}{2}$ would result in what new equation? Will the new graph be wider or narrower than the original?

540

5. On the set of axes below, graph and label the equations $y = |x|$ and $y = 3|x|$ for the interval $-3 \leq x \leq 3$. Explain how changing the coefficient of the absolute value from 1 to 3 affects the graph.

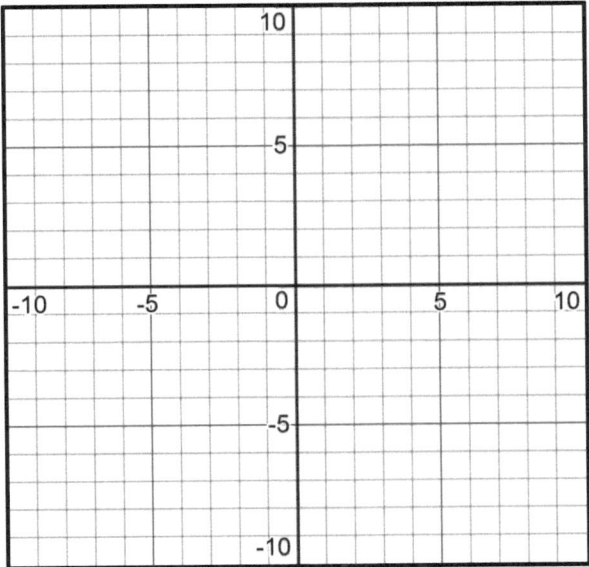

Regents Questions

MULTIPLE CHOICE

1. The graph of the equation $y = ax^2$ is shown below.

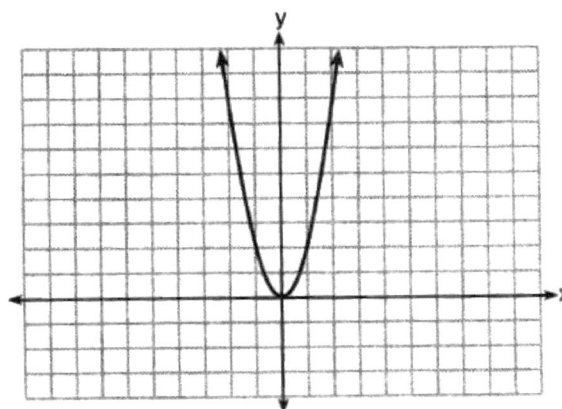

If a is multiplied by $-\frac{1}{2}$, the graph of the new equation is

(1)   wider and opens downward   (3)   narrower and opens downward

(2)   wider and opens upward   (4)   narrower and opens upward

2. How does the graph of $f(x) = 3(x - 2)^2 + 1$ compare to the graph of $g(x) = x^2$?

    (1) The graph of $f(x)$ is wider than the graph of $g(x)$, and its vertex is moved to the left 2 units and up 1 unit.

    (2) The graph of $f(x)$ is narrower than the graph of $g(x)$, and its vertex is moved to the right 2 units and up 1 unit.

    (3) The graph of $f(x)$ is narrower than the graph of $g(x)$, and its vertex is moved to the left 2 units and up 1 unit.

    (4) The graph of $f(x)$ is wider than the graph of $g(x)$, and its vertex is moved to the right 2 units and up 1 unit.

3. When the function $f(x) = x^2$ is multiplied by the value a, where $a > 1$, the graph of the new function, $g(x) = ax^2$

    (1) opens upward and is wider   (3) opens downward and is wider

    (2) opens upward and is narrower   (4) opens downward and is narrower

4. What would be the order of these quadratic functions when they are arranged from the narrowest graph to the widest graph?

    $$f(x) = -5x^2 \qquad g(x) = 0.5x^2 \qquad h(x) = 3x^2$$

    (1) $f(x), g(x), h(x)$ (3) $h(x), f(x), g(x)$

    (2) $g(x), h(x), f(x)$ (4) $f(x), h(x), g(x)$

5. Caitlin graphs the function $f(x) = ax^2$, where a is a positive integer. If Caitlin multiplies a by -2, when compared to $f(x)$, the new graph will become

    (1) narrower and open downward   (3) wider and open downward

    (2) narrower and open upward   (4) wider and open upward

6. The function $f(x) = |x|$ is multiplied by k to create the new function $g(x) = k|x|$. Which statement is true about the graphs of $f(x)$ and $g(x)$ if $k = \frac{1}{2}$?

    (1) $g(x)$ is a reflection of $f(x)$ over the y-axis.

    (2) $g(x)$ is a reflection of $f(x)$ over the x-axis.

    (3) $g(x)$ is wider than $f(x)$.

    (4) $g(x)$ is narrower than $f(x)$.

CHAPTER 17.   DISCONTINUOUS FUNCTIONS

17.1   Piecewise Functions

We often need to define functions in parts. These are called **piecewise functions**. We define these functions using two or more pieces, each for a different part of the function's domain, using a large brace symbol, {.

Example: The graph below is made up of two pieces. For values of $x < 0$, we see part of the parabola whose equation is $f(x) = x^2$. But for values of $x \geq 0$ the graph shows the square root function $f(x) = \sqrt{x}$. The function is defined as:

$$f(x) = \begin{cases} x^2 & x < 0 \\ \sqrt{x} & x \geq 0 \end{cases}$$

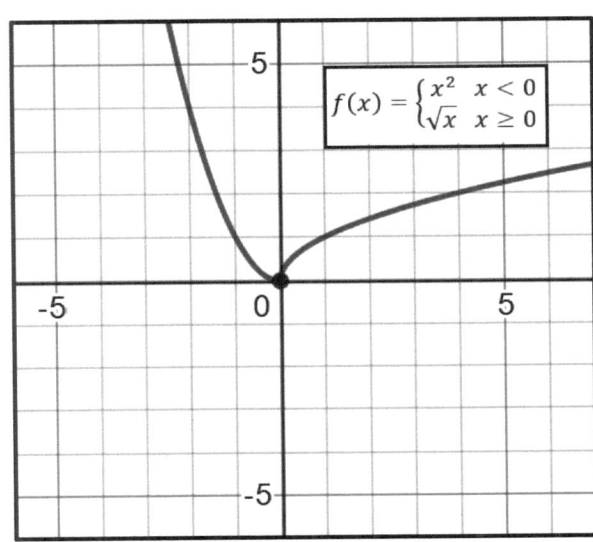

CALCULATOR TIP

To graph a piecewise function on most calculator models, you'll need to use a workaround.

1. Press MODE and change the graphing mode from Connected to Dot (or from Thick to Dot-Thick on some models). Just remember to change it back when you're done.

2. Press Y= and enter the first piece next to Y1 using the format (*piece* 1)(*condition* 1) and the second piece next to Y2 as (*piece* 2)(*condition* 2), and so on.

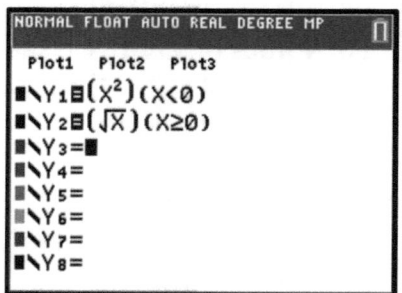
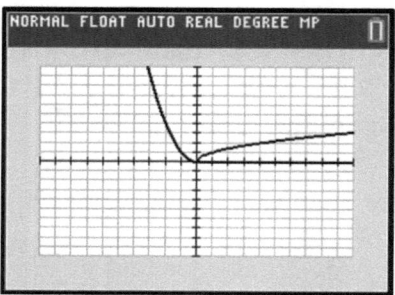

This is not exactly graphed as a piecewise function. It is graphed as two (or more) separate functions but with $f(x) = 0$ for values of x within the function's domain for which the condition is false. For these values of x outside of the condition, the points are drawn along the x-axis; be aware that these points are *not* part of the piecewise function.

To type a relational operator in a condition, press 2nd[TEST] followed by the appropriate number. To type a compound condition such as $0 < x < 1$, type it as two conditions with the logical operator "and" between them, as in 0 < X and X < 1. The logical operators, including "and," are listed under 2nd[TEST]<LOGIC>.

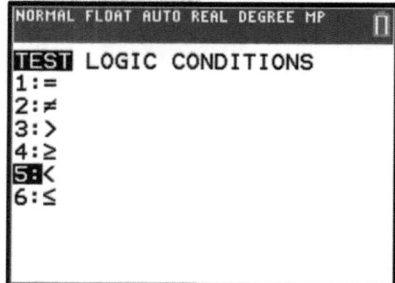
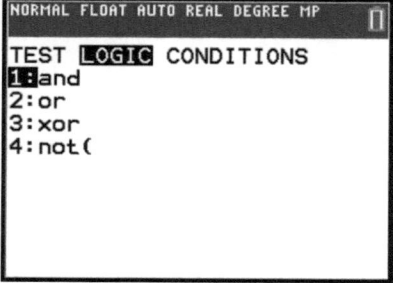

■■■□ CALCULATOR TIP

On some TI-84 models, a piecewise function is available, so the workaround isn't needed.

1. Press $\boxed{\text{Y=}}$ and, while the cursor is next to Y1, press $\boxed{\text{MATH}}$$\boxed{\text{ALPHA}}$[B].
2. Select the number of pieces using the $\boxed{◀}$ and $\boxed{▶}$ keys and select OK.
3. Enter the expressions and condtions into the template, and press $\boxed{\text{GRAPH}}$.

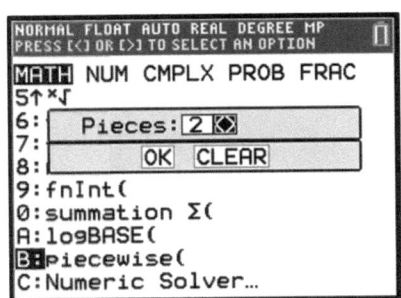 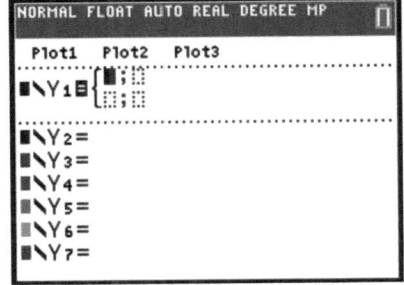 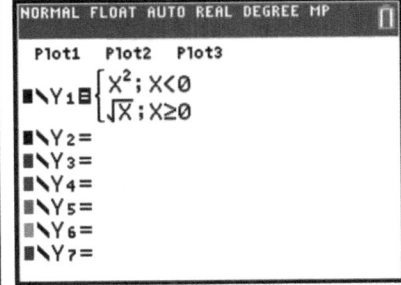

A **discontinuous** function is a function that is not continuous which, by a loose definition, means that you would not be able to draw or trace the function on a plane without lifting your pencil off the paper. The funcion above is *continuous*, since the two parts meet at and include (0,0), allowing us to draw the graph without lifting our pencil. However, piecewise functions are often discontinuous.

Example: The graph to the right shows the function defined as pieces of two lines:

$$f(x) = \begin{cases} x + 1 & x < 1 \\ x - 1 & x \geq 1 \end{cases}$$

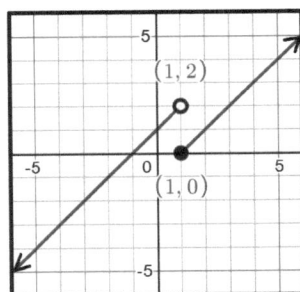

Note that for $f(-1)$ we apply the $x + 1$ piece of the definition, $f(-1) = (-1) + 1 = 0$, giving the point $(-1,0)$. But for $f(1)$, we apply the $x - 1$ piece of the definition, giving $f(1) = 1 - 1 = 0$ and the point $(1,0)$. Also note that the point $(1,2)$ is an open circle. That is because the top piece of the definition, $f(x) = x + 1$, is only for $x < 1$ and not for $x = 1$. When $x \geq 1$, the bottom piece of the definition, $f(x) = x - 1$, applies.

The **absolute value functions** can also be piecewise defined.

Example: $f(x) = |x|$ can be defined as $f(x) = \begin{cases} -x & x < 0 \\ x & x \geq 0 \end{cases}$

MODEL PROBLEM

Graph the piecewise function:

$$f(x) = \begin{cases} x^2 & x < 2 \\ 6 & x = 2 \\ 10 - x & 2 < x \le 6 \end{cases}$$

Solution:

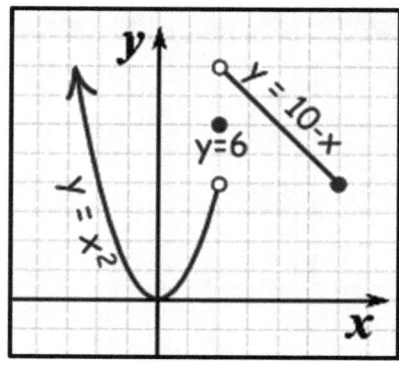

Explanation:

Graph each piece separately, using closed circles at the closed ends of the intervals (where x is $=$, $\le$, or $\ge$ some value) and open circles at the open ends (where x is $<$ or $>$ some value).
[Graph the top piece as part of the parabola $y = x^2$, but ending at an open circle at (2,4) to show that $x = 2$ ($y = x^2 = 2^2 = 4$) is not part of that piece. The middle piece defines (2,6) as a point on the graph. The bottom piece defines a line segment graphed as $y = 10 - x$ starting at but not including (2,8) – hence the open circle – and ending at and including (6,4).]

PRACTICE PROBLEMS

1. Given $f(x) = \begin{cases} -x & x < 0 \\ x + 1 & x \ge 0 \end{cases}$
   find $f(-3)$, $f(0)$, and $f(2)$.

2. Graph $f(x) = \begin{cases} -x & x < 0 \\ x + 1 & x \ge 0 \end{cases}$
   Is this a continuous function?

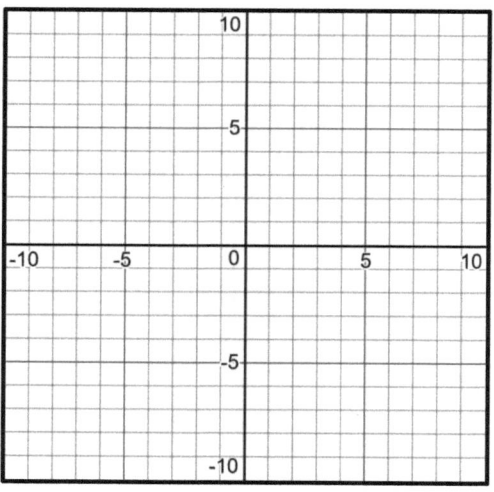

3. Graph $f(x) = \begin{cases} -1 & x < 1 \\ 1 & x = 1 \\ 2x - 2 & x > 1 \end{cases}$

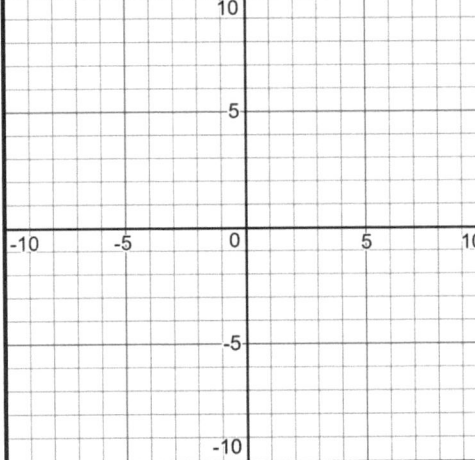

4. Graph $f(x) = \begin{cases} -x + 1 & x < 0 \\ 2^x & x \geq 0 \end{cases}$

    Is this a continuous function?

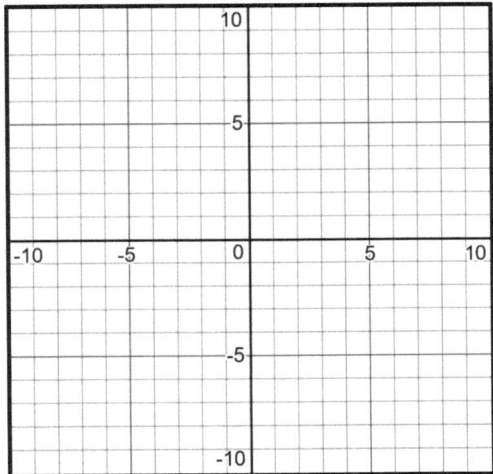

5. Graph $f(x) = |x + 1| + 1$. Is this a continuous function? Write an equivalent function using a piecewise definition instead of an absolute value symbol.

6. A garage charges the following rates for parking (with an 8 hour limit):

    $4 per hour for the first 2 hours
    $2 per hour for the next 4 hours
    No charge for the next 2 hours

    Write a piecewise function that gives the parking cost c (in dollars) in terms of the time t (in hours) that a car is parked.

Regents Questions

MULTIPLE CHOICE

1. A function is graphed on the set of axes below.

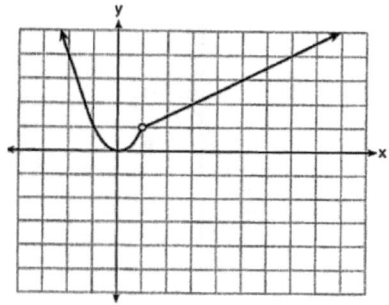

Which function is related to the graph?

(1) $f(x) = \begin{cases} x^2, & x < 1 \\ x - 2, & x > 1 \end{cases}$

(2) $f(x) = \begin{cases} x^2, & x < 1 \\ \frac{1}{2}x + \frac{1}{2}, & x > 1 \end{cases}$

(3) $f(x) = \begin{cases} x^2, & x < 1 \\ 2x - 7, & x > 1 \end{cases}$

(4) $f(x) = \begin{cases} x^2, & x < 1 \\ \frac{3}{2}x - \frac{9}{2}, & x > 1 \end{cases}$

2. Which graph represents $f(x) = \begin{cases} |x| & x < 1 \\ \sqrt{x} & x \geq 1 \end{cases}$?

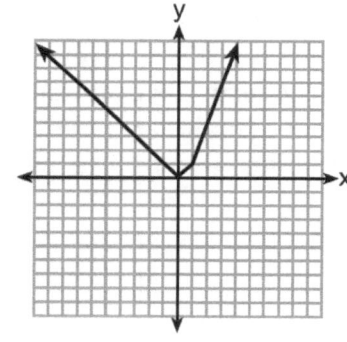

(1)

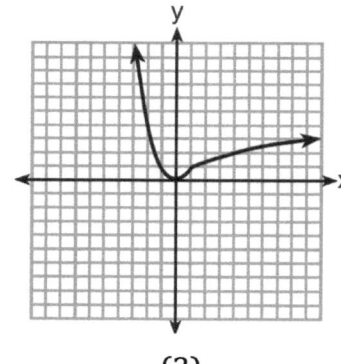

(3)

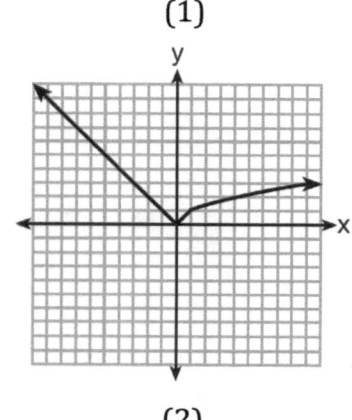

(2)

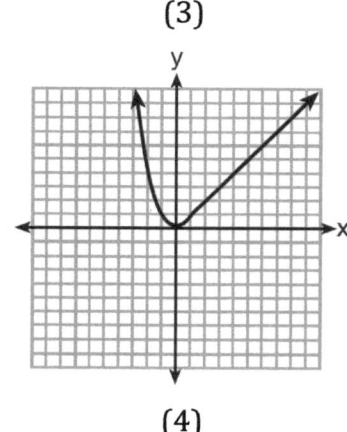

(4)

3. When the function $g(x) = \begin{cases} 5x, & x \leq 3 \\ x^2 + 4, & x > 3 \end{cases}$ is graphed correctly, how should the points be drawn on the graph for an x-value of 3?

    (1) open circles at (3,15) and (3,13)

    (2) closed circles at (3,15) and (3,13)

    (3) an open circle at (3,15) and a closed circle at (3,13)

    (4) a closed circle at (3,15) and an open circle at (3,13)

4. Which relation is a function?

x	y
−1	1
0	0
1	1
1	2
2	4
3	9

(1)

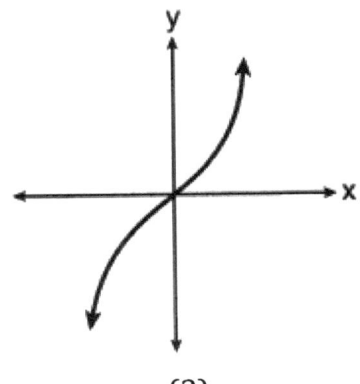

(3)

$y = \begin{cases} x, & -1 < x \leq 2 \\ x^2, & 2 \leq x < 4 \end{cases}$

(2)

$\{(0,1), (2,3), (3,2), (3,4)\}$

(4)

CONSTRUCTED RESPONSE

5. At an office supply store, if a customer purchases fewer than 10 pencils, the cost of each pencil is $1.75. If a customer purchases 10 or more pencils, the cost of each pencil is $1.25. Let c be a function for which $c(x)$ is the cost of purchasing x pencils, where x is a whole number.

$$c(x) = \begin{cases} 1.75x, & \text{if } 0 \le x \le 9 \\ 1.25x, & \text{if } x \ge 10 \end{cases}$$

Create a graph of c on the axes below.

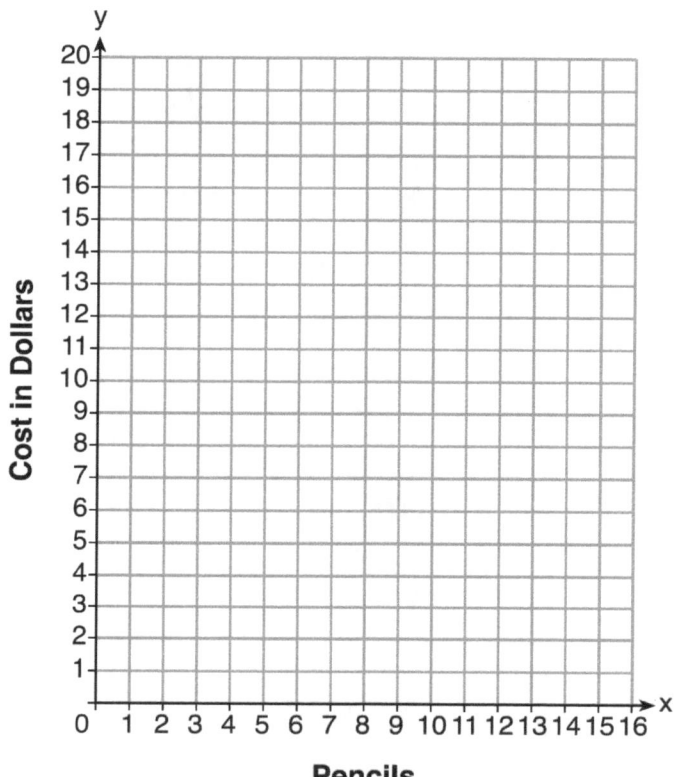

A customer brings 8 pencils to the cashier. The cashier suggests that the total cost to purchase 10 pencils would be less expensive. State whether the cashier is correct or incorrect. Justify your answer.

6.   Graph the following function on the set of axes below.

$$f(x) = \begin{cases} |x|, & -3 \le x < 1 \\ 4, & 1 \le x \le 8 \end{cases}$$

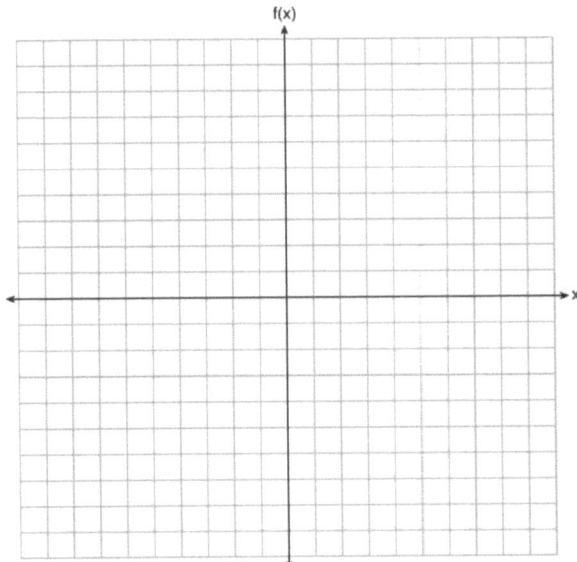

7.   The equation to determine the weekly earnings of an employee at The Hamburger Shack is given by $w(x)$, where x is the number of hours worked.

$$w(x) = \begin{cases} 10x, & 0 \le x \le 40 \\ 15(x - 40) + 400, & x > 40 \end{cases}$$

Determine the difference in salary, *in dollars*, for an employee who works 52 hours versus one who works 38 hours.

Determine the number of hours an employee must work in order to earn $445. Explain how you arrived at this answer.

8.   On the set of axes below, graph

$$g(x) = \frac{1}{2}x + 1 \quad \text{and} \quad f(x) = \begin{cases} 2x + 1, & x \leq -1 \\ 2 - x^2, & x > -1 \end{cases}$$

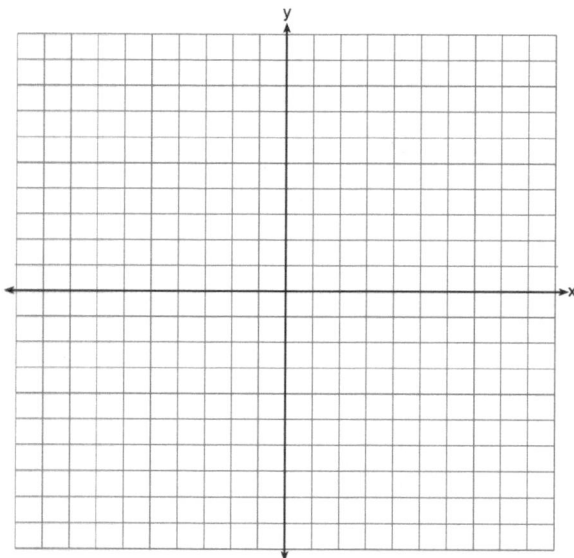

How many values of x satisfy the equation $f(x) = g(x)$? Explain your answer, using evidence from your graphs.

9.   On the set of axes below, graph the piecewise function:

$$f(x) = \begin{cases} -\frac{1}{2}x, & x < 2 \\ x, & x \geq 2 \end{cases}$$

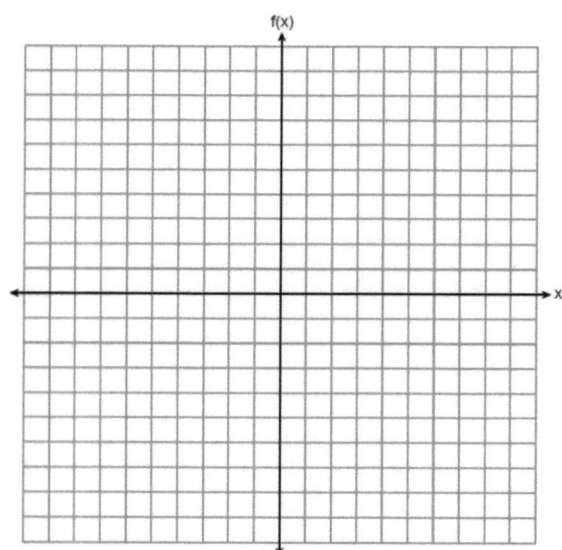

10. Graph the following piecewise function on the set of axes below.

$$f(x) = \begin{cases} |x|, & -5 \le x < 2 \\ -2x + 10, & 2 \le x \le 6 \end{cases}$$

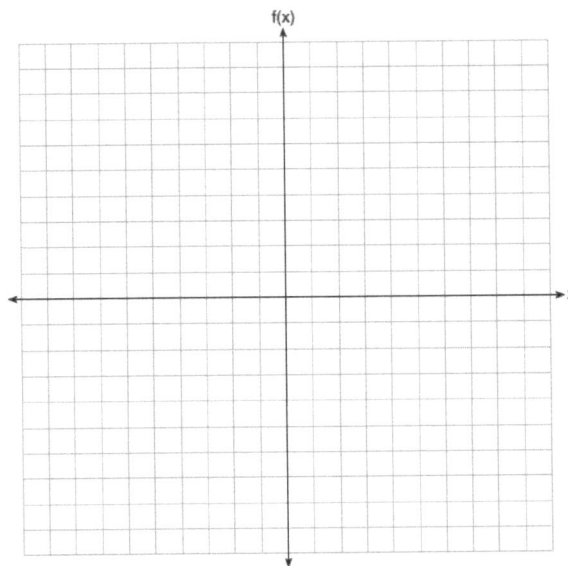

11. Graph the function: $h(x) = \begin{cases} 2x - 3, & x < 0 \\ x^2 - 4x - 5, & 0 \le x \le 5 \end{cases}$

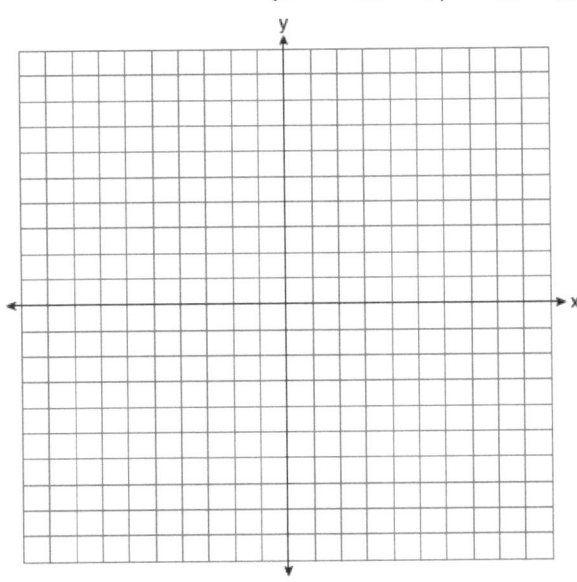

12. The function g is defined as $g(x) = \begin{cases} |x + 3|, & x < -2 \\ x^2 + 1, & -2 \leq x \leq 2 \end{cases}$

On the set of axes below, graph $g(x)$.

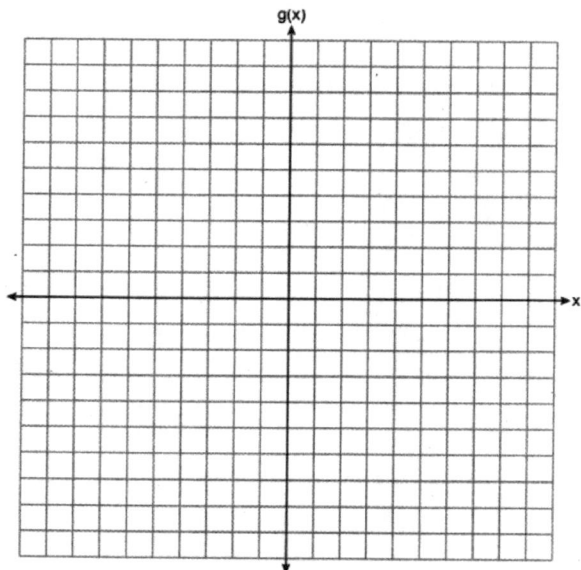

13. The two relations shown below are *not* functions.

Relation I: Relation II:

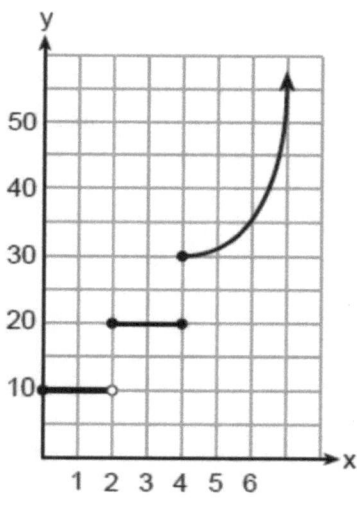 $\{(-5, -2), (-4, 0), (-2, 1), (-1, 3), (-4, 4)\}$

Explain how you could change each relation so that they each become a function.

14. Bryan said that the piecewise function graphed below has a domain of all real numbers.

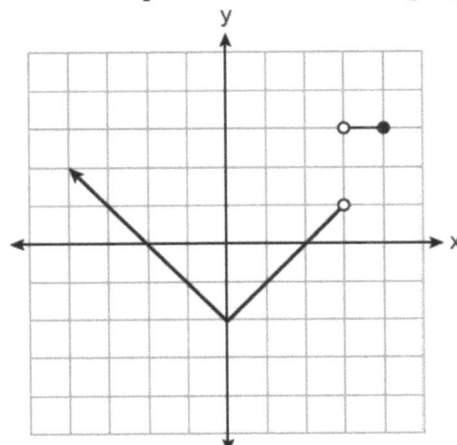

   State *two* reasons why Bryan is *incorrect*.

15. The piecewise function $f(x)$ is given below.

$$f(x) = \begin{cases} 2x - 3, & x > 3 \\ -x^2 + 15, & x \le 3 \end{cases}$$

   State the value of $f(3)$. Justify your answer.

17.2   Step Functions

A **step function** is a discontinuous function that, when graphed, appears as a series of disconnected line segments resembling steps on a staircase.

Two common step functions are called the floor and ceiling functions. The **floor function** uses special bracket symbols ⌊ ⌋ with serifs only at the bottom, which represents the greatest integer less than or equal to the value inside the brackets. The **ceiling function** uses similar bracket symbols ⌈ ⌉ but with serifs only at the top, which represents the least integer that is greater than or equal to the value inside the brackets.

Examples: (a) For the floor function $f(x) = \lfloor x \rfloor$, $f(2.9) = 2$ but $f(3) = 3$.
               (b) For the ceiling function $g(x) = \lceil x \rceil$, $g(5.3) = 6$ and $g(6) = 6$.

The graphs of floor and ceiling functions are shown below. The graph is made up of disconnected line segments with open or closed circles at their ends. Just as we saw when graphing inequalities, an **open circle** means the point is *excluded*, but a **closed circle** means the point is *included*.

Example: On the graph of $y = \lfloor x \rfloor$ to the left below, $y = \lfloor x \rfloor = 1$ for all real values of x between 1 and 2, including 1 but excluding 2. In other words, $y = 1$ for all values of x such that $1 \leq x < 2$, even for values very close to 2 such as 1.99999, but not including 2. For $x = 2$, $y = \lfloor 2 \rfloor = 2$, so the point (2,2) is closed while the point (2,1) is open.

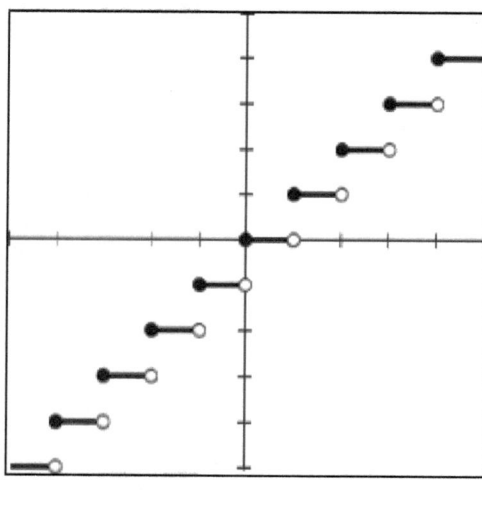

$$y = \lfloor x \rfloor$$

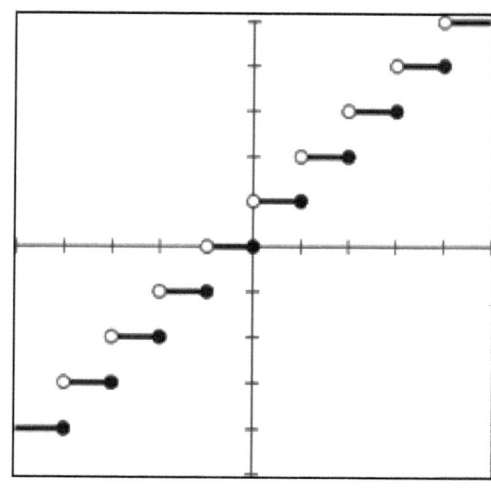

$$y = \lceil x \rceil$$

Step functions are used in various real world situations.

Example: The graph below shows the cost of mailing a letter way back in 2006. For letters of up to 1 ounce in weight, the postage cost 39 cents. The cost was 41 cents for letters that were more than 1 ounce but up to 2 ounces in weight. For weights of x ounces in the interval $2 < x \leq 3$, the cost was 43 cents. This pattern continued so that each additional ounce, or fraction of an ounce, cost an additional 2 cents.

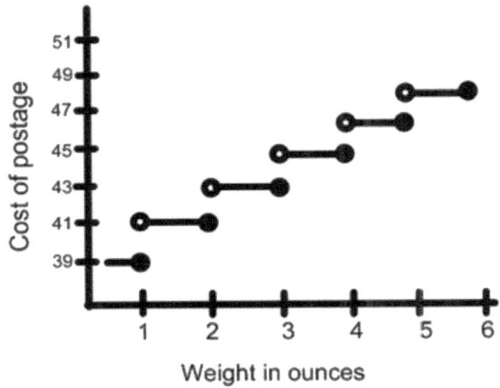

This function is a type of ceiling function, defined as $y = 2\lceil x \rceil + 37$ for all $x > 0$.

Although it is not continuous, it is still a function, as we can see by applying the vertical line test, shown below.

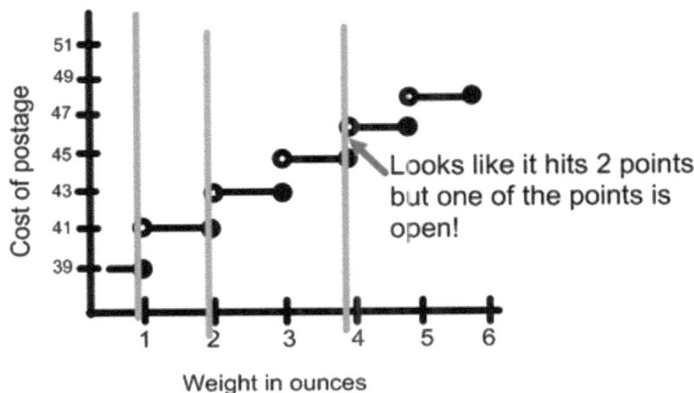

557

MODEL PROBLEM

A famous step function, called the Heaviside step function, is defined as:

$$H(x) = \begin{cases} 0 & x < 0 \\ 0.5 & x = 0 \\ 1 & x > 0 \end{cases}$$

Graph the function, $H(x)$.

Solution:

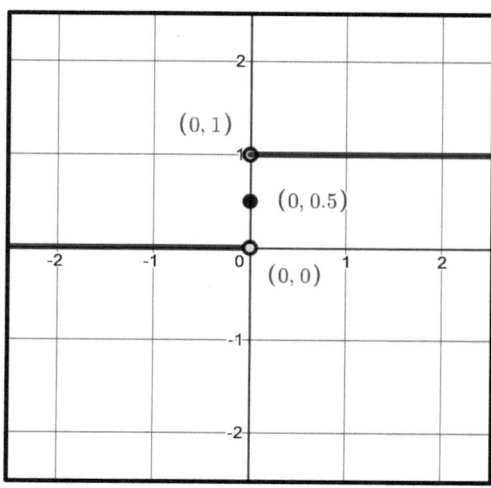

Explanation of steps:
Graph each piece separately. *[Graph $y = 0$ for $x < 0$ and $y = 1$ for $x > 0$. For both of these pieces, use an open dot at $x = 0$ because $H(0)$ does not equal 0 or 1. Instead, $H(0) = 0.5$, so graph the point (0,0.5).]*

PRACTICE PROBLEMS

1. Given $f(x) = 3[x] + 5$, find $f(6.25)$.	2. Graph $f(x) = \lfloor x \rfloor + 1$.
	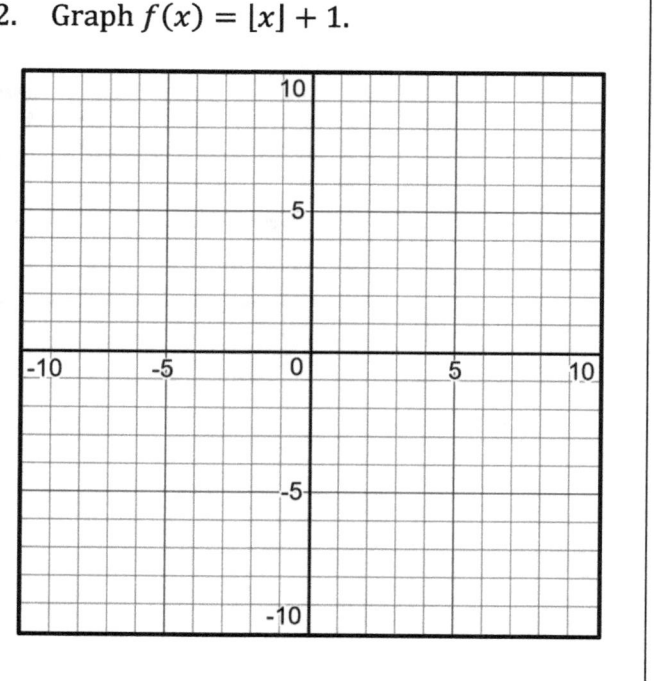

Regents Questions

MULTIPLE CHOICE

1. Morgan can start wrestling at age 5 in Division 1. He remains in that division until his next odd birthday when he is required to move up to the next division level. Which graph correctly represents this information?

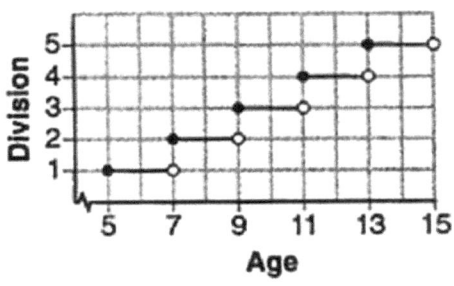

(1)

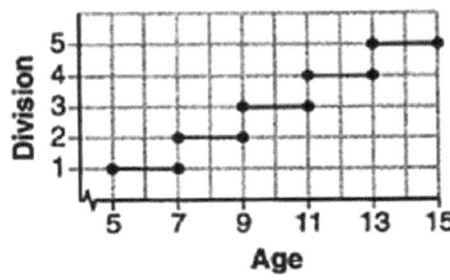

(3)

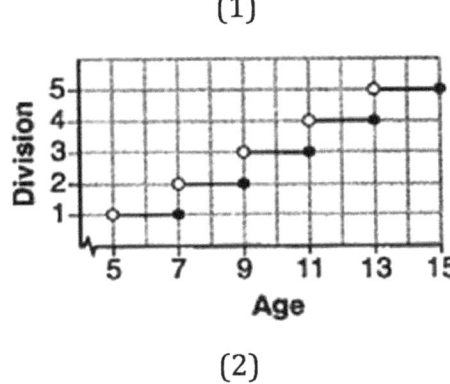

(2)

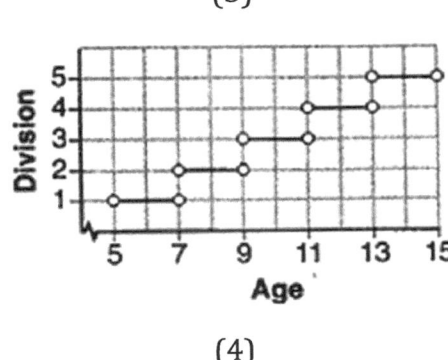

(4)

CONSTRUCTED RESPONSE

2.   The table below lists the total cost for parking for a period of time on a street in Albany, N.Y. The total cost is for any length of time up to and including the hours parked. For example, parking for up to and including 1 hour would cost $1.25; parking for 3.5 hours would cost $5.75.

Hours Parked	Total Cost
1	1.25
2	2.50
3	4.00
4	5.75
5	7.75
6	10.00

Graph the step function that represents the cost for the number of hours parked.

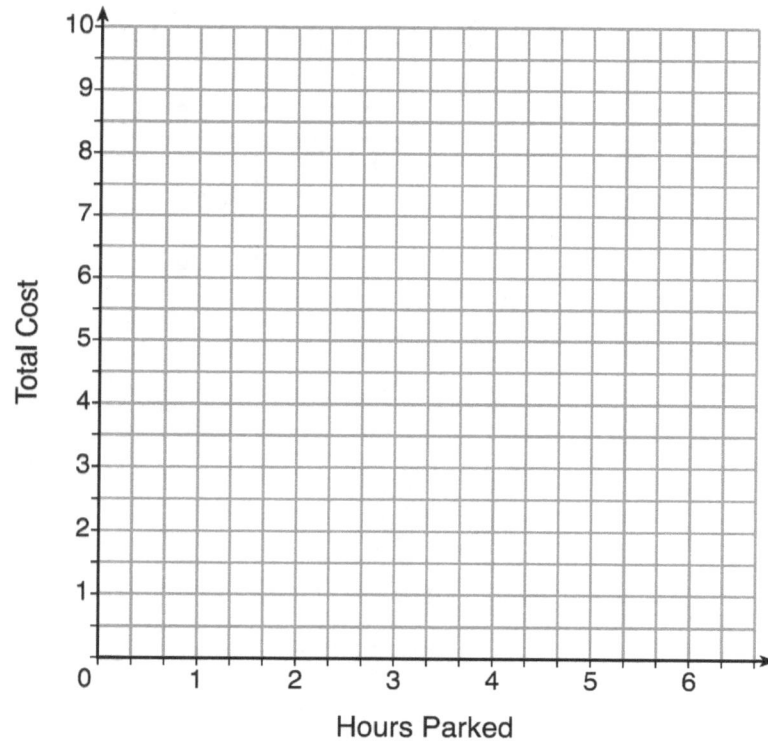

Explain how the cost per hour to park changes over the six-hour period.

CHAPTER 18.   UNIVARIATE DATA

18.1   Types of Data

Statistics is the practice or science of collecting, describing, analyzing, and making inferences about data in order to make decisions. **Data** are items of information. Each data value (*datum* is the singular) may be called an **observation** or **data point**. A collection of data is called a **data set** (or *dataset*). A data set may be described or analyzed using summary values or graphs. These descriptions and analyses are often used to make **inferences** (generalizations or predictions) about a larger data set. A data set may be quantitative or qualitative.

Data is **quantitative** (from the word, "quantity") if it is recorded as numeric values. Quantitative data allows for numerical analysis of the results, such as finding the average (mean) of the data. Quantitative data is also known as **numerical**.
Example: If a survey asks for SAT scores, the responses are quantitative.

Data is **qualitative** (from the word, "quality") if it is recorded using category names, such as labels or characteristics. Qualitative data may have values that are written as numbers, such as zip codes, but if we would never apply arithmetic operations on the data (such as adding zip codes), then it is not quantitative. Qualitative data is also known as **categorical**.
Example: If a survey asks for a favorite soda, the responses are qualitative.

Data is considered **univariate** if it can be recorded using a **single variable**. The **frequency** of each data value represents the number of times that value occurs in the data. Univariate data can be represented by various charts, such as a bar graph, histogram, dot plot, pie chart, or a box-and-whisker plot.
Example: A survey asks a sampling of people for their heights in inches.

Data is considered **bivariate** if it is recorded using **two variables**. Each variable may be categorical or quantitative. If the bivariate data is quantitative, it can be represented by plotting points on a coordinate graph or scatter plot. For each result or response, an ordered pair represents the values of the two variables, usually x and y, for that point.
Example: A survey asks people for their heights in inches and their weights in pounds.

One advantage of quantitative data is that it can be entered into the calculator as a list. As we'll see later, this will allow us to calculate statistics such as the mean or standard deviation of the data. Instructions for entering data into the calculator are provided below.

CALCULATOR TIP

To enter a set of quantitative data into the calculator:

1. Press $\boxed{\text{STAT}}\boxed{1}$ to select Edit.
2. If any values already appear in the L1 column (L stands for List), select the column heading and press $\boxed{\text{CLEAR}}\boxed{\text{ENTER}}$. Also, if the L1 column is missing, you can add it by selecting an empty column heading and then pressing $\boxed{\text{2nd}}\boxed{\text{L1}}\boxed{\text{ENTER}}$.
3. Enter the data values into the L1 column.
4. Press $\boxed{\text{2nd}}\boxed{\text{QUIT}}$ when you're done to return to the home screen.

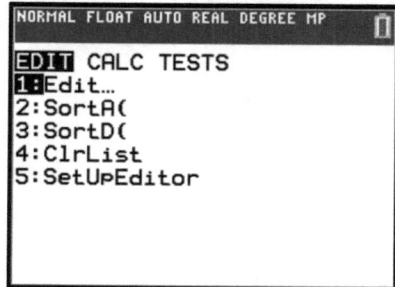 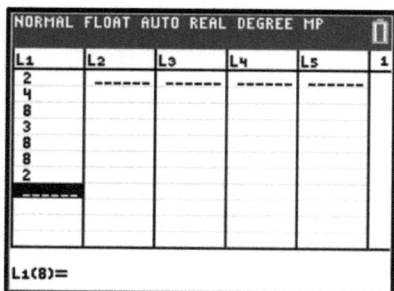

It is often helpful to sort a list of data in the calculator. Directions are provided below.

 CALCULATOR TIP

To sort and view your data as a horizontal list:

1. Press $\boxed{\text{STAT}}\boxed{2}$ to select SortA. [Or, press $\boxed{\text{STAT}}\boxed{3}$ select SortD for descending order.]
2. Press $\boxed{\text{2nd}}\boxed{\text{L1}}\boxed{)}\boxed{\text{ENTER}}$. The word "Done" will appear.
3. View your data by pressing $\boxed{\text{2nd}}\boxed{\text{L1}}\boxed{\text{ENTER}}$ and it will now be sorted. You will also see the data sorted in L1 if you press $\boxed{\text{STAT}}\boxed{1}$.

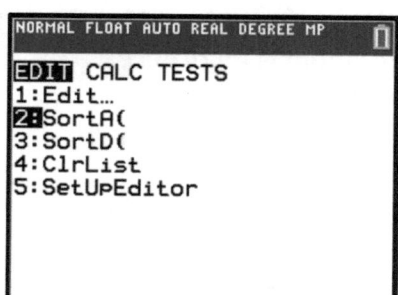 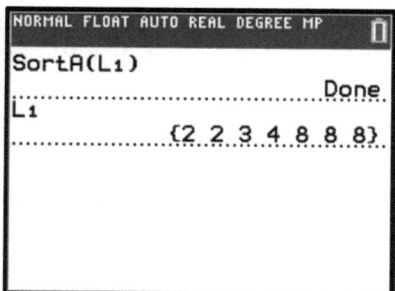

MODEL PROBLEM

For each of the following graphs, identify whether the data is (a) qualitative or quantitative, and (b) univariate or bivariate.

Types of Pitches in a Game

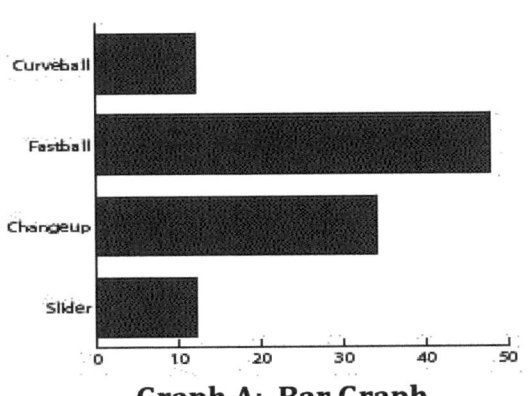

Graph A: Bar Graph

Revenue for an Ice Cream Shop Dependent on Temperature

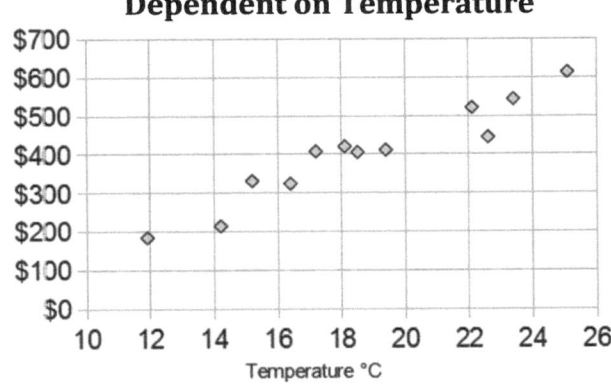

Graph B: Scatterplot

Solution:
    (A)   Graph A is qualitative and univariate.
    (B)   Graph B is quantitative and bivariate.

Explanation:
    (A)   Data is qualitative if their values are non-numerical *[curveball, fastball, etc.]* and univariate if it involves only one variable *[specifying which type of pitch]*.
            [The numerical values on the x-axis of the bar graph represent the frequency of the data; for example, nearly 50 of the pitches were fastballs.]
    (B)   Data is quantitative if their values are numerical *[a number of degrees and a number of dollars]* and bivariate if it involves two variables *[specifying the temperature in degrees Celsius and the revenue in dollars]*.

PRACTICE PROBLEMS

1. An art studio posts information about each sculpture that is for sale. Which data would *not* be classified as quantitative?
 	
    (1)   cost   (3)   artist
    (2)   height   (4)   weight
	2. Determine whether the data are quantitative or qualitative.
	
	a) body temperatures
	
	b) hair colors
	
	c) telephone numbers
	
	d) annual salaries

3. Which is an example of bivariate data?

    (1)   shoe sizes of the players on a high school basketball team
    (2)   goals scored in hockey games over the course of a season
    (3)   Calories consumed per day by a track athlete for one month
    (4)   hours spent studying compared to test scores in a science course

4. Specify whether the data used to create this chart is univariate or bivariate.

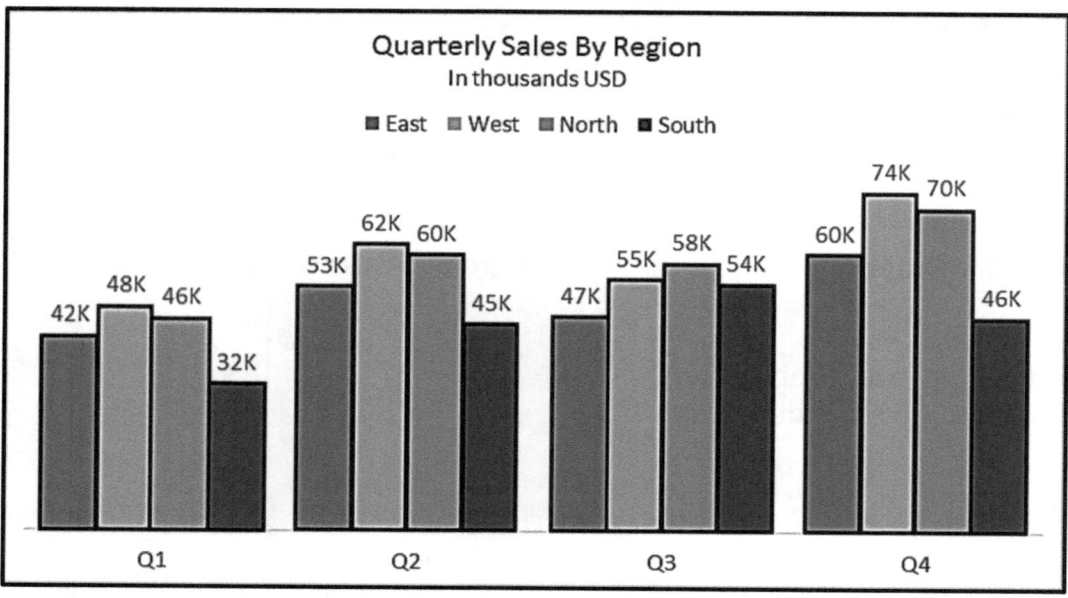

Regents Questions

There are no Regents exam questions on this topic.

18.2 Frequency Tables

Frequency is the number of times that a particular result or value occurs in a list of data. These frequencies may be compiled into a **frequency table**.

Example: Students were asked to record the color of all of the cars parked in the school lot. Their results are shown in the frequency table below.

Color	Frequency (f)
Black	7
Gray	10
White	8
Red	4
Blue	5
Green	2
Brown	4

We can find the number of observations, n, by finding the sum of the frequencies.

Example: By adding the frequencies in the table above, we can see that there was a sum of 40 cars in the parking lot, so $n = 40$.

To enter large amounts of quantitative data into the calculator, we can use a frequency table to enter each value and its frequency as L1 and L2, as shown below.

Example: The table below shows a distribution of 30 data values ($n = 30$). Instead of entering 30 values into L1, we can enter each of the six possible values once into L1 but specify its frequency in L2. Instructions for entering a frequency table are provided below.

Value (x)	Frequency (f)
0	4
1	3
2	5
3	5
4	6
5	7

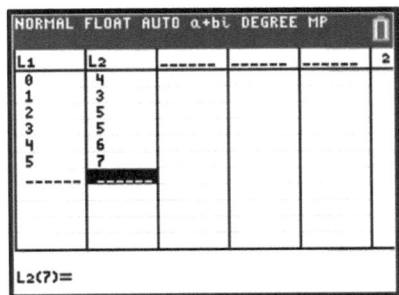

████▉▌ CALCULATOR TIP

To enter a frequency table:

1. Press STAT 1 to select Edit.
2. If any values already appear in the L1 or L2 column, select the column heading and press CLEAR ENTER.
3. Enter each data value into the L1 column and its corresponding frequency into the L2 column.
4. Press 2nd QUIT.

The **relative frequency** of each observation, usually written as *rf*, is the ratio of its frequency to the total number of observations. The formula is $rf = \frac{f}{n}$, where *f* is the frequency and *n* is the number of observations. Relative frequencies may be expressed as decimals, percents, or fractions.

Example: For a sample set of 200 observations, if a certain result occurs 30 times, then its relative frequency is $\frac{30}{200} = 0.15$. This may also be expressed as a percent, as in 15%.

A **relative frequency table** includes the relative frequency of each observation. When we need to find the sum of a column in the table, it is helpful to add a row to the bottom of the table, as shown below.

Example: For the table below, we can calculate the relative frequencies by (a) finding the sum of the frequencies, $n = 30$, and then (b) dividing each frequency by this sum: $rf = \frac{f}{n}$. So, for the first row, the relative frequency is $\frac{4}{30} \approx 0.13$, or about 13%.

Value (x)	Frequency (f)	Relative Frequency (rf)
0	4	0.13
1	3	0.10
2	5	0.17
3	5	0.17
4	6	0.20
5	7	0.23
Total	**30**	**1.00**

The sum of the relative frequencies *before rounding* will always equal 1, or 100%.

We can use the calculator to find the relative frequencies.

Example: With the frequencies already stored in L2, we can enter a formula to calculate the relative frequencies as $rf = \frac{f}{n}$ and store them in L3. Directions for adding this column on the calculator are provided below.

CALCULATOR TIP

To add relative frequencies to a frequency table:

1. Press $\boxed{\text{STAT}}\boxed{1}$ to select Edit. Be sure L1 and L2 contain the values and frequencies.
2. Move the cursor to the column heading for L3 and type the formula for the relative
    frequencies: $\boxed{\text{2nd}}\boxed{\text{L2}}\boxed{\div}\boxed{\text{2nd}}\boxed{\text{LIST}}<\text{MATH}>\boxed{5}\boxed{\text{2nd}}\boxed{\text{L2}}\boxed{)}\boxed{\text{ENTER}}$.

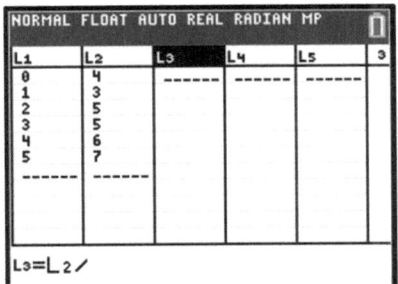

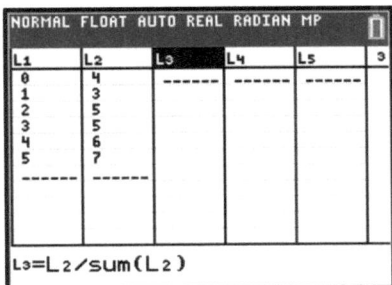

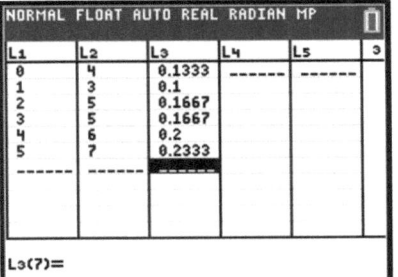

The **cumulative frequency** is the number of observations less than or equal to a value, and is usually denoted by cf. Since the phrase "less than or equal to" requires an order to the values, cumulative frequency would only be applied to quantitative data or to categorical data that have a meaningful order (known as *ordinal* data).

We can derive the cumulative frequencies by (a) setting the cf of the first row equal to its frequency, f, and then (b) for all other rows, adding the frequency of the current row to the previous row's cumulative frequency.

Example: In the cumulative frequency table below,
             for the first row, $cf = f = 4$,
             for the second row, $cf = 4 + 3 = 7$,
             for the third row, $cf = 7 + 5 = 12$, and so on.
         Note that the cumulative frequency of the last row will equal the total number of data values, n.

Value (x)	Frequency (f)	Cumulative Frequency (cf)
0	4	4
1	3	7
2	5	12
3	5	17
4	6	23
5	7	30

We can also add a column for cumulative frequencies into the calculator by using the **cumSum** function. Directions for adding this column on the calculator are provided below.

▓▓▒░ CALCULATOR TIP

To add cumulative frequencies to a frequency table:
1. Press ⌈STAT⌉⌈1⌉ to select Edit. Be sure L1 and L2 contain the values and frequencies.
2. Move the cursor to the column heading for L3 and enter the function for the cumulative frequencies: ⌈2nd⌉⌈LIST⌉<OPS>⌈6⌉⌈2nd⌉⌈L2⌉⌈ᵀ⌉⌈ENTER⌉.

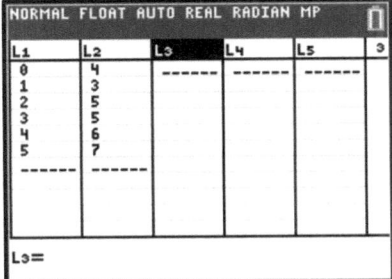   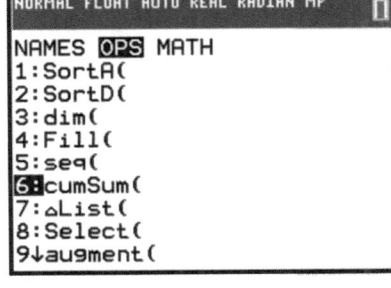

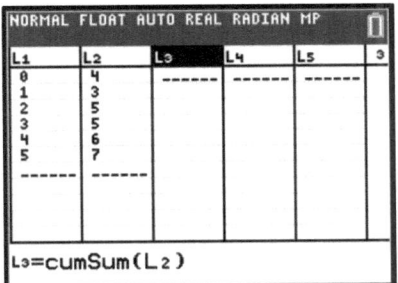

Quantitative data values are often organized into **intervals** (or **classes**) of equal sizes. The frequencies may be compiled into a **grouped frequency table**. Relative frequencies may also be included in the table.

Example: The frequency table below shows the ages of 30 people. The data is distributed across four intervals. For example, the table shows that 10 of the ages are between 40 and 49, inclusive. The table also shows the relative frequencies of each age group, rounded to the *nearest thousandths*.

Age Group	Frequency (f)	Relative Frequency (rf)
20-29	9	0.300
30-39	7	0.233
40-49	10	0.333
50-59	4	0.133

Each interval (or *class*) has a lower limit and an upper limit. The **class width** is the difference between the lower (or upper) limits of any two consecutive intervals. The class width must be the same for each interval.

Example: For the table above, the lower limits are 20, 30, 40, and 50, and the upper limits are 29, 39, 49, and 59, as given in the first column. We can find the class width by subtracting, for example, the first two lower limits:
$30 - 20 = 10$, so the class width is 10.

When creating a cumulative frequency table for grouped data, each interval should be relabeled as spanning from the start of the *first* interval to the end of the current interval. The cumulative frequencies are the sums of the frequencies from the *first* interval up to and including that interval. *The last cumulative frequency should equal the number of data items.*

Example: Below is a frequency table and its corresponding cumulative frequency table. Note that each Age Group in the cumulative frequency table starts with 20, the lower limit of the first interval.

Age Group	Frequency (f)	Age Group	Cumulative Frequency (cf)
20-29	9	20-29	9
30-39	7	20-39	16
40-49	10	20-49	26
50-59	4	20-59	30

MODEL PROBLEM

A pair of dice are rolled 100 times and the sums of the two dice are recorded in the frequency table shown (below left). Extend the table by adding columns for the relative frequencies and the cumulative frequencies.

Solution:

Result (x)	Frequency (f)
2	3
3	9
4	9
5	4
6	14
7	16
8	18
9	11
10	9
11	5
12	2

(A) Relative Frequency (rf)	(B) Cumulative Frequency (cf)
0.03	3
0.09	12
0.09	21
0.04	25
0.14	39
0.16	55
0.18	73
0.11	84
0.09	93
0.05	98
0.02	100

Explanation of steps:

(A) Divide each value in the (f) column by the total number of data values (the sum of the frequencies). [$n = 100$, so $\frac{3}{100} = 0.03$, $\frac{9}{100} = 0.09$, etc.]

(B) Start with the first frequency [3], then add each next value in the (f) column to get the next (cf) value. [$3 + 9 = 12$, $12 + 9 = 21$, $21 + 4 = 25$, etc.]

PRACTICE PROBLEMS

1. A 6-sided die is rolled 20 times and the results are recorded below. Complete the column for relative frequencies.

Result (x)	Frequency (f)	Relative Frequency (rf)
1	5	
2	3	
3	1	
4	5	
5	4	
6	2	

2. A 6-sided die is rolled 20 times and the results are recorded below. Complete the column for cumulative frequencies.

Result (x)	Frequency (f)	Cumulative Frequency (cf)
1	5	
2	3	
3	1	
4	5	
5	4	
6	2	

3. The table below shows the number of bases reached on hits in 2021 in major league baseball. Complete the columns for relative frequencies and cumulative frequencies. Round the values in the last two columns to the *nearest thousandths*.

Bases (x)	Frequency (f)	Relative Frequency (rf)	Cumulative Frequency (cf)
1	25,006		
2	7,863		
3	671		
4	5,944		

4. The following table shows the number of Nobel Prize recipients by category from 1901 to 2021. Complete the column for the relative frequencies, rounded to the *nearest hundredths*.

Category	Nobel Prizes	Relative Frequency
Physics	219	
Chemistry	186	
Medicine	224	
Literature	118	
Peace	137	
Economics	89	

5. The table below shows a cumulative frequency distribution of runners' ages. According to the table, how many runners are in their forties?

Cumulative Frequency Distribution of Runners' Ages

Age Group	Total
20–29	8
20–39	18
20–49	25
20–59	31
20–69	35

6. Complete the cumulative frequency table that corresponds to the data represented by the frequency table below.

Test Score	Frequency
41 – 55	8
56 – 70	12
71 – 85	26
86 – 100	14

Test Score	Cumulative Frequency

Regents Questions

There are no Regents exam questions on this topic.

18.3   Histograms

A **frequency histogram** is a type of graph where the height of each bar represents the number of data items in that interval. To create a frequency histogram:

1. First create a **frequency table** by counting the number of data values for each interval and writing this count in the Frequency column. It may be helpful to count the frequencies using **tally** marks, with every 5 tally marks written as ⦀⦀.
2. Create the frequency histogram by first labeling each interval along the horizontal axis using a consistent class width. Then create an appropriate scale for the vertical axis, using equally spaced labels for the frequencies. If the lowest interval does not start at 0, the symbol ⌁ may be drawn on the horizontal axis to represent the gap.
3. For each interval, draw a shaded rectangular bar as high as its corresponding frequency. There should be no spaces between the bars.

Example: The data from the frequency table below may be graphed using a frequency histogram, as shown to the right.

Age Group	Frequency (f)
20-29	9
30-39	7
40-49	10
50-59	4

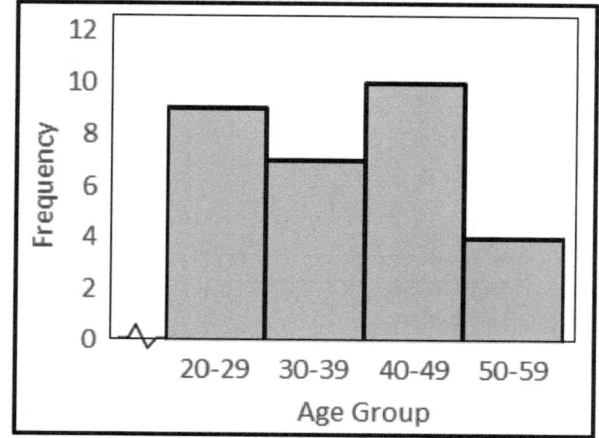

We can use the calculator to view a histogram of our data. When entering data from a grouped frequency table like the one above, you may enter the lower limits of each interval into L1 and the frequencies into L2. The calculator will only show the bars of the histogram, not the labelled axes.

Example: The directions below show how to create a histogram from the above table.

CALCULATOR TIP

To view a frequency histogram for a frequency table of data:

1. Press [STAT][1] to enter the frequency table data. For grouped data, you may enter the lower limits into L1. Enter the frequencies into L2.
2. Press [2nd][STAT PLOT][1] to select Plot 1.
3. Select <On> and press [ENTER].
4. Select the third Type of graph (histogram) and press [ENTER].
5. For Xlist, enter L1 by pressing [2nd][L1].
6. For Freq, enter L2 by pressing [2nd][L2].
7. Press [WINDOW]. For Xmin, enter the lowest value (or the lower limit of the first interval). For Xmax, enter 1 more than the highest value (or 1 more than the upper limit of the last interval). For Xscl, enter the class width of each interval. Ymin should be 0 and Ymax should be at least 1 more than the largest frequency. Yscl may be adjusted depending on the value of Ymax.
8. Press [GRAPH].

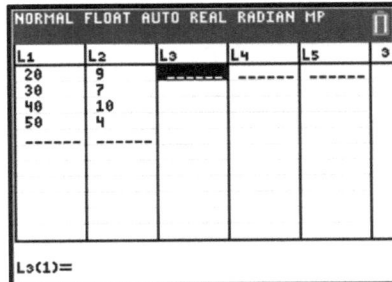 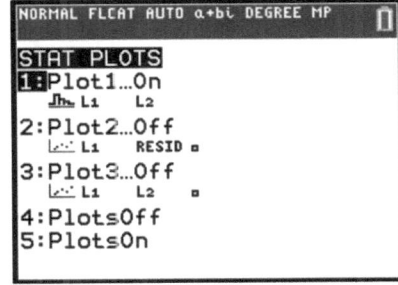 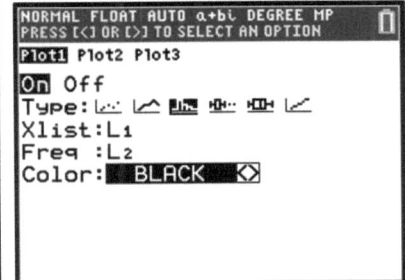

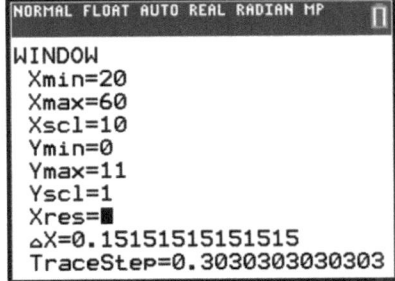 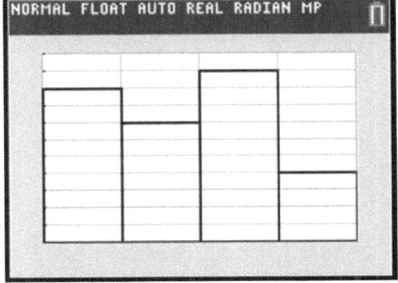

A **cumulative frequency histogram** is a type of graph where the height of each bar represents the sum of all the frequencies up to and including that interval. To create a cumulative frequency histogram:

1. Create a **cumulative frequency table**.
2. Create the cumulative frequency histogram by first labeling each interval along the horizontal axis and equally spaced cumulative frequencies along the vertical axis.
3. For each interval, draw a shaded rectangular bar as high as its corresponding cumulative frequency. There should be no spaces between the bars. *Each bar should be at least as tall as the bar to its left.*

Example: The same data used in the frequency histogram of the previous example may be displayed using a cumulative frequency histogram, as shown below.

Age Group	Cumulative Frequency (cf)
20-29	9
20-39	16
20-49	26
20-59	30

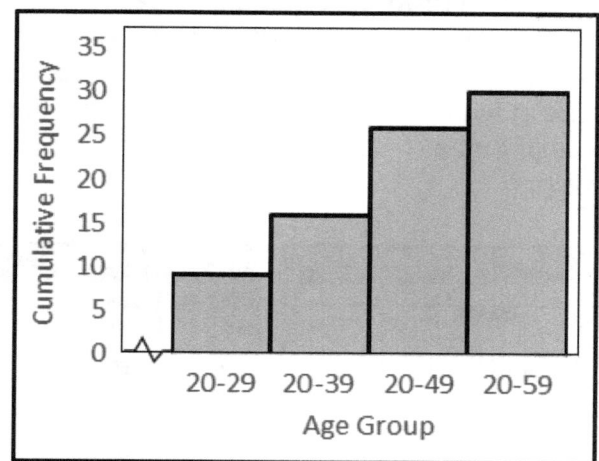

The directions on the next page show how to view a cumulative frequency histogram on the calculator.

CALCULATOR TIP

To view a cumulative frequency histogram for a frequency table of data:

1. Press [STAT][1] to enter the frequency table data. For grouped data, you may enter the lower limits into L1. Enter the frequencies into L2.

2. Move to the L3 column heading and enter [2nd][LIST]<OPS>[6][2nd][L2][)][ENTER] to fill the column with cumulative frequencies using the cumSum function.

3. Press [2nd][STAT PLOT][1] to select Plot 1.

4. Select <On> and press [ENTER].

5. Select the third Type of graph (histogram) and press [ENTER].

6. For Xlist, enter L1 by pressing [2nd][L1].

7. For Freq, enter L3 (*the cumulative frequencies*) by pressing [2nd][L3].

8. Press [WINDOW]. For Xmin, enter the lowest value (or the lower limit of the first interval). For Xmax, enter 1 more than the highest value (or 1 more than the upper limit of the last interval). For Xscl, enter the class width of each interval. Ymin should be 0 and Ymax should be at least 1 more than the number of data values. Yscl may be adjusted depending on the value of Ymax.

9. Press [GRAPH].

NORMAL FLOAT AUTO REAL RADIAN MP					3
L₁	L₂	L₃	L₄	L₅	
20	9	------	------	------	
30	7				
40	10				
50	4				
------	------				

L₃=cumSum(L₂)

NORMAL FLOAT AUTO REAL RADIAN MP					3
L₁	L₂	L₃	L₄	L₅	
20	9	9	------	------	
30	7	16			
40	10	26			
50	4	30			
------	------	------			

L₃={9,16,26,30}

```
NORMAL FLOAT AUTO REAL RADIAN MP
WINDOW
 Xmin=20
 Xmax=60
 Xscl=10
 Ymin=0
 Ymax=31
 Yscl=5
 Xres=1
 ΔX=0.15151515151515
 TraceStep=0.3030303030303
```

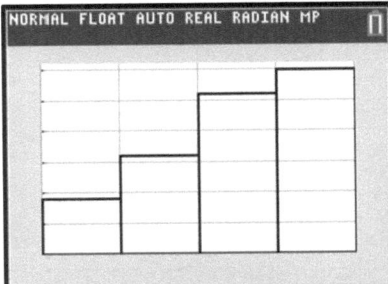

MODEL PROBLEM

Create a frequency histogram and cumulative frequency histogram for the following set of 36 test scores, using intervals of 10 points.

14, 17, 28, 28, 30, 36, 45, 52, 58, 58, 61, 64, 64, 68, 68, 77, 77,
81, 81, 81, 81, 87, 87, 94, 94, 95, 95, 95, 95, 95, 95, 97, 97, 97, 100, 100

Solution:

(A) Interval	(B) Tally	(C) Frequency
0-10		0
11-20	||	2
21-30	|||	3
31-40	|	1
41-50	|	1
51-60	|||	3
61-70	̶⊥̶⊥̶⊤	5
71-80	||	2
81-90	̶⊥̶⊥̶⊤ |	6
91-100	̶⊥̶⊥̶⊤ ̶⊥̶⊥̶⊤ |||	13

(D)

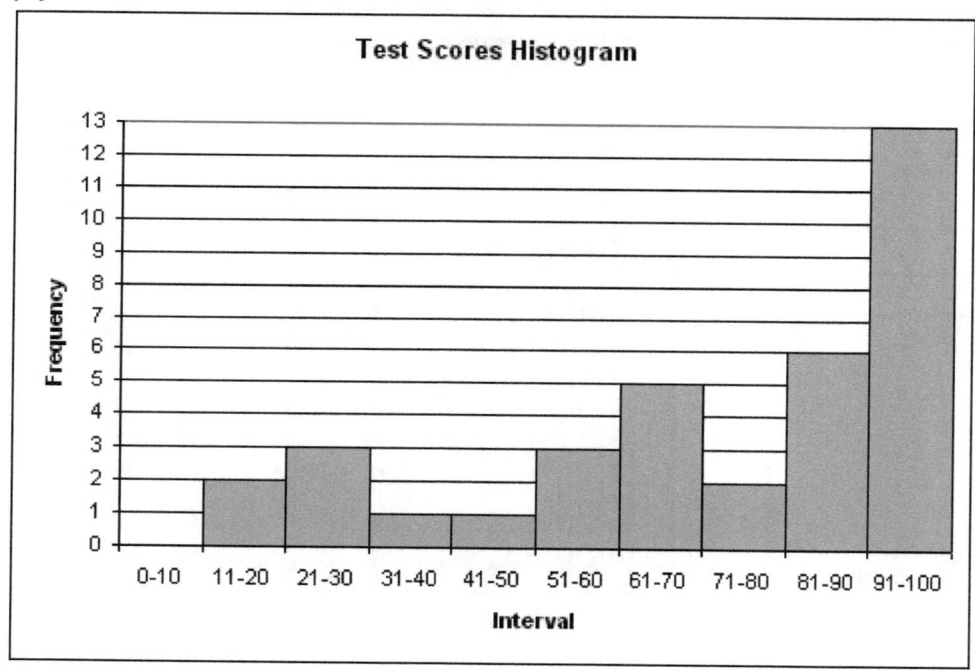

(Continued on next page...)

(Continued from previous page)

(E) Interval	(F) Cumulative Frequency
0-10	0
0-20	2
0-30	5
0-40	6
0-50	7
0-60	10
0-70	15
0-80	17
0-90	23
0-100	36

(G)

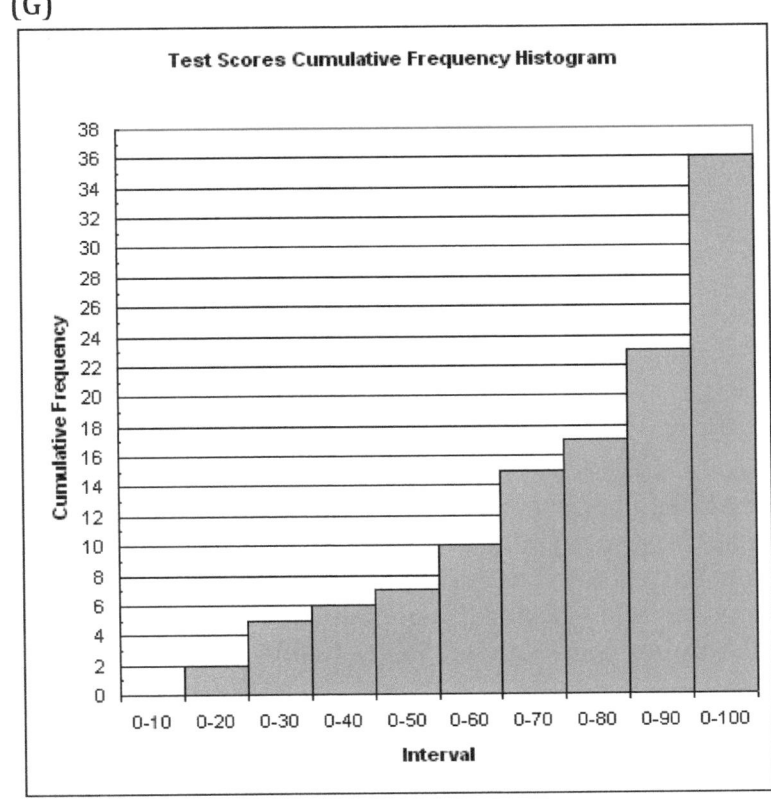

Explanation of steps:
(A) Create the frequency table using equally wide intervals. *[Each is 10 points wide.]*
(B) Tally each data item by writing a tally mark for the corresponding interval.
(C) Count the tally marks and write each count as a number in the Frequency column.
(D) Draw the frequency histogram.
(E) Create the cumulative frequency table based on the frequency table. Re-label the intervals so that they all start with the *first* interval's minimum value *[0]*.
(F) Enter the cumulative frequencies. The first interval's cumulative frequency is equal to the first interval's frequency. The cumulative frequency for each interval after the first is the sum of the preceding cumulative frequency plus the corresponding interval's frequency. *[For example, the cumulative frequency for 0-30 is the sum of the cumulative frequency for 0-20, which is 2, plus the frequency of the 21-30 interval, which is 3.]*
(G) Draw the cumulative frequency histogram.

PRACTICE PROBLEMS

1. The accompanying histogram shows the heights of the students in Kyra's health class. What is the total number of students in the class?

    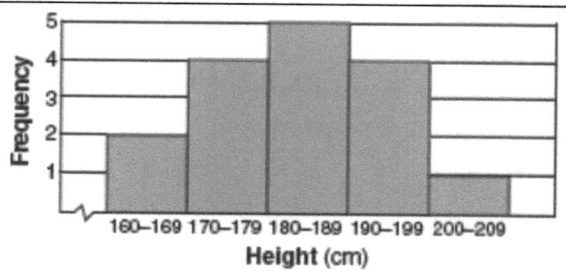

2. Casey talked to everyone in his apartment building to find out how many hours of television each person watched each day. The results are shown in the histogram below. Using the histogram, determine the total number of people in Casey's building.

    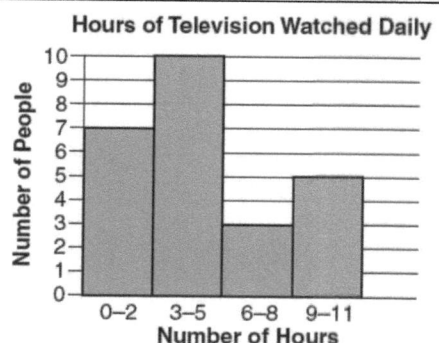

3. Mr. Trapp recorded the height, in inches, of every student in his class. From the data collected, he created the cumulative frequency histogram shown below. Based on the histogram, how many students are in the class?

    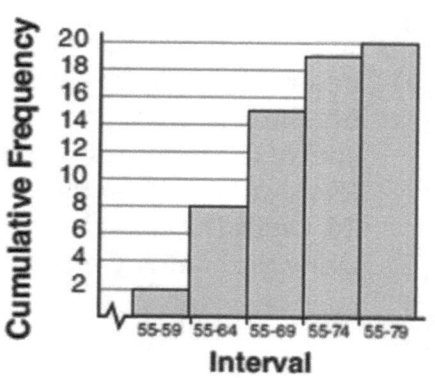

4. The cumulative frequency histogram below shows the distances swimmers completed in a swim team practice.

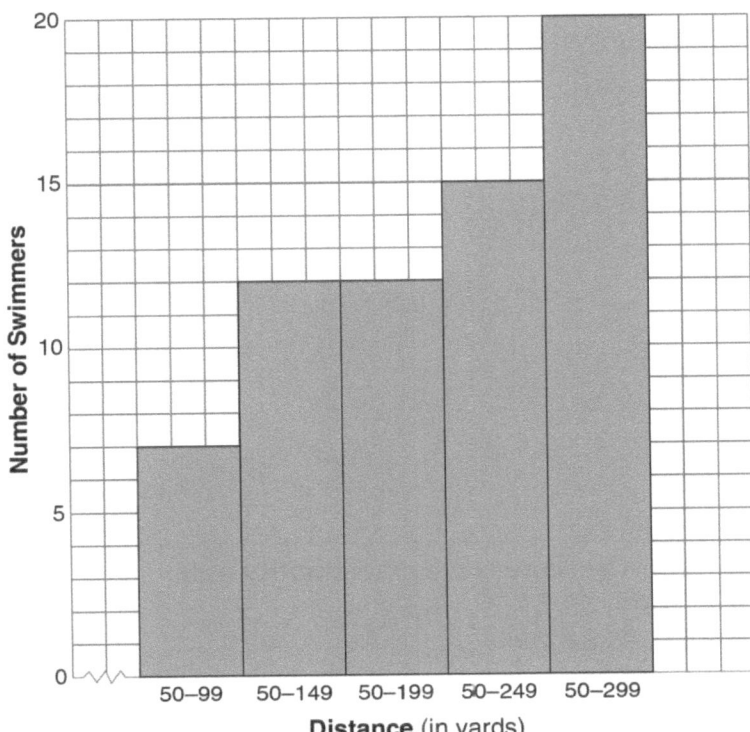

a) Based on the cumulative frequency histogram, determine the number of swimmers who swam between 200 and 249 yards.

b) Determine the number of swimmers who swam between 150 and 199 yards.

c) Determine the number of swimmers who participated in the swim team practice.

5. In the time trials for the 400-meter run at the state sectionals, the 15 runners recorded the times shown in the table below. Using the data from the frequency column, draw a frequency histogram on the grid below.

400-Meter Run

Time (sec)	Frequency
50.0–50.9	
51.0–51.9	||
52.0–52.9	ＨＴ |
53.0–53.9	|||
54.0–54.9	||||

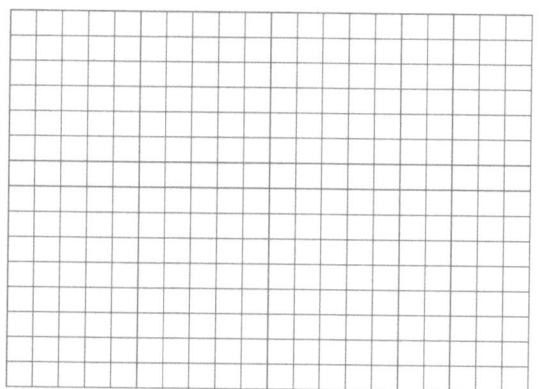

6. The following set of data represents the scores on a mathematics quiz:
       58, 79, 81, 99, 68, 92, 76, 84, 53, 57,
       81, 91, 77, 50, 65, 57, 51, 72, 84, 89

    a) Complete the frequency table below.

    b) On the grid below, draw and label a frequency histogram of these scores.

Mathematics Quiz Scores

Interval	Tally	Frequency
50–59		
60–69		
70–79		
80–89		
90–99		

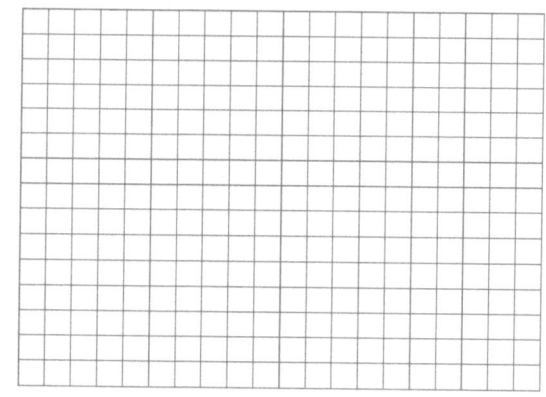

7. Ms. Hopkins recorded her students' final exam scores in the frequency table below.

    On the accompanying grid, construct a frequency histogram based on the table.

Interval	Tally	Frequency
61–70	‎TH+	5
71–80	IIII	4
81–90	‎TH+ IIII	9
91–100	‎TH+ I	6

8. The test scores for 18 students in Ms. Rom's class are listed below.
        86, 81, 79, 71, 58, 87, 52, 71, 87,
        87, 93, 64, 94, 81, 76, 98, 94, 68

    Complete the frequency table below. Then, draw and label a frequency histogram on the accompanying grid.

Interval	Tally	Frequency
51–60		
61–70		
71–80		
81–90		
91–100		

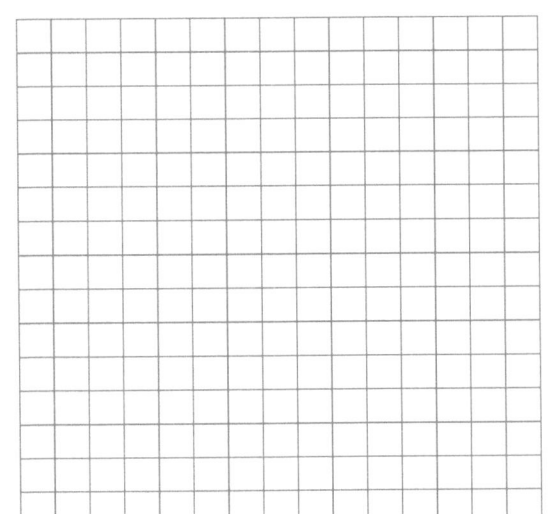

9. Twenty students were surveyed about the number of days they played outside in one week. The results of this survey are shown below.

6, 5, 4, 3, 0, 7, 1, 5, 4, 4, 3, 2, 2, 3, 2, 4, 3, 4, 0, 7

Complete the frequency table and cumulative frequency table below for these data.

Number of Days Outside

Interval	Tally	Frequency
0–1		
2–3		
4–5		
6–7		

Number of Days Outside

Interval	Cumulative Frequency
0–1	
0–3	
0–5	
0–7	

On the grid below, create a cumulative frequency histogram based on the table.

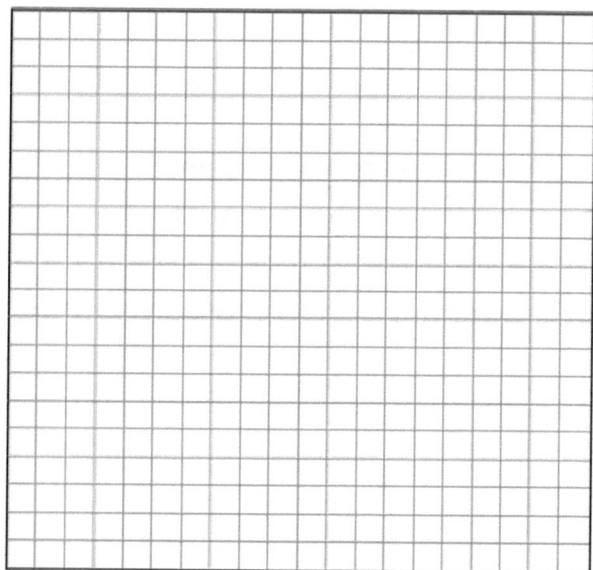

Regents Questions

There are no Regents exam questions on this topic.

18.4 Central Tendency

An **average** is a measure of **central tendency**, which means that it describes the **center** of a distribution. An average is a single number meant to typify a list of values. The most commonly used measures are the mean, median, and mode.

For the definitions below, we will refer to this sample set of quantitative data: 2, 4, 8, 3, 8, 8, 2

Each data value may be called an **observation**. The number of data values in a sample set of data is represented by the letter, n.

Example: For the sample data above, there are 7 observations, so $n = 7$.

The **mean** (represented by the symbol, $\bar{x}$) is calculated by *adding* all the numbers in the data and *dividing* by how many data values (or *observations*) there are.

Example: For the sample set of data, $\bar{x} = \dfrac{2+4+8+3+8+8+2}{7} = \dfrac{35}{7} = 5.$

To find the mean from a frequency table:
1. Multiply each data value, x, by its frequency, f, to get xf.
2. Find the sum of these products.
3. Divide this sum by the sum of the frequencies, n.

Example: To find the mean of the data represented by the table below,
(a) fill in the (xf) column by multiplying each data value by its frequency,
(b) find the sum of this column, 41, then (c) divide this by the sum of the frequency column, $n = 20$, to get a mean of $\dfrac{41}{20} \approx 2.05$.

x	f	xf
0	3	0
1	4	4
2	6	12
3	4	12
4	2	8
5	1	5
Total	20	41

The **median** is the value that lies in the middle of an ordered set of data. To find the median for a set of data, we must first arrange the data in ascending order.

For an _odd_ number of data values, the median is the _middle number_.

Example: The sample set of data is arranged as: 2, 2, 3, 4, 8, 8, 8.
                      There are 7 data values. The median, or middle number, is 4.

For an _even_ number of values, the median is the _mean_ (average) of the _two middle numbers._

Example: Suppose we add another 2 to the data: 2, 2, 2, 3, 4, 8, 8, 8.
                      Now there are 8 values. The median is the average of 3 and 4, or 3.5.

When finding the location of the median, you may use the formula $\frac{n+1}{2}$ to determine which value(s) to choose.

Examples: a) For 7 ordered data values, $\frac{7+1}{2} = 4$, so we should pick the fourth value.

                        b) For 8 ordered data values, $\frac{8+1}{2} = 4.5$, so we should find the average of the fourth and fifth values.

To find the median from a frequency table:

1. find the sum of the frequencies, n.
2. find the middle data value. (For an odd n, this will be a specific data value, but for an even n, this _may_ be the mean of two different data values.)

Example: To find the median of the data represented by the table below, (a) find the sum of the frequencies, $n = 20$, and (b) find the middle data value. In this case, $\frac{(20+1)}{2} = 10.5$, so we need the average of the tenth and eleventh smallest values. These values are both 2, so the median is 2.

x	f
0	3
1	4
2	6
3	4
4	2
5	1
Total	20

← median lies here

The **mode** is the data value that appears *most often*. For numerical data, it is helpful to arrange the data in ascending order before determining the mode.

Example: Given a sample set of data arranged in ascending order: 2, 2, 3, 4, 8, 8, 8.
                    Since the value 8 appears most often (three times), the mode is 8.

It is possible to have more than one mode. A set of data with two modes is called **bimodal**.

Example: Suppose we add another 2 to the sample data: 2, 2, 2, 3, 4, 8, 8, 8.
                    Now there are two modes, 2 and 8, since both numbers appear three times.
                    Therefore, this set of data is bimodal.

If each number appears with the same frequency, then there is **no mode**.

Examples: a) This set of data has no mode: 2, 2, 3, 3, 4, 4.
                      b) This set of data has no mode: 2, 3, 4, 7, 8, 9.

Unlike the other measures of central tendency, the mode may also be determined for a set of qualitative data.

Example: If a set of data represents the favorite flavors of ice cream for a group of
                    respondents, then we can have a mode (for example, "vanilla" occurs most often),
                    but a mean or median would not be possible.

For some sets of data, there are one or a few data values, called **outliers**, which are much larger or much smaller than the rest of the data. The median is a better measure than the mean at characterizing data that is skewed by outliers. For this reason, the median is considered a more **resistant** measure of center than the mean.

Examples: Consider these two sets of data.

2, 3, 5, 5, 7, 8, 33   2, 3, 5, 5, 25, 25, 33
Mean = 9   Mean = 14
Median = 5   Median = 5

In the first set of data, an outlier (33) skews the mean so that the mean (9) is actually larger than 6 of the 7 data values. In this case, the median may be a better indicator of central tendency. In the second set of data, the mean (14) may be more appropriate, since nearly half the values are at least 25.

CALCULATOR TIP

To find the measures of central tendency:

1. First, press [STAT][1] and enter the data into the L1 column, or enter the data as a frequency table into L1 and L2. *[For this example, the small sample 55, 60, 62, 70, 70 is stored in L1.]*

2. Press [STAT]<CALC>[1] for 1-Var Stats. Be sure that L1 is specified as the List, or press [2nd][L1] to enter it. If L2 contains frequencies, enter [2nd][L2] as FreqList; otherwise, leave it blank. Then, select Calculate. *[On the TI-83, you need to enter L1 after the 1-Var Stats prompt by pressing [2nd][L1]. If L2 contains frequencies, press [,][2nd][L2]. Then, press [ENTER].]*

3. The mean is the value of $\bar{x}$. Press the down arrow to Med for the median. Unfortunately, the calculator does not show the mode.

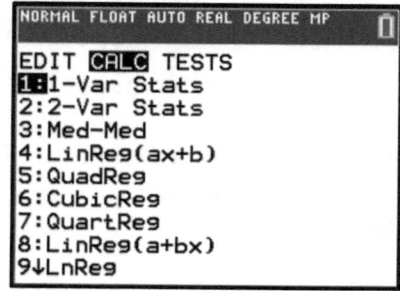
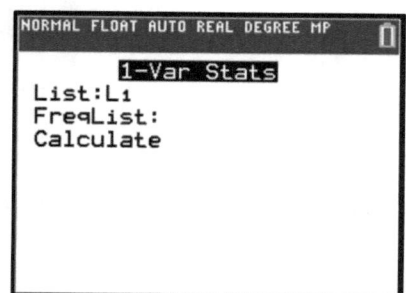
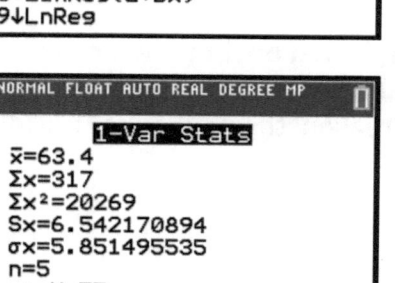
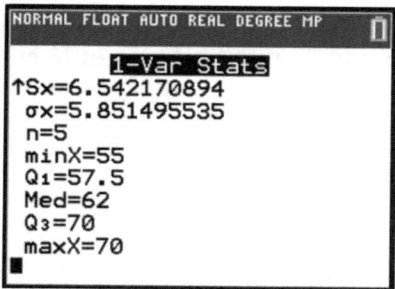

CALCULATOR TIP

To find the mean or median using the List Math functions:

You can also find the mean (or median) on the calculator by pressing [2nd][STAT]<MATH>. Then, press [3] for mean (or [4] for median), followed by [2nd][L1][)][ENTER].

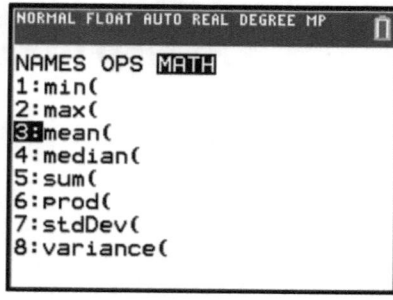
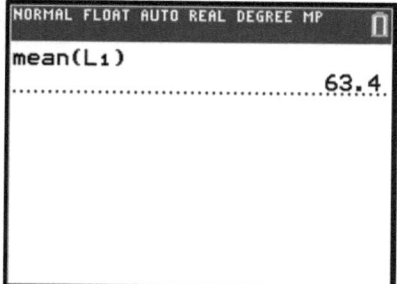

The lowest value in the data set is called the **minimum**, the highest value is called the **maximum**, and the **range** is the *difference* between the *maximum* and *minimum* values. Range is a measure of **spread**, not central tendency, in that it measures how spread out the data set is.
Example: For the sample data, 2 is the minimum, 8 is the maximum, and the range is 6.

If we **add the same constant to each value** in a set of data, this will add that constant to the mean, median, and mode. The range will not change.
Example: Test scores in a class are originally 55, 60, 62, 70, 70.
                    The mean = 63.4, median = 62, and mode = 70.
                    The teacher decides to "curve" the test by adding 10 points to each score.
                    The data set becomes 65, 70, 72, 80, 80. Each new measure of central tendency is
                    now 10 points higher: mean = 73.4, median = 72, and mode = 80.

The same is true if we **multiply each data value by the same constant** (other than zero). The mean, median and mode will also be multiplied by the same constant factor as a result. However, in this case the range will also multiplied by the same factor.

MODEL PROBLEM

For the given set of data, find the mean, median, and mode: 1, 3, 5, 7, 9, 13, 21, 25, 25, 31.

Solution:

(A)   Mean: $\dfrac{1+3+5+7+9+13+21+25+25+31}{10} = \dfrac{140}{10} = 14$

(B)   Median: $\dfrac{9+13}{2} = 11$

(C)   Mode: 25

Explanation of steps:
  (A)   Find the mean by adding all the values and dividing this sum *[140]* by the number of data values *[10]*.
  (B)   For an even number of data values *[there are 10 values in this set]*, we need to take the mean (numerical average) of the middle two numbers *[9 and 13]*.
  (C)   The mode is the value that appears most often. *[Only 25 appears more than once.]*

PRACTICE PROBLEMS

1. For a school report, Luke contacted a car dealership to collect data on recent sales. He asked, "What color do buyers choose most often for their car?" White was the response. What statistical measure does the response "white" represent?	2. The weights of all the students in grade 9 are arranged from least to greatest. Which statistical measure separates the top half of this set of data from the bottom half?
3. Which of the following sets of data is bimodal? (1) 1, 1, 2, 5, 5, 6 (3) 1, 2, 3, 4, 5, 6 (2) 1, 1, 2, 2, 2, 3 (4) 1, 1, 2, 2, 3, 3	4. Each value in a set of data is divided by two. How does this affect the mean, median and mode for this set of data?

5. Mr. Swift raised all his students' scores on a recent test by five points. How were the mean and the range of the scores affected?

6. The table below shows math quiz scores of seven students.

   5, 12, 7, 15, 20, 14, 7

   a) Determine the mean, median, and mode of the student scores, to the *nearest tenth*.

   b) Describe the effect on the mean, median, and mode if 5 bonus points are added to each of the students' scores.

7. The table shows the high and low temperatures for five cities. Which city had the greatest temperature range?

   TEMPERATURES ON OCTOBER 1ST
   FOR FIVE CITIES (in °F)

	High	Low
City A	72	50
City B	90	75
City C	83	72
City D	50	37
City E	92	72

8. Sara's test scores in mathematics were 64, 80, 88, 78, 60, 92, 84, 76, 86, 78, 72, and 90. Determine the mean, the median, and the mode of Sara's test scores.

9. Which statement is true about the data set
    4, 5, 6, 6, 7, 9, 12?

    (1)   mean = mode
    (2)   mode = median
    (3)   mean < median
    (4)   mode > mean

10. What is the relationship between the
    measures of central tendency of these
    data?

                22, 14, 19, 22, 8, 17

    (1)   mode > median > mean
    (2)   median > mode > mean
    (3)   mean > median > mode
    (4)   mode > mean > median

11. Based on the frequency table of student
    test grades below, which statement is true
    for the data?

    (1)   mean > median > mode
    (2)   mean > mode > median
    (3)   mode > median > mean
    (4)   median > mean > mode

Score	Frequency
96	2
92	5
88	3
84	2
78	4
60	1

12. Two social studies classes took the same
    current events examination that was
    scored on the basis of 100 points.

    Mr. Wong's class had a median score of 78
    and a range of 4 points, while Ms. Rizzo's
    class had a median score of 78 and a range
    of 22 points.

    Explain how these classes could have the
    same median score while having very
    different ranges.

13. The accompanying graph shows the high temperatures in Elmira, New York, for a 5-day period in January. Find the mean, median, and mode.

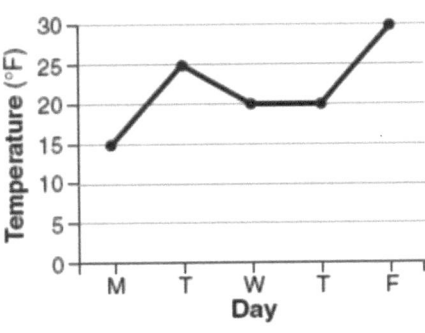

14. The table below shows the distribution of bowling scores. In which interval does the median lie?

Interval	Frequency
91–110	10
111–130	11
131–150	8
151–170	4
171–190	6
191–210	5

15. The values of 11 houses on West St. are shown in the table below.

Value per House	Number of Houses
$100,000	1
$175,000	5
$200,000	4
$700,000	1

Find the mean value and the median value of these houses in dollars.

State which measure of central tendency, the mean or the median, *best* represents the values of these 11 houses. Justify your answer.

16. The cumulative frequency table below shows the number of minutes students spent texting on a weekend.

Text-Use Interval (minutes)	Cumulative Frequency
41–50	2
41–60	5
41–70	10
41–80	19
41–90	31

Which 10-minute interval contains the median? Justify your choice.

Regents Questions

MULTIPLE CHOICE

1. Isaiah collects data from two different companies, each with four employees. The results of the study, based on each worker's age and salary, are listed in the tables below.

Company 1

Worker's Age in Years	Salary in Dollars
25	30,000
27	32,000
28	35,000
33	38,000

Company 2

Worker's Age in Years	Salary in Dollars
25	29,000
28	35,500
29	37,000
31	65,000

Which statement is true about these data?
   (1) The median salaries in both companies are greater than $37,000.
   (2) The mean salary in company 1 is greater than the mean salary in company 2.
   (3) The salary range in company 2 is greater than the salary range in company 1.
   (4) The mean age of workers at company 1 is greater than the mean age of workers at company 2.

2. The table below shows the annual salaries for the 24 members of a professional sports team in terms of millions of dollars.

0.5	0.5	0.6	0.7	0.75	0.8
1.0	1.0	1.1	1.25	1.3	1.4
1.4	1.8	2.5	3.7	3.8	4
4.2	4.6	5.1	6	6.3	7.2

The team signs an additional player to a contract worth 10 million dollars per year. Which statement about the median and mean is true?
   (1) Both will increase.
   (2) Only the median will increase.
   (3) Only the mean will increase.
   (4) Neither will change.

3.   The 15 members of the French Club sold candy bars to help fund their trip to Quebec. The
      table below shows the number of candy bars each member sold.

Number of Candy Bars Sold				
0	35	38	41	43
45	50	53	53	55
68	68	68	72	120

      When referring to the data, which statement is *false*?
      (1)   The mode is the best measure of central tendency for the data.
      (2)   The data have two outliers.
      (3)   The median is 53.
      (4)   The range is 120.

CONSTRUCTED RESPONSE

4.   Student scores on a recent test are shown in the table below.

85	96	92	82	90
90	88	95	85	88
90	87	96	82	85
92	96	85	92	87

      On the number line below, create a dot plot to model the data.

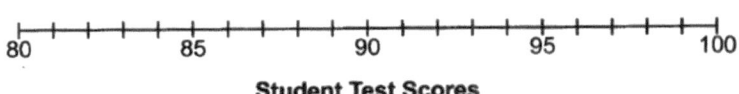

Student Test Scores

      State the median test score for the data set.

18.5   Distribution

A **dot plot** may be used to provide a graphic representation of a small set of data. In a dot plot, each data value is represented by a dot, or by a similar symbol such as a plus sign (+) or asterisk (*). The dots are stacked on top of a number line. The height of a stacked column of dots represents the frequency for that data value, just like the height of a bar represents the frequency for a data value in a frequency histogram.

Example: 29 students were asked to time their trip to school one morning and to round their results to the nearest multiple of 5 minutes. Their responses were plotted below. From the dot plot, we can see that two students took about 5 minutes each, one student took about 10 minutes, and so on.

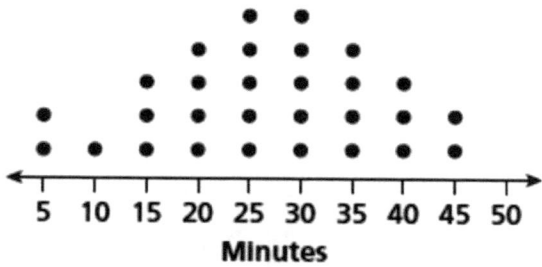

The **shape** of a dot plot can show the type of **distribution** of the data. If the data values are somewhat equally distributed, we call it a **uniform distribution**. In a **symmetrical distribution**, one could draw a vertical line on the dot plot that would divide it into two parts that are approximate mirror images of each other. A **skewed distribution** is neither uniform nor symmetric; the data is stacked mostly on the low end or on the high end of the dot plot.

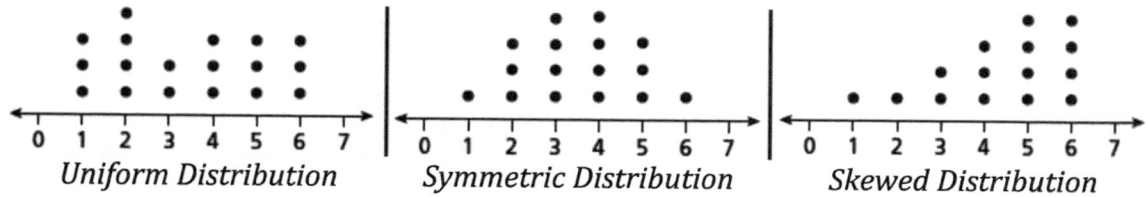

When the data tends to gather around a central value in a *symmetrical* dot plot, it is said to have a **normal distribution**, and the shape is often called a *bell curve*. The middle dot plot above is an example of a normal distribution of data; notice the bell-shaped curve as shown below:

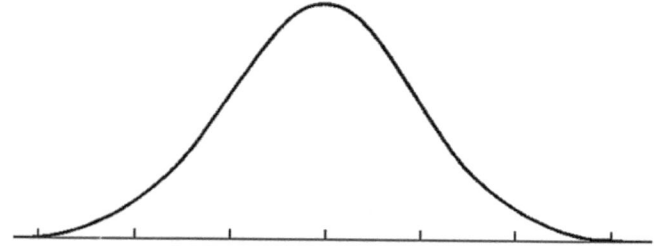

In a **skewed distribution**, if the graph appears to have a tail to the right (and more of the data to the left), it is **skewed to the right**, or *positively skewed*. If it appears to have its data stretched to the left like a tail, it is considered **skewed to the left**, or *negatively skewed*.
Example: The following dot plot is skewed to the left.

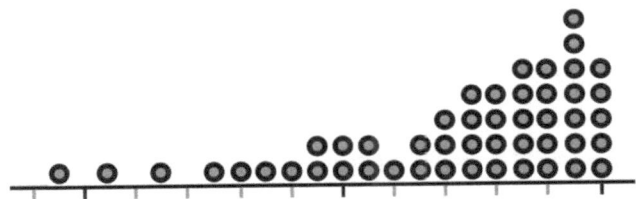

A *skewed distribution* can easily be seen in a frequency histogram. The histogram on the left shows a somewhat *symmetrical distribution*. We can draw a vertical line down the middle so that the left and right sides would have approximately the same shape. The histogram on the right is skewed to the right (*positively skewed*), in that it has a "tail" to the right and more of the data to the left.

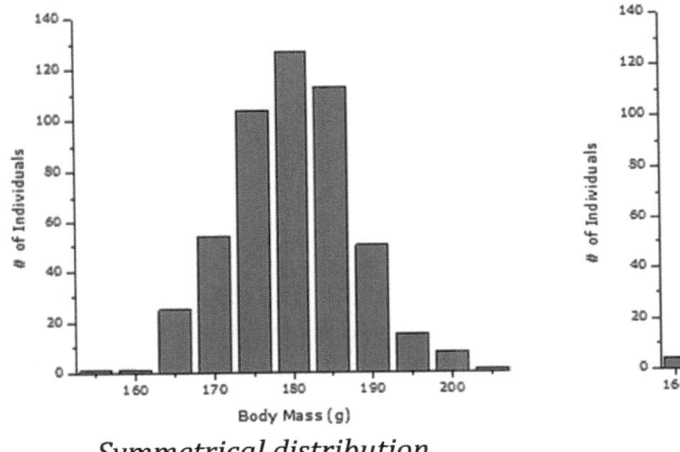

Symmetrical distribution

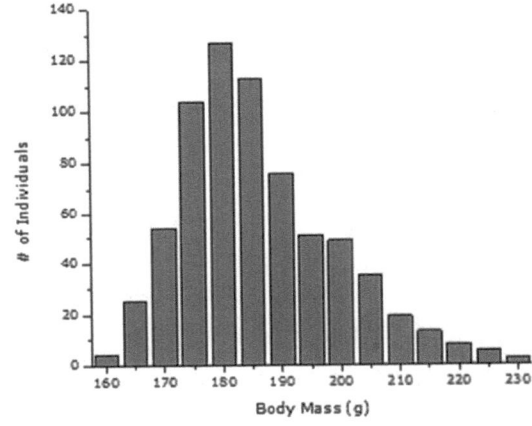

Skewed distribution

When the data has a symmetrical distribution, the mean and median tend to be approximately equal (at the center of the bell) and either can be used as measures of central tendency. However, when the data is skewed, the median is probably a better indicator. The mean will be greater than the median when the data is skewed right (positive skew) or less than the median when the data is skewed left (negative skew).

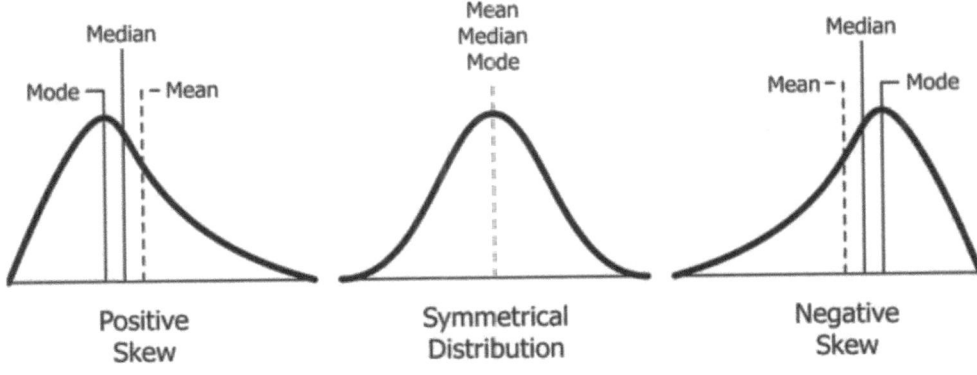

The mode is always at the highest point on the graph (the "peak"). Usually, the mode will be less than the median when data is skewed right or greater than the median when data is skewed left.

For a more informative analysis, the *standard deviation* can be calculated with the mean, or the *quartiles* may be calculated with the median; these will be presented in later sections.

MODEL PROBLEM 1:   *READING DOT PLOTS*

The following dot plot shows the fuel economy for a number of cars in miles per gallon (mpg). How many cars have a fuel economy between 30 and 40 mpg inclusive?

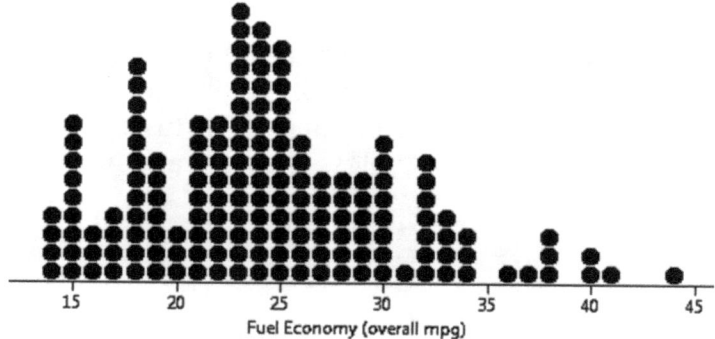

Solution:

   30 cars

Explanation:

   Count the dots stacked vertically. *[There are 8 dots at the 30 mpg marker, 2 dots at the 40 mpg marker, and 20 more dots stacked between these two markers.]*

PRACTICE PROBLEMS

1. Given the dot plot below, what percent of the data values are less than 5?

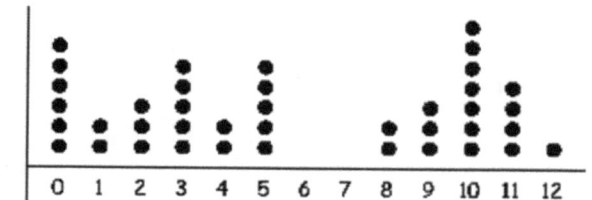

2. Given the dot plot below, what are the mean, median, and mode of the data?

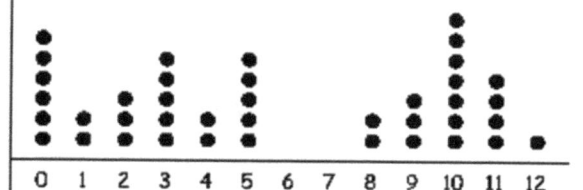

MODEL PROBLEM 2:   *DISTRIBUTION AND SHAPE*

The dot plot below shows the number of televisions owned by each family on a city block.

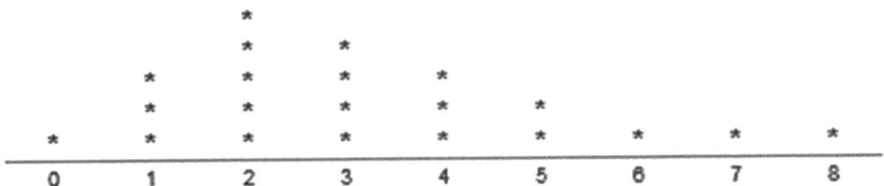

Which of the following statements are true?
   (1) The distribution is right-skewed with no outliers.
   (2) The distribution is right-skewed with one outlier.
   (3) The distribution is left-skewed with no outliers.
   (4) The distribution is left-skewed with one outlier.
   (5) The distribution is symmetric.

Solution:
The correct answer is (1).

Explanation:
Most of the observations are on the left side of the distribution, so the distribution is right-skewed. And none of the observations is extreme, so there are no outliers.

PRACTICE PROBLEMS

3. Identify the distribution of the data represented by the dot plot below. 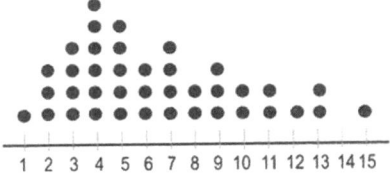	4. Describe the distribution. 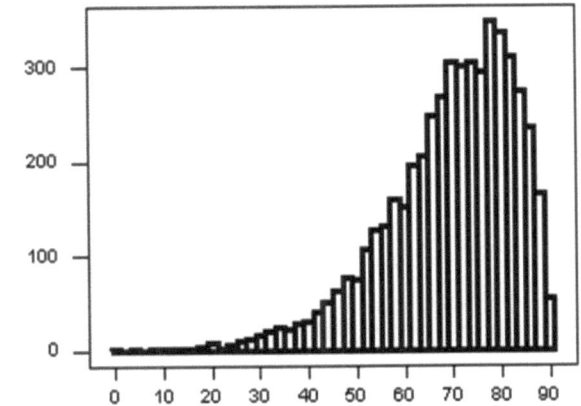

5. Describe the distribution.

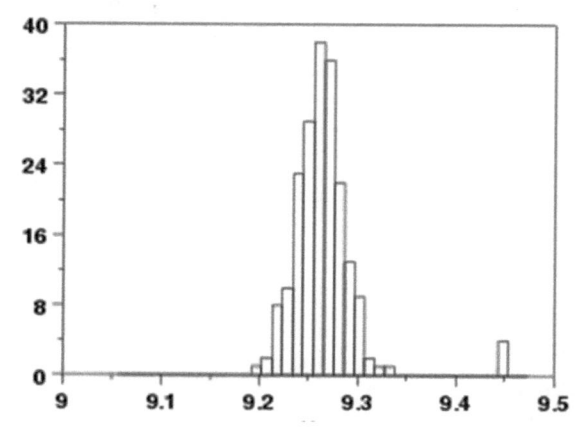

6. Draw a dot plot of data that has a symmetrical but *not* a normal distribution.

Regents Questions

MULTIPLE CHOICE

1. Noah conducted a survey on sports participation. He created the following two dot plots to represent the number of students participating, by age, in soccer and basketball.

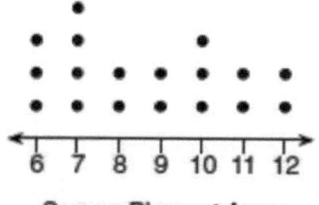

Soccer Players' Ages

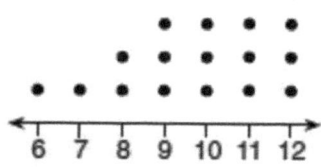

Basketball Players' Ages

Which statement about the given data sets is correct?
    (1) The data for soccer players are skewed right.
    (2) The data for soccer players have less spread than the data for basketball players.
    (3) The data for basketball players have the same median as the data for soccer players.
    (4) The data for basketball players have a greater mean than the data for soccer players.

18.6 Standard Deviation

The **standard deviation** is a measure of the **spread** (also known as *variability* or *dispersion*) of a set of data. A *low* standard deviation means the data values tend to be *close to the mean*, while a *high* standard deviation means the values are *more spread out*. In other words, more **consistent** data values should result in a **lower** standard deviation.

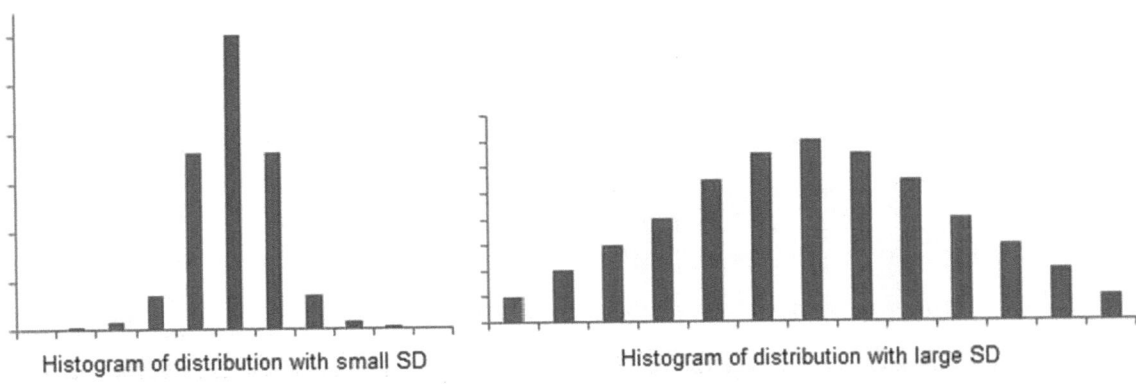

Histogram of distribution with small SD   Histogram of distribution with large SD

For the purposes of this course, we will calculate the **sample standard deviation**, which is represented by the letter, *s*.

The sample standard deviation is preferred when we are attempting to generalize from a subset of the data. For example, suppose we want to find the average age among *all registered U.S. voters*. This group represents the **population**, estimated at over 150 million people (which is still only half of the estimated general population of the U.S.). To determine the average age, we might take a **sample** of, say, one million registered voters. Even if this sample is random and unbiased, the average age of the sample may be close to, but not necessarily equal to, the average age of the population. After all, even this very large sample is still less than 1% of the entire population of registered voters.

Now, let's calculate a sample standard deviation.

The difference between each value and the mean is called the **deviation** of that data value. That is, for sample data, the deviation of each x is $x - \bar{x}$.

We find the **variance** by finding the sum of the squared deviations of all the values and dividing this sum by $n - 1$. That is, the variance, s^2, is calculated as

$s^2 = \frac{\Sigma(x - \bar{x})^2}{n-1}$ (where Σ means the *sum of each*).

The standard deviation is the square root of the variance: $s = \sqrt{\frac{\Sigma(x - \bar{x})^2}{n-1}}$.

To calculate the standard deviation of n data values:
1. Calculate the mean, $\bar{x}$.
2. Calculate the *deviation* of each value from the mean, $x - \bar{x}$.
3. Square each deviation from the mean, $(x - \bar{x})^2$.
4. The *variance* is the sum of these squares divided by $n - 1$.
5. The *standard deviation* is the square root of the variance.

When calculating a standard deviation by hand, it is helpful to use a table.

Example: To calculate the sample standard deviation for $12, 15, 36, 49, 62, 84$:

a) the sample mean $\bar{x} = \dfrac{258}{6} = 43$,

b) use a table to find each deviation and their squares,

x	$(x - \bar{x})$	$(x - \bar{x})^2$
12	−31	961
15	−28	784
36	−7	49
49	6	36
62	19	361
84	41	1681
258		**3872**

c) the variance is the sum of the squares (the sum of the last column) divided by $n - 1$. Therefore, $s^2 = \dfrac{3872}{5} = 774.4$.

d) the sample standard deviation is the square root of the variance: $s = \sqrt{774.4} \approx 27.8$.

Thankfully, we can have the calculator find the standard deviation for us, as directed below.

CALCULATOR TIP

To find the standard deviation on the calculator:

1. First, press [STAT][1] and enter the data into the L1 column, or enter the data as a frequency table into L1 and L2.

2. Press [STAT]<CALC>[1] for 1-Var Stats. Be sure that L1 is specified as the List, or press [2nd][L1] to enter it. If L2 contains frequencies, enter [2nd][L2] as FreqList; otherwise, leave it blank. Then, select Calculate.

   [On the TI-83, you need to enter L1 after the 1-Var Stats prompt by pressing [2nd][L1]. If L2 contains frequencies, press [,][2nd][L2]. Then, press [ENTER].]

3. The sample standard deviation is the value of Sx.

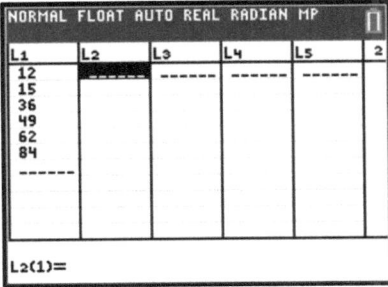

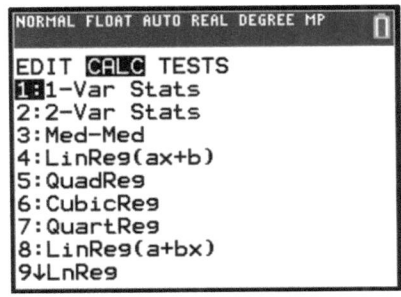

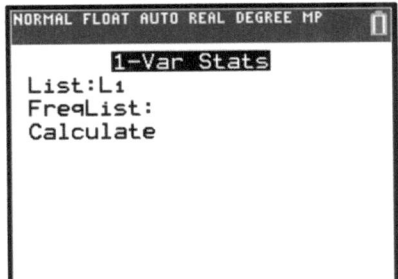

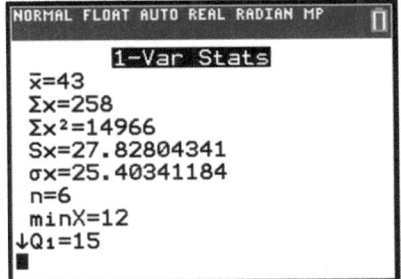

603

MODEL PROBLEM 1:   *POPULATION AND SAMPLE*

A survey of 1353 American households found that 18% of the households own a computer. Identify the population and sample.

Solution:

> population = all American households
> sample = the 1353 American houeholds that were surveyed

Explanation:

> The population is the entire set we want to study. The sample is the subset that we actually survey.

PRACTICE PROBLEMS

1. Identify the population and the sample: A manufacturer received a large shipment of bolts. The bolts must meet certain specifications to be useful. Before accepting shipment, 100 bolts were selected, and it was determined whether or not each met specifications.	2. Identify the population and the sample: The average weight of every sixth person entering the mall within a 3 hour period was 146 lb.

MODEL PROBLEM 2:   *CONCEPTUALIZING STANDARD DEVIATION*

In the following chart, stock prices for companies UCX and SWC are recorded at the start of each month over a seven month period. Both stocks have the same mean price of $70 over that time. Which stock's monthly prices have a greater standard deviation, according to the graph?

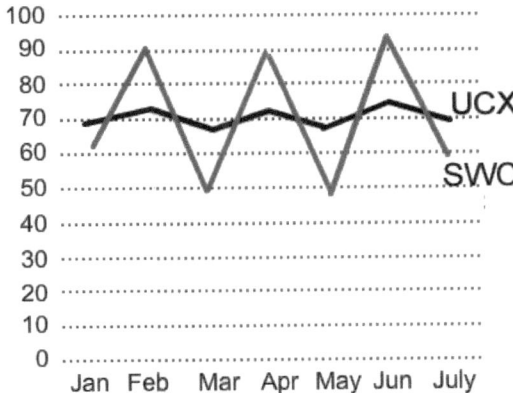

Solution:
SWC stock has a larger standard deviation.

Explanation:
The data set with the greater variability will have the higher standard deviation.
[The UCX prices are much more consistent, with prices closer to the mean, whereas the SWC prices fluctuate greatly, with monthly prices that are more distant from the mean.]

PRACTICE PROBLEMS

3. Jason and Eric discovered that the means of their grades for the first marking period in their math class were identical. They also noticed that the standard deviation of Jason's grades is 20.7, while the standard deviation of Eric's grades is 2.7. Which statement must be true?

   (1) In general, Eric's grades were lower than Jason's grades.
   (2) Eric's grades are more consistent than Jason's grades
   (3) Eric had more failing grades during the marking period than Jason had.
   (4) The median for Eric's grades is lower than the median for Jason's grades.

4. Billie's scores on five science tests were 98, 97, 99, 98, and 96. Her scores on five history tests were 78, 84, 95, 72, and 79. Which statement is true about the standard deviations for the scores?

   (1) The standard deviation for the history scores is greater than the standard deviation for the science scores.
   (2) The standard deviation for the science scores is greater than the standard deviation for the history scores.
   (3) The standard deviations for both sets of scores are equal.
   (4) More information is needed to determine the relationship between the standard deviations.

5. The mean for a set of data is 8.9 and the standard deviation is 1. The mean for a second set of data is 8.9 and the standard deviation is 2. In which data set do the values cluster closer to the mean?

6. On their college admission exams, Quincy College had a mean score of 875 and a standard deviation of 12. McCrane College had a mean score of 855 and a standard deviation of 20. In which school was there greater variability in the scores? Explain how you arrived at your answer.

MODEL PROBLEM 3:   CALCULATING STANDARD DEVIATION

Find the standard deviation of this set of data.
    $2, 4, 4, 4, 5, 5, 5, 7, 9$

Solution:

(A) $\dfrac{2+4+4+4+5+5+5+7+9}{9} = \dfrac{45}{9} = 5$

(B)

x	$(x - \bar{x})^2$
2	$(2 - 5)^2 = 9$
4	$(4 - 5)^2 = 1$
4	$(4 - 5)^2 = 1$
4	$(4 - 5)^2 = 1$
5	$(5 - 5)^2 = 0$
5	$(5 - 5)^2 = 0$
5	$(5 - 5)^2 = 0$
7	$(7 - 5)^2 = 4$
9	$(9 - 5)^2 = 16$
45	**32**

Explanation of steps:

(A) Calculate the mean *[5]*.

(B) For each data value, find the square of the difference from the mean.

(C) Find the variance by dividing the sum of the squares by $n - 1$ *[32 ÷ 8]*.

(D) Take the square root of the variance *[2]*.

(C) $\dfrac{9+1+1+1+0+0+0+4+16}{9-1} = \dfrac{32}{8} = 4$

(D) The standard deviation $= \sqrt{4} = 2$.

PRACTICE PROBLEMS

7. Using the calculator, find the mean and standard deviation to the *nearest tenth*. 22, 99, 102, 33, 57, 75, 100, 81, 62, 29	8. Using the calculator, find the mean and standard deviation to the *nearest tenth*. 35, 50, 60, 60, 75, 65, 80
9. The scores on a mathematics test are: 42, 51, 58, 64, 70, 76, 76, 82, 84, 88, 88, 90, 94, 94, 94, 97 For this set of data, find the standard deviation to the *nearest tenth*.	10. The following Apgar scores were recorded on one day at a local hospital: 9, 8, 10, 9, 8, 10, 9, 10, 8, 10. Find the standard deviation of the scores, to the *nearest hundredth*.
11. A New York weather bureau recorded snowfalls of more than 6 inches over a four year period. The snowfall amounts, in inches, were as follows: 7.1, 9.2, 8.0, 6.1, 14.4, 8.5, 6.1, 6.8, 7.7, 21.5, 6.7, 9.0, 8.4, 7.0, 11.5, 14.1, 9.5, 8.6 Find the mean and standard deviation to the *nearest hundredth*.	12. Use the method shown in the model problem above to find the mean and standard deviation to the *nearest tenth*. 51, 48, 47, 46, 45, 43, 41, 40, 40, 39 Then verify your answer using the calculator.

13. This chart shows the weekly salary of five employees. Find the mean and standard deviation of this data.

Employee Number	Salary
3201	$612
2734	$588
2461	$604
3582	$625
3144	$621

14. This table shows the age at inauguration of ten presidents of the United States.

President	Age at Inauguration
Harry Truman	60
Dwight D. Eisenhower	62
John F. Kennedy	43
Lyndon B. Johnson	55
Richard M. Nixon	56
Gerald R. Ford	61
Jimmy Carter	52
Ronald Reagan	69
George Bush	64
Bill Clinton	46

Find, to the *nearest tenth*, the standard deviation of the age at inauguration of these ten presidents.

Regents Questions

MULTIPLE CHOICE

1. The two sets of data below represent the number of runs scored by two different youth baseball teams over the course of a season.

    Team *A*:   4, 8, 5, 12, 3, 9, 5, 2
    Team *B*:   5, 9, 11, 4, 6, 11, 2, 7

    Which set of statements about the mean and standard deviation is true?

    (1) mean *A* < mean *B*
         standard deviation *A* > standard deviation *B*

    (2) mean *A* > mean *B*
         standard deviation *A* < standard deviation *B*

    (3) mean *A* < mean *B*
         standard deviation *A* < standard deviation *B*

    (4) mean *A* > mean *B*
         standard deviation *A* > standard deviation *B*

2. The following table shows the heights, in inches, of the players on the opening-night roster of the 2015-2016 New York Knicks.

84	80	87	75	77	79	80	74	76	80	80	82	82

    The population standard deviation of these data is approximately

    (1) 3.5   (3) 79.7
    (2) 13   (4) 80

CONSTRUCTED RESPONSE

3. Santina is considering a vacation and has obtained high-temperature data from the last two weeks for Miami and Los Angeles.

Miami	76	75	83	73	60	66	76
	81	83	85	83	87	80	80

Los Angeles	74	63	65	67	65	65	65
	62	62	72	69	64	64	61

    Which location has the least variability in temperatures? Explain how you arrived at your answer.

18.7   Percentiles and Quartiles

Quantiles divide an ordered set of data into adjacent subgroups of equal sizes. There is always one fewer quantile than the number of subgroups created.

Examples: A median is a quantile that cuts a set of data into two equal halves. Quartiles and percentiles divide data into 4 or 100 equal parts, respectively. The three quartiles are represented by Q_1, Q_2, and Q_3 and the 99 percentiles are represented by P_1 through P_{99}.

A value's **percentile** is the percent of values in a set of data that lie below the given value. The percent is generally rounded to the nearest whole percent and the given data value is stated as being "at the pth percentile."

Unfortunately, there is no general consensus as to how a percentile should be calculated, but one commonly used formula is $p = \dfrac{b}{n}$, where b is the number of values that are below the given value, and n is the number of data values in the set.

Examples: (1) Jake is the third tallest in a group of ten students. Since 7 out of the 10 students are shorter than Jake, his height is at the 70th percentile.

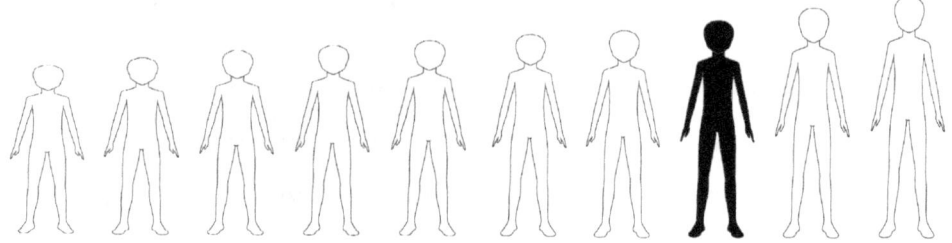

(2) Andy scores an 85 on an exam, which is better than 300 of the 325 exam scores. Andy's score is at the 92nd percentile. $(300 \div 325 \approx 92.3\%.)$

Quartiles divide data into quarters. For a set of data, the value at the 25th percentile is called the **first quartile** or **lower quartile** (and is often denoted as Q_1). The value at the 50th percentile is called the **second quartile** or **median** (or Q_2), and the value at the 75th percentile is called the **third quartile** or **upper quartile** (or Q_3).

To identify the quartiles for a set of data:
1. Arrange the data in ascending order.
2. First find the *median (or second quartile)*, and label it Q2. It is the middle value (for an odd number of data values), or the average of the two middle values (for an even number of data values).

   > **Use this tip:** For an *odd number* of data values, *circle* the middle value, but for an *even number* of data values, *draw a vertical line* between the two middle values. When we find the quartiles in the next steps, we will ignore a circled value but include values to the left or right of a line.

3. Find the *lower (first) quartile* by looking only at the subgroup of values that are to the *left* of the middle (that is, to the left of a circled value or line). The median of this subgroup is the lower quartile.
4. Find the *upper (third) quartile* by looking only at the subgroup of values that are to the *right* of the middle (to the right of a circled value or line). The median of this subgroup is the upper quartile.

The difference between the upper and lower quartiles, $Q_3 - Q_1$, is called the **interquartile range** (or IQR). The interquartile range is another measure of spread, along with range and standard deviation. We can think of it as the range of the middle 50% of the data.

The IQR can be used to identify *outliers*, which are extremely high or low values in the data. A data value should be considered an outlier if it is more than $1.5 \times$ IQR below the lower quartile, $Q1$, or $1.5 \times$ IQR above the upper quartile, $Q3$.

Example: For a set of data, $Q_1 = 60$, $Q_2 = 80$, and $Q_3 = 100$. The IQR is $100 - 60 = 40$. Since $1.5 \times 40 = 60$, a value that is above $100 + 60 = 160$, such as 180, would be considered an outlier.

CALCULATOR TIP

To find the quartiles for a set of data:

1. First, press STAT 1 and enter the data into the L1 column, or enter the data as a frequency table into L1 and L2.

2. Press STAT <CALC> 1 for 1-Var Stats. Be sure that L1 is specified as the List, or press 2nd[L1] to enter it. If L2 contains frequencies, enter 2nd[L2] as FreqList; otherwise, leave it blank. Then, select Calculate.

   [On the TI-83, you need to enter L1 after the 1-Var Stats prompt by pressing 2nd[L1]. If L2 contains frequencies, press , 2nd[L2]. Then, press ENTER.]

3. Scroll down to find the number of data values, n; the minimum, minX; the first quartile, Q1; the median, Med; the third quartile, Q3; and the maximum, maxX.

   [This example uses the data set, 12, 15, 36, 49, 62, 84.]

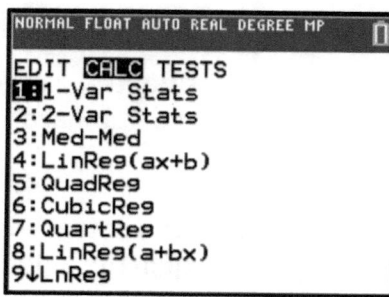 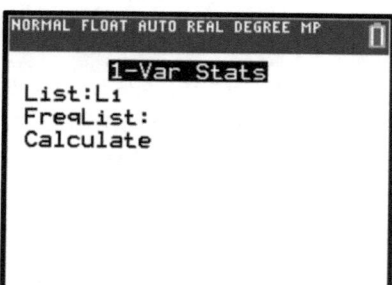 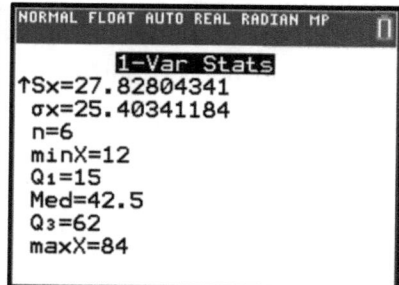

Once these statistics are calculated, they may be used in an expression by pressing VARS 5. minX and maxX are listed under <XY>, while Q1, Med, and Q3 are listed under <PTS>.

Example: We can calculate the IQR by typing VARS 5 <PTS> 9 − VARS 5 <PTS> 7 ENTER.

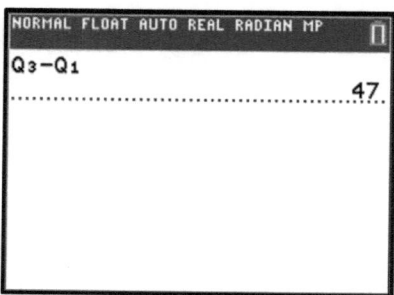

MODEL PROBLEM

Identify the first, second and third quartiles and calculate the interquartile range for the following set of ten data values:

86, 72, 85, 89, 86, 92, 73, 71, 91, 82

Solution:

(A)

71, 72, ⓐ73ⓐ 82, 85, | 86, 86, ⓐ89ⓐ 91, 92

(B) Q_2 (median) = 85.5
(C) $Q_1 = 73$ and $Q_3 = 89$
(D)   IQR = 89 − 73 = 16

Explanation of steps:

(A)   Arrange the data in ascending order.
(B)   Find the median. *[Since there is an even number of values, we draw a line between the two middle values, 85 and 86, and calculate the median (or second quartile) as the average of these two middle values.]*
(C)   Find the lower and upper quartiles. *[For the subgroup of 5 values to the left of the line, the middle number (73) is the lower quartile. For the subgroup of 5 values to the right of the line, the middle number (89) is the upper quartile.]*
(D)   Calculate the interquartile range (IQR) as the difference between Q_3 and Q_1.

PRACTICE PROBLEMS

1. The weights of 40 students were recorded. If the 75th percentile of their weights was 150 pounds, what is the total number of students who weighed *more than* 150 pounds?	2. Brian's score on a college entrance exam exceeded the scores of 95,000 of the 125,000 students who took the exam. What was his score's percentile?

3. Dawn scored higher than 22 out of the 30 students in her class. What was her score's percentile?

4. For the given set of data, what percentile is the value 70?
   25, 90, 87, 58, 42, 95, 64, 75, 39, 70

5. The students at Adams High School held a canned food drive for 12 weeks. The results are summarized in the table below.

Canned Food Drive Results

Week	1	2	3	4	5	6	7	8	9	10	11	12
Number of Cans	20	35	32	45	58	46	28	23	31	79	65	62

Find the second quartile of the number of cans of food collected.

6. Find the first, second and third quartiles for the following set of data:
   5, 6, 7, 8, 12, 14, 17, 17, 18, 19, 19

7. Find the first, second and third quartiles for the following set of data:
   3, 6, 7, 7, 8, 9, 9, 9, 10, 12, 13, 15

8. Find the first, second and third quartiles, and the interquartile range, for the following set of data:
33, 28, 45, 21, 32, 53, 41, 28, 50

9. The heights, in inches, of 10 high school varsity basketball players are given below. Find the interquartile range.
78, 79, 79, 72, 75, 71, 74, 74, 83, 71

10. For the set of test scores shown by the frequency table below, find the first, second and third quartiles.

Score	Frequency
60	1
70	9
80	8
90	2
100	5

11. The cumulative frequency table below shows the length of time that 30 students spent texting on a weekend.

Minutes Used	Cumulative Frequency
31–40	2
31–50	5
31–60	10
31–70	19
31–80	30

Which 10-minute interval contains the first quartile?

Regents Questions

MULTIPLE CHOICE

1. Christopher looked at his quiz scores shown below for the first and second semester of his Algebra class.

    Semester 1:   78, 91, 88, 83, 94
    Semester 2:   91, 96, 80, 77, 88, 85, 92

    Which statement about Christopher's performance is correct?
    (1)   The interquartile range for semester 1 is greater than the interquartile range for semester 2.
    (2)   The median score for semester 1 is greater than the median score for semester 2.
    (3)   The mean score for semester 2 is greater than the mean score for semester 1.
    (4)   The third quartile for semester 2 is greater than the third quartile for semester 1.

2. The dot plot shown below represents the number of pets owned by students in a class.

    Which statement about the data is *not* true?
    (1)   The median is 3.   (3)   The mean is 3.
    (2)   The interquartile range is 2.   (4)   The data contain no outliers.

3. The heights, in inches, of 12 students are listed below.

    61, 67, 72, 62, 65, 59, 60, 79, 60, 61, 64, 63

    Which statement best describes the spread of these data?
    (1)   The set of data is evenly spread.
    (2)   The median of the data is 59.5.
    (3)   The set of data is skewed because 59 is the only value below 60.
    (4)   79 is an outlier, which would affect the standard deviation of these data.

4. Donna and Andrew compared their math final exam scores from grade 8 through grade 12. Their scores are shown below.

Donna	
8th	90
9th	92
10th	87
11th	94
12th	95

Andrew	
8th	78
9th	96
10th	87
11th	94
12th	93

Which statement about their final exam scores is correct?

(1) Andrew has a higher mean than Donna.

(2) Donna and Andrew have the same median.

(3) Andrew has a larger interquartile range than Donna.

(4) The 3rd quartile for Donna is greater than the 3rd quartile for Andrew.

CONSTRUCTED RESPONSE

5. The heights, in feet, of former New York Knicks basketball players are listed below.

6.4	6.9	6.3	6.2	6.3	6.0	6.1	6.3	6.8	6.2
6.5	7.1	6.4	6.3	6.5	6.5	6.4	7.0	6.4	6.3
6.2	6.3	7.0	6.4	6.5	6.5	6.5	6.0	6.2	

Using the heights given, complete the frequency table below.

Based on the frequency table created, draw and label a frequency histogram on the grid below.

Interval	Frequency
6.0 – 6.1	
6.2 – 6.3	
6.4 – 6.5	
6.6 – 6.7	
6.8 – 6.9	
7.0 – 7.1	

Determine and state which interval contains the upper quartile. Justify your response.

6. The students in Mrs. Lankford's 4th and 6th period Algebra classes took the same test. The results of the scores are shown in the following table:

	$\bar{x}$	σ_x	n	min	Q_1	med	Q_3	max
4th Period	77.75	10.79	20	58	69	76.5	87.5	96
6th Period	78.4	9.83	20	59	71.5	78	88	96

Based on these data, which class has the larger spread of test scores? Explain how you arrived at your answer.

7. The ages of the last 16 United States presidents on their first inauguration day are shown in the table below.

51	54	51	60
62	43	55	56
61	52	69	64
46	54	47	70

Determine the interquartile range for this set of data.

18.8 Box Plots

A **box plot** (or **box-and-whisker plot**) graphically summarizes statistics for a set of data. The "box" marks the values of the *first, second,* and *third quartiles* of the data set. The "whiskers" show the *minimum* (lowest) and *maximum* (highest) values of the data set. These five values are called the *five-number summary*.

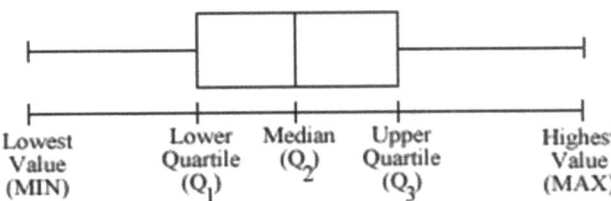

A boxplot gives us one measure of center – the *median* – and also provides information that allows us to calculate two measures of spread – the *range* (which is the difference between the maximum and minimum values) and the *interquartile range* or IQR (which is the difference between the upper and lower quartiles).

By the definition of quartiles, each section of a box plot represents 25% (one quarter) of the data values. 25% of the data is at or below the lower quartile, 50% is at or below the median, and 75% is at or below the upper quartile.

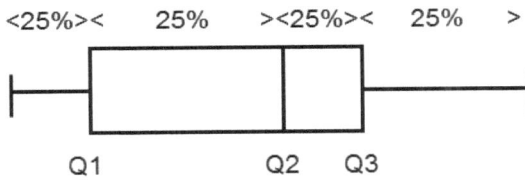

The width of each part of the boxplot represents how spread out the data is within that quarter.

Example: For the above boxplot, although there are the same number of data values between Q_1 and Q_2 as there are between Q_2 and Q_3 (25% of the data in each), the boxplot shows that data between Q_1 and Q_2 are more spread out than the data between Q_2 and Q_3.

If the data has a **symmetrical distribution**, the whiskers will be about the same length and the median will be at about the center of the box. For a **skewed distribution**, the whisker on the side of the "tail" will tend to be longer and the median within the box will tend to be closer to the side with the shorter whisker.

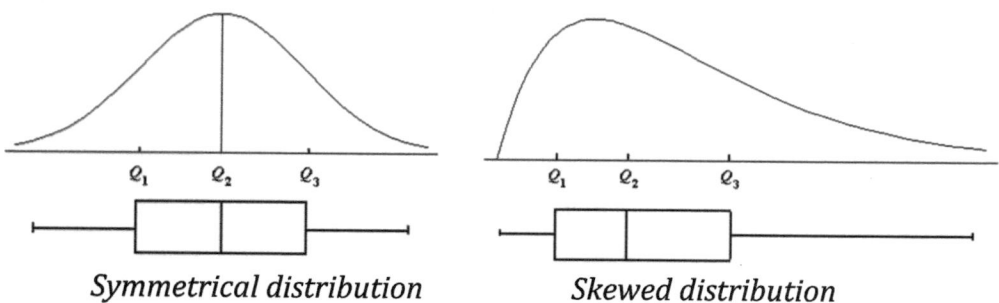

Symmetrical distribution   *Skewed distribution*

This does *not* imply that two distributions with the same boxplot will have the same shape.

Example: The three distributions below all have the same five-number summary (min = 1, $Q_1 = 2$, $Q_2 = 3.5$, $Q_3 = 5$, and max = 6), resulting in the same boxplot, shown below. Clearly, however, they have very different shapes.

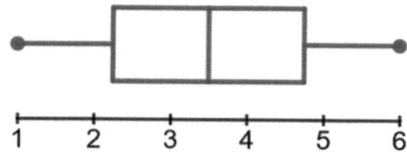

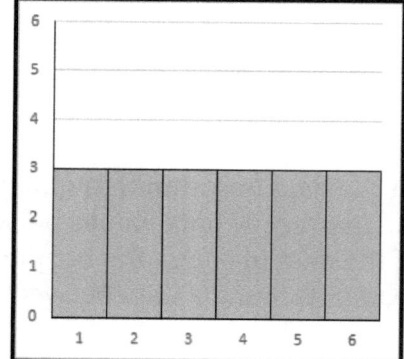   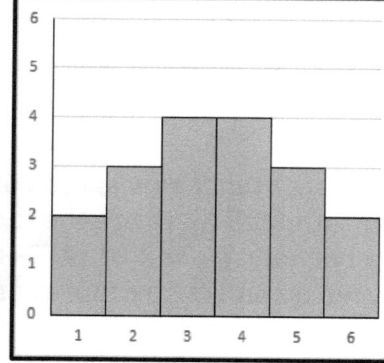   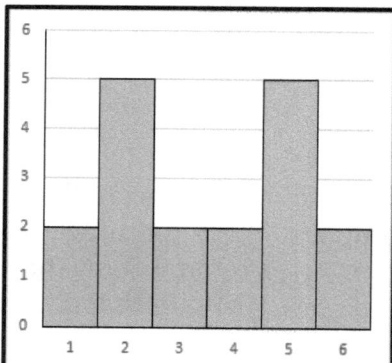

To create a boxplot:

1. Arrange the data from smallest to largest. Create a five-number summary.
2. Draw a number line from below the minimum to above the maximum.
3. Draw a box from the lower to the upper quartile (Q_1 to Q_3).
4. Draw a line through the box at the median.
5. Draw the minimum and maximum as smaller lines and connect them as whiskers.

 CALCULATOR TIP

To display summary data as a box plot.

1. First, press STAT 1 and enter the data into the L1 column, or enter the data as a frequency table into L1 and L2.
2. Press 2nd STAT PLOT 1 to select Plot1. Select <On> ENTER. For the Type, select the fifth chart type, ⊡⊢. For Xlist, if L1 isn't already selected, press 2nd L1 to enter it. For Freq, enter 1 if all of your data is in L1, or enter 2nd L2 if your data is entered as a frequency table with the frequencies stored in L2.
3. Press ZOOM 1 for ZBox. You may need to adjust the WINDOW settings, as shown below, to see the entire box plot clearly, depending on the range of values.
4. Press TRACE and then the ◁▷ keys to see the five summary numbers: minX, Q1, Med, Q2, and maxX.

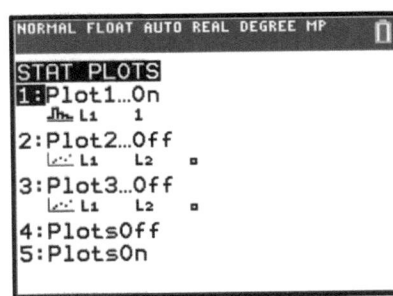 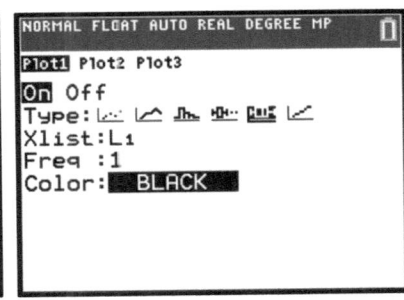 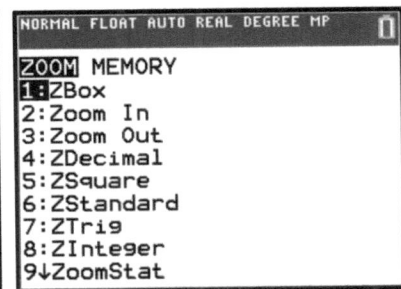

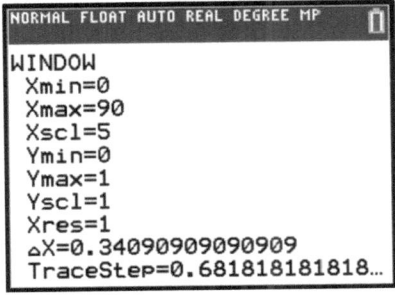 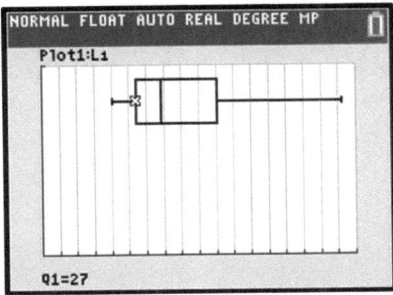

Box plots with outliers: Sometimes, outliers are drawn separately as dots (or similar symbols such as asterisks) on a box plot and excluded from the lengths of the whiskers. Remember, an *outlier* is considered to be any data value that is more than $1.5 \times IQR$ above the upper quartile or more than $1.5 \times IQR$ below the lower quartile.

Example: Consider the data set, {20, 25, 25, 27, 28, 31, 33, 34, 36, 37, 44, 50, 59, 85, 86}. The quartiles are 27, 34, and 50, so the box plot would look like this:

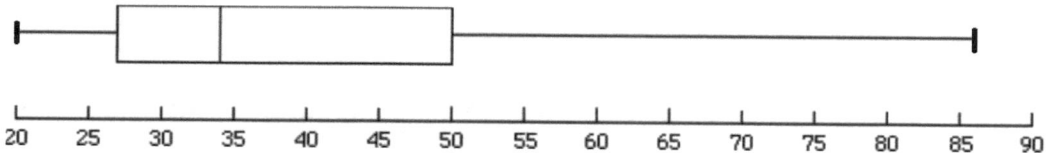

Notice the long right whisker due to the fact that two outliers, 85 and 86, are included. We know these are outliers because they are more than $1.5 \times IQR$ above the upper quartile.

$$IQR = Q_3 - Q_1 = 50 - 27 = 23$$
$$Q_3 + (1.5 \times IQR) = 50 + (1.5 \times 23) = 84.5$$

So, any values above 84.5 are outliers. Therefore, 85 and 86 are outliers.

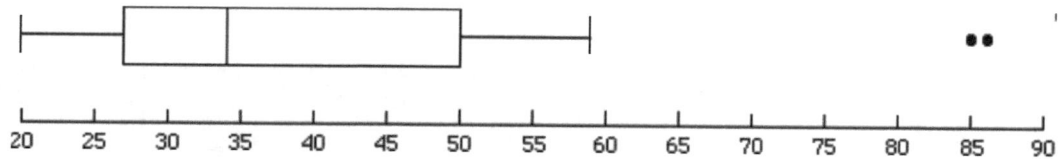

If we show these outliers as dots, the box plot would appear as above. The right whisker is shortened in length to the next largest (*adjacent*) value below the outliers, which happens to be 59. This now shows a big gap below the outliers.

We can check that there are no outliers at the low end in the same way:
$$Q_1 - (1.5 \times IQR) = 27 - (1.5 \times 23) = -7.5$$
There are no data values below –7.5, so there are no outliers on the low end.

CALCULATOR TIP

To show outliers on boxplots:

After pressing [2nd][STAT PLOT][1] to select Plot1, you can choose the fourth chart type, ⊡⋯, instead of the fifth chart type, ⊡⊢. Then, choose the Mark you prefer to use for outliers. This will show the outliers in your boxplot as marks separated from the tails.

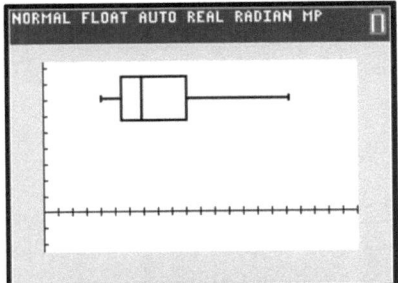

standard boxplot

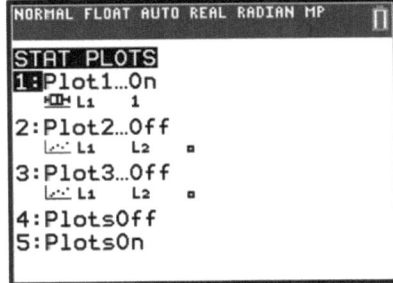

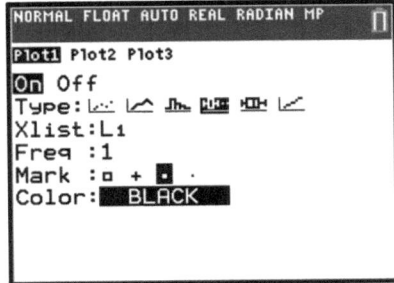

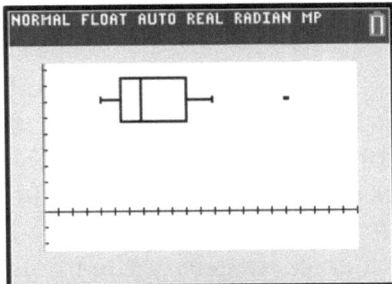

boxplot with outliers shown

MODEL PROBLEM

Create a box plot for the following data:
        2, 4, 8, 10, 14, 18, 20, 22, 30, 32, 38

Solution:

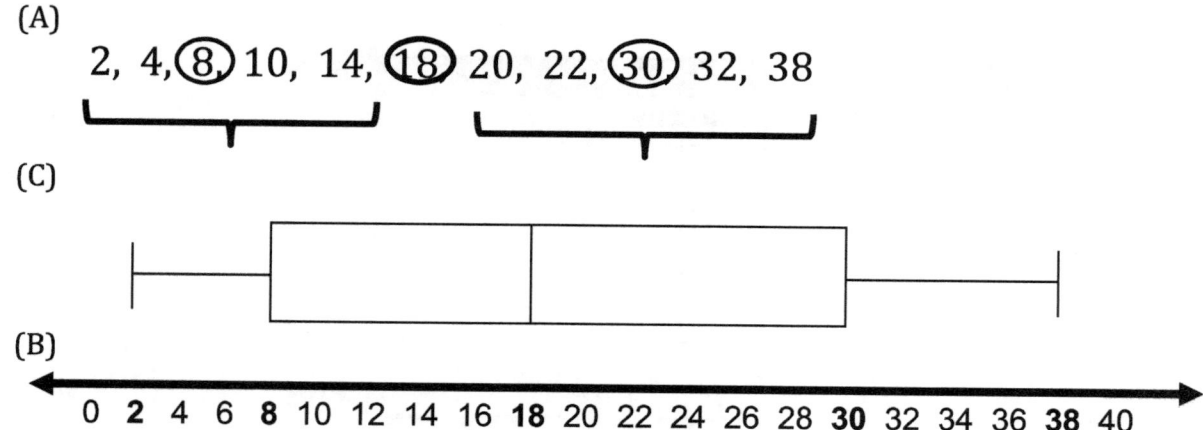

Explanation of steps:

(A) Once the data is arranged in ascending order *[as this data already is]*, create a five-number summary. *[Min = 2, Q_1 = 8, Q_2 = Med = 18, Q_3 = 30, Max = 38.]*

(B) draw a number line from below the minimum *[2]* to above the maximum *[38]*, and label the intervals. *[Our number line can go from 0 to 40 in 2-unit intervals.]*

(C) Draw a box from the lower to the upper quartile *[8 to 30]*. Draw a line through the box at the median *[18]*. Add shorter lines over the minimum *[2]* and maximum *[38]* and connect them to the box as whiskers.

PRACTICE PROBLEMS

1. The accompanying diagram shows a box plot of student test scores. What is the median score?	2. The accompanying box plot represents the scores earned on a science test. What is the median score?

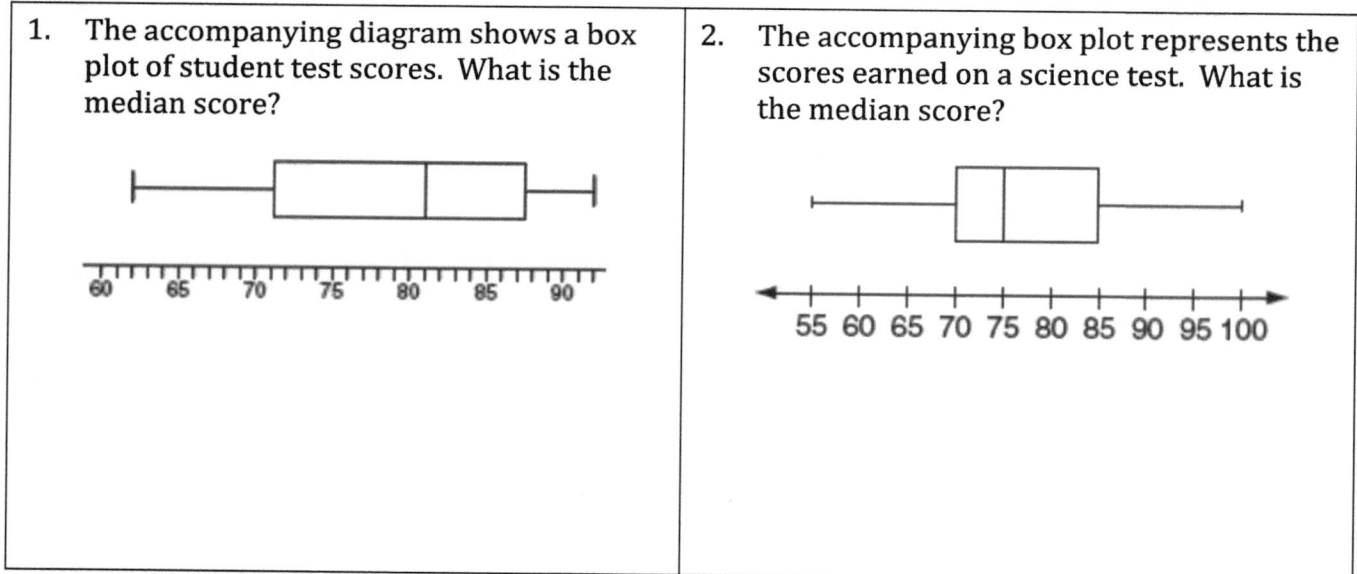

3. What is the value of the third quartile shown on the box-and-whisker plot below?

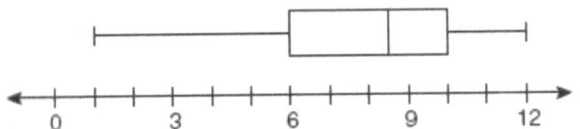

4. The box-and-whisker plot below represents students' scores on a recent test. What is the value of the upper quartile?

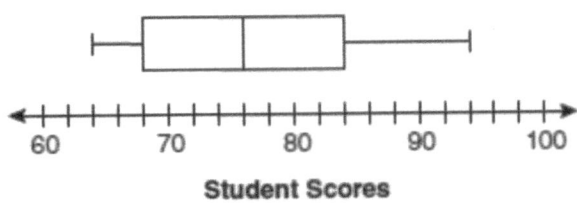

Student Scores

5. In the box-and-whisker plot below, what is the 2nd quartile?

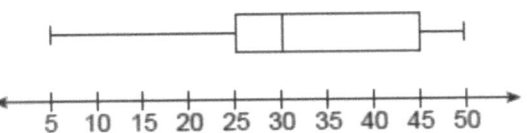

6. In the box-and-whisker plot below, what is the 1st quartile?

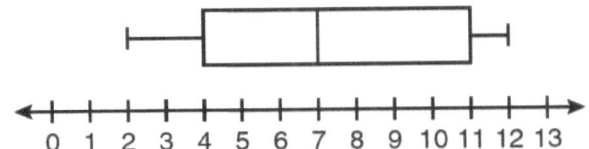

7. What is the range of the data represented in the box-and-whisker plot shown below?

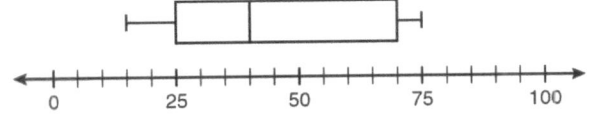

8. The box-and-whisker plot below represents the math test scores of 20 students.

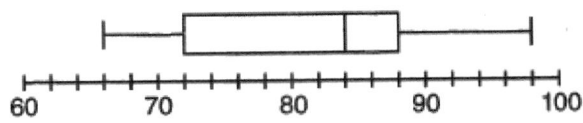

What percentage of the test scores are *less than* 72?

9. The box-and-whisker plot below represents a set of grades in an algebra class.

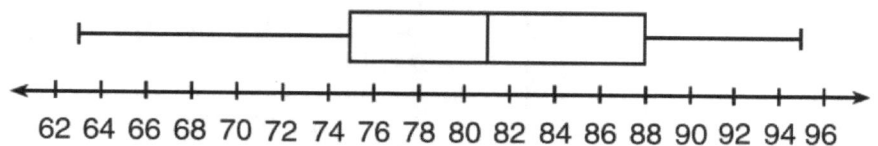

62 64 66 68 70 72 74 76 78 80 82 84 86 88 90 92 94 96

Which interval contains exactly 50% of the grades?

(1)   63-88   (3)   75-81

(2)   63-95   (4)   75-88

10. Match the boxplots with the type of distribution.

(a)   (b)   (c)

(1) left skewed

(2) right skewed

(3) no skew

11. Create a box plot for the following data:
       89, 73, 84, 91, 87, 77, 94

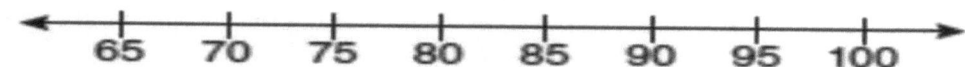

65   70   75   80   85   90   95   100

12. Create a box plot for the following data:
    65, 75, 92, 84, 62, 96, 88, 79, 82

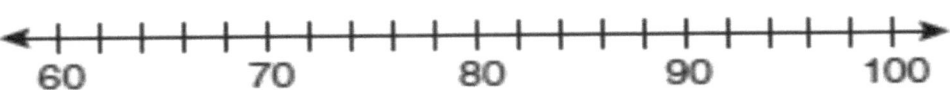

13. Create a box plot for the following data:
    72, 73, 66, 71, 82, 85, 95, 85, 86, 89, 91, 92

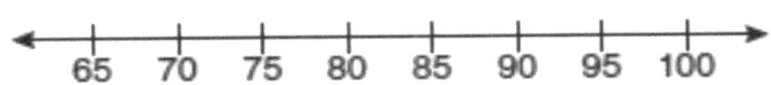

14. Using the line provided, construct a box-and-whisker plot for the 12 scores below.
    26, 32, 19, 65, 57, 16, 28, 42, 40, 21, 38, 10

Regents Questions

MULTIPLE CHOICE

1. Corinne is planning a beach vacation in July and is analyzing the daily high temperatures for her potential destination. She would like to choose a destination with a high median temperature and a small interquartile range. She constructed box plots shown in the diagram below.

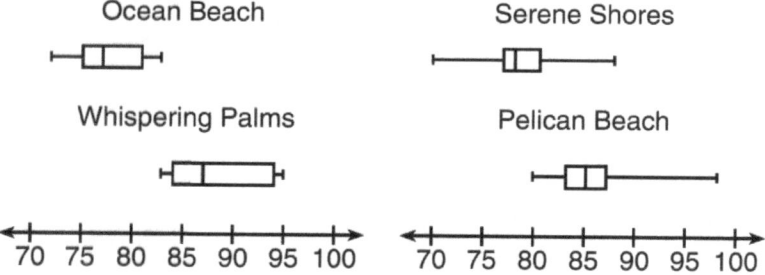

Which destination has a median temperature above 80 degrees and the smallest interquartile range?

    (1)   Ocean Beach   (3)   Serene Shores

    (2)   Whispering Palms   (4)   Pelican Beach

2. Which statistic can *not* be determined from a box plot representing the scores on a math test in Mrs. DeRidder's algebra class?

    (1)   the lowest score   (3)   the highest score

    (2)   the median score   (4)   the score that occurs most frequently

3. The box plot below summarizes the data for the average monthly high temperatures in degrees Fahrenheit for Orlando, Florida.

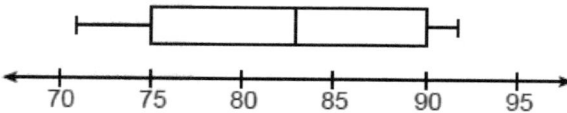

The third quartile is

    (1)   92   (3)   83

    (2)   90   (4)   71

4. What is the range of the box plot shown below?

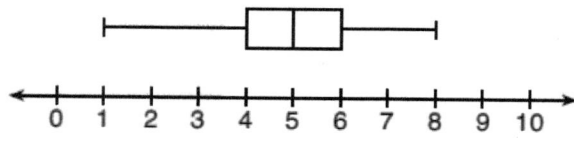

    (1)   7   (3)   3

    (2)   2   (4)   4

5.   Given the following data set:

65, 70, 70, 70, 70, 80, 80, 80, 85, 90, 90, 95, 95, 95, 100

Which representations are correct for this data set?

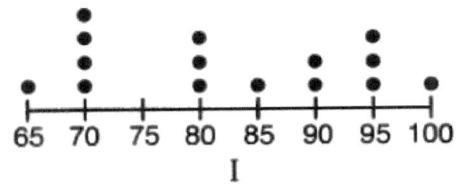

I

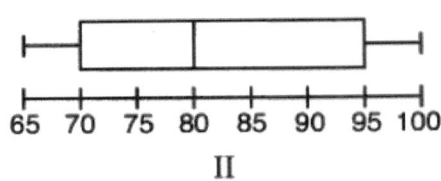

II

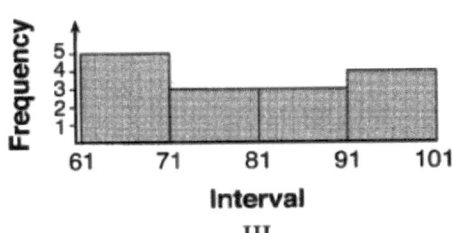

III

   (1)   I and II, only   (3)   II and III, only

   (2)   I and III, only   (4)   I, II, and III

6.   Below are two representations of data.

A: 2, 5, 5, 6, 6, 6, 7, 8, 9

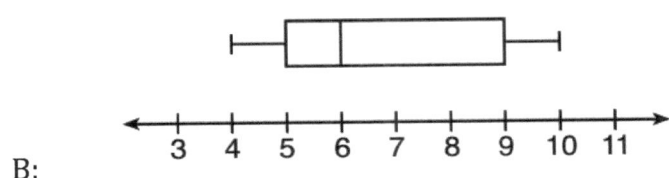

B:

Which statement about A and B is true?

   (1)   median of A > median of B (3)   upper quartile of A < upper quartile of B

   (2)   range of A < range of B (4)   lower quartile of A > lower quartile of B

7.   Different ways to represent data are shown below.

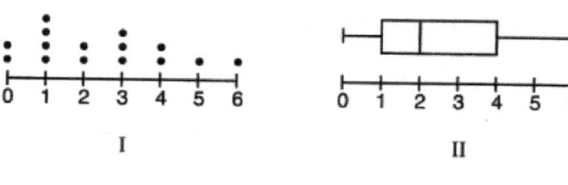

I

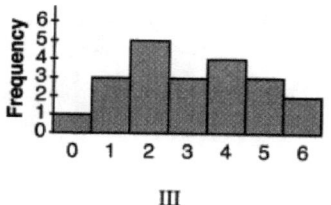

II

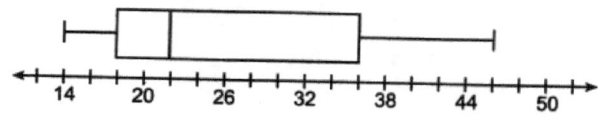

III

Which data representations have a median of 2?

(1)   I and II, only   (3)   II and III, only
(2)   I and III, only   (4)   I, II, and III

8.   What is the value of the third quartile in the box plot shown below?

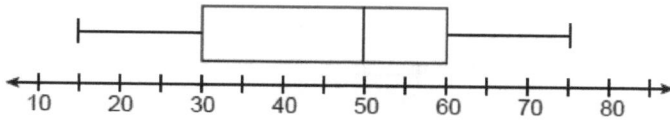

(1)   18   (3)   36
(2)   22   (4)   46

9.   A box plot is shown below.

Which number represents the third quartile?

(1)   30   (3)   60
(2)   50   (4)   75

CONSTRUCTED RESPONSE

10. Robin collected data on the number of hours she watched television on Sunday through Thursday nights for a period of 3 weeks. The data are shown in the table below.

	Sun	Mon	Tues	Wed	Thurs
Week 1	4	3	3.5	2	2
Week 2	4.5	5	2.5	3	1.5
Week 3	4	3	1	1.5	2.5

Using an appropriate scale on the number line below, construct a box plot for the 15 values.

11. The data set 20, 36, 52, 56, 24, 16, 40, 4, 28 represents the number of books purchased by nine book club members in a year. Construct a box plot for these data on the number line below.

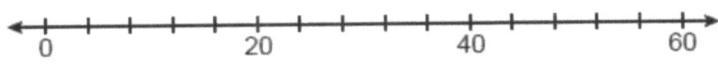

CHAPTER 19.   BIVARIATE DATA

19.1   Two-Way Frequency Tables

Bivariate statistics deals with the relationship between two variables. A **two-way table** (also known as a *contingency table* or *pivot table*) shows frequencies for bivariate data.

Example: A bakery holds a taste test in which participants select their favorite cup cake icing flavor. The two-way table below shows the results for 50 adults - 20 women and 30 men. In this survey, only 2 out of 20 women preferred vanilla, but 16 out of the 30 men chose vanilla. In this table, the *row variable* is the participant's gender and the *column variable* is the favorite flavor.

	Vanilla	Chocolate	Strawberry	Total
Women	2	10	8	20
Men	16	6	8	30
Total	18	16	16	50

Entries in the body of the table are called **joint frequencies**. Entries in the "Total" row and "Total" column are called **marginal frequencies**, since they are written in the right and bottom margins. The marginal frequencies are the sums of the joint frequencies on that row or column of the table. The **grand total** in the lower right hand cell is the total number of data points. It should equal the sum of each set of marginal frequencies.

Sometimes we may prefer to show ratios (in *percent, decimal,* or *fraction* format). In this case, each entry is a **relative frequency**; that is, a ratio of the frequency for that cell to the total number of data points. The total percentage written in the lower right hand cell is always 100%.

Example: We could have represented the above table using relative frequencies, below.

	Vanilla	Chocolate	Strawberry	Total
Women	4%	20%	16%	40%
Men	32%	12%	16%	60%
Total	36%	32%	32%	100%

Two-way tables can help us to find conditional relative frequencies. A **conditional relative frequency** is one that is calculated given that a certain condition (row or column) is true. Within the table, we set the given condition to be true by isolating a row or column.
Example:

	Vanilla	Chocolate	Strawberry	Total
Women	2	10	8	20
Men	16	6	8	30
Total	18	16	16	50

(a) Using the table above, we could say that among *women*, 2 out of 20, or 10% of the women, prefer *vanilla*. The given condition, "among women," restricts us to the top row.

(b) Among those who prefer *chocolate*, 6 out of 16, or 37.5%, are *men*. In this case, the given condition restricts us to the column labelled "Chocolate."

MODEL PROBLEM

A public opinion survey explored the relationship between age and support for increasing the minimum wage. The results are summarized in the two-way table below.

	For	Against	No opinion	Total
21 – 40	25	20	5	50
41 – 60	20	35	20	75
Over 60	55	15	5	75
Total	100	70	30	200

In the 21 to 40 age group, what percent supports increasing the minimum wage?

Solution: 50%

Explanation:
A total of 50 people in the 21 to 40 age group were surveyed (isolate the first row). Of those, 25 were "for" increasing the minimum wage. Therefore, 50% (25 ÷ 50) of the respondents in this age group supported the increase.

PRACTICE PROBLEMS

1. In a survey of eighth and ninth grade students, participants were asked what grade they
    were in and whether they planned to watch the Super Bowl. Results are shown in the table
    below. Round your answers to the *nearest tenth of a percent.*

	Watching	Not Watching	Undecided	Total
8th Grade	25	20	8	53
9th Grade	31	22	7	60
Total	56	42	15	113

a) What percent of the students are undecided?

b) What percent of the ninth graders are watching?

2. The first table shows the number of books sold at a library sale. Complete this *joint
    frequency table* by writing the *marginal frequencies* in the blank cells. Then, using the same
    data, create an equivalent two-way *relative frequency table* in the second table below.

	Fiction	Nonfiction	Total
Hardcover	28	52	
Paperback	94	36	
Total			

	Fiction	Nonfiction	Total
Hardcover			
Paperback			
Total			

3. You go to a dance and help clean up afterwards. To help, you collect the soda cans, Coca-Cola and Sprite, and organize them. Some cans were on the table and some were in the garbage. 72 total cans were found. 42 total cans were found in the garbage and 50 total cans were Coca-Cola. 14 Sprite cans were found on the table. From the given information, complete the two-way joint frequency table below.

	Coca-Cola	Sprite	Total
Table			
Garbage			
Total			

Regents Questions

MULTIPLE CHOICE

1. A public opinion poll was taken to explore the relationship between age and support for a candidate in an election. The results of the poll are summarized in the table below.

Age	For	Against	No Opinion
21 – 40	30	12	8
41 – 60	20	40	15
Over 60	25	35	15

What percent of the 21–40 age group was for the candidate?

(1) 15   (3) 40

(2) 25   (4) 60

2. A radio station did a survey to determine what kind of music to play by taking a sample of middle school, high school, and college students. They were asked which of three different types of music they prefer on the radio: hip-hop, alternative, or classic rock. The results are summarized in the table below.

	Hip-Hop	Alternative	Classic Rock
Middle School	28	18	4
High School	22	22	6
College	16	20	14

What percentage of college students prefer classic rock?

(1) 14%   (3) 33%

(2) 28%   (4) 58%

3. Students were asked to name their favorite sport from a list of basketball, soccer, or tennis. The results are shown in the table below.

	Basketball	Soccer	Tennis
Girls	42	58	20
Boys	84	41	5

What percentage of the students chose soccer as their favorite sport?

(1) 39.6%   (3) 50.4%

(2) 41.4%   (4) 58.6%

4. An outdoor club conducted a survey of its members. The members were asked to state their preference between skiing and snowboarding. Each member had to pick one. Of the 60 males, 45 stated they preferred to snowboard. Twenty-two of the 60 females preferred to ski. What is the relative frequency that a male prefers to ski?

(1) 0.125   (3) $0.\overline{333}$

(2) 0.25   (4) $0.\overline{405}$

5. Jenna took a survey of her senior class to see whether they preferred pizza or burgers. The results are summarized in the table below.

	Pizza	Burgers
Male	23	42
Female	31	26

Of the people who preferred burgers, approximately what percentage were female?

(1) 21.3   (3) 45.6

(2) 38.2   (4) 61.9

6. A survey was given to 12th-grade students of West High School to determine the location for the senior class trip. The results are shown in the table below.

	Niagara Falls	Darien Lake	New York City
Boys	56	74	103
Girls	71	92	88

To the *nearest percent*, what percent of the boys chose Niagara Falls?
(1) 12   (3) 44
(2) 24   (4) 56

7. A middle school conducted a survey of students to determine if they spent more of their time playing games or watching videos on their tablets. The results are shown in the table below.

	Playing Games	Watching Videos	Total
Boys	138	46	184
Girls	54	142	196
Total	192	188	380

Of the students who spent more time playing games on their tablets, approximately what percent were boys?
(1) 41   (3) 72
(2) 56   (4) 75

8. At Berkeley Central High School, a survey was conducted to see if students preferred cheeseburgers, pizza, or hot dogs for lunch. The results of this survey are shown in the table below.

	Cheeseburgers	Pizza	Hot Dogs
Females	32	44	24
Males	36	30	34

Based on this survey, what percent of the students preferred pizza?
(1) 30   (3) 44
(2) 37   (4) 74

9.   Some adults were surveyed to find out if they would prefer to buy a sports utility vehicle (SUV) or a sports car. The results of the survey are summarized in the table below.

	SUV	Sports Car	Totals
Male	21	38	59
Female	135	46	181
Totals	156	84	240

Of the number of adults that preferred sports cars, approximately what percent were males?

(1) 15.8   (3) 64.4

(2) 45.2   (4) 82.6

10. Mrs. Smith's math class surveyed students to determine their favorite flavors of soft ice cream. The results are shown in the table below.

	Chocolate	Vanilla	Twist
Juniors	42	27	45
Seniors	67	42	21

Of the students who preferred chocolate, approximately what percentage were seniors?

(1) 27.5   (3) 51.5

(2) 44.7   (4) 61.5

CONSTRUCTED RESPONSE

11. The school newspaper surveyed the student body for an article about club membership. The table below shows the number of students in each grade level who belong to one or more clubs.

	1 Club	2 Clubs	3 or More Clubs
9th	90	33	12
10th	125	12	15
11th	87	22	18
12th	75	27	23

If there are 180 students in ninth grade, what percentage of the ninth grade students belong to more than one club?

12. A statistics class surveyed some students during one lunch period to obtain opinions about television programming preferences. The results of the survey are summarized in the table below.

Programming Preferences

	Comedy	Drama
Male	70	35
Female	48	42

Based on the sample, predict how many of the school's 351 males would prefer comedy. Justify your answer.

13. A survey of 100 students was taken. It was found that 60 students watched sports, and 34 of these students did not like pop music. Of the students who did *not* watch sports, 70% liked pop music. Complete the two-way frequency table.

	Watch Sports	Don't Watch Sports	Total
Like Pop			
Don't Like Pop			
Total			

14. The sixth-grade classes at West Road Elementary School were asked to vote on the location of their class trip. The results are shown in the table below.

	Playland	Splashdown	Fun Central
Boys	38	53	25
Girls	39	46	37

Determine, to the *nearest percent*, the percentage of girls who voted for Splashdown.

15. Julia surveyed 150 of her classmates at City Middle School to determine their favorite animals. Of the 150 students, 46% were male. Forty-two students said their favorite animal was a horse, and of those students were female. Of the 60 students who said dolphins were their favorite animal, 30% were male. Using this information, complete the two-way frequency table below.

	Horse	Dolphin	Penguin	Total
Male				
Female				
Total				

639

19.2 Scatter Plots

A **scatter plot** is a graph used to plot pairs of bivariate quantitative data values in a coordinate plane. They are often used for gathering experimental data to determine whether a correlation exists between the two variables. The **independent** variable is represented by *x*-**values** in a horizontal axis and the **dependent** variable is represented by *y*-**values** in a vertical axis. Only points are plotted; the points are not connected by lines.

The bivariate data can be written using a table. The *x*-**values** are always written first (the top row in a horizontal table or the left column in a vertical table), followed by the corresponding *y*-**values**. Each pair of values in the table can be plotted as a single point.

Example: Park administrators use a table to keep track of daily temperatures and the number of daily visitors to the beach. The points are then plotted on a scatter plot, using the temperatures as *x*-values and visitors as *y*-values. The first data column is plotted as the point $(84, 225)$ on the scatter plot.

Temp (*x*)	84	86	82	87	86	92	88	89	94	96	94
Visitors (*y*)	225	350	100	125	300	450	455	525	600	565	510

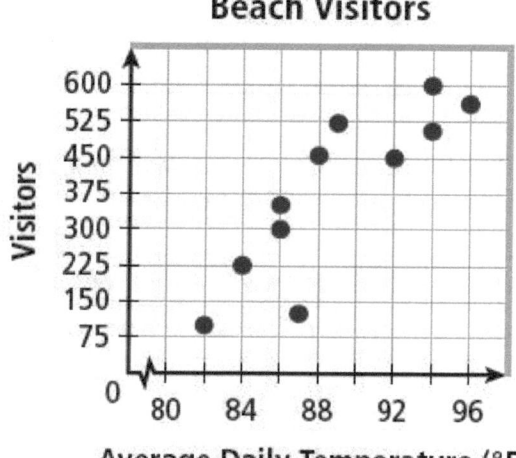

Beach Visitors

Each axis must include **labels** in equal intervals to cover all of the *x* or *y* values in the data, and should include an axis **title** describing what the axis labels represent.

Example: In the scatter plot above, the *x*-axis is labeled from 80 to 96 in intervals of 4, where each grid square is 2 units wide. To save room, you may omit labels between 0 and the first tick by using a ⌄ symbol, as shown between 0 and 80 on the *x*-axis. To cover values as high as 600 visitors, intervals of 75 visitors are used on the *y*-axis.

CALCULATOR TIP

To enter bivariate data:

8. Press [STAT][1] to select Edit.
9. If any values already appear in the L1 or L2 columns, select the column heading and press [CLEAR][ENTER].
10. Enter the *x* values into the L1 column and the corresponding *y* values into the L2 column.

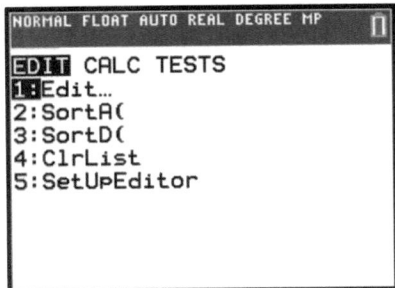   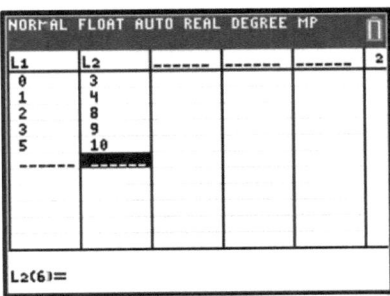

If any columns other than L1 or L2 appear in the table, you can either leave them there or delete them by selecting the column heading and pressing [DEL]. Also, if either the L1 or L2 column is missing, you can add it by selecting an empty column heading and then pressing [2nd][L1][ENTER] or [2nd][L2][ENTER].

CALCULATOR TIP

To view a scatter plot:

1. Enter the *x* and *y* values into L1 and L2 as described above.

2. Press [2nd][STAT PLOT][1] for Plot1. Select <On>[ENTER]. Be sure the Type is set at the first option ⊡ , Xlist is L1, and Ylist is L2.

3. Press [ZOOM][9] for ZoomStat. You may need to adjust the dimensions and scale of the scatter plot by pressing [WINDOW], as shown below.

4. To view the coordinates of each point, press [TRACE] and the arrow keys, [◄][►].

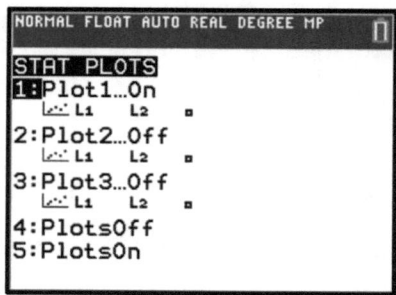
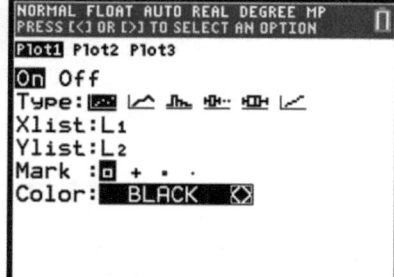
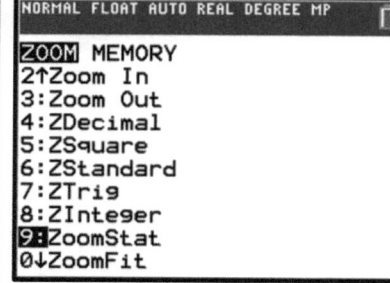

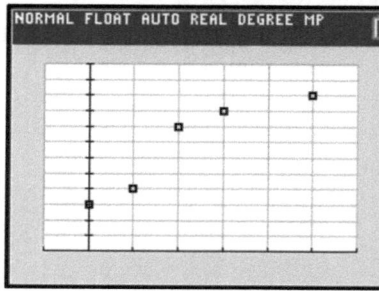
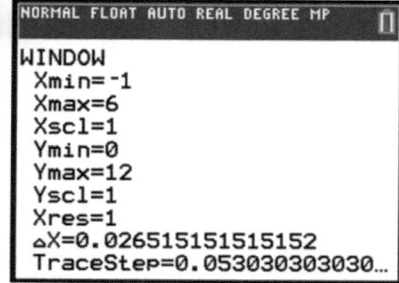

CALCULATOR TIP

To view summary statistics for bivariate data:
1. Enter the x and y values into L1 and L2 as described above.
2. Press [STAT]\<CALC\>[2] for 2-Var Stats.
3. Enter [2nd][L1] for Xlist and [2nd][L2] for Ylist, and press [ENTER] to Calculate.

Summary statistics, including the mean and standard deviation for both x and y, are shown. Scroll down with the [▼] key to see the rest of the statistics.

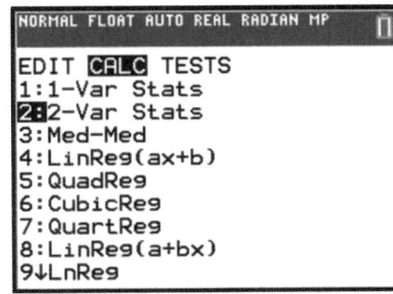  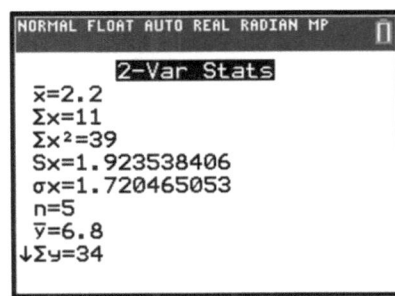

 **CALCULATOR TIP**

To use statistics variables for bivariate data:
Once the 2-Var Stats have been calculated, you may now use these statistics in an expression by pressing [VARS][5] for Statistics Variables.

Example: To find the difference between the means of x and y, we can type
              [VARS][5][2][–][VARS][5][5][ENTER].

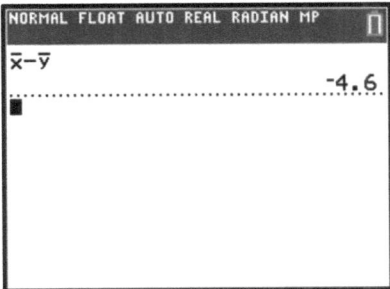

MODEL PROBLEM

A teacher records how many hours her students studied during the week leading up to their state exam and the scores they received. The data is shown in the following table. Create a scatter plot for the data.

Hours	3	5	2	6	7	1	2	7	1	7
Score	80	90	75	80	90	50	65	85	40	100

Solution:

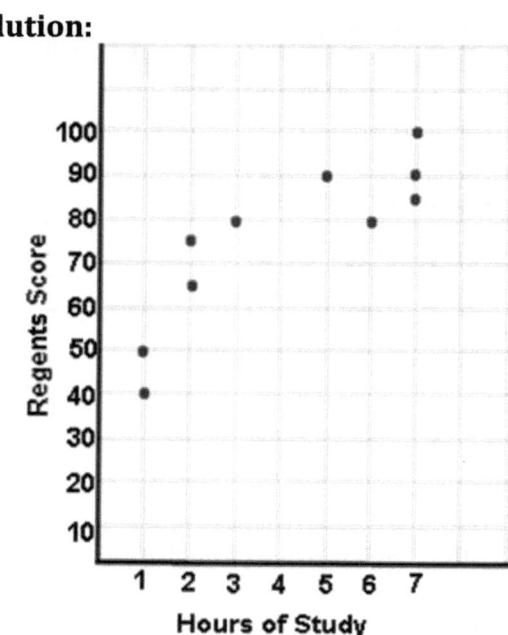

Explanation of steps:

(A) Draw a grid. Label the *x*-axis using equal intervals covering the range of x values in the table *[1 through 7, with intervals of 1]*.

(B) Label the *y*-axis similarly to cover the range of y values *[10 through 100, with intervals of 10]*. Add appropriate axis titles.

(C) Plot each pair of data values as a point on the grid *[the first point in the table is (3, 80)]*.

To check that your scatter plot is correct, you can enter the table into the calculator and create a scatter plot according to the instructions described in this section.

PRACTICE PROBLEMS

1. For 10 days, a real estate agent kept a record of the number of hours she spent showing homes to potential buyers. The information is shown in the table below.

Day	1	2	3	4	5	6	7	8	9	10
Hours	9	3	2	6	8	6	10	4	5	2

Which scatter plot shows the agent's data graphically?

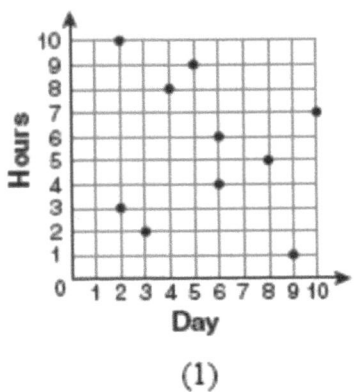

(1)

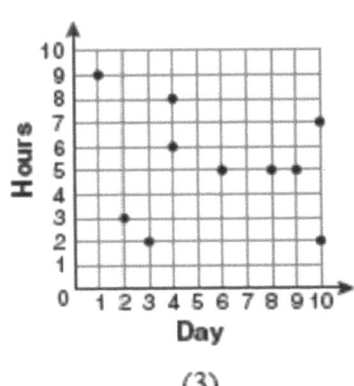

(3)

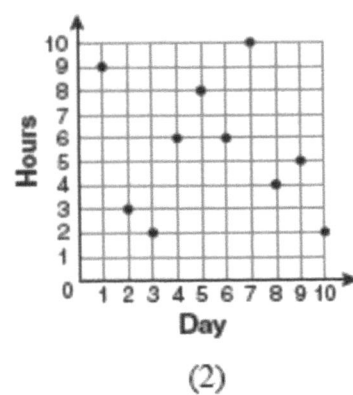

(2)

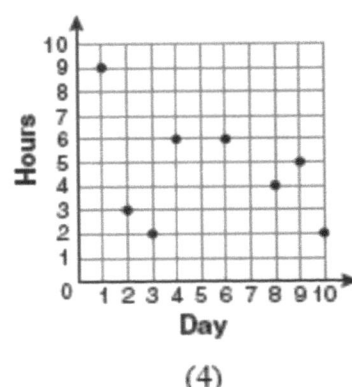

(4)

2. The table shows the height (in inches) and the weight (in pounds) of five starters on a high school basketball team. Create the corresponding scatter plot.

Height (x)	67	72	77	74	69
Weight (y)	155	220	240	195	175

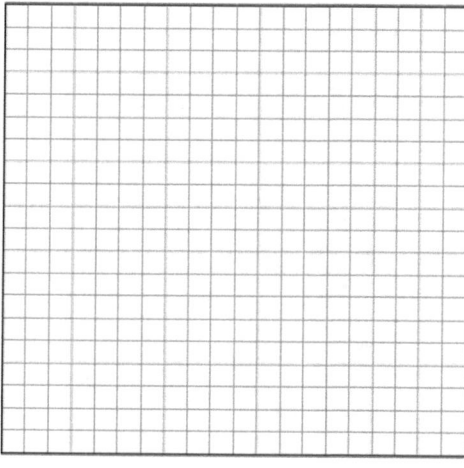

3. The following table lists weights (in hundreds of pounds) and highway fuel usage rates (in mpg) for a sample of domestic cars. Create the corresponding scatter plot.

Weight (x)	29	35	28	44	25	34	30	33	28	24
Fuel (y)	31	27	29	25	31	29	28	28	28	33

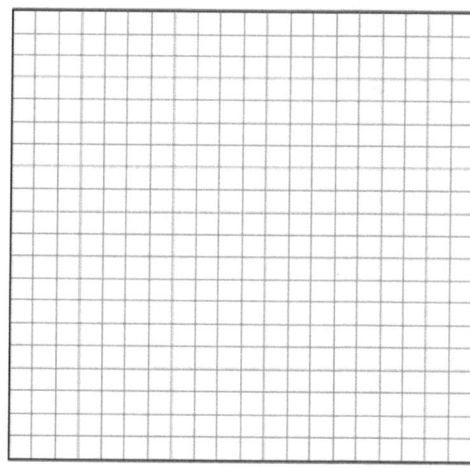

Regents Questions

There are no Regents exam questions on this topic.

19.3   Correlation and Causality

Correlation describes an association (a statistical relationship) between two quantitative variables.

If the two variables are *x* and *y*, the data is said to have a **positive correlation** (or *direct correlation*) if, as *x* increases, *y* also increases, and as *x* decreases, *y* also decreases.
Example: As the temperature *increases, more* ice cream cones are sold. As the temperature *decreases, less* ice cream cones are sold. This shows a positive correlation.

There is a **negative correlation** (or *indirect correlation*) when, as *x* increases, *y* decreases, and as *x* decreases, *y* increases.
Example: As the temperature *increases, less* cups of hot chocolate are sold. As the temperature *decreases, more* cups of hot chocolate are sold. This shows a negative correlation.

In some cases, a change in one variable is directly responsible for the change in the second variable. These are called **causal** relationships.
Example: When boiling a pot of water, the relationship between the time spent on the stove and the temperature of the water is causal.

However, the existence of a correlation does not necessarily mean that the relationship is causal. An alternate explanation may be that the two variables of interest are related to a third variable not being considered, which is called a **confounding factor**.
Example: A study of middle school students found a positive correlation between shoe sizes and reading comprehension scores. Clearly, a larger shoe size does not cause an increase in reading comprehension, or vice-versa, so this is *not a causal relationship*. In fact, a missing confounding factor in this research is the age of the students: older children tend to have larger feet, and also tend to have higher reading scores.

Almost all real life relationships would have some hidden factors, but for our purposes we will define a "cause" as a primary factor responsible.
Example: In the above example about boiling water, a number of other factors could be partially responsible for the rise in water temperature. But certainly, placing a pot of water on a hot stove for a period of time will directly cause the temperature of the water to increase.

MODEL PROBLEM

A number of children between the ages of 5 and 15 are measured for height. How would you describe the relationship between the ages and heights? Is there a correlation, and if so, is it positive or negative? Is it a causal relationship?

Solution:
Children grow as they get older, so there is a positive correlation and a causal relationship.

Explanation of steps:
If an increase or decrease in the first variable *causes* a change in the second variable, there is a causal relationship. *[Yes, there are other hidden factors, such as nutrition, genetics, etc., but it safe to say that aging is a primary cause of growth within this age group.]* If an increase in one variable results in an increase in the other, it is a positive correlation, but if an increase in one variable results in a decrease in the other, it is a negative correlation.

PRACTICE PROBLEMS

1. Which of the following best describes the relationship between the distance driven and the amount of gasoline used? (1) causal, but not correlated (2) correlated, but not causal (3) both correlated and causal (4) neither correlated nor causal	2. A study showed that a decrease in the cost of a quart of milk led to an increase in the number of quarts of milk sold. Which statement best describes this relationship? (1) positive correlation and causal (2) negative correlation and causal (3) positive correlation and not causal (4) negative correlation and not causal

3. Identify the correlation you would expect to see (*positive, negative,* or *none*) between each pair of data sets. Explain.

a) children's ages and their weights

b) the volume of water poured into a container and the amount of empty space left in the container

c) a woman's shoe size and the length of her hair

d) the outside temperature and the number of people at the beach

4. For each research finding below, (a) determine whether there is a *positive or negative correlation*; (b) decide if there is a *causal relationship*; and (c) if a correlation is *not* causal, state what *confounding factors*, if any, there may be.

a) In a balloon, as the volume of air increases, the diameter increases.

b) As ice cream sales increase within the state, the number of forest fires also increases.

c) As the number of workers increases, the number of days required to complete a job decreases.

d) The more firefighters sent to a fire, the longer it takes to put out the fire.

e) Over the past few centuries, the number of pirates worldwide has decreased while the level of CO_2 in the atmosphere has increased.

f) As snow shovel sales decrease within the state, the number of shark attacks increases.

Regents Questions

MULTIPLE CHOICE

1. What type of relationship exists between the number of pages printed on a printer and the amount of ink used by that printer?
    (1) positive correlation, but not causal
    (2) positive correlation, and causal
    (3) negative correlation, but not causal
    (4) negative correlation, and causal

2. Which situation does *not* describe a causal relationship?
    (1) The higher the volume on a radio, the louder the sound will be.
    (2) The faster a student types a research paper, the more pages the research paper will have.
    (3) The shorter the time a car remains running, the less gasoline it will use.
    (4) The slower the pace of a runner, the longer it will take the runner to finish the race.

3. The data obtained from a random sample of track athletes showed that as the foot size of the athlete decreased, the average running speed decreased. Which statement is best supported by the data?
    (1) Smaller foot sizes cause track athletes to run slower.
    (2) The sample of track athletes shows a causal relationship between foot size and running speed.
    (3) The sample of track athletes shows a correlation between foot size and running speed.
    (4) There is no correlation between foot size and running speed in track athletes.

4. Which correlation shows a causal relationship?
    (1) The more minutes an athlete is on the playing field, the more goals he scores.
    (2) The more gasoline that you purchase at the pump, the more you pay.
    (3) The longer a shopper stays at the mall, the more purchases she makes.
    (4) As the price of a gift increases, the size of the gift box increases.

19.4   Identify Correlation in Scatter Plots

Although the points in a scatter plot may not be collinear (i.e., they may not lie exactly in a straight line), one can visually determine whether a *correlation* between the *independent* (x) and *dependent* (y) variables exists.

When a correlation exists, the nature of the correlation may be *linear* (best described by a straight line, despite some possible variation), or *nonlinear* (best described by some other type of function). If there is a linear correlation, then the type of correlation (positive or negative) will match the sign of the slope of the related line. It's also possible that the points show **no correlation** between the variables.

Example: The following diagrams show that (a) there is *no correlation* between a person's arm length and his or her results on an exam; (b) there is a *positive correlation* between the time a person spends revising the exam essay and the results on the exam; and (c) there is a *negative correlation* between the number of absences from school and the exam results.

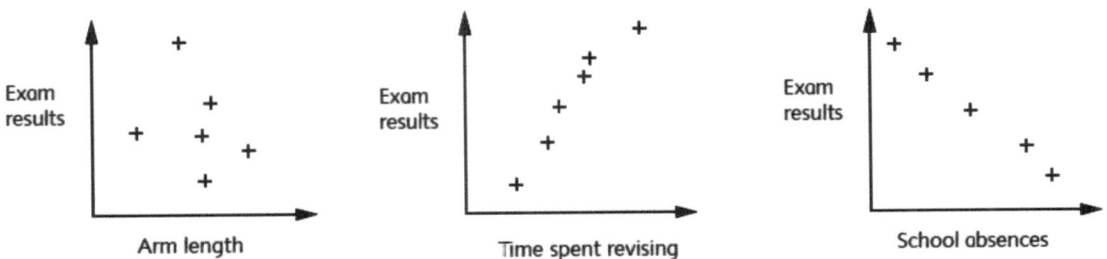

A linear correlation is considered **strong** if the points closely approximate a straight line. The strength of a correlation is represented by the *correlation coefficient*, which we will learn about in Section 19.6.

Example: The first graph below shows a stronger positive correlation than the second. The correlation between two variables is stronger when the data points come closer to forming a straight line when plotted.

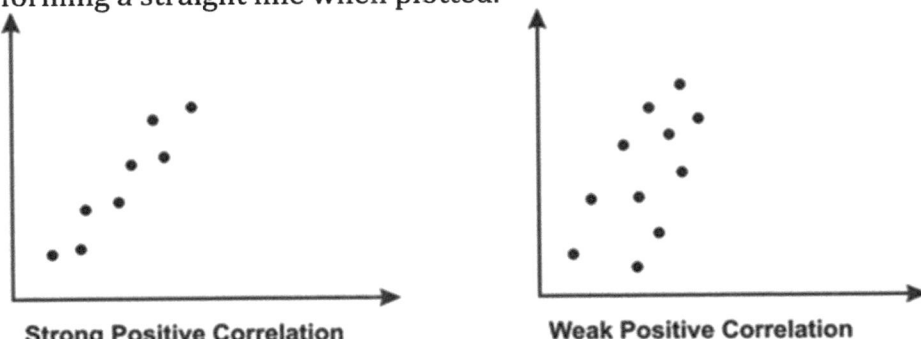

MODEL PROBLEM

Which diagram shows a negative correlation?

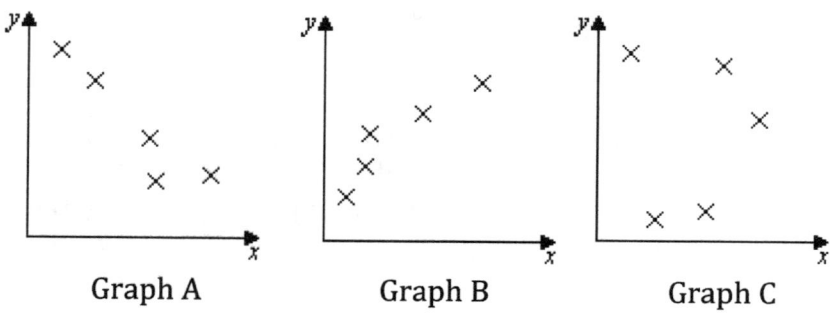

Graph A   Graph B   Graph C

Solution: Graph A

Explanation of steps:

A graph has a negative correlation if, as the *x*-values increase, the *y*-values tend to decrease. *[Graph B shows a positive correlation and Graph C shows no correlation.]*

PRACTICE PROBLEMS

1. Which diagram shows the strongest positive correlation?

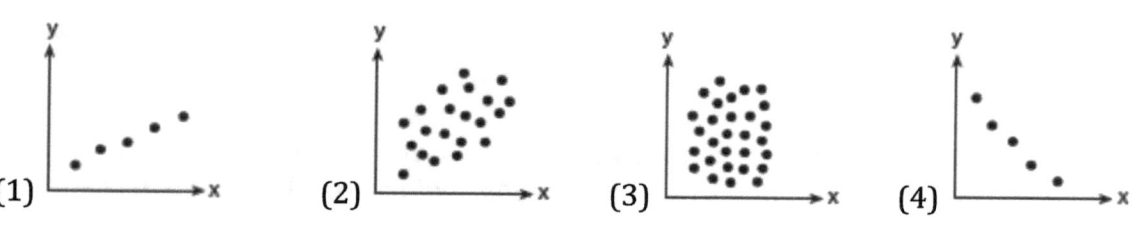

2. State whether the diagram below shows a positive, negative, or no correlation.

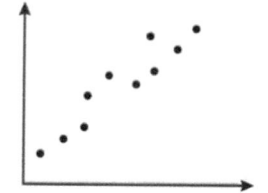

3. State whether the diagram below shows a positive, negative, or no correlation.

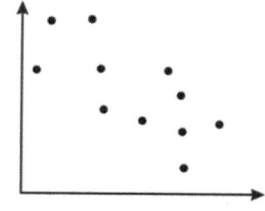

4. State whether the scatter plot below shows a positive, negative, or no correlation.

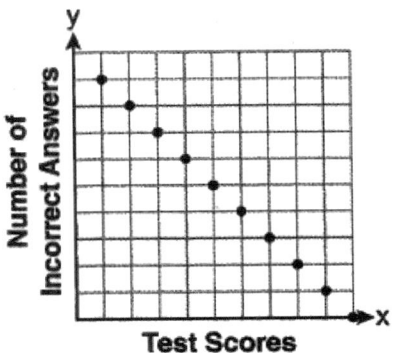

5. State whether the scatter plot below shows a positive, negative, or no correlation.

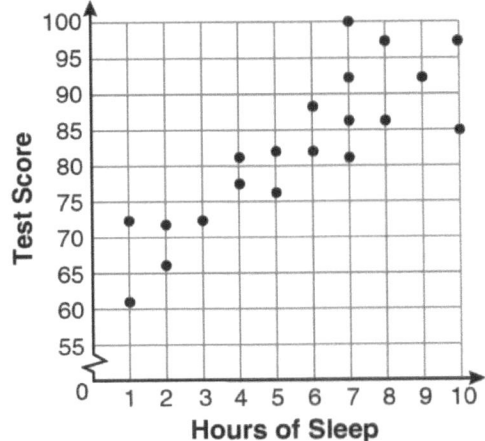

6. State whether the scatter plot below shows a positive, negative, or no correlation.

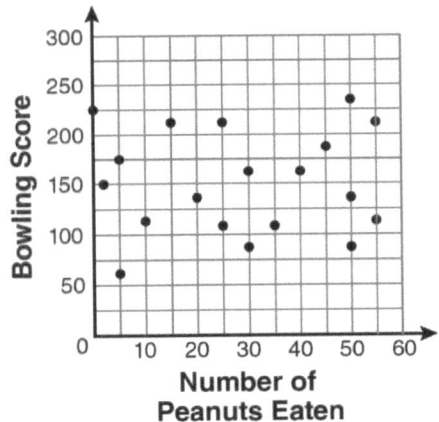

7. State whether the scatter plot below shows a positive, negative, or no correlation.

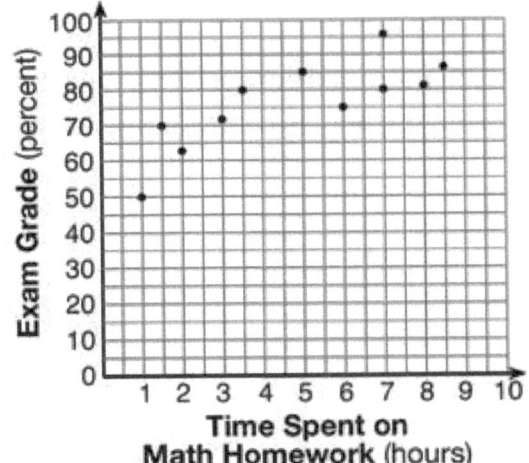

Regents Questions

MULTIPLE CHOICE

1.   The scatterplot below compares the number of bags of popcorn and the number of sodas sold at each performance of the circus over one week.

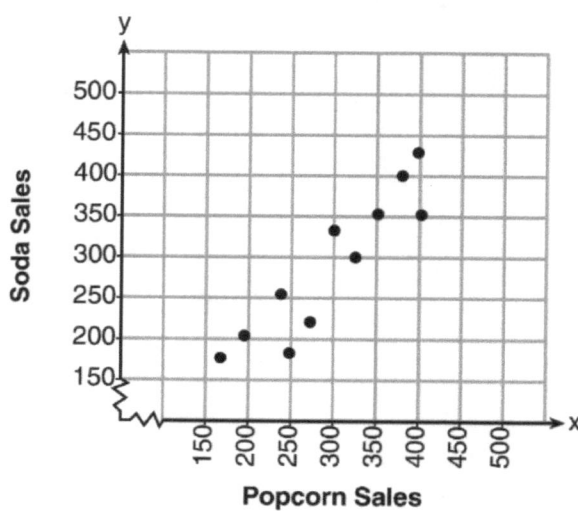

Which conclusion can be drawn from the scatterplot?
   (1)   There is a negative correlation between popcorn sales and soda sales.
   (2)   There is a positive correlation between popcorn sales and soda sales.
   (3)   There is no correlation between popcorn sales and soda sales.
   (4)   Buying popcorn causes people to buy soda.

19.5   Lines of Fit

When a scatter plot shows a linear correlation, a line which approximates this relationship is called a **line of fit** (or *trend line*).

Example: Suppose a set of data, given below, compares students' hours worked, x, with their report grades, y. The data is graphed on a scatterplot and a line of fit is drawn. Because the graph shows a positive correlation, the line will have a positive slope.

x	1	1	2	2.5	3	3	3.5	4	4.5	5
y	60	70	70	75	75	80	80	85	90	90

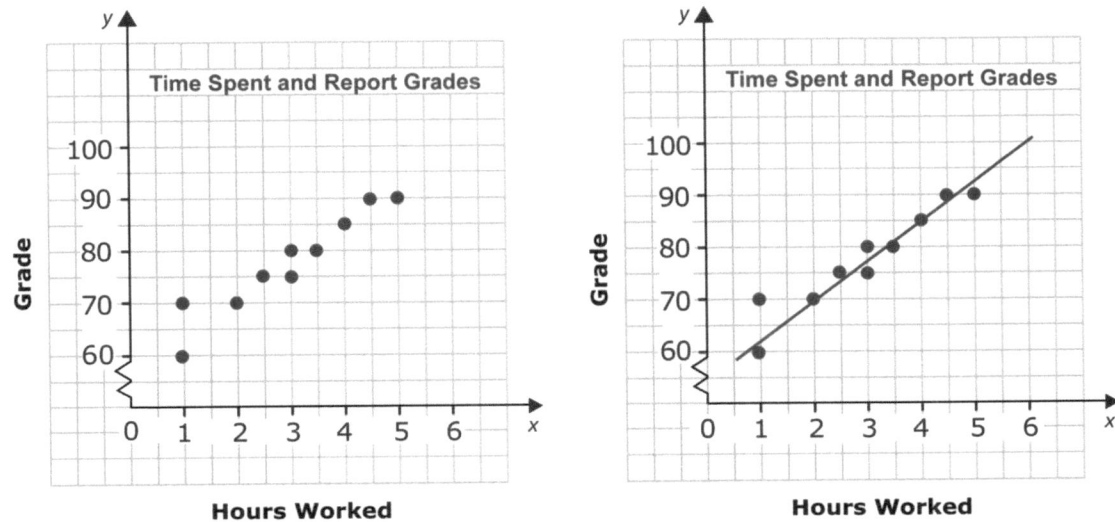

We can create a **line of fit** and find its equation by one of two methods: either by (a) *drawing* an appropriate line by hand (using a straight edge) or by (b) using the *calculator*.

The calculator finds the **linear regression**, which passes through the mean point $(\overline{x}, \overline{y})$, where $\overline{x}$ is the mean of the x's and $\overline{y}$ is the mean of the y's. This line minimizes the combined distances of the points from the line, and so it is usually called a **line of best fit**.

The regression line is also called the **least-squares line** because it is calculated by minimizing the sum of the squares of the residuals. The *residuals*, as we shall see in Section **Error! Reference source not found.**, are the differences between the y values of the given points and the points on the line.

Drawing a Line of Fit:

When drawing a straight line, we try to have the points lie as close to the line as possible, and preferably with as many points above the line as below it. We can then determine its equation by finding two points which appear to lie directly on the line. Try not to pick points that are too close to each other. From these two points, we can determine a slope and y-intercept.

Example: The line above appears to run through points $(2, 70)$ and $(4, 85)$.

Using these points, the slope $m = \dfrac{85 - 70}{4 - 2} = \dfrac{15}{2} = 7.5$.

Substituting point $(2, 70)$ for x and y and the slope 7.5 for m in the general equation $y = mx + b$, we can find b:

$$70 = 7.5(2) + b$$
$$70 = 15 + b$$
$$b = 55$$

Therefore, the equation for this line of fit is $y = 7.5x + 55$.

An easier and more accurate way to find a line of fit is to allow the calculator to find the equation of the linear regression for us. These directions are given below.

Example: If we enter the points as they appear in the graph on the previous page, the calculator will create the equation Y1 = 6.651718984X + 57.877429. Rounding to the *nearest tenth*, the equation would be $y = 6.7x + 57.9$.

 **CALCULATOR TIP**

To find the equation of the regression line using the LinReg function:

1. Press $\boxed{STAT}\boxed{1}$ and enter the x and y coordinates of the points as L1 and L2.

2. Press $\boxed{STAT}$ <CALC> $\boxed{4}$ to select LinReg(ax+b).

3. On the next screen, select L1 for Xlist, L2 for Ylist, and Y1 (by pressing $\boxed{ALPHA}\boxed{F4}\boxed{1}$) for Store RegEQ in order to store the equation in Y1.

   [On the TI-83, this screen is skipped; instead, after the LinReg(ax+b) *prompt, press* $\boxed{VARS}$ *<Y-VARS>* $\boxed{1}\boxed{1}$ *to select Y1, and press* $\boxed{ENTER}$.*]*

4. The screen will show the equation $y = ax + b$ along with the values of a (the slope) and b (the y-intercept).

5. To view the line, now stored in Y1, press $\boxed{ZOOM}\boxed{9}$ for ZoomStat.

6. To see the equation of the line, press $\boxed{Y=}$.

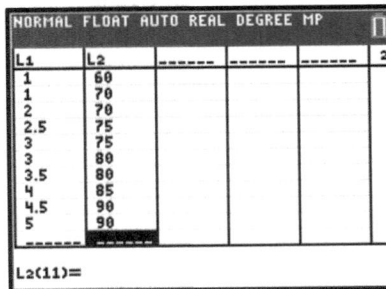 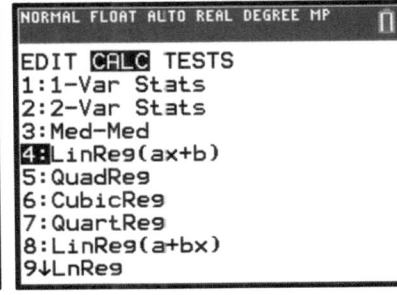 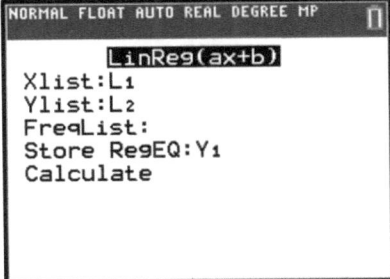

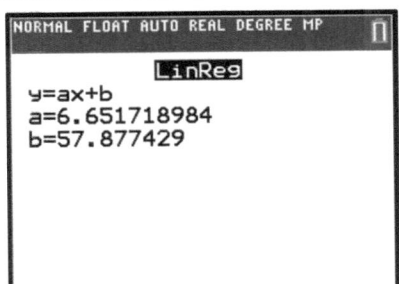 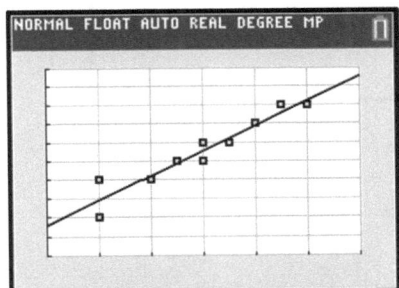

The line of fit helps us to **predict** values not included in the original data. We can **extrapolate** about data that is outside (but near) the given x-values, or **interpolate** about data that is within the x-values but not already included in the data.

Example: Using the graph above, if a student submits a project on which he has worked 1.5 hours, we can predict his grade. This would be *interpolation*, since 1.5 is within the span of given x values. Either of the following predictions would be acceptable:

a) Substituting 1.5 for x in the equation of the drawn line, we get
   $$y = 7.5(1.5) + 55 = 66.25 \approx 66.$$

b) Using the equation of the calculator's linear regression, we get
   $$y = 6.7(1.5) + 57.9 = 67.95 \approx 68.$$

 CALCULATOR TIP
_____

To predict values based on a linear regression equation

For part b) of the example above, since we have already stored the line of best fit into Y1, we could also have the calculator find this value by entering [ALPHA][F4][1][(][1][.][5][)][ENTER]
[or on the TI-83, [VARS]<Y-VARS>[1][1][(][1][.][5][)][ENTER]].

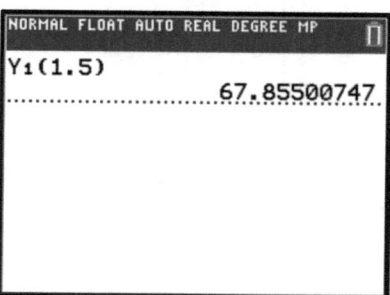

MODEL PROBLEM

Given the data table below, create a scatter plot. State the equation of the regression line and draw the line on the graph. Use the equation to extrapolate the next _y_ value for an _x_ value of 7, rounded to the _nearest whole_.

x	0	1	2	3	4	5	6
y	2	5	9	11	13	18	20

Solution:

(A)   (B)   (C)

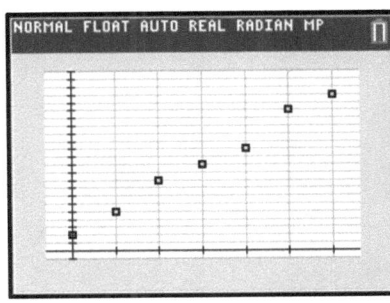   

(D) $y = 3x + 2.14$ (E)

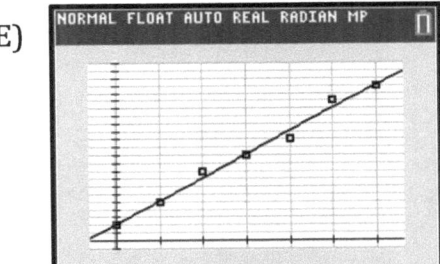

(F) $y = 3(7) + 2.14 \approx 23$

Explanation of steps:

(A)   On the calculator, press [STAT][1] to enter the table values into L1 and L2.

(B)   Press [2nd][STAT PLOT][1] to turn on and set up the scatter plot (the first Type) using L1 as the Xlist and L2 as the Ylist. Display the graph by pressing [ZOOM][9] for ZoomStat.

(C)   Press [STAT]<CALC>[4] for LinReg(ax+b). Again, use L1 as Xlist and L2 as Ylist. For Store RegEq, press [VARS]<Y-VARS>[1][1] for Y1 and Calculate.

(D)   Write the equation by substituting the given values for _a_ and _b_ into $y = ax + b$, or press [Y=] to see the equation stored in Y1.

(E)   Press [ZOOM][9] to view the regression line on the scatter plot.

(F)   To extrapolate, use the new given value of _x [7]_ and substitute it into the equation to find the value of _y_. _[Or, enter_ [VARS]_<Y-VARS>[1][1][(][7][)][ENTER] on the calculator.]_

PRACTICE PROBLEMS

1. The scatter plots below display the same data about the ages of eight health club members and their heart rates during exercises. Which line is a better fit for the data? Explain your reasoning.

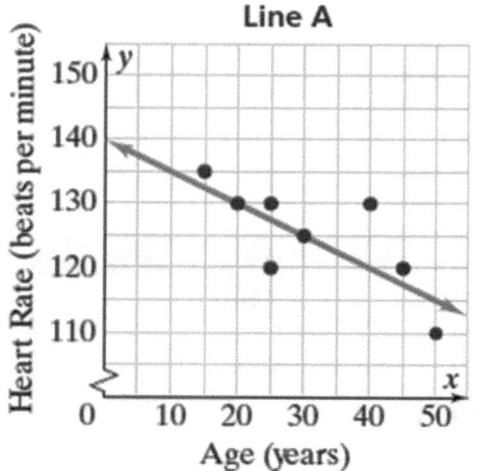

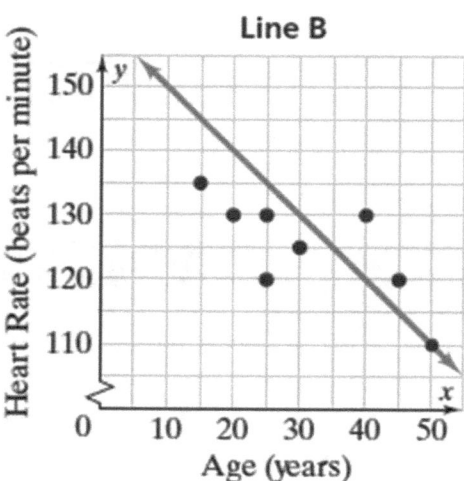

2. The following chart shows students' typing speeds in words per minute (wpm) after a certain number of weeks of practice.

a) Based on the line of fit shown, approximately how fast would you expect a student to type after 8 weeks of practice?

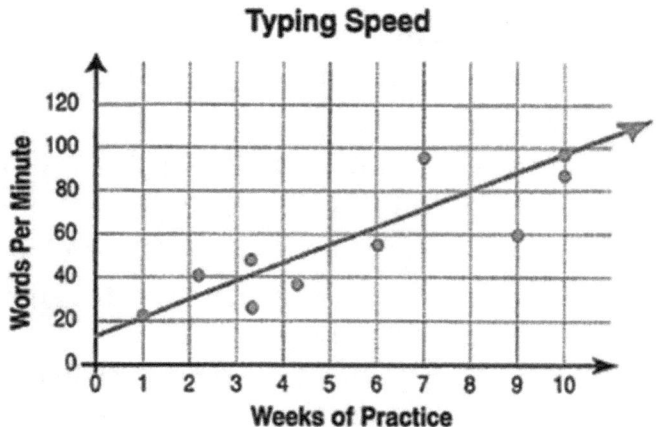

b) By how many words per minute, approximately, can an employee expect to increase her or his speed for each additional week of practice?

3. The local ice cream shop keeps track of daily sales (in dollars) and the temperature (in Celsius) on that day. On the scatter plot below, draw a line of fit.

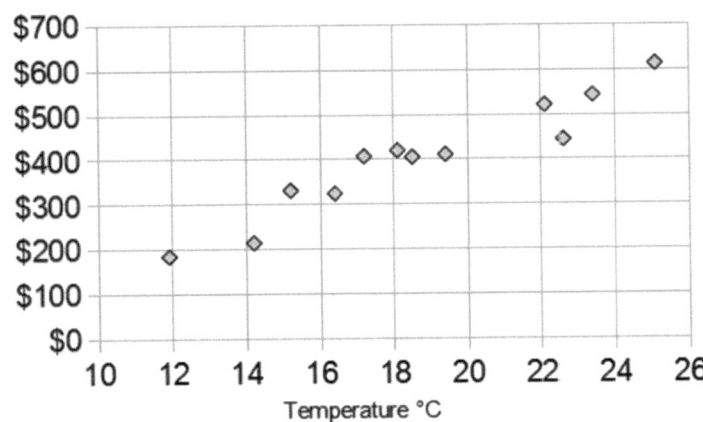

4. Write an equation for the line of fit shown in the scatter plot below.

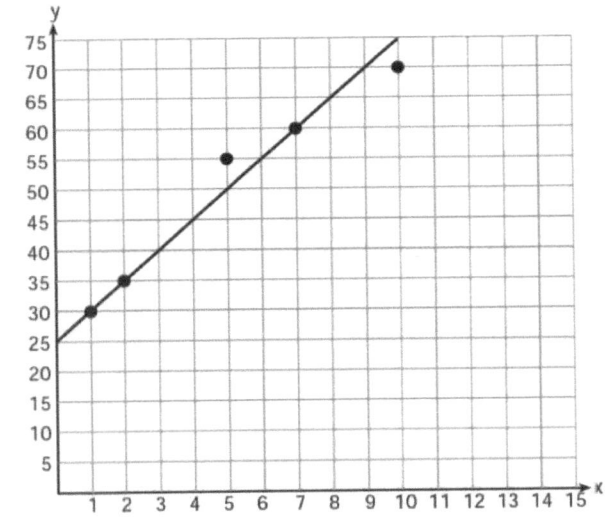

5. Write an equation for the line of fit shown in the scatter plot below.

6. The data below shows hours spent researching the stock market per week and the percent gain for an investor.

Hours	6	8	10	12	14	16	18
% Gain	17	20.5	26.5	29	32.5	37.5	41

Find an equation of the line of best fit for gain with respect to hours of study.

7. A random sample of graduates from a particular college program reported their ages and incomes in response to a survey. Each point on the scatter plot below represents the age and income of a different graduate. Of the following equations, which best fits the data?

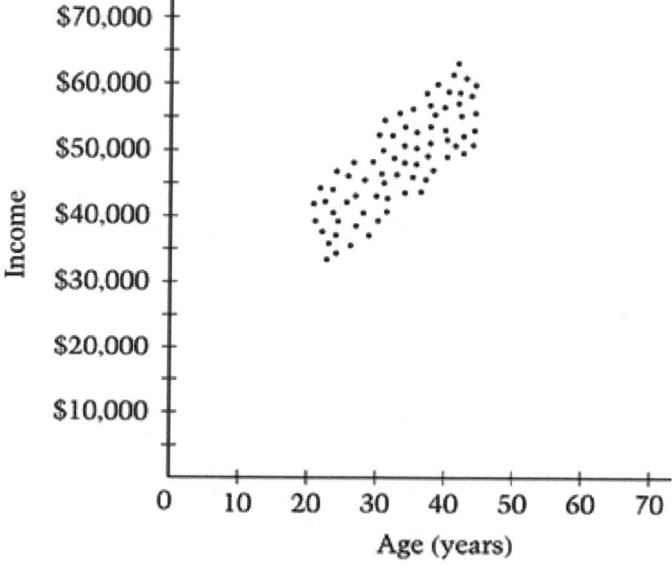

(1) $y = -1{,}000x + 15{,}000$ (3) $y = 1{,}000x + 15{,}000$

(2) $y = 1{,}000x$ (4) $y = 10{,}000x + 15{,}000$

8. Based on the data in the scatter plot of the previous question, predictions can be made about the income of a 35 year old and the income of a 65 year old. For which age is the prediction more likely to be accurate? Justify your answer.

9. Which equation most closely represents the line of best fit for the scatter plot below?

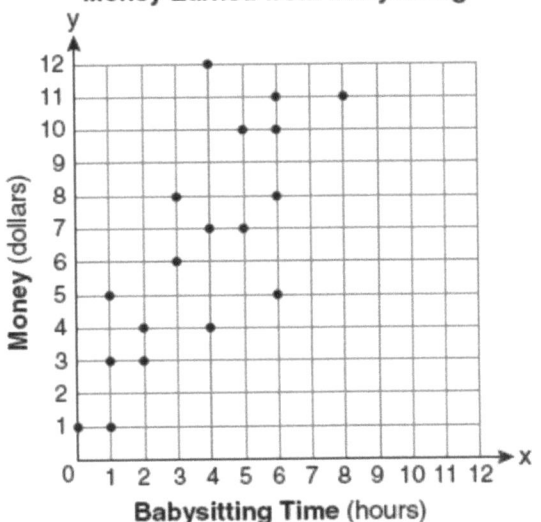

Money Earned from Babysitting

(1) $y = x$

(2) $y = \frac{2}{3}x + 1$

(3) $y = \frac{3}{2}x + 4$

(4) $y = \frac{3}{2}x + 1$

10. Based on a line of best fit for the scatter plot below, which exam grade is the best prediction for a student who spends 4 hours on math homework?

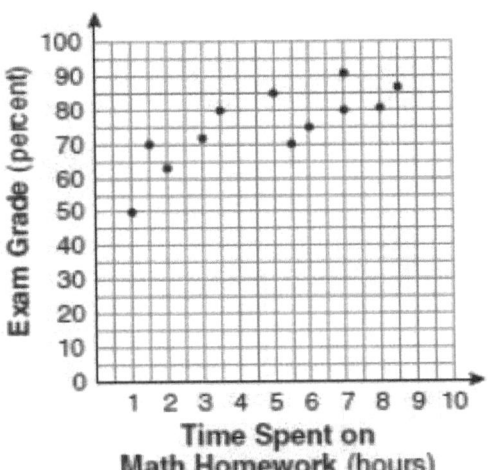

Time Spent on Math Homework (hours)

(1) 62

(2) 72

(3) 82

(4) 92

11. Based on the line of best fit drawn below, which value could be expected for the data in June 2015?

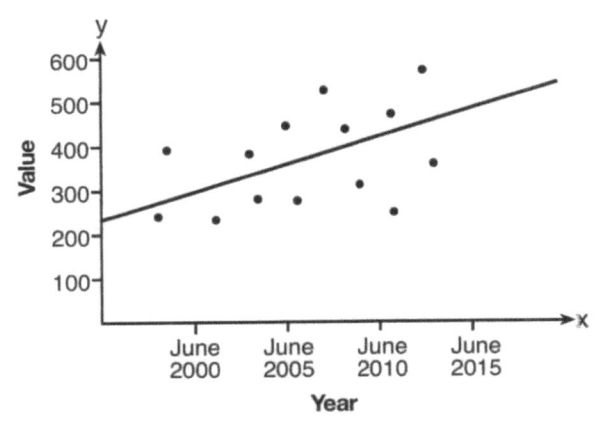

(1) 230

(2) 310

(3) 480

(4) 540

12. Based on the line of best fit drawn below, what is the best estimate for profit in the 18th month?

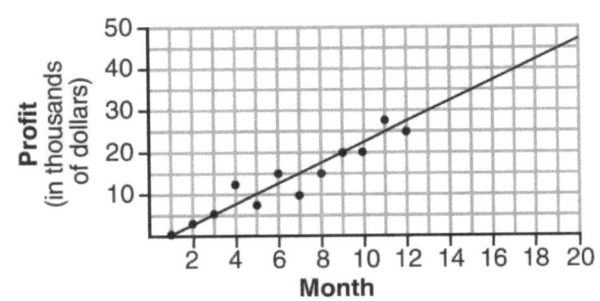

(1) $35,000

(2) $37,750

(3) $42,500

(4) $45,000

13. A new business' goal is to reach a profit of $20,000 in its 18th month of business. The table and scatter plot below represent the profit, *P*, in thousands of dollars, that the business made during the first 12 months.

t (months)	P (profit, in thousands of dollars)
1	3.0
2	2.5
3	4.0
4	5.0
5	6.5
6	5.5
7	7.0
8	6.0
9	7.5
10	7.0
11	9.0
12	9.5

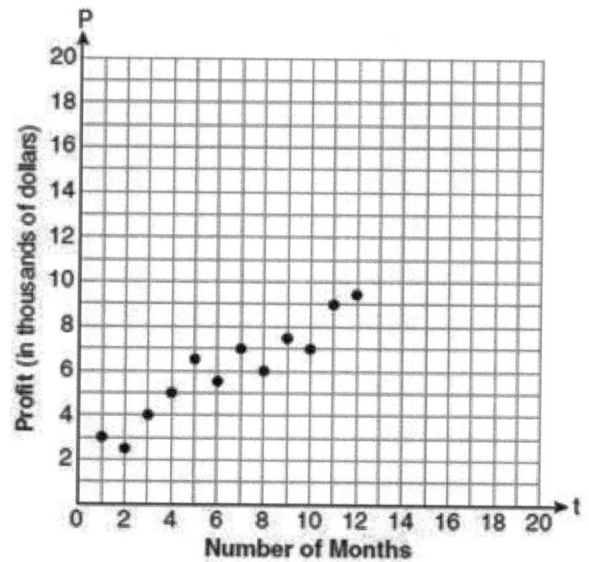

Draw the regression line. Using the line, predict whether the business will reach its goal in the 18th month. Justify your answer.

14. The table shows the temperature, *t* degrees Fahrenheit, displayed on an oven while it was heating as a function of the amount of time, *s* seconds, since it was turned on. Create a scatter plot for this data. Find the equation for its linear regression.

s	t
31	175
61	200
104	225
158	250
202	275
250	300
285	325
327	350
380	375
428	400

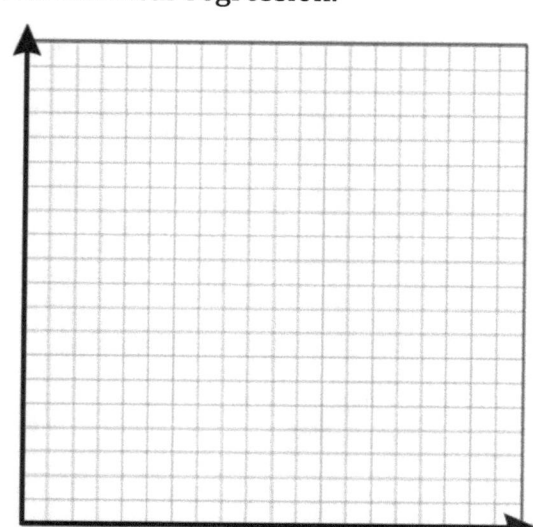

15. The table below shows ocean temperatures at various depths.

Water Depth (x) (meters)	Temperature (y) (°C)
50	18
75	15
100	12
150	7
200	1

a) Write the linear regression equation for this set of data, rounding all values to the *nearest thousandth*.

b) Using this equation, predict the temperature, to the *nearest integer*, at a water depth of 255 meters.

16. The table below shows the number of new crime cases reported in a city over a period of four years.

Year (x)	New Cases (y)
1999	440
2000	457
2001	369
2002	351

a) Write the linear regression equation for this set of data. (Let $x = 0$ represent 1999.)

b) Using this equation, find the projected number of new cases for 2009, rounded to the *nearest whole number*.

Regents Questions

MULTIPLE CHOICE

1. The table below shows the number of grams of carbohydrates, x, and the number of Calories, y, of six different foods.

Carbohydrates (x)	Calories (y)
8	120
9.5	138
10	147
6	88
7	108
4	62

    Which equation best represents the line of best fit for this set of data?
    (1) $y = 15x$ (3) $y = 0.1x - 0.4$
    (2) $y = 0.07x$ (4) $y = 14.1x + 5.8$

2. The scatter plot below shows the relationship between the number of members in a family and the amount of the family's weekly grocery bill.

    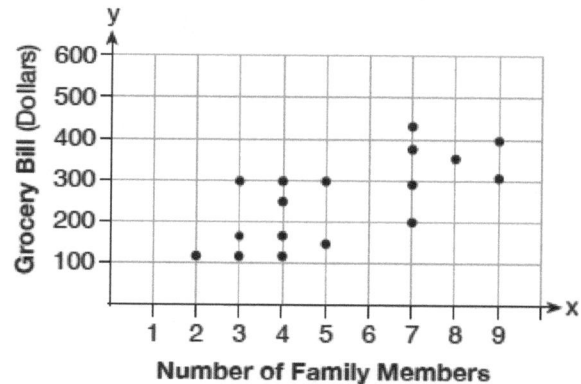

    The most appropriate prediction of the grocery bill for a family that consists of six members is
    (1) $100   (3) $400
    (2) $300   (4) $500

666

CONSTRUCTED RESPONSE

3.   Emma recently purchased a new car. She decided to keep track of how many gallons of gas she used on five of her business trips. The results are shown in the table below.

Miles Driven	Number of Gallons Used
150	7
200	10
400	19
600	29
1000	51

Write the linear regression equation for these data where miles driven is the independent variable. (Round all values to the *nearest hundredth*.)

4.   The data table below shows the median diameter of grains of sand and the slope of the beach for 9 naturally occurring ocean beaches.

Median Diameter of Grains of Sand, in Millimeters (x)	0.17	0.19	0.22	0.235	0.235	0.3	0.35	0.42	0.85
Slope of Beach, in Degrees (y)	0.63	0.7	0.82	0.88	1.15	1.5	4.4	7.3	11.3

Write the linear regression equation for this set of data, rounding all values to the *nearest thousandth*.

Using this equation, predict the slope of a beach, to the *nearest tenth of a degree*, on a beach with grains of sand having a median diameter of 0.65 mm.

5.   Omar has a piece of rope. He ties a knot in the rope and measures the new length of the rope. He then repeats this process several times. Some of the data collected are listed in the table below.

Number of Knots	4	5	6	7	8
Length of Rope (cm)	64	58	49	39	31

State, to the *nearest tenth*, the linear regression equation that approximates the length, *y*, of the rope after tying *x* knots.

Explain what the *y*-intercept means in the context of the problem.

Explain what the slope means in the context of the problem.

19.6   Correlation Coefficients

The **correlation coefficient**, represented by the letter r, tells us the degree of correlation for a set of bivariate quantitative data. In other words, it tells us how close the entire set of points is to the regression line. It is a value between –1 and 1, with *negative* values used for lines with *negative* slopes and *positive* values used for lines with *positive* slopes. A correlation coefficient of 0 means there is no correlation. A value close to 0 represents a weak correlation. A value of –1 or 1 would represent data points that are collinear (all points are on the regression line). A value close to –1 or 1 represents a strong correlation.

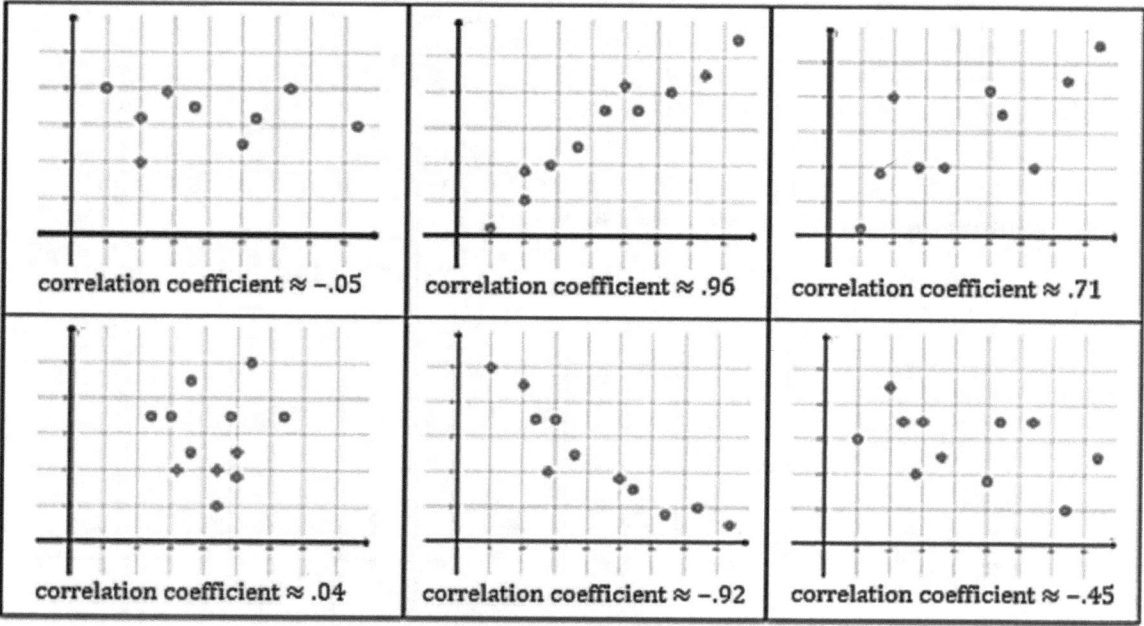

Although there are no specific cutoff numbers for what may constitute strong or weak correlations, we will use $|r| < 0.3$ to represent a weak correlation and $|r| \geq 0.7$ to represent a strong correlation.

One way to calculate the correlation coefficient is to use the formula,

$$r = b\left(\frac{s_x}{s_y}\right)$$

where b is the slope of the regression line and s_x and s_y are the sample standard deviations of the x and y values, respectively.

Fortunately, there is an easier way to find the correlation coefficient on the calculator using the LinReg function, as explained below.

 CALCULATOR TIP

To find the correlation coefficient, *r*, using the LinReg function:
If you turn Diagnostics on, then any time you calculate a linear regression, you will be told the correlation coefficient *r* along with the values of *a* and *b*.

- On newer TI models, you can turn Diagnostics on by pressing [MODE], scrolling down to Stat Diagnostics and selecting On.
- For other models, you can turn Diagnostics on by pressing [2nd][CATALOG], scrolling down to <DiagnosticOn> and then pressing [ENTER] twice.

Now, when you find a linear regression using LinReg, instead of showing the screen to the left below, the calculator will display a screen like the one to the right, including *r*.

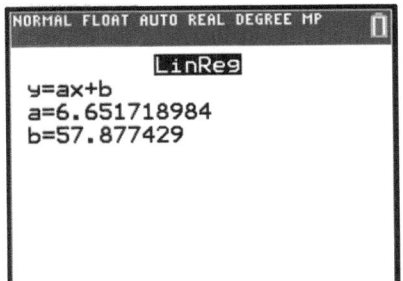

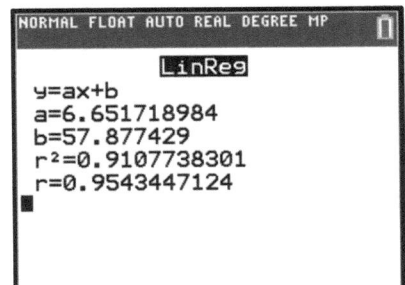

MODEL PROBLEM

Use the calculator to find the correlation coefficient for the data below, rounded to the *nearest thousandth*.

x	3	4	5	6	7	8	9	10
y	6.25	8.0	10.5	13.75	15.5	17.75	19.0	20.75

Solution:

(A)
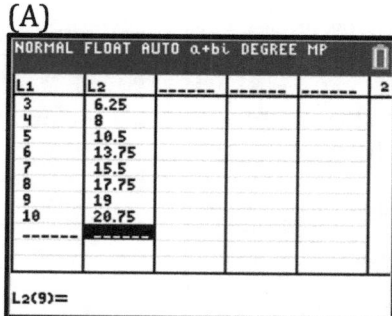

(B)
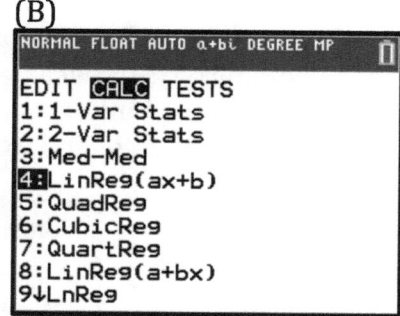

(C)
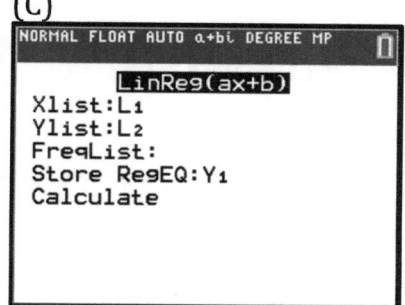

(D)
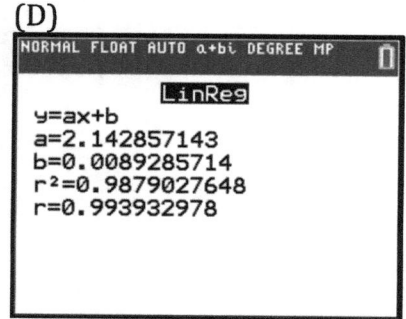

$r \approx 0.994$

Explanation of steps:

(A) Press the MODE key and turn Stat Diagnostics on. Enter the data into L1 and L2.

(B) Press the STAT key and select CALC for the linear regression (LinReg) function.

(C) Select L1 and L2 for the Xlist and Ylist, and store the equation in Y1.

(D) The correlation coefficient is calculated as r
[$r \approx 0.994$ *and the equation is approximately* $y = 2.143x + 0.009$].

PRACTICE PROBLEMS

1. Which value of r represents data with a strong positive linear correlation between two variables?

   (1) 0.89   (3) 1.04

   (2) 0.34   (4) 0.01

2. What could be the approximate value of the correlation coefficient for the accompanying scatter plot?

   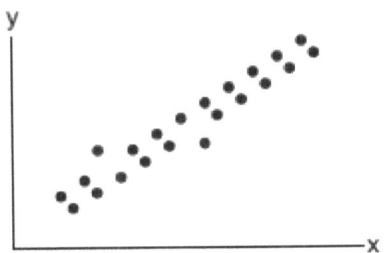

   (1) –0.85   (3) 0.21

   (2) –0.16   (4) 0.90

3. The relationship between the variables x and y is modeled by the linear regression equation $y = 12.5 + 6.1x$. Based om this model, the correlation coefficient could be

   (1) between –1 and 0   (3) equal to –1

   (2) between 0 and 1   (4) equal to 0

4. The relationship between x and y is modeled by a linear regression equation with a negative correlation coefficient. Which of the following could be the regression equation?

   (1) $y = -15.2 + 6x$

   (2) $y = 8.3x + 7.4$

   (3) $y = 18.9 - 3.4x$

   (4) $y = 2.3x - 1.7$

5.   Which graph would have a linear correlation coefficient closest to −1?

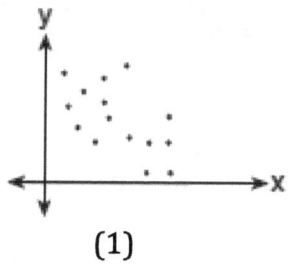

(1)

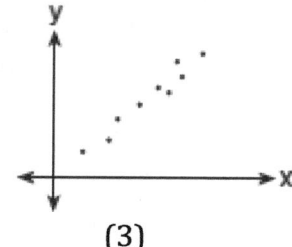

(3)

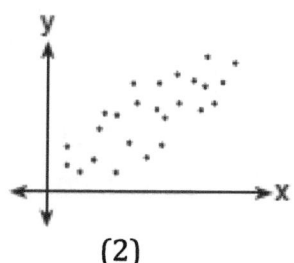

(2)

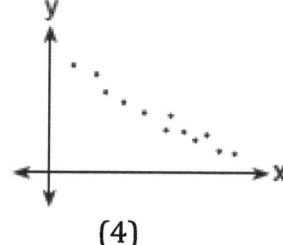

(4)

6.   What could be the approximate value of the correlation coefficient for the accompanying scatter plot?

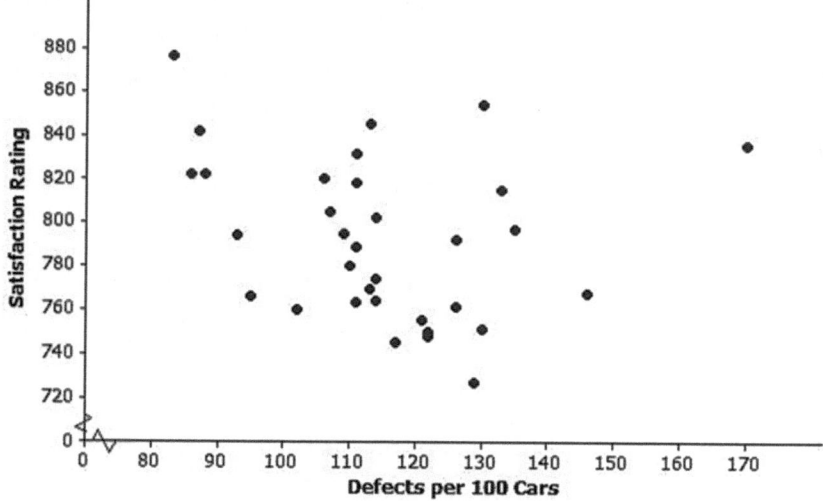

(1) −0.95   (3) 0.83

(2) −0.24   (4) 1.00

7. The correlation coefficients for the six scatter plots shown below are
        -0.85, -0.40, 0, 0.50, 0.90, 0.99
    Match each scatter plot with the correct correlation coefficient.

a.
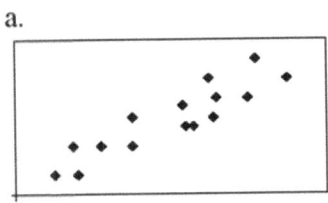

b.
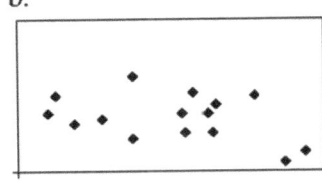

c.
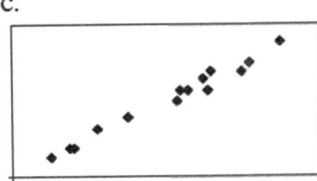

d.
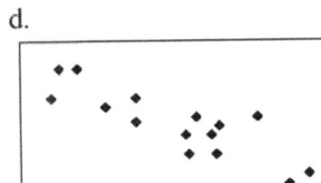

e.
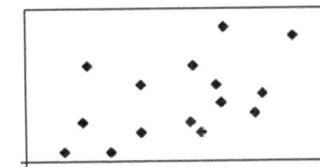

f.
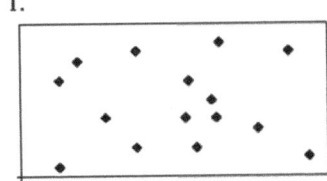

8. Find the correlation coefficient for the data below.

Woman's Shoe Size	5	6	7	8
Foot Length (in)	9.00	9.25	9.50	9.75

9. Find the correlation coefficient for the data below, rounded to the *nearest thousandth*.

x	1	3	4	8	13	16	24	37	42
y	64	37	18	28	56	43	92	44	60

10. Find the correlation coefficient for the data below, rounded to the *nearest thousandth*.

x	15	14	10	9	8	8	7	6	4	2
y	4	6	4	8	7	8	10	9	14	12

11. Use the calculator to find the correlation coefficient for the data below, rounded to the *nearest thousandth*.

Exit #	2	39	48	57	67	75	91	110
Toll	1.50	2.50	3.00	3.50	4.25	4.50	5.50	6.50

12. Use the calculator to find the correlation coefficient for the data below, rounded to the *nearest thousandth*.

Age (years)	Target Heart Rate (beats per minute)
20	135
25	132
30	129
35	125
40	122
45	119
50	115

Regents Questions

MULTIPLE CHOICE

1. What is the correlation coefficient of the linear fit of the data shown below, to the *nearest hundredth?*

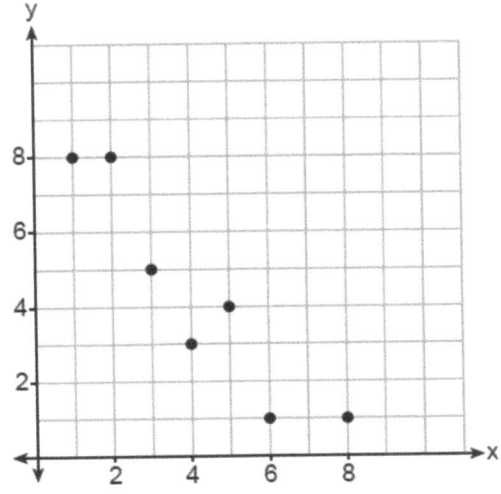

    (1) 1.00   (3) −0.93

    (2) 0.93   (4) −1.00

2. Beverly did a study this past spring using data she collected from a cafeteria. She recorded data weekly for ice cream sales and soda sales. Beverly found the line of best fit and the correlation coefficient, as shown in the diagram below.

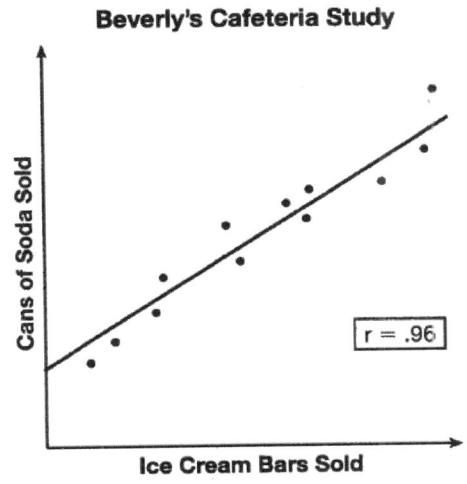

Given this information, which statement(s) can correctly be concluded?

    I. Eating more ice cream causes a person to become thirsty.

    II. Drinking more soda causes a person to become hungry.

    III. There is a strong correlation between ice cream sales and soda sales.

    (1) I, only   (3) I and III

    (2) III, only   (4) II and III

3.   The table below shows 6 students' overall averages and their averages in their math class.

Overall Student Average	92	98	84	80	75	82
Math Class Average	91	95	85	85	75	78

If a linear model is applied to these data, which statement best describes the correlation coefficient?

(1)   It is close to −1.   (3)   It is close to 0.
(2)   It is close to 1.   (4)   It is close to 0.5.

4.   Analysis of data from a statistical study shows a linear relationship in the data with a correlation coefficient of −0.524. Which statement best summarizes this result?

(1)   There is a strong positive correlation between the variables.
(2)   There is a strong negative correlation between the variables.
(3)   There is a moderate positive correlation between the variables.
(4)   There is a moderate negative correlation between the variables.

5.   Bella recorded data and used her graphing calculator to find the equation for the line of best fit. She then used the correlation coefficient to determine the strength of the linear fit. Which correlation coefficient represents the strongest linear relationship?

(1)   0.9   (3)   −0.3
(2)   0.5   (4)   −0.8

6.   The results of a linear regression are shown below.

$y = ax + b$
$a = -1.15785$
$b = 139.3171772$
$r = -0.896557832$
$r^2 = 0.8038159461$

Which phrase best describes the relationship between x and y?

(1)   strong negative correlation   (3)   weak negative correlation
(2)   strong positive correlation   (4)   weak positive correlation

7. The table below shows the time, in hours, spent by students on electronic devices and their math test scores. The data collected model a linear regression.

Time Spent on an Electronic Device (hours)	Math Test Score
3	85
1	99
4	81
0	98
3	90
7	65
5	78
2	90

What is the correlation coefficient, to the *nearest hundredth*, for these data?

(1) –0.98   (3) 0.98

(2) –0.95   (4) 0.95

CONSTRUCTED RESPONSE

8. A nutritionist collected information about different brands of beef hot dogs. She made a table showing the number of Calories and the amount of sodium in each hot dog.

Calories per Beef Hot Dog	Milligrams of Sodium per Beef Hot Dog
186	495
181	477
176	425
149	322
184	482
190	587
158	370
139	322

Write the correlation coefficient for the line of best fit. Round your answer to the *nearest hundredth*.

Explain what the correlation coefficient suggests in the context of this problem.

9.   The table below shows the attendance at a museum in select years from 2007 to 2013.

Attendance at Museum

Year	2007	2008	2009	2011	2013
Attendance (millions)	8.3	8.5	8.5	8.8	9.3

State the linear regression equation represented by the data table when $x = 0$ is used to represent the year 2007 and y is used to represent the attendance. Round all values to the *nearest hundredth*.

State the correlation coefficient to the *nearest hundredth* and determine whether the data suggest a strong or weak association.

10.   Erica, the manager at Stellarbeans, collected data on the daily high temperature and revenue from coffee sales. Data from nine days this past fall are shown in the table below.

	Day 1	Day 2	Day 3	Day 4	Day 5	Day 6	Day 7	Day 8	Day 9
High Temperature, t	54	50	62	67	70	58	52	46	48
Coffee Sales, f(t)	$2900	$3080	$2500	$2380	$2200	$2700	$3000	$3620	$3720

State the linear regression function, $f(t)$, that estimates the day's coffee sales with a high temperature of t. Round all values to the *nearest integer*.

State the correlation coefficient, r, of the data to the *nearest hundredth*. Does r indicate a strong linear relationship between the variables? Explain your reasoning.

11.   At Mountain Lakes High School, the mathematics and physics scores of nine students were compared as shown in the table below.

Mathematics	55	93	89	60	90	45	64	76	89
Physics	66	89	94	52	84	56	66	73	92

State the correlation coefficient, to the *nearest hundredth*, for the line of best fit for these data.

Explain what the correlation coefficient means with regard to the context of this situation.

12. The percentage of students scoring 85 or better on a mathematics final exam and an English final exam during a recent school year for seven schools is shown in the table below.

Percentage of Students Scoring 85 or Better	
Mathematics, x	English, y
27	46
12	28
13	45
10	34
30	56
45	67
20	42

Write the linear regression equation for these data, rounding all values to the *nearest hundredth*.

State the correlation coefficient of the linear regression equation, to the *nearest hundredth*. Explain the meaning of this value in the context of these data.

13. The data given in the table below show some of the results of a study comparing the height of a certain breed of dog, based upon its mass.

Mass (kg)	4.5	5	4	3.5	5.5	5	5	4	4	6	3.5	5.5
Height (cm)	41	40	35	38	43	44	37	39	42	44	31	30

Write the linear regression equation for these data, where *x* is the mass and *y* is the height. Round all values to the *nearest tenth*.

State the value of the correlation coefficient to the *nearest tenth*, and explain what it indicates.

14. The table below shows the number of hours ten students spent studying for a test and their scores.

Hours Spent Studying (x)	0	1	2	4	4	4	6	6	7	8
Test Scores (y)	35	40	46	65	67	70	82	88	82	95

Write the linear regression equation for this data set. Round all values to the *nearest hundredth*.

State the correlation coefficient of this line, to the *nearest hundredth*. Explain what the correlation coefficient suggests in the context of the problem.

15. Stephen collected data from a travel website. The data included a hotel's distance from Times Square in Manhattan and the cost of a room for one weekend night in August. A table containing these data appears below.

Distance From Times Square (city blocks) (x)	0	0	1	1	3	4	7	11	14	19
Cost of a Room (dollars) (y)	293	263	244	224	185	170	219	153	136	111

Write the linear regression equation for this data set. Round all values to the *nearest hundredth*.

State the correlation coefficient for this data set, to the *nearest hundredth*. Explain what the sign of the correlation coefficient suggests in the context of the problem.

16. The following table represents a sample of sale prices, in thousands of dollars, and number of new homes available at that price in 2017.

Sale Price, p (in thousands of dollars)	160	180	200	220	240	260	280
Number of New Homes Available f(p)	126	103	82	75	82	40	20

State the linear regression function, $f(p)$, that estimates the number of new homes available at a specific sale price, p. Round all values to the *nearest hundredth*.

State the correlation coefficient of the data to the *nearest hundredth*. Explain what this means in the context of the problem.

17.   Joey recorded his heart rate, in beats per minute (bpm), after doing different numbers of jumping jacks. His results are shown in the table below.

Number of Jumping Jacks x	Heart Rate (bpm) y
0	68
10	84
15	104
20	100
30	120

State the linear regression equation that estimates the heart rate per number of jumping jacks.

State the correlation coefficient of the linear regression equation, rounded to the *nearest hundredth*. Explain what the correlation coefficient suggests in the context of this problem.

18.   An insurance agent is looking at records to determine if there is a relationship between a driver's age and percentage of accidents caused by speeding. The table below shows his data.

Age (x)	17	18	21	25	30	35	40	45	50	55	60	65
Percentage of Accidents Caused by Speeding (y)	49	49	48	38	31	33	24	25	16	10	5	6

State the linear regression equation that models the relationship between the driver's age, x, and the percentage of accidents caused by speeding, y. Round all values to the *nearest hundredth*. State the value of the correlation coefficient to the *nearest hundredth*. Explain what this means in the context of the problem.

19.   The table below shows the number of math classes missed during a school year for nine students, and their final exam scores.

Number of Classes Missed (x)	2	10	3	22	15	2	20	18	9
Final Exam Score (y)	99	72	90	35	60	80	40	43	75

Write the linear regression equation for this data set. Round all values to the *nearest hundredth*.

State the correlation coefficient for your linear regression. Round your answer to the *nearest hundredth*. State what the correlation coefficient indicates about the linear fit of the data.

20. Suzanna collected information about a group of ponies and horses. She made a table showing the height, measured in hands (hh), and the weight, measured in pounds (lbs), of each pony and horse.

Height (hh) x	Weight (lbs) y
11	264
12	638
13	700
14	850
15	1000
16	1230
17	1495

Write the linear regression equation for this set of data. Round all values to the *nearest hundredth*.

State the correlation coefficient for the linear regression. Round your answer to the *nearest hundredth*.

Explain what the correlation coefficient indicates about the linear fit of the data in the context of the problem.

21.   A software company kept a record of their annual budget for advertising and their profit for each of the last eight years. These data are shown in the table below.

Annual Advertising Budget (in thousands, $) (x)	Profit (in millions, $) (y)
10	2.2
13	2.4
14	3.2
16	4.6
19	5.7
24	6.9
24	7.9
28	9.3

Write the linear regression equation for this set of data.

State, to the *nearest hundredth*, the correlation coefficient of these linear data.

State what this correlation coefficient indicates about the linear fit of the data.

22.   The table below shows the number of SAT prep classes five students attended and the scores they received on the test.

Number of Prep Classes Attended (x)	3	1	6	7	6
Math SAT Score (y)	500	410	620	720	500

State the linear regression equation for this data set, rounding all values to the *nearest hundredth*.

State the correlation coefficient, rounded to the *nearest hundredth*. State what this correlation coefficient indicates about the linear fit of the data.

Reference Sheet for Algebra I (NGLS)

Conversions

1 mile = 5280 feet

1 mile = 1760 yards

1 pound = 16 ounces

1 ton = 2000 pounds

Conversions Across Measurement Systems

1 inch = 2.54 centimeters

1 meter = 39.37 inches

1 mile = 1.609 kilometers

1 kilometer = 0.6214 mile

1 pound = 0.454 kilogram

1 kilogram = 2.2 pounds

Quadratic Equation	$y = ax^2 + bx + c$	Exponential Equation	$y = ab^x$
Quadratic Formula	$x = \dfrac{-b \pm \sqrt{b^2 - 4ac}}{2a}$	Annual Compound Interest	$A = P(1 + r)^n$
Equation of the Axis of Symmetry	$x = -\dfrac{b}{2a}$	Arithmetic Sequence	$a_n = a_1 + d(n - 1)$
Slope	$m = \dfrac{y_2 - y_1}{x_2 - x_1}$	Geometric Sequence	$a_n = a_1 r^{n-1}$
Linear Equation Slope Intercept	$y = mx + b$	Interquartile Range (IQR)	$IQR = Q_3 - Q_1$
Linear Equation Point Slope	$y - y_1 = m(x - x_1)$	Outlier	Lower Outlier Boundary $= Q_1 - 1.5(IQR)$ Upper Outlier Boundary $= Q_3 + 1.5(IQR)$

maximum, 589
mean, 585
measurement equivalent, 82
median, 586, 610
minimum, 589
mode, 587
monetary values, 47
monomial, 196

N

negative correlation, 647
negative slope, 102
no slope, 102
normal distribution, 596
numerical data, 561

O

observation, 561, 585
open circle, 247, 556
operands, order of, 46
ordered pair, 92, 215
outlier, 587, 622
overlap, 159

P

parabola, 449, 457, 461, 470, 496
   graphing, 470
   y-intercept, 449
parallel, 96
parent function, 524
percentile, 610
plus-minus symbol, 374, 435
points, 92
point-slope form, 118
polynomial, 196
polynomials
   adding, 200
   dividing, 212
   subtracting, 200
population, 601
positive correlation, 647
positive slope, 102
power rules for exponents, 342
prime factorization, 375
prime trinomial, 397
properties, 10
properties of equality, 11

Q

quadratic equation
   solving, 428
quadratic equations
   solving, 410, 417
quadratic formula, 435
quadratic regression, 472
qualitative, 561
quantiles, 610
quantitative, 561
quartile, 610, 619
quartiles, 610

R

radical, 374
radical sign, 374
radicals
   combining, 378
   dividing, 378
   like, 378
   multiplying, 378
   simplifying, 375
radicand, 374
range, 245, 589
rate of change, 276
ratio, 47
rational, 10, 389
rational roots, 430
rationalizing the denominator, 385
real numbers, 10
rectangle diagram, 204, 205
regression, 655
relation, 215
relative frequency, 566, 632
relative frequency table, 566
relative maximum, 240
relative minimum, 240
resistant, 587
rise, 102
roots, 410, 421, 437, 449, 506, 507
run, 102

S

sample, 601
scale, 271
scatter plot, 640
secant line, 282
second differences, 410
sequence, 361, 369